CONTINUITY EQUATION FOR ANY FLUID

$$\rho_1 A_1 v_1 = \rho_2 A_2 v_2$$

(6–4)

CONTINUITY EQUATION FOR LIQUIDS

$$A_1 v_1 = A_2 v_2$$

(6–5)

BERNOULLI'S EQUATION

$$\frac{p_1}{\gamma} + z_1 + \frac{v_1^2}{2g} = \frac{p_2}{\gamma} + z_2 + \frac{v_2^2}{2g}$$

(6–9)

TORRICELLI'S THEOREM

$$v_2 = \sqrt{2gh}$$

(6–16)

TIME REQUIRED TO DRAIN A TANK

$$t_2 - t_1 = \frac{2(A_t/A_j)}{\sqrt{2g}} (h_1^{1/2} - h_2^{1/2})$$

(6–26)

GENERAL ENERGY EQUATION

$$\frac{p_1}{\gamma} + z_1 + \frac{v_1^2}{2g} + h_A - h_R - h_L = \frac{p_2}{\gamma} + z_2 + \frac{v_2^2}{2g}$$

(7–3)

POWER ADDED TO A FLUID BY A PUMP

$$P_A = h_A \gamma A$$

(7–5)

PUMP EFFICIENCY

$$e_M = \frac{\text{Power delivered to fluid}}{\text{Power put into pump}} = \frac{P_A}{P_I}$$

(7–6)

POWER REMOVED FROM A FLUID BY A MOTOR

$$P_R = h_R W = h_R \gamma Q$$

(7–8)

MOTOR EFFICIENCY

$$e_M = \frac{\text{Power output from motor}}{\text{Power delivered by fluid}} = \frac{P_O}{P_R}$$

(7–9)

REYNOLDS NUMBER — CIRCULAR SECTIONS

$$N_R = \frac{vD\rho}{\mu} = \frac{vD}{v}$$

(8–1)

HYDRAULIC RADIUS

$$R = \frac{A}{WP} = \frac{\text{area}}{\text{wetted perimeter}}$$

(8–5)

REYNOLDS NUMBER — NONCIRCULAR SECTIONS

$$N_R = \frac{v(4R)\rho}{\mu} = \frac{v(4R)}{v}$$

(8–6)

DARCY'S EQUATION FOR ENERGY LOSS

$$h_L = f \times \frac{L}{D} \times \frac{v^2}{2g}$$

(9–1)

HAGEN-POISEUILLE EQUATION

$$h_L = \frac{32\mu L v}{\gamma D^2}$$

(9–2)

FRICTION FACTOR FOR LAMINAR FLOW

$$f = \frac{64}{N_R}$$

(9–3)

Applied Fluid Mechanics

Fourth Edition

Applied Fluid Mechanics

Fourth Edition

Robert L. Mott
University of Dayton

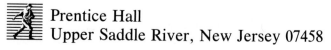

Prentice Hall
Upper Saddle River, New Jersey 07458

Library of Congress Cataloging-in-Publication Data

Mott, Robert L.
    Applied fluid mechanics / Robert L. Mott. - 4th ed.
      p. cm.
    Includes bibliographical references and index.
    ISBN 0-02-384231-8
    1. Fluid mechanics.   I. Title.
TA357.M67   1994
620.1′06--dc20
                                  93-15111
                                            CIF

Cover art: Rendering artist Larry Hewitt, Modern Medium Inc., Portland
OR for EDA, Berkeley, CA
Editor: Stephen Helba
Developmental Editor: Monica Ohlinger
Production Editor: Louise N. Sette
Art Coordinator: Lorraine Woost
Text Designer: Anne Flanagan
Cover Designer: Thomas Mack
Production Buyer: Patricia A. Tonneman
Illustrations: Academy ArtWorks, Inc.

This book was set in Times Roman by Bi-Comp, Inc.

 © 1994 by Prentice-Hall, Inc.
A Simon & Schuster Company
Upper Saddle River, New Jersey 07458

Printed in the United States of America

10

ISBN   0-02-384231-8

Prentice-Hall International (UK) Limited, London
Prentice-Hall of Australia Pty. Limited, Sydney
Prentice-Hall Canada Inc., Toronto
Prentice-Hall Hispanoamericana, S.A., Mexico
Prentice-Hall of India Private Limited, New Delhi
Prentice-Hall of Japan, Inc., Tokyo
Simon & Schuster Asia Pte. Ltd., Singapore
Editora Prentice-Hall do Brasil, Ltda., Rio de Janeiro

To my wife Marge;
our children Lynné, Robert Jr., and Stephen;
and my Mother and Father.

Preface

Introduction

The objective of this book is to present the principles of fluid mechanics and the application of these principles to practical, applied problems. Primary emphasis is on fluid properties, fluid statics, flow of fluids in pipes and noncircular conduits, pump selection and application, open channel flow, forces developed by fluids in motion, the design and analysis of HVAC ducts, the flow of gases, and the measurement of pressure, density, viscosity, and flow.

Applications are shown in the mechanical field including fluid power, heating, ventilation, and air conditioning (HVAC); the chemical field including flow in materials processing systems; and the civil and environmental fields as applied to water and waste water systems, fluid storage and distribution systems, and open channel flow. This book is directed to anyone in an engineering field where the ability to apply the principles of fluid mechanics is the primary goal.

Those using this book are expected to have an understanding of algebra, trigonometry, and physics mechanics. After completing the book, the student should have the ability to design and analyze practical fluid flow systems and to continue learning in the field. Other applied courses, such as fluid power, HVAC, and civil hydraulics could be taught following this course. Alternatively, this book could be used to teach selected fluid mechanics topics within such courses.

Approach

The approach used in this book encourages the student to become involved in the learning of the principles of fluid mechanics at five levels:

1. The understanding of concepts.
2. The recognition of the logical approach to problem solutions.
3. The ability to perform the details required in the solutions.
4. The ability to critique the design of a given system and recommend improvements.
5. The ability to design practical, efficient fluid systems.

This multilevel approach has been successful in building students' confidence in their ability to analyze and design fluid systems.

Concepts are presented in clear language and illustrated by reference to physical systems with which the reader should be familiar. An intuitive justification as well as a mathematical basis is given for each concept. The methods of solution to many types of complex problems are presented in step-by-step procedures. The importance of recognizing the relationships among what is known, what is to be found, and the choice of a solution procedure is emphasized.

Many practical problems in fluid mechanics require relatively long solution procedures. It has been my experience that students often have difficulty in accomplishing the details of the solution. For this reason, each example problem is worked in complete detail including the manipulation of units in equations. In the more complex examples, a programmed instruction format is used in which the student is asked to perform a small segment of the solution before being shown the correct result. The programs are of the linear type in which one panel presents a concept and then either poses a question or asks that a certain operation be performed. The following panel gives the correct result and the details of how it was obtained. The program then continues. Computer programming assignments are included in nine chapters to encourage students to use the computer in solving the longer, more computationally intensive problems. Forty-two programming assignments are included.

The International System (Système International) of units (SI) and the U.S. Customary System units are used approximately equally. The SI notation in this book follows the guidelines set forth by the American National Metric Council. Other references used are

International Standard ISO 1000, *SI Units and Recommendations for the Use of their Multiples and of Certain Other Units*, International Organization for Standardization.

ASME Orientation and Guide for the Use of SI (Metric) Units, 5th ed., The American Society of Mechanical Engineers.

SI Units in Fluid Mechanics, The American Society of Mechanical Engineers.

Features New to the Fourth Edition

The fourth edition continues the pattern of earlier editions in refining the presentation of several topics, enhancing the visual attractiveness and usability of the book, updating data and analysis techniques, and adding selective new material. The following list highlights some of the changes in this edition:

- The use of color obviously makes the book more attractive. But, more importantly, it enhances student learning and the rate of comprehension. Fluids in figures stand out in blue. Curves in graphs are printed in blue to make them stand out from the black grid lines. This is especially effective in Moody's diagram in Chapter 9 and in the pump performance curves in Chapter 15.
- Throughout the book, primary formulas are highlighted and annotated with labels in the left margin so readers can easily find them for review or use in problem solutions.

- New information is included in Chapter 1 on density and specific gravity measurement with references to updated standards of the American Society for Testing and Materials (ASTM).

- New photographs of viscosity measurement devices and instruments have been included in Chapter 2 along with new, extensively revised standards for the SAE grades of lubricating oils and the ISO Viscosity Grades. SAE now uses the calibrated glass capillary viscometer as a standard rather than the Saybolt viscometer.

- New photographs of commercially available pressure measuring instruments are included in Chapter 3.

- A large-format version of the Moody diagram for friction factors has been packaged with the book, allowing the user to lay it out on a table during the solution of problems.

- Chapter 9 includes a completely new approach to calculating the friction factor for the flow of fluids in pipes and tubes based on the work of Swamee and Jain that allows for direct, one-step computation.

- The Hazen-Williams method for analyzing the flow of water in pipes, very popular with the civil engineering and environmental fields, has been included in Chapter 9.

- Minor loss data in Chapter 10 have been refined with new information adapted from the latest version of Technical Paper 410 of the Crane Company and *Cameron Hydraulic Data* from Ingersoll-Rand. Minor loss coefficients are now referred to as K rather than C_L.

- The analysis of series pipe line problems in Chapter 11 has been refined with critiques of system designs and the inclusion of computed friction factors in problems of Classes II and III.

- The solution of parallel flow problems in Chapter 12 has been refined using the computed friction factors.

- Chapter 14 on *Flow Measurement* now includes a discussion of mass flow rate and several new photographs.

- Performance curves for six new centrifugal pumps have been added to Chapter 15. Covering a wide range of capacities and total head capabilities, they allow instructors to create a virtually unlimited set of assignments on pump selection and application.

- Data for drag coefficients in Chapter 17 have been updated.

- The former Chapter 18 of the 3rd edition has been split into two chapters placing the discussion of fans and compressors and the analysis of the flow of gases in the new Chapter 18 and the design and analysis of ducts in the new Chapter 19.

- Several Appendix tables have been completely reworked. Data for both SI and U.S. Customary dimensions of pipes and tubes have been integrated into the same tables. Data for ductile iron pipes, used frequently for water systems, have been added. A table on the areas of circles has been added. The table of typical properties of lubricating oils has been completely changed. Data on the properties of the atmosphere now include both SI and U.S. Customary units and a new graph of the variation of pressure and temperature with altitude has been added.

- The solutions manual has been completely reworked and set in type for ease of use and to enhance its appearance.
- The solutions manual includes a computer disk with useful software for analyzing linear fluid flow systems and networks. The versions of *HYDROFLO*™ and *HYDRONET*™ included are functional, smaller versions of the popular commercially available software produced by Engineering Software of Truckee, California. Instructors can use the disks for classroom and laboratory demonstrations. Unlimited copies can be made for distribution to students for their own use.

Acknowledgments

I would like to thank all who helped and encouraged me in the writing of this book, including users of the third edition and the several reviewers who provided detailed suggestions: John Fletcher, Hudson Valley Community College; Peggy R. Geier, Lawrence Technological University; Matthew Kovac, Butler County Community College; Thomas J. Lepore, Hartford State Technical College; Paul Ricketts, New Mexico State University; David J. Unger; Steve Wells, Old Dominion University; and Donald L. Young, California State Polytechnic University. Special thanks go to my colleagues, Jesse H. Wilder , David H. Myszka, and Joseph A. Untener of the University of Dayton, who used the third edition of this book in class many times and offered helpful suggestions. Robert L. Wolff, also of the University of Dayton, has provided much help in the use of the SI system of units, based on his long experience in metrication through the American Society for Engineering Education. Comments from students who have used the book are also appreciated, since the book is written for them.

Robert L. Mott

Contents

CHAPTER 9
ENERGY LOSSES DUE TO FRICTION   237

CHAPTER 10
MINOR LOSSES   267

CHAPTER 11
SERIES PIPE LINE SYSTEMS   293

■ ■ ■ ■ 1 The Nature of Fluids

1.1
OBJECTIVES OF THIS BOOK

The term *fluid mechanics* refers to the study of the behavior of fluids, either at rest or in motion. Fluids can be either *liquids* (such as water, oil, gasoline, or glycerine) or *gases* (such as air, oxygen, nitrogen, or helium). The behavior of fluids affects your daily life in many ways.

When you turn on a faucet, water is delivered to you by a distribution system composed of pumps, valves, and pipes. The source of the water may be a storage tank, reservoir, river, lake, or well. The flow of water from its source to your faucet is controlled by the principles of fluid mechanics. These principles must be understood in order to select properly the size and type of pumps and pipes, to design the storage tanks, to select flow control valves, and to monitor the performance of the system.

The performance of an automated manufacturing machine, which is controlled by fluid power systems, is dependent on the flow of the hydraulic oil and the pressure at the actuators. A typical system is shown in Fig. 1.1. The greater the pressure of the oil in a cylinder, the greater the force it can exert. The greater the flow rate of oil into the cylinder, the faster it will move. You will learn how to analyze such a system using the material in this book.

A buoy marking a boating channel seems like a very simple device, and it is. However, the material from which the buoy is made and its geometry must be specified according to the laws of buoyancy and stability of floating bodies, which you will learn in Chapter 5 of this book.

In your automobile, fuel is delivered to the carburetor from the tank by a fuel pump. How much power is being supplied by the engine to drive the pump? The material in Chapter 7 will help you to calculate this.

A highway sign giving directions to motorists may have to withstand high winds. In order to determine the forces exerted on the sign due to the winds, the principle of impulse-momentum as presented in Chapter 16 must be understood.

Automation equipment for manufacturing systems often employs high-pressure compressed air to actuate pneumatic cylinders and air motors. The flow of air in piping systems must be analyzed to ensure that the air pressure at the point of use is sufficient.

Heating, ventilating, and air-conditioning systems deliver air at low pressure to living and working spaces to improve the comfort of the occupants. The methods presented in Chapter 19 will help you analyze such systems.

FIGURE 1.1   Typical piping system for fluid power.

These are just a few of the many practical problems you are likely to encounter that require an understanding of the principles of fluid mechanics for their solution. The objective of this book is to help you solve these kinds of problems. Included in each chapter are problems representing situations from many fields of technology. Your ability to solve these problems will be a measure of how well the objective of this book has been accomplished.

1.2 OBJECTIVES OF THIS CHAPTER

In order to understand the behavior of fluids, it is necessary to understand the nature of the fluids themselves. This chapter defines the properties of fluids, introduces the symbols and units involved, and discusses the kinds of calculations required in the study of fluid mechanics.

After completing this chapter, you should be able to:

1. Differentiate between a gas and a liquid.
2. Identify the units for the basic quantities of time, length, force, and mass in the SI system (metric unit system).
3. Identify the units for the basic quantities of time, length, force, and mass in the U.S. Customary System.
4. Properly set up equations to ensure consistency of units.
5. Define *compressibility* and *bulk modulus*.
6. Define *pressure*.
7. Define the relationship between force and mass.
8. Define *density*.
9. Define *specific weight*.
10. Define *specific gravity*.
11. Identify the relationships between specific weight, specific gravity, and density, and solve problems using these relationships.

1.3 DIFFERENCE BETWEEN LIQUIDS AND GASES

When a liquid is held in a container, it tends to take the shape of the container, covering the bottom and sides. The top surface, in contact with the atmosphere above it, maintains a uniform level. As the container is tipped, the liquid tends to pour out; the rate of pouring is dependent on a property called *viscosity,* which will be defined later.

When a gas is held in a closed container, it tends to expand and completely fill the container. If the container is opened, the gas tends to continue to expand and escape from the container.

In addition to these familiar differences between gases and liquids, another difference is important in the study of fluid mechanics:

- Liquids are only slightly compressible.
- Gases are readily compressible.

Compressibility refers to the change in volume of a substance when the pressure on it changes. These distinctions are sufficient for most purposes.

The continuing discussion of pressure, compressibility, and other properties of fluids requires an understanding of the units in which these quantities are measured as presented in the following sections.

1.4
FORCE AND MASS

An understanding of fluid properties requires a careful distinction between *mass* and *weight*. The following definitions apply.

- *Mass* is the property of a body of fluid that is a measure of its inertia or resistance to a change in motion. It is also a measure of the quantity of fluid.

We will use the symbol *m* for mass.

- *Weight* is the amount that a body weighs, that is, the force with which a body is attracted toward the earth by gravitation.

We will use the symbol *w* for weight.

Weight is related to mass and the acceleration due to gravity, *g,* by Newton's law of gravitation,

▷ **WEIGHT-MASS RELATIONSHIP**

$$w = mg \tag{1-1}$$

In this book, we will use $g = 9.81$ m/s^2 in the SI system and $g = 32.2$ ft/s^2 in the U.S. Customary System. These are the standard values for g to three significant digits. To a greater degree of precision, the standard values of $g = 9.806\ 65$ m/s^2 or $g = 32.1740$ ft/s^2. For high precision work and at high elevations (such as aerospace operations) where the actual value of g is different from the standard, the local value should be used.

1.5
THE INTERNATIONAL
SYSTEM OF UNITS (SI)

In any technical work the units in which physical properties are measured must be stated. A system of units specifies the units of the basic quantities of length, time, force, and mass. The units of other terms are then derived from these.

Le Système International d'Unités, or the International System of Units (abbreviated SI), is used in this book. The units for the basic quantities are

$$length = meter* \text{ (m)}$$
$$time = second \text{ (s)}$$
$$mass = kilogram \text{ (kg)}$$
$$force = newton \text{ (N) or } kg \cdot m/s^2$$

An equivalent unit for force is the $kg \cdot m/s^2$, as indicated above. This is derived from the relationship between force and mass from physics,

$$F = ma$$

where a is acceleration expressed in units of m/s^2. Therefore, the derived unit for force is

$$F = ma = kg \cdot m/s^2 = newton$$

Thus a force of 1.0 N would give a mass of 1.0 kg an acceleration of 1.0 m/s^2. This means that either newtons (N) or $kg \cdot m/s^2$ may be used as the units for force. In fact, some calculations in this book require that you be able to use both or to convert from one to the other.

For example, we may say that a rock with a mass of 5.60 kg is suspended by a wire. Then, in order to determine what force is exerted on the wire, Newton's law of gravitation ($w = mg$) should be used:

$$w = mg = \text{mass} \times \text{acceleration of gravity}$$

But, under standard conditions at sea level, $g = 9.81 \, m/s^2$. Then we have

$$w = 5.60 \text{ kg} \times 9.81 \, m/s^2 = 54.9 \, kg \cdot m/s^2 = 54.9 \text{ N}$$

Thus, 5.60 kg of the rock weighs 54.9 N.

Because the actual size of physical quantities in the study of fluid mechanics covers a wide range, prefixes are added to the basic quantities. Table 1.1 shows these prefixes. Standard usage in the SI system calls for only those prefixes varying in steps of 10^3 as shown. Results of calculations should normally be adjusted so that the number is between 0.1 and 10 000 times some multiple of 10^3.† Then the proper unit with a prefix can be specified. Some examples follow.

---

* The American National Metric Council (ANMC) recommends the spellings "meter" and "liter" instead of "metre" and "litre." We will use the preferred notation of the ANMC throughout this book.

† Because commas are used as decimal markers in many countries, we will not use commas to separate groups of digits. We will separate the digits into groups of three, counting both to the left and to the right from the decimal point, and use a space to separate the groups of three digits. We will not use a space if there are only four digits to the left or right of the decimal point unless required in tabular matter.

TABLE 1.1 SI unit prefixes

Prefix	SI Symbol	Factor
giga	G	$10^9 = 1\ 000\ 000\ 000$
mega	M	$10^6 = 1\ 000\ 000$
kilo	k	$10^3 = 1\ 000$
milli	m	$10^{-3} = 0.001$
micro	μ	$10^{-6} = 0.000\ 001$

Computed Result	Reported Result
0.004 23 m	4.23×10^{-3} m, or 4.23 mm (millimeters)
15 700 kg	15.7×10^3 kg, or 15.7 Mg (megagrams)
86 330 N	86.33×10^3 N, or 86.33 kN (kilonewtons)

1.6 THE U.S. CUSTOMARY SYSTEM

Sometimes called the *English gravitational unit system* or the *pound-foot-second* system, the U.S. Customary System defines the basic quantities as follows:

$$\text{length} = \text{foot (ft)}$$
$$\text{time} = \text{second (s)}$$
$$\text{force} = \text{pound (lb)}$$
$$\text{mass} = \text{slug (lb-s}^2\text{/ft)}$$

Probably the most difficult of these units to understand is the slug, since we are familiar with measuring in terms of pounds, seconds, and feet. It may help to note the relationship between force and mass from physics,

$$F = ma$$

where a is acceleration expressed in units of ft/s². Therefore, the derived unit for mass is

$$m = \frac{F}{a} = \frac{\text{lb}}{\text{ft/s}^2} = \frac{\text{lb-s}^2}{\text{ft}} = \text{slug}$$

This means that you may use either slugs or lb-s²/ft for the unit of mass. In fact, some calculations in this book require that you be able to use both or to convert from one to the other.

Mass Expressed as lbm (Pounds-Mass)

Some professionals, in the analysis of fluid flow systems, have used the unit lbm (pounds-mass) for the unit of mass instead of the unit of slugs. In this system, an object or a quantity of fluid having a weight of 1.0 lb would have a mass of 1.0 lbm. The pound-force is then sometimes designated lbf. It must be noted that the numerical equivalence of lbf and lbm applies *only* when the value of g is equal to the standard value.

This system is avoided in this book because it is not a coherent system. When one tries to relate force and mass units using Newton's law, the following is obtained.

$$F = m \cdot a = \text{lbm(ft/s}^2) = \text{lbm-ft/s}^2$$

This is *not* the same as the lbf.

To overcome this difficulty, a conversion constant, commonly called g_c, is defined having both a numerical value and units. That is,

$$g_c = \frac{32.2 \text{ lbm}}{\text{lbf/(ft/s}^2)} = \frac{32.2 \text{ lbm-ft/s}^2}{\text{lbf}}$$

Then to convert from lbm to lbf, a modified form of Newton's law is used,

$$F = m(a/g_c)$$

Letting the acceleration $a = g$,

$$F = m(g/g_c)$$

For example, to determine the weight of material in lbf that has a mass of 100 lbm, and assuming that the local value of g is equal to the standard value of 32.2 ft/s²,

$$w = F = m\frac{g}{g_c} = 100 \text{ lbm} \frac{32.2 \text{ ft/s}^2}{\dfrac{32.2 \text{ lbm-ft/s}^2}{\text{lbf}}} = 100 \text{ lbf}$$

This shows that weight in lbf is numerically equal to mass in lbm *provided* $g = 32.2$ ft/s².

But if the analysis were to be done for an object or fluid on the earth's moon where g is approximately ⅙ that on the earth, say, 5.4 ft/s²,

$$w = F = m\frac{g}{g_c} = 100 \text{ lbm} \frac{5.4 \text{ ft/s}^2}{\dfrac{32.2 \text{ lbm-ft/s}^2}{\text{lbf}}} = 16.8 \text{ lbf}$$

This is a dramatic difference.

In summary, because of the cumbersome nature of the relationship between lbm and lbf, we avoid the use of lbm in this book. Mass will be expressed in the unit of slugs when problems are in the U.S. Customary System of units.

**1.7
CONSISTENT UNITS IN
AN EQUATION**

The analyses required in fluid mechanics involve the algebraic manipulation of several terms. The equations are often complex, and it is extremely important that the results be dimensionally correct. That is, they must have the proper units. Indeed, answers will have the wrong numerical value if the units in the equation are not consistent.

A simple straightforward procedure called *unit cancellation* will ensure proper units in any kind of calculation, not only in fluid mechanics but also in virtually all your technical work. The six steps of the procedure are listed below.

UNIT CANCELLATION PROCEDURE

1. Solve the equation algebraically for the desired term.
2. Decide on the proper units for the result.
3. Substitute known values, including units.
4. Cancel units that appear in both the numerator and denominator of any term.
5. Use conversion factors to eliminate unwanted units and obtain the proper units as decided in Step 2.
6. Perform the calculation.

This procedure, properly executed, will work for any equation. It is really very simple, but some practice may be required to use it. We are going to borrow some material from elementary physics with which you should be familiar to illustrate the method. However, the best way to learn how to do something is to do it. The following example problems are presented in a form called *programmed instruction.* You will be guided through the problems in a step-by-step fashion with your participation required at each step.

To proceed with the program you should cover all material below the heading ''Programmed Example Problem,'' using a heavy piece of paper. You should have a blank piece of paper handy on which to perform the requested operations. Then successively uncover one panel at a time down to the heavy line that runs across the page. The first panel presents a problem and asks you to perform some operation or to answer a question. After doing what is asked, uncover the next panel, which will contain information that you can use to check your result. Then continue with the next panel, and so on through the program.

Remember, the purpose of this is to help you learn how to get correct answers using the unit cancellation method. You may want to refer to the table of conversion factors in Appendix K.

PROGRAMMED EXAMPLE PROBLEM

☐ **EXAMPLE PROBLEM 1.1** Imagine you are traveling in a car at a constant speed of 80 kilometers per hour (km/h). How many seconds (s) would it take to travel 1.5 km?

For the solution, use the equation

$$s = vt$$

where s is the distance traveled, v is the speed, and t is the time. Using the unit cancellation procedure outlined above, what is the first thing to do?

The first step is to solve for the desired term. Since you were asked to find time, you should have written

$$t = \frac{s}{v}$$

Now perform Step 2 of the procedure described above.

Step 2 is to decide on the proper units for the result, in this case time. From the problem statement the proper unit is seconds. If no specification had been given for units, you could choose any acceptable time unit such as hours.

Proceed to Step 3.

The result should look something like this:

$$t = \frac{s}{v} = \frac{1.5 \text{ km}}{80 \text{ km/h}}$$

For the purpose of cancellation it is not convenient to have the units in the form of a compound fraction as we have above. To clear this to a simple fraction, write it in the form

$$t = \frac{\dfrac{1.5 \text{ km}}{1}}{\dfrac{80 \text{ km}}{h}}$$

This can be reduced to

$$t = \frac{1.5 \text{ km} \cdot h}{80 \text{ km}}$$

After some practice, equations may be written in this form directly. Now perform Step 4 of the procedure.

The result should now look like this:

$$t = \frac{1.5 \cancel{\text{km}} \cdot h}{80 \cancel{\text{km}}}$$

This illustrates that units can be cancelled just as numbers can if they appear in both the numerator and denominator of a term in an equation.

Now do Step 5.

The answer looks like this:

$$t = \frac{1.5 \cancel{\text{km}} \cdot \cancel{h}}{80 \cancel{\text{km}}} \times \frac{3600 \text{ s}}{1 \cancel{h}}$$

The equation in the preceding panel showed the result for time in hours after the kilometer units were cancelled. Although hours is an acceptable time unit, our desired unit is seconds as determined in Step 2. Thus the conversion factor 3600 s/1 h is required.

How did we know to multiply by 3600 instead of dividing?

The units determine this. Our objective in using the conversion factor was to eliminate the hour unit and obtain the second unit. Since the unwanted hour unit was in the numerator of the original equation, the hour unit in the conversion factor must be in the denominator in order to cancel.

Now that we have the time unit of seconds we can proceed with Step 6.

The correct answer is $t = 67.5$ s.

■

1.8
THE DEFINITION
OF PRESSURE

▷ **PRESSURE**

Pressure is defined as the amount of force exerted on a unit area of a substance. This can be stated by the equation

$$p = \frac{F}{A} \tag{1–2}$$

Two important principles about pressure were described by Blaise Pascal, a seventeenth-century scientist.

■ Pressure acts uniformly in all directions on a small volume of a fluid.
■ In a fluid confined by solid boundaries, pressure acts perpendicular to the boundary.

These principles, sometimes called *Pascal's laws,* are illustrated in Figs. 1.2 and 1.3.

FIGURE 1.2 Pressure acting uniformly in all directions on a small volume of fluid.

Fluid surface

FIGURE 1.3   Direction of fluid
pressure on boundaries.

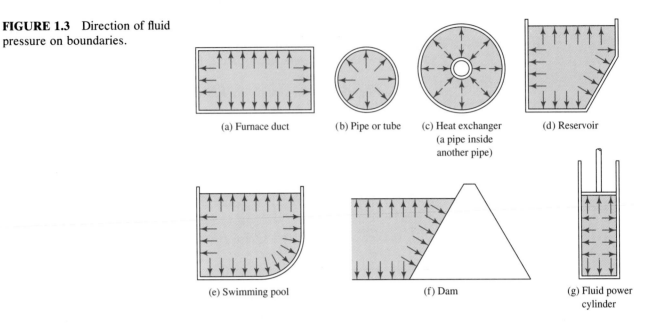

(a) Furnace duct   (b) Pipe or tube   (c) Heat exchanger   (d) Reservoir
                                                    (a pipe inside
                                                    another pipe)

(e) Swimming pool   (f) Dam   (g) Fluid power
                                                                        cylinder

Using Eq. (1–2) and the second of Pascal's laws, we can compute the
magnitude of the pressure in a fluid if we know the amount of force exerted
on a given area.

☐ **EXAMPLE PROBLEM 1.2**   Figure 1.4 shows a container of liquid with a movable piston supporting a load.
Compute the magnitude of the pressure in the liquid under the piston if the total
weight of the piston and the load is 500 N and the area of the piston is 2500 mm².

Solution   It is reasonable to assume that the entire surface of the fluid under the piston is
sharing in the task of supporting the load. The second of Pascal's laws states that the
fluid pressure acts perpendicular to the piston. Then, using Eq. (1–2),

$$p = \frac{F}{A} = \frac{500 \text{ N}}{2500 \text{ mm}^2} = 0.20 \text{ N/mm}^2$$

The standard unit of pressure in the SI system is the N/m², called the *pascal* (Pa) in
honor of Blaise Pascal. The conversion can be made by using the factor 10^3 mm =
1 m.

$$p = \frac{0.20 \text{ N}}{\text{mm}^2} \times \frac{(10^3 \text{ mm})^2}{\text{m}^2} = 0.20 \times 10^6 \text{ N/m}^2 = 0.20 \text{ MPa}$$

Note that pressure in N/mm² is numerically equal to pressure in MPa. It is not
unusual to encounter pressure in the range of several megapascals (MPa) or several
hundred kilopascals (kPa).

Pressure in the U.S. Customary System is illustrated in the following example
problem.

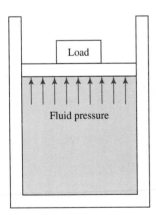

Load

Fluid pressure

FIGURE 1.4   Illustration of fluid
pressure supporting a load.

☐ **EXAMPLE PROBLEM 1.3**   A load of 200 pounds (lb) is exerted on a piston confining an oil in a circular cylinder with an inside diameter of 2.50 inches (in). Compute the pressure in the oil at the piston. See Fig. 1.4.

Solution   Using Eq. (1–2), we must compute the area of the piston.

$$A = \pi D^2/4 = \pi(2.50 \text{ in})^2/4 = 4.91 \text{ in}^2$$

Then,

$$p = \frac{F}{A} = \frac{200 \text{ lb}}{4.91 \text{ in}^2} = 40.7 \text{ lb/in}^2$$

Although the standard unit for pressure in the U.S. Customary System is pounds per square foot (lb/ft^2), it is not often used because it is inconvenient. Length measurements are more conveniently made in inches, and pounds per square inch (lb/in^2), abbreviated psi, is used most often for pressure in this system. The pressure in the oil is 40.7 psi. This is a fairly low pressure; it is not unusual to encounter pressures of several hundred or several thousand psi.

The *bar* is another unit used by some people working in fluid mechanics and thermodynamics. The bar is defined as 10^5 Pa or 10^5 N/m^2. Another way of expressing the bar is 100×10^3 N/m^2, which is equivalent to 100 kPa. Since atmospheric pressure near sea level is very nearly the same, the bar has a convenient point of physical reference. This, plus the fact that pressures expressed in bars yield smaller numbers, makes this unit attractive to some practitioners. You must realize, however, that the bar is not a part of the coherent SI system and that you must carefully convert to N/m^2 (pascals) in problem solving.

1.9 COMPRESSIBILITY

Compressibility refers to the change in volume (*V*) of a substance that is subjected to a change in pressure on it. The usual quantity used to measure this phenomenon is the *bulk modulus of elasticity* or, simply, *bulk modulus, E.*

⇨ BULK MODULUS

$$E = \frac{-\Delta p}{(\Delta V)/V} \tag{1–3}$$

Because the quantities ΔV and V would have the same units, the denominator of Eq. (1–3) is dimensionless. Therefore, the units for E are the same as those for the pressure.

As stated before, liquids are very slightly compressible, indicating that it would take a very large change in pressure to produce a small change in volume. Thus, the magnitudes of E for liquids, as shown in Table 1.2, are very high. For this reason, liquids will be considered incompressible in this book, unless stated otherwise.

The term *bulk modulus* is not usually applied to gases, and the principles of thermodynamics must be applied to determine the change in volume of a gas with a change in pressure.

TABLE 1.2 Values for bulk modulus for selected liquids

Liquid	Bulk Modulus	
	(psi)	(MPa)
Ethyl alcohol	130 000	896
Benzene	154 000	1 062
Machine oil	189 000	1 303
Water	316 000	2 179
Glycerine	654 000	4 509
Mercury	3 590 000	24 750

☐ **EXAMPLE PROBLEM 1.4**

Compute the change in pressure that must be applied to water to change its volume by 1.0 percent.

Solution

The 1.0-percent volume change indicates that $\Delta V/V = -0.01$. Then, the required change in pressure is

$$\Delta p = -E[(\Delta V)/V] = [-316\ 000\ \text{psi}][-0.01] = 3160\ \text{psi}$$

**1.10
DENSITY, SPECIFIC
WEIGHT, AND
SPECIFIC GRAVITY**

Because the study of fluid mechanics typically deals with a continuously flowing fluid or with a small amount of a fluid at rest, it is most convenient to relate the mass and weight of the fluid to a given volume of the fluid. Thus, the properties of density and specific weight are defined as follows.

Density *is the amount of mass per unit volume of a substance.*

Therefore, using the greek letter ρ (rho) for density,

⇨ DENSITY

$$\rho = m/V \tag{1-4}$$

where V is the volume of the substance having a mass m. The units for density are kilograms per cubic meter in the SI system and slugs per cubic foot in the U.S. Customary System.

The American Society for Testing and Materials (ASTM) has published several standard test methods for measuring density that describe vessels having precisely known volumes, called *pycnometers*. The proper filling, handling, temperature control, and reading of these devices are prescribed. Two types are known as the *Bingham pycnometer* and the *Lipkin bicapillary pycnometer*. The standards also call for the precise determination of the mass of the fluids in the pycnometers to the nearest 0.1 mg using an analytical balance. See References 2, 3, 5, and 6.

Specific weight *is the amount of weight per unit volume of a substance.*

Using the Greek letter γ (gamma) for specific weight,

⇨ SPECIFIC WEIGHT

$$\gamma = w/V \tag{1-5}$$

where V is the volume of a substance having the weight w. The units for specific weight are newtons per cubic meter (N/m^3) in the SI system and pounds per cubic foot (lb/ft^3) in the U.S. Customary System.

It is often convenient to indicate the specific weight or density of a fluid in terms of its relationship to the specific weight or density of a common fluid. When the term *specific gravity* is used in this book, the reference fluid is pure water at 4°C. At that temperature, water has its greatest density. Then, specific gravity can be defined in either of two ways:

a. *Specific gravity* is the ratio of the density of a substance to the density of water at 4°C.

b. *Specific gravity* is the ratio of the specific weight of a substance to the specific weight of water at 4°C.

These definitions for specific gravity can be shown mathematically as

▷ **SPECIFIC GRAVITY**

$$sg = \frac{\gamma_s}{\gamma_w \ @ \ 4°C} = \frac{\rho_s}{\rho_w \ @ \ 4°C} \tag{1–6}$$

where the subscript s refers to the substance whose specific gravity is being determined and the subscript w refers to water. The properties of water at 4°C are constant, having the values shown below.

$$\gamma_w \ @ \ 4°C = 9.81 \ kN/m^3 \qquad \gamma_w \ @ \ 4°C = 62.4 \ lb/ft^3$$
$$\text{or}$$
$$\rho_w \ @ \ 4°C = 1000 \ kg/m^3 \qquad \rho_w \ @ \ 4°C = 1.94 \ slugs/ft^3$$

Therefore, the mathematical definition of specific gravity can be written

$$sg = \frac{\gamma_s}{9.81 \ kN/m^3} = \frac{\rho_s}{1000 \ kg/m^3} \ \text{or} \ sg = \frac{\gamma_s}{62.4 \ lb/ft^3} = \frac{\rho_s}{1.94 \ slugs/ft^3} \tag{1–7}$$

This definition holds regardless of the temperature at which the specific gravity is being determined.

The properties of fluids do, however, vary with temperature. In general, the density (and therefore, the specific weight and specific gravity) decreases with increasing temperature. The properties of water at various temperatures are listed in Appendix A. The properties of other liquids at a few selected temperatures are listed in Appendix B and Appendix C.

You should seek other references for data on specific gravity at specified temperatures if they are not reported in the appendix and if high precision is desired. One estimate that gives reasonable accuracy for petroleum oils, as presented more fully in References 8 and 10, is that the specific gravity of oils decreases approximately 0.036 for a 100°F (37.8°C) rise in temperature. This applies for nominal values of s.g. from 0.80 to 1.00 and for temperatures in the range from approximately 32°F to 400°F (0°C to 204°C).

Some industry sectors prefer modified definitions for specific gravity. Instead of using the properties of water at 4°C (39.2°F) as the basis, the petroleum industry and others use water at 60°F (15.6°C). This makes little difference for typical design and analysis. Although the density of water at 4°C is 1000.00 kg/m³, at 60°F it is 999.04 kg/m³. The difference is less than 0.1 percent. References 2, 3, 5, 6, 7, and 10 contain more extensive tables of the properties of water at temperatures from 0°C to 100°C (32°F to 212°F).

Specific gravity in the Baumé and API scales is discussed in Section 1.10.2. We will continue to use water at 4°C as the basis for specific gravity in this book.

The ASTM also refers to the property of specific gravity as *relative density*. See References 2–6.

1.10.1
Relation Between Density and Specific Weight

Quite often the specific weight of a substance must be found when its density is known and vice versa. The conversion from one to the other can be made by the following equation:

▷ **γ-ρ RELATION**

$$\gamma = \rho g \qquad (1\text{–}8)$$

where g is the acceleration due to gravity. This equation can be justified by referring to the definitions of density and specific gravity and by using the equation relating mass to weight, $w = mg$.

The definition of specific weight is

$$\gamma = \frac{w}{V}$$

By multiplying both the numerator and the denominator of this equation by g we obtain

$$\gamma = \frac{wg}{Vg}$$

But $m = w/g$. Therefore, we have

$$\gamma = \frac{mg}{V}$$

Since $\rho = m/V$, we get

$$\gamma = \rho g$$

The following problems illustrate the definitions of the basic fluid properties presented above and the relationships between the various properties.

☐ **EXAMPLE PROBLEM 1.5**   Calculate the weight of a reservoir of oil if it has a mass of 825 kg.

Solution   Since $w = mg$,

$$w = 825 \text{ kg} \times 9.81 \text{ m/s}^2 = 8093 \text{ kg·m/s}^2$$

Substituting the newton for the unit kg·m/s², we have

$$w = 8093 \text{ N} = 8.093 \times 10^3 \text{ N} = 8.093 \text{ kN}$$

□ **EXAMPLE PROBLEM 1.6**   If the reservoir from Example Problem 1.5 has a volume of 0.917 m³, compute the density, specific weight, and the specific gravity of the oil.

Solution   Density:

$$\rho = \frac{m}{V} = \frac{825 \text{ kg}}{0.917 \text{ m}^3} = 900 \text{ kg/m}^3$$

Specific weight:

$$\gamma = \frac{w}{V} = \frac{8.093 \text{ kN}}{0.917 \text{ m}^3} = 8.83 \text{ kN/m}^3$$

Specific gravity:

$$sg = \frac{\rho_o}{\rho_w @ 4°C} = \frac{900 \text{ kg/m}^3}{1000 \text{ kg/m}^3} = 0.90$$

□ **EXAMPLE PROBLEM 1.7**   Glycerine at 20°C has a specific gravity of 1.263. Compute its density and specific weight.

Solution   Density:

$$\rho_g = (sg)_g(1000 \text{ kg/m}^3) = (1.263)(1000 \text{ kg/m}^3) = 1263 \text{ kg/m}^3$$

Specific weight:

$$\gamma_g = (sg)_g(9.81 \text{ kN/m}^3) = (1.263)(9.81 \text{ kN/m}^3) = 12.39 \text{ kN/m}^3$$

□ **EXAMPLE PROBLEM 1.8**   A pint of water weighs 1.041 lb. Find its mass.

Solution   Since $w = mg$, the mass is

$$m = \frac{w}{g} = \frac{1.041 \text{ lb}}{32.2 \text{ ft/s}^2} = \frac{1.041 \text{ lb-s}^2}{32.2 \text{ ft}}$$
$$= 0.0323 \text{ lb-s}^2/\text{ft} = 0.0323 \text{ slugs}$$

Remember that the units of slugs and lb-s²/ft are the same.

□ **EXAMPLE PROBLEM 1.9**   One gallon of mercury has a mass of 3.51 slugs. Find its weight.

Solution

$$w = mg = 3.51 \text{ slugs} \times 32.2 \text{ ft/s}^2 = 113 \text{ slug-ft/s}^2$$

This is correct, but the units may seem confusing since weight is normally expressed in pounds. The units of mass may be rewritten as lb-s²/ft.

$$w = mg = 3.51 \frac{\text{lb-s}^2}{\text{ft}} \times \frac{32.2 \text{ ft}}{\text{s}^2} = 113 \text{ lb}$$

1.10.2
Specific Gravity in Degrees
Baumé or Degrees API

The reference temperature for measurements in the Baumé or API scale is 60°F rather than 4°C as defined before. To emphasize this difference, the API or Baumé specific gravity is often reported as

$$\text{Specific gravity } \frac{60°}{60°} \text{ F}$$

This notation indicates that both the reference fluid (water) and the oil are at 60°F.

Specific gravities of crude oils vary widely depending on where they are found. Those from the western U.S. range from approximately 0.87 to 0.92. Eastern U.S. oil fields produce oil at about 0.82 specific gravity. Mexican crude oil is among the highest at 0.97. A few heavy asphaltic oils have sg > 1.0. (See Reference 7.)

Most oils are distilled before use to enhance their quality of burning. The resulting gasolines, kerosenes, and fuel oils have specific gravities ranging from about 0.67 to 0.98.

The equation used to compute specific gravity when the degrees Baumé are known is different for fluids lighter than water and fluids heavier than water. For liquids heavier than water,

$$\text{sg} = \frac{145}{145 - \text{deg Baumé}} \tag{1–9}$$

Or, to compute the degrees Baumé for a given specific gravity,

$$\text{deg Baumé} = 145 - \frac{145}{\text{sg}} \tag{1–10}$$

For liquids lighter than water,

$$\text{sg} = \frac{140}{130 + \text{deg Baumé}} \tag{1–11}$$

$$\text{deg Baumé} = \frac{140}{\text{sg}} - 130 \tag{1–12}$$

The American Petroleum Institute (API) has developed the API scale, slightly different from the Baumé scale, for liquids lighter than water. The formulas are

$$\text{sg} = \frac{141.5}{131.5 + \text{deg API}} \tag{1–13}$$

$$\text{deg API} = \frac{141.5}{\text{sg}} - 131.5 \tag{1–14}$$

Degrees API for oils may range from 10 to 80. Most fuel grades will fall in the range of API 20 to 70, corresponding to specific gravities from 0.93 to 0.70. Note that the heavier oils have the lower values of degrees API.

ASTM Standards D287 and D1298 (References 1 and 4) describe standard test methods for determining API gravity using a *hydrometer*. Figure

1.5 shows a sketch of a typical hydrometer incorporating a weighted glass bulb with a smaller diameter stem at the top that is designed to float upright in the test liquid. Based on the principles of buoyancy (see Chapter 5), the hydrometer rests at a position that is dependent on the density of the liquid. The stem is marked with a calibrated scale from which the direct reading of density, specific gravity, or API gravity can be made. Because of the importance of temperature to an accurate measurement of density, some hydrometers, called *thermohydrometers,* have a built-in precision thermometer.

FIGURE 1.5 Hydrometer with built-in thermometer (thermohydrometer).

Direct reading scale

Precision thermometer

Ballast

REFERENCES

1. American Society for Testing and Materials (ASTM). 1987. *Standard D287-82 (Reaffirmed 1987): Standard Test Method for API Gravity of Crude Petroleum and Petroleum Products (Hydrometer Method).* Philadelphia: Author.

2. ———. 1988. *Standard D941-88: Standard Test Method for Density and Relative Density (Specific Gravity) of Liquids by Lipkin Bicapillary Pycnometer.* Philadelphia: Author.

3. ———. 1986. *Standard D1217-86: Standard Test*

Method for Density and Relative Density (Specific Gravity) of Liquids by Bingham Pycnometer. Philadelphia: Author.

4. ———. 1985. *Standard D1298-85: Standard Test Method for Density, Relative Density (Specific Gravity), or API Gravity of Crude Petroleum and Liquid Petroleum Products by Hydrometer Method.* Philadelphia: Author.

5. ———. 1986. *Standard D1480-86: Standard Test Method for Density and Relative Density (Specific Gravity) of Viscous Materials by Bingham Pycnometer.* Philadelphia: Author.

6. ———. 1986. *Standard D1481-86: Standard Test Method for Density and Relative Density (Specific Gravity) of Viscous Materials by Lipkin Bicapillary Pycnometer.* Philadelphia: Author.

7. Avallone, Eugene A., and Theodore Baumeister III, eds. 1987. *Marks' Standard Handbook for Mechanical Engineers.* 9th ed. New York: McGraw-Hill.

8. Bolz, Ray E., and George L. Tuve, eds. 1973. *CRC Handbook of Tables for Applied Engineering Science.* 2nd ed. Boca Raton, Florida: CRC Press, Inc.

9. Cheremisinoff, N. P., ed. 1986. *Encyclopedia of Fluid Mechanics.* Vol. 1, *Flow Phenomena and Measurement.* Houston, Texas: Gulf Publishing Co.

10. Heald, C. C., ed. 1988. *Cameron Hydraulic Data.* 17th ed. Woodcliff Lake, New Jersey: Ingersoll-Rand.

11. Miller, R. W. 1983. *Flow Measurement Engineering Handbook.* New York: McGraw-Hill.

PRACTICE PROBLEMS

Conversion Factors

1.1 Convert 1250 millimeters to meters.

1.2 Convert 1600 square millimeters to square meters.

1.3 Convert 3.65×10^3 cubic millimeters to cubic meters.

1.4 Convert 2.05 square meters to square millimeters.

1.5 Convert 0.391 cubic meters to cubic millimeters.

1.6 Convert 55.0 gallons to cubic meters.

1.7 An automobile is moving at 80 kilometers per hour. Calculate its speed in meters per second.

1.8 Convert a length of 25.3 feet to meters.

1.9 Convert a distance of 1.86 miles to meters.

1.10 Convert a length of 8.65 inches to millimeters.

1.11 Convert a distance of 2580 feet to meters.

1.12 Convert a volume of 480 cubic feet to cubic meters.

1.13 Convert a volume of 7390 cubic centimeters to cubic meters.

1.14 Convert a volume of 6.35 liters to cubic meters.

1.15 Convert 6.0 feet per second to meters per second.

1.16 Convert 2500 cubic feet per minute to cubic meters per second.

(*Note:* In all Practice Problems sections in this book, the problems will use both SI and U.S. Customary System units. When SI units are used, "M" will follow the problem number and the problem will be printed in *italic*. When U.S. Customary System units are used, "E" will follow the problem number. When both systems of units are combined in a problem, "C" will follow the problem number.)

Consistent Units in an Equation

A body moving with constant velocity obeys the relationship $s = vt$, where s = distance, v = velocity, and t = time.

1.17M *A car travels 0.50 km in 10.6 s. Calculate its average speed in m/s.*

1.18M *In an attempt at a land speed record, a car travels 1.50 km in 5.2 s. Calculate its average speed in km/h.*

1.19E A car travels 1000 ft in 14 s. Calculate its average speed in mi/h.

1.20E In an attempt at a land speed record, a car travels a mile in 5.7 s. Calculate its average speed in mi/h.

A body starting from rest with constant acceleration moves according to the relationship $s = \frac{1}{2}at^2$, where s = distance, a = acceleration, and t = time.

1.21M *If a body moves 3.2 km in 4.7 min while moving with constant acceleration, calculate the acceleration in m/s².*

1.22M *An object is dropped from a height of 13 m. Neglecting air resistance, how long would it take for the body to strike the ground? Use $a = g = 9.81$ m/s².*

1.23C If a body moves 3.2 km in 4.7 min while moving with constant acceleration, calculate the acceleration in ft/s^2.

1.24E An object is dropped from a height of 53 in. Neglecting air resistance, how long would it take for the body to strike the ground? Use $a = g = 32.2$ ft/s^2.

The formula for kinetic energy is $KE = \frac{1}{2}mv^2$, where $m =$ mass and $v =$ velocity.

1.25M *Calculate the kinetic energy in N·m of a 15-kg mass if it has a velocity of 1.20 m/s.*

1.26M *Calculate the kinetic energy in N·m of a 3600-kg truck moving at 16 km/h.*

1.27M *Calculate the kinetic energy in N·m of a 75-kg box moving on a conveyor at 6.85 m/s.*

1.28M *Calculate the mass of a body in kg if it has a kinetic energy of 38.6 N·m when moving at 31.5 km/h.*

1.29M *Calculate the mass of a body in g if it has a kinetic energy of 94.6 mN·m when moving at 2.25 m/s.*

1.30M *Calculate the velocity in m/s of a 12-kg object if it has a kinetic energy of 15 N·m.*

1.31M *Calculate the velocity in m/s of a 175-g body if it has a kinetic energy of 212 mN·m.*

1.32E Calculate the kinetic energy in ft-lb of a 1-slug mass if it has a velocity of 4 ft/s.

1.33E Calculate the kinetic energy in ft-lb of an 8000-lb truck moving 10 mi/h.

1.34E Calculate the kinetic energy in ft-lb of a 150-lb box moving on a conveyor at 20 ft/s.

1.35E Calculate the mass of a body in slugs if it has a kinetic energy of 15 ft-lb when moving at 2.2 ft/s.

1.36E Calculate the weight of a body in lb if it has a kinetic energy of 38.6 ft-lb when moving at 19.5 mi/h.

1.37E Calculate the velocity in ft/s of a 30-lb object if it has a kinetic energy of 10 ft-lb.

1.38E Calculate the velocity in ft/s of a 6-oz body if it has a kinetic energy of 30 in-oz.

One measure of a baseball pitcher's performance is his earned run average, or ERA. It is the average number of earned runs he has allowed if all the innings he has pitched were converted to equivalent nine-inning games. Therefore, the units for ERA are runs per game.

1.39 If a pitcher has allowed 39 runs during 141 innings, calculate his ERA.

1.40 A pitcher has an ERA of 3.12 runs/game and has pitched 150 innings. How many earned runs has he allowed?

1.41 A pitcher has an ERA of 2.79 runs/game and has allowed 40 earned runs. How many innings has he pitched?

1.42 A pitcher has allowed 49 earned runs during 123 innings. Calculate his ERA.

The Definition of Pressure

1.43E Compute the pressure produced in the oil in a closed cylinder by a piston exerting a force of 2500 lb on the enclosed oil. The piston has a diameter of 3.00 in.

1.44E A hydraulic cylinder must be able to exert a force of 8700 lb. The piston diameter is 1.50 in. Compute the required pressure in the oil.

1.45M *Compute the pressure produced on the oil in a closed cylinder by a piston exerting a force of 12.0 kN on the enclosed oil. The piston has a diameter of 75 mm.*

1.46M *A hydraulic cylinder must be able to exert a force of 38.8 kN. The piston diameter is 40 mm. Compute the required pressure in the oil.*

1.47E The hydraulic lift for an automobile service garage has a cylinder with a diameter of 8.0 in. What pressure must the oil have to be able to lift 6000 lb?

1.48E A coining press is used to produce commemorative coins with the likenesses of all the U.S. presidents. The coining process requires a force of 18 000 lb. The hydraulic cylinder has a diameter of 2.50 in. Compute the required oil pressure.

1.49M *The maximum pressure that can be developed for a certain fluid power cylinder is 20.5 MPa. Compute the force it can exert if its piston diameter is 50 mm.*

1.50E The maximum pressure that can be developed for a certain fluid power cylinder is 6000 psi. Compute the force it can exert if its piston diameter is 2.00 in.

1.51E The maximum pressure that can be developed for a certain fluid power cylinder is 5000 psi. Compute the required diameter for the piston if the cylinder must exert a force of 20 000 lb.

1.52M *The maximum pressure that can be developed for a certain fluid power cylinder is 15.0 MPa. Compute* *the required diameter for the piston if the cylinder must exert a force of 30 kN.*

1.53E A line of fluid power cylinders has a range of diameters in 1.00-in increments from 1.00 in to 8.00 in. Compute the force that could be exerted by each cylinder with a fluid pressure of 500 psi. Draw a graph of the force versus cylinder diameter.

1.54E A line of fluid power cylinders has a range of diameters in 1.00-in increments from 1.00 in to 8.00 in. Compute the pressure required by each cylinder if it must exert a force of 5000 lb. Draw a graph of the pressure versus cylinder diameter.

1.55C Determine your own weight in newtons. Then, compute the pressure in pascals that would be created in an oil if you stood on a piston in a cylinder with a 20-mm diameter. Convert the resulting pressure to psi.

1.56C For the pressure you computed in Problem 1.55, compute the force in newtons that could be exerted on a piston with 250-mm diameter. Then, convert the resulting force to pounds.

Bulk Modulus

1.57C Compute the pressure change required to cause a decrease in the volume of ethyl alcohol by 1.00 percent. Express the result in both psi and MPa.

1.58C Compute the pressure change required to cause a decrease in the volume of mercury by 1.00 percent. Express the result in both psi and MPa.

1.59C Compute the pressure change required to cause a decrease in the volume of machine oil by 1.00 percent. Express the result in both psi and MPa.

1.60E For the conditions described in Problem 1.59, assume that the 1.00 percent volume change occurred in a cylinder with an inside diameter of 1.00 in and a length of 12.00 in. Compute the axial distance the piston would travel as the volume change occurs.

1.61E A certain hydraulic system operates at 3000 psi. Compute the percentage change in the volume of the oil in the system as the pressure is increased from zero to 3000 psi if the oil is similar to the machine oil listed in Table 1.2.

1.62M *A certain hydraulic system operates at 20.0 MPa. Compute the percentage change in the volume of the oil in the system if the oil is similar to the machine oil listed in Table 1.2.*

1.63E A measure of the stiffness of a linear actuator system is the amount of force required to cause a certain linear deflection. For an actuator that has an inside diameter of 0.50 in and a length of 42.0 in and that is filled with a machine oil, compute the stiffness in lb/in.

1.64E Repeat Problem 1.63 but change the length of the cylinder to 10.0 in. Compare the results.

1.65E Repeat Problem 1.63 but change the cylinder diameter to 2.00 in. Compare the results.

1.66E Using the results of Problems 1.63, 1.64, and 1.65 together, generate a statement about the general design approach to achieve a very stiff system.

Force and Mass

1.67M *Calculate the mass of a can of oil if it weighs 610 N.*

1.68M *Calculate the mass of a tank of gasoline if it weighs 1.35 kN.*

1.69M *Calculate the weight of 1 m³ of kerosene if it has a mass of 825 kg.*

1.70M *Calculate the weight of a jar of castor oil if it has a mass of 450 g.*

1.71E Calculate the mass of 1 gal of oil if it weighs 7.8 lb.

1.72E Calculate the mass of 1 ft³ of gasoline if it weighs 42.0 lb.

1.73E Calculate the weight of 1 ft³ of kerosene if it has a mass of 1.58 slugs.

1.74E Calculate the weight of 1 gal of water if it has a mass of 0.258 slug.

1.75C Assume that a man weighs 160 lb (force).
  a. Compute his mass in slugs.
  b. Compute his weight in N.
  c. Compute his mass in kg.

1.76C In the United States, hamburger and other meats are sold by the pound. Assuming that this is 1.00-lb force, compute the mass in slugs, the mass in kg, and the weight in N.

1.77M *The metric ton is 1000 kg (mass). Compute the force in Newtons required to lift it.*

1.78C Convert the force found in Problem 1.77M to lb.

1.79C Determine your own weight in lb and N and your own mass in slugs and kg.

Density, Specific Weight, and Specific Gravity

1.80M *The specific gravity of benzene is 0.876. Calculate its specific weight and its density in SI units.*

1.81M *Air at 16°C and standard atmospheric pressure has a specific weight of 12.02 N/m³. Calculate its density.*

1.82M *Carbon dioxide has a density of 1.964 kg/m³ at 0°C. Calculate its specific weight.*

1.83M *A certain medium lubricating oil has a specific weight of 8.860 kN/m³ at 5°C and 8.483 kN/m³ at 50°C. Calculate its specific gravity at each temperature.*

1.84M *At 100°C mercury has a specific weight of 130.4 kN/m³. What volume of the mercury would weigh 2.25 kN?*

1.85M *A cylindrical can, 150 mm in diameter, is filled to a depth of 100 mm with a fuel oil. The oil has a mass of 1.56 kg. Calculate its density, specific weight, and specific gravity.*

1.86M *Glycerine has a specific gravity of 1.258. How much would 0.50 m³ of glycerine weigh? What would be its mass?*

1.87M *The fuel tank of an automobile holds 0.095 m³. If it is full of gasoline having a specific gravity of 0.68, calculate the weight of the gasoline.*

1.88M *The density of muriatic acid is 1200 kg/m³. Calculate its specific weight and its specific gravity.*

1.89M *Liquid ammonia has a specific gravity of 0.826. Calculate the volume of ammonia that would weigh 22.0 N.*

1.90M *Vinegar has a density of 1080 kg/m³. Calculate its specific weight and its specific gravity.*

1.91M *Methyl alcohol has a specific gravity of 0.789. Calculate its density and its specific weight.*

1.92M *A cylindrical container is 150 mm in diameter and weighs 2.25 N when empty. When filled to a depth of 200 mm with a certain oil, it weighs 35.4 N. Calculate the specific gravity of the oil.*

1.93M *A storage vessel for gasoline (sg = 0.68) is a vertical cylinder 10 m in diameter. If it is filled to a depth of 6.75 m, calculate the weight and mass of the gasoline.*

1.94M *What volume of mercury (sg = 13.54) would weigh the same as 0.020 m³ of castor oil, which has a specific weight of 9.42 kN/m³?*

1.95M *A rock has a specific gravity of 2.32 and a volume of 1.42×10^{-4} m³. How much does it weigh?*

1.96E The specific gravity of benzene is 0.876. Calculate its specific weight and its density in U.S. Customary System units.

1.97E Air at 59°F and standard atmospheric pressure has a specific weight of 0.0765 lb/ft³. Calculate its density.

1.98E Carbon dioxide has a density of 0.003 81 slug/ft³ at 32°F. Calculate its specific weight.

1.99E A certain medium lubricating oil has a specific weight of 56.4 lb/ft³ at 40°F and 54.0 lb/ft³ at 120°F. Calculate its specific gravity at each temperature.

1.100E At 212°F mercury has a specific weight of 834 lb/ft³. What volume of the mercury would weigh 500 lb?

1.101E One gal of a certain fuel oil weighs 7.50 lb. Calculate its specific weight, its density, and its specific gravity.

1.102E Glycerine has a specific gravity of 1.258. How much would 50 gal of glycerine weigh?

1.103E The fuel tank of an automobile holds 25.0 gal. If it is full of gasoline having a density of 1.32 slugs/ft³, calculate the weight of the gasoline.

1.104C The density of muriatic acid is 1.20 g/cm³. Calculate its density in slugs/ft³, its specific weight in lb/ft³, and its specific gravity. (Note that specific gravity and density in g/cm³ are numerically equal.)

1.105C Liquid ammonia has a specific gravity of 0.826. Calculate the volume in cm³ that would weigh 5.0 lb.

1.106C Vinegar has a density of 1.08 g/cm³. Calculate its specific weight in lb/ft³.

1.107C Alcohol has a specific gravity of 0.79. Calculate its density both in slugs/ft³ and g/cm³.

1.108E A cylindrical container has a 6.0-in diameter and weighs 0.50 lb when empty. When filled to a depth of 8.0 in with a certain oil, it weighs 7.95 lb. Calculate the specific gravity of the oil.

1.109E A storage vessel for gasoline (sg = 0.68) is a vertical cylinder 30 ft in diameter. If it is filled to a depth of 22 ft, calculate the number of gallons in the tank and the weight of the gasoline.

1.110E How many gal of mercury (sg = 13.54) would

weigh the same as 5 gal of castor oil, which has a specific weight of 59.69 lb/ft³?

1.111E A rock has a specific gravity of 2.32 and a volume of 8.64 in³. How much does it weigh?

COMPUTER PROGRAMMING ASSIGNMENT

1. Write a program that computes the specific weight of water for a given temperature using the data from Appendix A. Such a program could be part of a more comprehensive program to be written later. The following options could be used:

   a. Enter the table data for specific weight as a function of temperature into an array. Then, for a specified temperature, search the array for the corresponding specific weight. Interpolate temperatures between values given in the table.

   b. Include data in both SI units and U.S. Customary System units.

   c. Include density.

   d. Include checks in the program to ensure that the specified temperature is within the range given in the tables (i.e., above freezing point and below boiling point).

   e. Instead of using the table look-up approach, use a curve-fit technique to obtain equations of the properties of water versus temperature. Then compute the desired property value for any specified temperature.

2 Viscosity of Fluids

■ ■ ■ ■

2.1 OBJECTIVES

The ease with which a fluid pours is an indication of its viscosity. Cold oil has a high viscosity and pours very slowly, whereas water has a relatively low viscosity and pours quite readily. We define *viscosity* as the property of a fluid that offers resistance to the relative motion of fluid molecules. The energy loss due to friction in a flowing fluid is due to its viscosity. Viscosity is used in problem solutions beginning in Chapter 8 of this book so you may choose to delay coverage of the material in this chapter until you are ready to cover that chapter. The material is placed here for those who desire to learn about all fluid properties at the same time.

After completing this chapter, you should be able to:

1. Define *dynamic viscosity*.
2. Define *kinematic viscosity*.
3. Identify the units of viscosity in both the SI system and the U.S. Customary System.
4. Describe the difference between a *newtonian fluid* and a *nonnewtonian fluid*.
5. Describe the methods of viscosity measurement using the *rotating drum viscometer,* the *capillary tube viscometer,* the *falling ball viscometer,* and the *Saybolt Universal viscometer.*
6. Describe the variation of viscosity with temperature for both liquids and gases.
7. Define *viscosity index*.
8. Describe the viscosity of lubricants using the SAE viscosity numbers and the ISO viscosity grades.

2.2 DYNAMIC VISCOSITY

As a fluid moves there is developed in it a shear stress, the magnitude of which depends on the viscosity of the fluid. *Shear stress,* denoted by the Greek letter τ (tau), can be defined as the force required to slide one unit area layer of a substance over another. Thus, τ is a force divided by an area and can be measured in the units of newtons per square meter or lb/ft^2. In a fluid such as water, oil, alcohol, or other common liquids we find that the magnitude of the shearing stress is directly proportional to the change of velocity between different positions in the fluid.

Figure 2.1 illustrates the concept of velocity change in a fluid by showing a thin layer of fluid between two surfaces, one of which is stationary while the other is moving. A fundamental condition that exists when a real

FIGURE 2.1 Velocity gradient in a moving fluid.

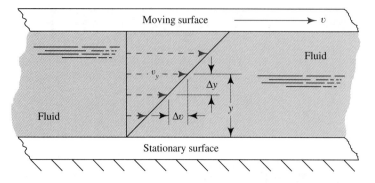

fluid is in contact with a boundary surface is that the fluid has the same velocity as the boundary. In Fig. 2.1, then, the fluid in contact with the lower surface has a zero velocity and that in contact with the upper surface has the velocity v. If the distance between the two surfaces is small, then the rate of change of velocity with position y is linear. That is, it varies in a straight-line manner. The *velocity gradient* is a measure of the velocity change and is defined as $\Delta v/\Delta y$. This is also called the *shear rate*.

The fact that the shear stress in the fluid is directly proportional to the velocity gradient can be stated mathematically as

$$\tau = \mu(\Delta v/\Delta y) \tag{2--1}$$

where the constant of proportionality μ (the Greek letter mu) is called the *dynamic viscosity* of the fluid.

You can visualize the physical interpretation of Eq. (2–1) by stirring a fluid with a rod. The action of stirring causes a velocity gradient to be created in the fluid. A greater force is required to stir a cold oil with a high viscosity (a high value of μ) than is required to stir water with a low viscosity. This is an indication of the higher shear stress in the cold oil.

The direct application of Eq. (2–1) is used in some types of viscosity measuring devices as will be explained later.

2.2.1
Units for Dynamic
Viscosity

Many different unit systems are used to express viscosity. The systems used most frequently are described here for dynamic viscosity and in the next section for kinematic viscosity. Summary tables listing many conversion factors are included in Appendix K.

The definition of dynamic viscosity can be derived from Eq. (2–1) by solving for μ.

▷ **DYNAMIC VISCOSITY**

$$\mu = \frac{\tau}{\Delta v/\Delta y} = \tau\left(\frac{\Delta y}{\Delta v}\right) \tag{2--2}$$

The units for μ can be derived by substituting the SI units only into Eq. (2–2) as follows.

$$\mu = \frac{N}{m^2} \times \frac{m}{m/s} = \frac{N \cdot s}{m^2}$$

Since Pa is another name for N/m^2, we can also express μ as

$$\mu = \text{Pa} \cdot \text{s}$$

Sometimes, when units for μ are being combined with other terms—especially density—it is convenient to express μ in terms of kg rather than N. Since $1\text{N} = 1 \text{ kg} \cdot \text{m/s}^2$, μ can be expressed as

$$\mu = \text{N} \times \frac{\text{s}}{\text{m}^2} = \frac{\text{kg} \cdot \text{m}}{\text{s}^2} \times \frac{\text{s}}{\text{m}^2} = \frac{\text{kg}}{\text{m} \cdot \text{s}}$$

Thus, either N·s/m^2, Pa·s, or kg/m·s may be used for μ in the SI system.

Table 2.1 lists the dynamic viscosity units in the three most widely used systems. The dimensions of force multiplied by time divided by length squared are evident in each system. The units of poise and centipoise are listed here because much published data are given in these units. They are part of the obsolete metric system called cgs, derived from its base units of centimeter, gram, and second. Conversion factors are given in Appendix K.

TABLE 2.1

Unit System	Dynamic Viscosity Units
International System (SI)	N·s/m^2, Pa·s, or kg/m·s
U.S. Customary System	lb-s/ft^2 or slug/ft-s
cgs system (obsolete)	poise = dyne·s/cm^2 = g/(cm·s) = 0.1 Pa·s
	centipoise = poise/100 = 0.001 Pa·s = 1.0 mPa·s

2.3
KINEMATIC VISCOSITY

⇨ KINEMATIC VISCOSITY

Many calculations in fluid mechanics involve the ratio of the dynamic viscosity to the density of the fluid. As a matter of convenience, the kinematic viscosity ν (the Greek letter nu) is defined as

$$\nu = \mu/\rho \tag{2–3}$$

Since μ and ρ are both properties of the fluid, ν is also a property.

2.3.1
Units for Kinematic Viscosity

We can derive the SI units for kinematic viscosity by substituting the previously developed units for μ and ρ:

$$\nu = \frac{\mu}{\rho} = \mu \left(\frac{1}{\rho}\right)$$

$$\nu = \frac{\text{kg}}{\text{m} \cdot \text{s}} \times \frac{\text{m}^3}{\text{kg}}$$

$$\nu = \text{m}^2/\text{s}$$

Table 2.2 lists the kinematic viscosity units in the three most widely used systems. The basic dimensions of length squared divided by time are evident in each system. The units of stoke and centistoke are listed because published data often employ these units. Appendix K lists conversion factors.

TABLE 2.2

Unit System	Kinematic Viscosity Units
International System (SI)	m²/s
U.S. Customary System	ft²/s
cgs system (obsolete)	stoke = cm²/s = 1×10^{-4} m²/s
	centistoke = stoke/100 = 1×10^{-6} m²/s

2.4
NEWTONIAN FLUIDS
AND NONNEWTONIAN
FLUIDS

The study of the deformation and flow characteristics of substances is called *rheology*, which is the field from which we learn about the viscosity of fluids. One important distinction you should understand is between a *newtonian fluid* and a *nonnewtonian fluid*. Any fluid that behaves in accordance with Eq. (2–1) is called a *newtonian fluid*. The viscosity μ is a function only of the condition of the fluid, particularly its temperature. The magnitude of the velocity gradient $\Delta v/\Delta y$ has no effect on the magnitude of μ. Most common fluids such as water, oil, gasoline, alcohol, kerosene, benzene, and glycerine are classified as newtonian fluids.

Conversely, a fluid that does not behave in accordance with Eq. (2–1) is called a *nonnewtonian fluid*. The difference between the two is shown in Fig. 2.2. The viscosity of the nonnewtonian fluid is dependent on the velocity gradient in addition to the condition of the fluid.

FIGURE 2.2   Newtonian and nonnewtonian fluids.

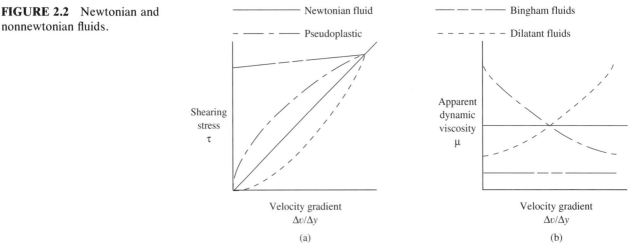

Note that in Fig. 2.2(a), the slope of the curve for shear stress versus the velocity gradient is a measure of the *apparent viscosity* of the fluid. The steeper the slope, the higher is the apparent viscosity. Because the newtonian fluids have a linear relationship between shear stress and velocity gradient, the slope is constant and, therefore, the viscosity is constant. The slope of the curves for nonnewtonian fluids varies. Figure 2.2(b) shows how viscosity changes with velocity gradient.

Two major classifications of nonnewtonian fluids are *time-independent* and *time-dependent*. As their name implies, time-independent fluids have a viscosity at any given shear stress that does not vary with time. The viscosity of time-dependent fluids, however, will change with time (see Reference [10]).

Three types of time-independent fluids can be defined:

- *Pseudoplastic*   The plot of shear stress versus velocity gradient lies above the straight, constant sloped line for newtonian fluids, as shown in Fig. 2.2. The curve begins steeply, indicating a high apparent viscosity. Then the slope decreases with increasing velocity gradient. Examples of such fluids are blood plasma, molten polyethylene, and water suspensions of clay.
- *Dilatant Fluids*   The plot of shear stress versus velocity gradient lies below the straight line for newtonian fluids. The curve begins with a low slope, indicating a low apparent viscosity. Then, the slope increases with increasing velocity gradient. Examples of dilatant fluids are corn starch in ethylene glycol, starch in water, and titanium dioxide.
- *Bingham Fluids*   Sometimes called *plug-flow fluids,* Bingham fluids require the development of a significant level of shear stress before flow will begin, as illustrated in Fig. 2.2. Once flow starts, there is an essentially linear slope to the curve indicating a constant apparent viscosity. Examples of Bingham fluids are chocolate, catsup, mustard, mayonnaise, toothpaste, paint, asphalt, some greases, and water suspensions of fly ash or sewage sludge.

Time-dependent fluids are very difficult to analyze because apparent viscosity varies with time as well as with velocity gradient and temperature. Examples of time-dependent fluids are some crude oils at low temperatures, printer's ink, nylon, some jellies, flour dough, and several polymer solutions. Such fluids are called *thixotropic* fluids.

Electrorheological fluids are being developed that offer very unique properties that are controllable by the application of an electric current. Sometimes referred to as ''ER fluids,'' they are suspensions of fine particles, such as starch, polymers, and ceramics, in a nonconducting oil, such as mineral oil or silicone oil. When no current is applied they behave like other liquids. But when a current is applied they turn to a gel and behave more like a solid. The change can occur in less than $1/1000$ of a second. Potential applications for such fluids are as a replacement for conventional valves, in clutches, in suspension systems for vehicles and machinery, and in automation actuators. (See Reference 11.)

2.5 VARIATION OF VISCOSITY WITH TEMPERATURE

You are probably familiar with some examples of the variation of fluid viscosity with temperature. Engine oil is generally quite difficult to pour when it is cold, indicating that it has a high viscosity. As the temperature of the oil is increased, its viscosity decreases noticeably.

All fluids exhibit this behavior to some extent. Appendix D gives two graphs of dynamic viscosity versus temperature for many common liquids.

TABLE 2.3

Fluid	Temperature (°C)	Dynamic Viscosity (N·s/m² or Pa·s)
Water	20	1.0×10^{-3}
Gasoline	20	3.1×10^{-4}
SAE 30 oil	20	3.5×10^{-1}
SAE 30 oil	80	1.9×10^{-2}

Notice that viscosity is plotted on a logarithmic scale because of the large range of numerical values. In order to check your ability to interpret these graphs, a few examples are listed in Table 2.3.

Gases behave differently from liquids in that the viscosity increases as the temperature increases. Also, the amount of change is generally smaller than that for liquids.

**2.5.1
Viscosity Index**

A measure of how greatly the viscosity of a fluid changes with temperature is given by its viscosity index, sometimes referred to as VI. This is especially important for lubricating oils and hydraulic fluids used in equipment that must operate at wide extremes of temperature.

> *A fluid with a high viscosity index exhibits a small change in viscosity with temperature. A fluid with a low viscosity index exhibits a large change in viscosity with temperature.*

Typical curves for oils with viscosity indexes of 50, 100, and 140 are shown in Fig. 2.3. Viscosity index is determined by measuring the viscosity of the

FIGURE 2.3 Typical viscosity index curves.

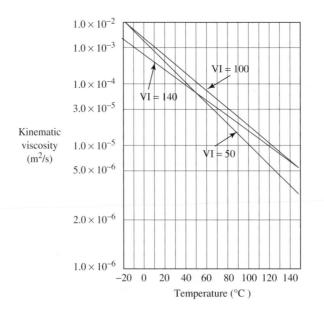

sample fluid at 40°C and 100°C and comparing these values with those of certain reference fluids.*

2.6
VISCOSITY
MEASUREMENT

Procedures and equipment for measuring viscosity are numerous. Some employ fundamental principles of fluid mechanics to indicate viscosity in its basic units. Others indicate only relative values for viscosity which can be used to compare different fluids. In this section we will describe several common methods used for viscosity measurement.

The American Society for Testing and Materials (ASTM) generates standards for viscosity measurement and reporting. Specific standards are cited in the sections that follow.

2.6.1
Rotating Drum Viscometer

The apparatus shown in Fig. 2.4(a) measures viscosity by the definition of dynamic viscosity given in Eq. (2–2):

$$\mu = \tau/(\Delta v/\Delta y) \qquad (2–2)$$

The outer drum is caused to rotate at a constant angular velocity ω while the inner drum is held stationary. Therefore, the fluid in contact with the rotating drum has a known linear velocity v while the fluid in contact with the inner drum has a zero velocity. If we know the thickness Δy of the fluid

---

* For a complete discussion of this method, see Standard ASTM D2270 of the American Society for Testing and Materials. (See Reference 3.)

FIGURE 2.4 Rotating drum viscometer. (Source of photo: Extech Instruments Corporation, Waltham, MA)

(a) Sketch of system components   (b) Commercially available viscometer sample, then we can calculate the term $\Delta v/\Delta y$ in Eq. (2–2). Special consideration is given to the fluid at the bottom of the drum since its velocity is not uniform at all points. Because of the fluid viscosity, a drag force exists on the surface of the inner drum, causing a torque to be developed which can be measured by a sensitive torquemeter. The magnitude of this torque is a measure of the shear stress τ in the fluid. Thus, the viscosity μ can be calculated from Eq. (2–2).

A commercially available device that uses a similar principle is shown in Fig. 2.4(b). The viscometer drives a special cylindrical rotor that is suspended in the fluid to be tested. The viscous drag on the cylinder causes the deflection of the meter with a scale calibrated in viscosity units. This device can be hand-held for in-plant operation or mounted on a stand for laboratory use.

A variant of the rotating drum viscometer is used in ASTM Standard D2602: *Standard Test Method for Apparent Viscosity of Engine Oils at Low Temperature Using the Cold-Cranking Simulator*. In this apparatus a universal motor drives a rotor which is closely fitted inside a stator. The test is run at 0°F (-17.78°C). The rotor speed is related to the viscosity of the test oil that fills the space between the stator and the rotor because of the viscous drag produced by the oil. Speed measurement is correlated to viscosity in centipoise (cP or mPa·s) by reference to a calibration chart prepared by running a set of at least five standard calibration oils of known viscosity on the particular apparatus being used. The resulting data are used by engine designers and users to ensure the proper operation of the engine at cold temperatures.

The Society of Automotive Engineers (SAE) specifies that a test to determine the maximum *borderline pumping temperature* (BPT) of engine oils be run. See Section 2.7 in this chapter. The test method and apparatus is described by ASTM Standard D3829: *Standard Test Method for Predicting the Borderline Pumping Temperature of Engine Oil*. The apparatus incorporates a rotary viscometer having a calibrated rotor-stator assembly. The time required for the rotor to make one revolution is measured with the test oil in the space between the rotor and stator and with the system at a known temperature. After a series of tests, the temperature at which the test oil has an apparent viscosity of 30 Pa·s (30 000 cP) is reported as the BPT. Another part of the standard method calls for the determination of a *critical yield stress*. Refer to the ASTM standard for this procedure. (See Reference 7).

2.6.2
Capillary Tube Viscometer

Figure 2.5 shows two reservoirs connected by a long, small-diameter tube called a *capillary tube*. As the fluid flows through the tube with a constant velocity, some energy is lost from the system, causing a pressure drop that can be measured by using manometers. The magnitude of the pressure drop is related to the fluid viscosity by the following equation, which is developed in a later chapter of this book:

$$\mu = \frac{(p_1 - p_2)D^2}{32vL} \tag{2–4}$$

FIGURE 2.5 Capillary tube viscometer.

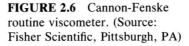

FIGURE 2.6 Cannon-Fenske routine viscometer. (Source: Fisher Scientific, Pittsburgh, PA)

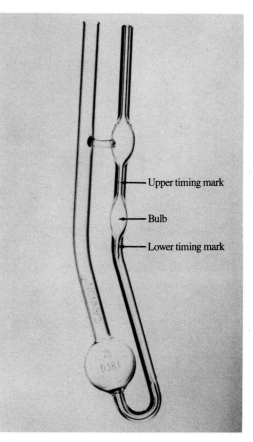

In Eq. (2–4), D is the inside diameter of the tube, v is the fluid velocity, and L is the length of the tube between points 1 and 2 where the pressure is measured.

2.6.3
Standard Calibrated Glass
Capillary Viscometers

ASTM Standards D445 and D446 (References 1 and 2) describe the use of standard glass capillary viscometers to measure the kinematic viscosity of transparent and opaque liquids. Figures 2.6 and 2.7 show two of the 17 types of viscometers discussed in the standard. Figure 2.8 shows a commercially available bath for holding the tubes and maintaining test temperatures within 0.01°C (0.02°F) throughout the bath.

In preparation for the viscosity test, the viscometer tube is charged with a specified quantity of test fluid.

After stabilizing at the test temperature, suction is used to draw fluid through the bulb and slightly above the upper timing mark. The suction is

FIGURE 2.7   Ubbelohde viscometer. (Source: Fisher Scientific, Pittsburgh, PA)

Upper timing mark

Bulb

Lower timing mark

FIGURE 2.8 Kinematic viscosity bath for holding standard calibrated glass capillary viscometers. (Source: Precision Scientific Petroleum Instruments Company, Bellwood, IL)

removed and the fluid is allowed to flow by gravity. The working section of the tube is the capillary below the lower timing mark. The time required for the leading edge of the meniscus to pass from the upper timing mark to the lower timing mark is recorded. The kinematic viscosity is computed by multiplying the flow time by the calibration constant of the viscometer supplied by the vendor. The viscosity unit used in these tests is the centistoke (cSt), which is equivalent to mm^2/s. This value must be multiplied by 10^{-6} to obtain the SI standard unit of m^2/s that is used for calculations in this book.

2.6.4
Falling Ball Viscometer

As a body falls in a fluid under the influence of gravity only, it will accelerate until the downward force (its weight) is just balanced by the buoyant force and the viscous drag force acting upward. Its velocity at that time is called the *terminal velocity*. The falling ball viscometer sketched in Fig. 2.9 uses this principle by causing a spherical ball to fall freely through the fluid and

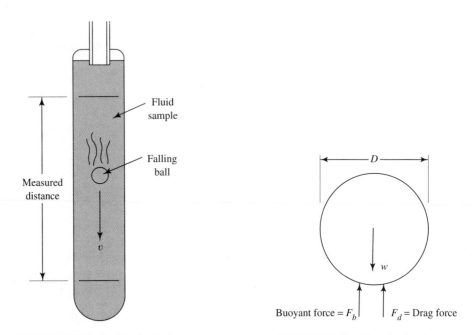

FIGURE 2.9   Falling ball viscometer.

FIGURE 2.10   Free body diagram of ball in falling ball viscometer.

measuring the time required for the ball to drop a known distance. Thus, the
velocity can be calculated. Figure 2.10 shows a free body diagram of the ball
where w is the weight of the ball, F_b is the buoyant force, and F_d is the
viscous drag force on the ball. When the ball has reached its terminal velocity, it is in equilibrium. Therefore, we have

$$w - F_b - F_d = 0. \qquad (2\text{--}5)$$

If γ_s is the specific weight of the sphere, γ_f is the specific weight of the fluid,
V is the volume of the sphere, and D is the diameter of the sphere, we have

$$w = \gamma_s V = \gamma_s \pi D^3/6 \qquad (2\text{--}6)$$
$$F_b = \gamma_f V = \gamma_f \pi D^3/6 \qquad (2\text{--}7)$$

For very viscous fluids and a small velocity, the drag force on the sphere is

$$F_d = 3\pi \mu v D \qquad (2\text{--}8)$$

(This will be discussed in Chapter 17.) Equation (2–5) then becomes

$$\gamma_s \pi D^3/6 - \gamma_f \pi D^3/6 - 3\pi \mu v D = 0$$

Solving for μ gives

$$\mu = \frac{(\gamma_s - \gamma_f)D^2}{18v} \qquad (2\text{--}9)$$

**2.6.5
Saybolt Universal
Viscometer**

The ease with which a fluid flows through a small-diameter orifice is an indication of its viscosity. This is the principle on which the Saybolt viscometer is based. The fluid sample is placed in an apparatus similar to that sketched in Fig. 2.11(a). After flow is established, the time required to collect 60 mL of the fluid is measured. The resulting time is reported as the viscosity of the fluid in Saybolt Seconds Universal (SSU or sometimes SUS). Since the measurement is not based on the basic definition of viscosity, the results are only relative. However, they do serve to compare the viscosities of different fluids. The advantage of this procedure is that it is simple and requires relatively unsophisticated equipment. Approximate conversion can be made from SSU to kinematic viscosity as shown in Appendix K. Figure 2.11(b) and (c) show a commercially available Saybolt viscometer and the 60 mL flask used to collect the sample. The use of the Saybolt viscometer was formerly covered by Standard ASTM D88. However, that standard is no longer supported by ASTM. Preference is now given to the use of the glass capillary viscometers described in ASTM D445 and D446 and discussed in Section 2.6.3 of this chapter.

**2.7
SAE VISCOSITY
GRADES**

The Society of Automotive Engineers (SAE) has developed a rating system for engine oils (Table 2.4) and gear and axle lubricants (Table 2.5) which indicates the viscosity of the oils at specified temperatures. Oils with a suffix W are based on maximum absolute viscosity at specified cold temperatures under conditions that simulate both the cranking of an engine and the pumping of the oil by the oil pump. They must also exhibit a kinematic viscosity above a specified minimum at 100°C. Those without the suffix W must have kinematic viscosities in the indicated ranges at 100°C. Multiviscosity oils, such as SAE 10W-30, must meet the standards at both the low- and high-temperature conditions.

The specification of maximum low-temperature viscosity values for oils is related to the ability of the oil to flow to the surfaces needing lubrication at the engine speeds encountered during starting at cold temperatures. The pumping viscosity indicates the ability of the oil to flow into the oil pump inlet of an engine. The high temperature viscosity range specifications relate to the ability of the oil to provide a satisfactory oil film to carry expected loads while not having an excessively high viscosity that would increase friction and energy losses generated by moving parts.

The following standards apply to the SAE classifications and the methods of testing:

SAE J300	*Engine Oil Viscosity Classification*
SAE J306	*Axle and Manual Transmission Lubricant Viscosity Classification*
ASTM D445	*Standard Test Method for Kinematic Viscosity of Transparent and Opaque Liquids*
ASTM D446	*Standard Specifications and Operating Instructions for Glass Capillary Kinematic Viscometers*

(a)

(b) Universal Saybolt viscometer

(c) 60 mL flask for collecting Saybolt sample

FIGURE 2.11 Saybolt viscometer. [Sources of photos: (b) Precision Scientific Petroleum Instruments Co., Bellwood, IL; (c) Corning, Inc., Corning, NY]

TABLE 2.4 SAE viscosity grades for engine oils

SAE Viscosity Grade	Low-Temperature—Dynamic Viscosity		High-Temperature Kinematic Viscosity at 100°C (cSt)[+]	
	Cranking Condition* (cP) Max. at (°C)	Pumping Condition[#] (cP) Max. at (°C)	Min.	Max.
0W	3250 at −30	30 000 at −35	3.8	—
5W	3500 at −25	30 000 at −30	3.8	—
10W	3500 at −20	30 000 at −25	4.1	—
15W	3500 at −15	30 000 at −20	5.6	—
20W	4500 at −10	30 000 at −15	5.6	—
25W	6000 at −5	30 000 at −10	9.3	—
20	—	—	5.6	<9.3
30	—	—	9.3	<12.5
40	—	—	12.5	<16.3
50	—	—	16.3	<21.9
60	—	—	21.9	<26.1

Source: Reprinted with permission from SAE J300 ©1991. Society of Automotive Engineers, Inc. Warrendale, PA. (See Reference 13.)

* Using modified ASTM Standard D2602

\# Using ASTM D4684

+ Using ASTM D445

TABLE 2.5 SAE viscosity grades for axle and manual transmission lubricants

SAE Viscosity Grade	Maximum Temperature for Dynamic Viscosity of 150 000 cP* (°C)	Kinematic Viscosity at 100°C (cSt)[#]	
		Min.	Max.
70W	−55	4.1	—
75W	−40	4.1	—
80W	−26	7.0	—
85W	−12	11.0	—
90	—	13.5	<24.0
140	—	24.0	<41.0
250	—	41.0	—

Source: Reprinted with permission from SAE J306 ©1986. Society of Automotive Engineers, Inc. Warrendale, PA. (See Reference 14.)

* Using ASTM D2983

\# Using ASTM D445

ASTM D2602	*Standard Test Method for Apparent Viscosity of Engine Oils at Low Temperature Using the Cold-Cranking Simulator*
ASTM D2983	*Method of Test for Apparent Viscosity at Low Temperature Using the Brookfield Viscometer*
ASTM D3829	*Standard Test Method for Predicting the Borderline Pumping Temperature of Engine Oil*
ASTM D4684	*Standard Test Method for Determination of Yield Stress and Apparent Viscosity of Engine Oils at Low Temperature*

Consult the latest revision of these standards.

See also Appendix C for typical properties of petroleum lubricating oils used in hydraulic systems and machine tool applications.

Note that oils designed to operate at wide ranges of temperatures have special additives to increase the viscosity index. An example is a multiviscosity engine oil that must meet stringent low-temperature viscosity limits while maintaining a sufficiently high viscosity at higher engine operating temperatures for effective lubrication. Also, automotive hydraulic system oils that must operate with similar performance in cold as well as warm climates and machine tool hydraulic system oils that must operate outdoors as well as indoors must have high viscosity indexes. See Appendix C for representative values.

Achieving a high viscosity index in an oil often calls for the blending of polymeric materials with the petroleum. The resulting blend may exhibit nonnewtonian characteristics, particularly at the lower temperatures.

**2.8
ISO VISCOSITY
GRADES**

Lubricants used in industrial applications must be available in a wide range of viscosities to meet the needs of production machinery, bearings, gear drives, electrical machines, fans and blowers, fluid power systems, mobile equipment, and many other devices. The designers of such systems must ensure that the lubricant can withstand the temperatures to be experienced while providing sufficient load-carrying ability. The result is a need for a wide range of viscosities.

To meet these requirements and still have an economical and manageable number of options, ASTM Standard D2422 *Standard Classification of Industrial Fluid Lubricants by Viscosity System* defines a set of 18 ISO viscosity grades. The standard designation includes the prefix ISO VG followed by a number representing the nominal viscosity in cSt (mm²/s) for a temperature of 40°C. Table 2.6 gives the data. The maximum and minimum values are +/− 10 percent from the nominal. Although the standard is voluntary, the intent is to encourage producers and users of lubricants to agree on the specification of viscosities from the list. This system is gaining favor throughout the world markets.

TABLE 2.6 ISO viscosity grades

Grade ISO VG	Kinematic Viscosity at 40°C (cSt) or (mm²/s)		
	Nominal	Minimum	Maximum
2	2.2	1.98	2.40
3	3.2	2.88	3.52
5	4.6	4.14	5.06
7	6.8	6.12	7.48
10	10	9.00	11.0
15	15	13.5	16.5
22	22	19.8	24.2
32	32	28.8	35.2
46	46	41.4	50.6
68	68	61.2	74.8
100	100	90.0	110
150	150	135	165
220	220	198	242
320	320	288	352
460	460	414	506
680	680	612	748
1000	1000	900	1100
1500	1500	1350	1650

Source: American Society for Testing and Materials. ASTM Standard D2422-86. Philadelphia, PA. (See Reference 4.) Copyright ASTM. Reprinted with permission.

REFERENCES

1. American Society for Testing and Materials (ASTM). 1988. *ASTM D445-88: Standard Test Method for Kinematic Viscosity of Transparent and Opaque Liquids.* Philadelphia: Author.

2. ———. 1989. *ASTM D446-89a: Standard Specifications and Operating Instructions for Glass Capillary Kinematic Viscometers.* Philadelphia: Author.

3. ———. 1986. *ASTM D2270-86: Standard Practice for Calculating Viscosity Index from Kinematic Viscosity at 40 and 100°C.* Philadelphia: Author.

4. ———. 1986. *ASTM D2422-86: Standard Classification of Industrial Lubricants by Viscosity System.* Philadelphia: Author.

5. ———. 1986. *ASTM D2602-86: Standard Test Method for Apparent Viscosity of Engine Oils at Low Temperature Using the Cold-Cranking Simulator.* Philadelphia: Author.

6. ———. 1987. *ASTM D2983-87: Standard Test Method for Low-Temperature Viscosity of Automotive Fluid Lubricants Measured by Brookfield Viscometer.* Philadelphia: Author.

7. ———. 1987. *ASTM D3829-87: Standard Test Method for Predicting the Borderline Pumping Temperature of Engine Oil.* Philadelphia: Author.

8. ———. 1989. *ASTM D4684-89: Standard Test Method for Determination of Yield Stress and Apparent Viscosity of Engine Oils at Low Temperature.* Philadelphia: Author.

9. Avallone, Eugene A., and Theodore Baumeister, eds. 1987. *Marks' Standard Handbook for Mechanical Engineers*. 9th ed. New York: McGraw-Hill.

10. Cheremisinoff, N. P., ed. 1986. *Encyclopedia of Fluid Mechanics*. Vol. 1, *Flow Phenomena and Measurement*. Houston, Texas: Gulf Publishing Co.

11. Goldstein, Gina. 1990 (October). "Electrorheological Fluids," *Mechanical Engineering Magazine* 112(10): 48–52.

12. Miller, R. W. 1983. *Flow Measurement Engineering Handbook*. New York: McGraw-Hill.

13. Society of Automotive Engineers (SAE). 1991. *SAE Standard J300: Engine Oil Viscosity Classification*. Pittsburgh: Author.

14. ———. 1986. *SAE Standard J306: Axle and Manual Transmission Lubricant Viscosity Classification*. Pittsburgh: Author.

PRACTICE PROBLEMS

2.1 Define *shear stress* as it applies to a moving fluid.

2.2 Define *velocity gradient*.

2.3 State the mathematical definition for *dynamic viscosity*.

2.4 Which would have the greater dynamic viscosity, a cold lubricating oil or fresh water? Why?

2.5 State the standard units for dynamic viscosity in the SI system.

2.6 State the standard units for dynamic viscosity in the U.S. Customary System.

2.7 State the equivalent units for *poise* in terms of the basic quantities in the cgs system.

2.8 Why are the units of poise and centipoise considered obsolete?

2.9 State the mathematical definition for *kinematic viscosity*.

2.10 State the standard units for kinematic viscosity in the SI system.

2.11 State the standard units for kinematic viscosity in the U.S. Customary System.

2.12 State the equivalent units for *stoke* in terms of the basic quantities in the cgs system.

2.13 Why are the units of stoke and centistoke considered obsolete?

2.14 Define a *newtonian fluid*.

2.15 Define a *nonnewtonian fluid*.

2.16 Give five examples of newtonian fluids.

2.17 Give four examples of the types of fluids that are nonnewtonian.

Appendix D gives dynamic viscosity for a variety of fluids as a function of temperature. Using Appendix D, give the value of the viscosity for the following fluids:

2.18M *Water at 40°C.*

2.19M *Water at 5°C.*

2.20M *Air at 40°C.*

2.21M *Hydrogen at 40°C.*

2.22M *Glycerine at 40°C.*

2.23M *Glycerine at 20°C.*

2.24E Water at 40°F.

2.25E Water at 150°F.

2.26E Air at 40°F.

2.27E Hydrogen at 40°F.

2.28E Glycerine at 60°F.

2.29E Glycerine at 110°F.

2.30E Mercury at 60°F.

2.31E Mercury at 210°F.

2.32E SAE 10 oil at 60°F.

2.33E SAE 10 oil at 210°F.

2.34E SAE 30 oil at 60°F.

2.35E SAE 30 oil at 210°F.

2.36 Define *viscosity index* (VI).

2.37 If you want to choose a fluid that exhibits a small change in viscosity as the temperature changes, would you choose one with a high VI or a low VI?

2.38 Which type of viscosity measurement method uses the basic definition of dynamic viscosity for direct computation?

2.39 In the *rotating drum viscometer*, describe how the velocity gradient is created in the fluid to be measured.

2.40 In the *rotating drum viscometer*, describe how the magnitude of the shear stress is measured.

2.41 What measurements must be taken to determine dynamic viscosity when using a *capillary tube viscometer*?

2.42 Define the term *terminal velocity* as it applies to a falling ball viscometer.

2.43 What measurements must be taken to determine dynamic viscosity when using the *falling ball viscometer*?

2.44 Describe the basic features of the *Saybolt Universal viscometer*.

2.45 Are the results of the Saybolt viscometer tests considered to be direct measurements of viscosity?

2.46 Does the Saybolt viscometer produce data related to a fluid's dynamic viscosity or kinematic viscosity?

2.47 On which type of viscometer is the SAE numbering system for viscosity at 100°C based?

2.48 Describe the difference between an SAE 20 oil and an SAE 20W oil.

2.49 What grades of SAE oil are suitable for lubricating the crankcases of engines?

2.50 What grades of SAE oil are suitable for lubricating gear-type transmissions?

2.51 If you were asked to check the viscosity of an oil that is described as SAE 40, at what temperature would you make the measurement?

2.52 If you were asked to check the viscosity of an oil that is described as SAE 10W, at what temperature would you make the measurement?

2.53 How would you determine the viscosity of an oil labeled SAE 5W-40 for comparison with SAE standards?

2.54C The viscosity of a lubricating oil is given as 500 SSU. Calculate the viscosity in m^2/s and ft^2/s.

2.55M *Using the data from Table 2.4, report the standard values for kinematic viscosity in m^2/s for SAE 10W-30 oil (sg = 0.88) at both the low and high temperature points.*

2.56C Convert a dynamic viscosity measurement of 4500 centipoises to Pa·s and lb-s/ft².

2.57C Convert a kinematic viscosity measurement of 5.6 centistokes to m^2/s and ft^2/s.

2.58C The viscosity of an oil is given as 80 SSU. Calculate the viscosity in m^2/s.

2.59C Convert a viscosity measurement of 6.5×10^{-3} Pa·s to the units of lb·s²/ft.

2.60C An oil container indicates that it has a viscosity of 0.12 poise at 60°C. Which oil in Appendix D has a similar viscosity?

2.61M *In a falling ball viscometer, a steel ball 1.6 mm in diameter is allowed to fall freely in a heavy fuel oil having a specific gravity of 0.94. Steel weighs 77 kN/m^3. If the ball is observed to fall 250 mm in 10.4 s, calculate the viscosity of the oil.*

2.62M *A capillary tube viscometer similar to that shown in Fig. 2.5 is being used to measure the viscosity of an oil having a specific gravity of 0.90. The following data apply:*

> *Tube inside diameter = 2.5 mm = D*
> *Length between manometer taps = 300 mm = L*
> *Manometer fluid is mercury*
> *Manometer deflection = 177 mm = h*
> *Velocity of flow = 1.58 m/s = v*

Determine the viscosity of the oil.

COMPUTER PROGRAMMING ASSIGNMENTS

1. Write a program to convert viscosity units from any given system to another system using the conversion factors and techniques from Appendix K. Note the special conditions for conversion of SSU data to kinematic viscosity in m^2/s when SSU < 100.

2. Write a program to determine the viscosity of water at a given temperature using data from Appendix A. This program could be joined with the one you wrote in Chapter 1, which used other properties of water. Use the same options described in Chapter 1.

3 Pressure Measurement

3.1 OBJECTIVES

In Chapter 1, fluid pressure, p, was defined as the amount of force, F, on a unit area, A, of a substance. Fluid pressure is computed from

$$p = F/A \qquad \textbf{(3–1)}$$

The standard unit for pressure in SI units is the pascal (Pa) or N/m^2. Whereas the standard unit for pressure in the U.S. Customary System is lb/ft^2, the unit lb/in^2 (psi) is more convenient and is used most often.

This chapter will focus on the measurement of fluid pressure. After completing this chapter, you should be able to:

1. Define the relationship between *absolute pressure, gage pressure*, and *atmospheric pressure*.
2. Describe the degree of variation of atmospheric pressure near the earth's surface.
3. Describe the properties of air at standard atmospheric pressure.
4. Describe the properties of the atmosphere at elevations from sea level to 30 000 m.
5. Define the relationship between a change in elevation and the change in pressure in a fluid.
6. Describe how a *manometer* works and how it is used to measure pressure.
7. Describe a U-tube manometer, a differential manometer, a well-type manometer, and an inclined well-type manometer.
8. Describe a *barometer* and how it indicates the value of the local atmospheric pressure.
9. Describe various types of pressure gages and pressure transducers.

3.2 ABSOLUTE AND GAGE PRESSURE

When making calculations involving pressure in a fluid, you must make the measurements relative to some reference pressure. Normally the reference pressure is that of the atmosphere, and the resulting measured pressure is called *gage pressure*. Pressure measured relative to a perfect vacuum is called *absolute pressure*. It is extremely important for you to know the difference between these two ways of measuring pressure and to be able to convert from one to the other.

A simple equation relates the two pressure measuring systems:

⇨ **ABSOLUTE AND GAGE PRESSURE**

$$p_{abs} = p_{gage} + p_{atm} \qquad \textbf{(3–2)}$$

where p_{abs} = absolute pressure

p_{gage} = gage pressure

p_{atm} = atmospheric pressure

Figure 3.1 shows an interpretation of this equation graphically. A few basic concepts may help you to understand the equation.

1. A perfect vacuum is the lowest possible pressure. Therefore, an absolute pressure will always be positive.
2. A gage pressure above atmospheric pressure is positive.
3. A gage pressure below atmospheric pressure is negative, sometimes called *vacuum*.
4. Gage pressure will be indicated in the units of Pa(gage) or psig.
5. Absolute pressure will be indicated in the units of Pa(abs) or psia.
6. The actual magnitude of the atmospheric pressure varies with location and with climatic conditions. The barometric pressure as broadcast in weather reports is an indication of the continually varying atmospheric pressure.

FIGURE 3.1   Comparison between absolute and gage pressure.

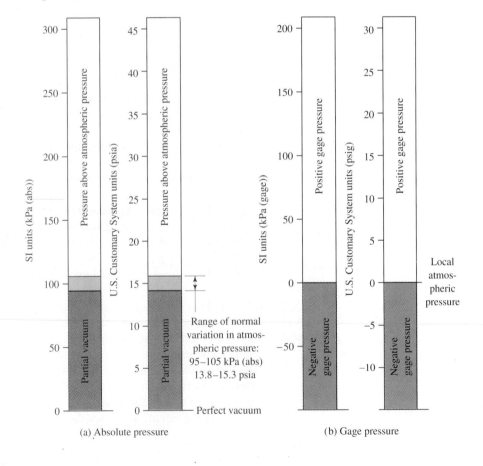

(a) Absolute pressure   (b) Gage pressure

7. The range of normal variation of atmospheric pressure near the earth's surface is approximately from 95 kPa(abs) to 105 kPa(abs) or from 13.8 psia to 15.3 psia. At sea level, the standard atmospheric pressure is 101.3 kPa(abs) or 14.69 psia. Unless the prevailing atmospheric pressure is given, we will assume it to be 101 kPa(abs) or 14.7 psia in this book.

☐ **EXAMPLE PROBLEM 3.1**   Express a pressure of 155 kPa(gage) as an absolute pressure. The local atmospheric pressure is 98 kPa(abs).

Solution

$$p_{abs} = p_{gage} + p_{atm}$$
$$p_{abs} = 155 \text{ kPa(gage)} + 98 \text{ kPa(abs)} = 253 \text{ kPa(abs)}$$

Notice that the units in this calculation are kilopascals (kPa) for each term and are consistent. The indication of gage or absolute is for convenience and clarity.

☐ **EXAMPLE PROBLEM 3.2**   Express a pressure of 225 kPa(abs) as a gage pressure. The local atmospheric pressure is 101 kPa(abs).

Solution

$$p_{abs} = p_{gage} + p_{atm}$$

Solving algebraically for p_{gage} gives

$$p_{gage} = p_{abs} - p_{atm}$$
$$p_{gage} = 225 \text{ kPa(abs)} - 101 \text{ kPa(abs)} = 124 \text{ kPa(gage)}$$

☐ **EXAMPLE PROBLEM 3.3**   Express a pressure of 10.9 psia as a gage pressure. The local atmospheric pressure is 15.0 psia.

Solution

$$p_{abs} = p_{gage} + p_{atm}$$
$$p_{gage} = p_{abs} - p_{atm}$$
$$p_{gage} = 10.9 \text{ psia} - 15.0 \text{ psia} = -4.1 \text{ psig}$$

Notice that the result is negative. This can also be read "4.1 psi below atmospheric pressure" or "4.1 psi vacuum."

☐ **EXAMPLE PROBLEM 3.4**   Express a pressure of −6.2 psig as an absolute pressure.

Solution

$$p_{abs} = p_{gage} + p_{atm}$$

Since no value was given for the atmospheric pressure, we will use $p_{atm} = 14.7$ psia:

$$p_{abs} = -6.2 \text{ psig} + 14.7 \text{ psia} = 8.5 \text{ psia}$$

3.3
RELATIONSHIP
BETWEEN PRESSURE
AND ELEVATION

You are probably familiar with the fact that as you go deeper in a fluid, such as in a swimming pool, the pressure increases. There are many situations in which it is important to know just how pressure varies with a change in depth or elevation.

In this book the term *elevation* means the vertical distance from some reference level to a point of interest and is called z. A *change* in elevation between two points is called h. Elevation will always be measured positively in the upward direction. In other words, a higher point has a larger elevation than a lower point.

The reference level can be taken at any level as illustrated in Fig. 3.2, which shows a submarine under water. In part (a) of the figure the sea bottom is taken as reference, while in (b) the position of the submarine is the reference level. Since fluid mechanics calculations usually consider differences in elevation, it is advisable to choose the lowest point of interest in a problem as the reference level to eliminate the use of negative values for z. This will be especially important in later work.

FIGURE 3.2 Illustration of reference level for elevation.

(a) (b)

The change in pressure in a homogeneous liquid at rest due to a change in elevation can be calculated from

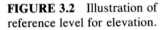

PRESSURE-ELEVATION
RELATIONSHIP

$$\Delta p = \gamma h \qquad (3\text{--}3)$$

where Δp = change in pressure

γ = specific weight of liquid

h = change in elevation

Some general conclusions from Eq. (3–3) will help you to apply it properly:

1. The equation is valid only for a homogeneous liquid at rest.
2. Points on the same horizontal level have the same pressure.

3. The change in pressure is directly proportional to the specific weight of the liquid.
4. Pressure varies linearly with the change in elevation or depth.
5. A decrease in elevation causes an increase in pressure. (This is what happens when you go deeper in a swimming pool.)
6. An increase in elevation causes a decrease in pressure.

Equation (3–3) does not apply to gases because the specific weight of a gas changes with a change in pressure. However, it requires a large change in elevation to produce a significant change in pressure in a gas. For example, an increase in elevation of 300 m (about 1000 ft) in the atmosphere causes a decrease in pressure of only 3.4 kPa (about 0.5 psi). *In this book we assume that the pressure in a gas is uniform unless otherwise specified.*

☐ **EXAMPLE PROBLEM 3.5** Calculate the change in water pressure from the surface to a depth of 5 m.

Solution Using Eq. (3–3), $\Delta p = \gamma h$, let $\gamma = 9.81$ kN/m³ for water and $h = 5$ m. Then we have

$$\Delta p = (9.81 \text{ kN/m}^3)(5.0 \text{ m}) = 49.05 \text{ kN/m}^2 = 49.05 \text{ kPa}$$

If the surface of the water is exposed to the atmosphere, the pressure there is 0 Pa(gage). Descending in the water (decreasing elevation) produces an increase in pressure. Therefore, at 5 m the pressure is 49.05 kPa(gage).

☐ **EXAMPLE PROBLEM 3.6** Calculate the change in water pressure from the surface to a depth of 15 ft.

Solution Using Eq. (3–3), $\Delta p = \gamma h$, let $\gamma = 62.4$ lb/ft³ for water and $h = 15$ ft. Then we have

$$\Delta p = \frac{62.4 \text{ lb}}{\text{ft}^3} \times 15 \text{ ft} \times \frac{1 \text{ ft}^2}{144 \text{ in}^2} = 6.5 \frac{\text{lb}}{\text{in}^2}$$

If the surface of the water is exposed to the atmosphere, the pressure there is 0 psig. Descending in the water (decreasing elevation) produces an increase in pressure. Therefore, at 15 ft the pressure is 6.5 psig.

☐ **EXAMPLE PROBLEM 3.7** Figure 3.3 shows a tank of oil with one side open to the atmosphere and the other side sealed with air above the oil. The oil has a specific gravity of 0.90. Calculate the gage pressure at points A, B, C, D, E, and F and the air pressure in the right side of the tank.

Solution At *point A* the oil is exposed to the atmosphere, and therefore

$$p_A = 0 \text{ Pa(gage)}$$

Point B: The change in elevation between point A and point B is 3.0 m, with B lower than A. To use Eq. (3–3) we need the specific weight of the oil:

$$\gamma_{oil} = (sg)_{oil}(9.81 \text{ kN/m}^3) = (0.90)(9.81 \text{ kN/m}^3) = 8.83 \text{ kN/m}^3$$

Then, we have

$$\Delta p_{A-B} = \gamma h = (8.83 \text{ kN/m}^3)(3.0 \text{ m}) = 26.5 \text{ kN/m}^2 = 26.5 \text{ kPa}$$

FIGURE 3.3   Tank for Example Problem 3.7.

Now, the pressure at B is

$$p_B = p_A + \Delta p_{A-B} = 0 \text{ Pa(gage)} + 26.5 \text{ kPa} = 26.5 \text{ kPa(gage)}$$

Point C: The change in elevation from point A to point C is 6.0 m, with C lower than A. Then, the pressure at point C is

$$\Delta p_{A-C} = \gamma h = (8.83 \text{ kN/m}^3)(6.0 \text{ m}) = 53.0 \text{ kN/m}^2 = 53.0 \text{ kPa}$$

$$p_C = p_A + \Delta p_{A-C} = 0 \text{ Pa(gage)} + 53.0 \text{ kPa} = 53.0 \text{ kPa(gage)}$$

Point D: Since point D is at the same level as point B, the pressure is the same. That is, we have

$$p_D = p_B = 26.5 \text{ kPa(gage)}$$

Point E: Since point E is at the same level as point A, the pressure is the same. That is, we have

$$p_E = p_A = 0 \text{ Pa(gage)}$$

Point F: The change in elevation between point A and point F is 1.5 m, with F higher than A. Then, the pressure at F is

$$\Delta p_{A-F} = -\gamma h = (-8.83 \text{ kN/m}^3)(1.5 \text{ m}) = -13.2 \text{ kN/m}^2 = -13.2 \text{ kPa}$$

$$p_F = p_A + \Delta p_{A-F} = 0 \text{ Pa(gage)} + (-13.2 \text{ kPa}) = -13.2 \text{ kPa}$$

Air pressure: Since the air in the right side of the tank is exposed to the surface of the oil where $p_F = -13.2$ kPa, the air pressure is also -13.2 kPa, or 13.2 kPa below atmospheric pressure.

**3.3.1
Summary of Observations
from Example Problem 3.7**

The results from Problem 3.7 illustrate the general conclusions listed below Eq. (3–3) on pages 46–47.

a. The pressure increases as the depth in the fluid increases. This result can be seen from $p_C > p_B > p_A$.
b. Pressure varies linearly with a change in elevation; that is, p_C is two times greater than p_B, and C is at twice the depth of B.
c. Pressure on the same horizontal level is the same. Note that $p_E = p_A$ and $p_D = p_B$.
d. The decrease in pressure from E to F occurs because point F is at a higher elevation than point E. Note that p_F is negative; that is, it is below the atmospheric pressure that exists at A and E.

**3.4
DEVELOPMENT
OF THE
PRESSURE-ELEVATION
RELATION**

The relationship between a change in elevation in a liquid, h, and a change in pressure, Δp, is

$$\Delta p = \gamma h \tag{3–3}$$

where γ is the specific weight of the liquid. This section presents the basis for this equation.

FIGURE 3.4 Small volume of fluid within a body of static fluid.

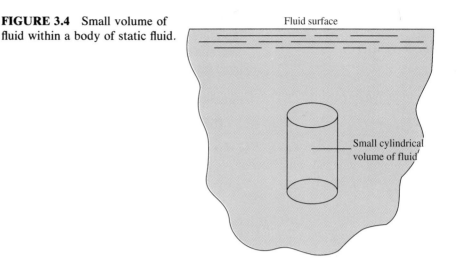

Figure 3.4 shows a body of static fluid with a specific weight γ. Consider a small volume of the fluid somewhere below the surface. In Fig. 3.4, the small volume is illustrated as a cylinder, but the actual shape is arbitrary.

Because the entire body of fluid is stationary and in equilibrium, the small cylinder of the fluid is also in equilibrium. From physics, we know that for a body in static equilibrium, the sum of the forces acting on it in all directions must sum to zero.

First consider the forces acting in the horizontal direction. A thin ring around the cylinder is shown at some arbitrary elevation in Fig. 3.5. The vectors acting on the ring represent the forces exerted on it by the fluid pressure. Recall from previous work that the pressure at any horizontal level

FIGURE 3.5 Pressure forces acting in a horizontal plane on a thin ring.

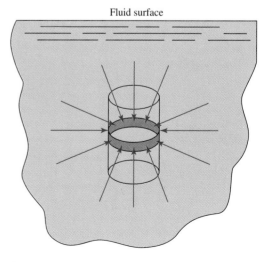

in a static fluid is the same. Also recall that the pressure at a boundary, and therefore the force due to the pressure, acts perpendicular to the boundary. Therefore, the forces are completely balanced around the sides of the cylinder.

Now consider Fig. 3.6. The forces acting on the cylinder in the vertical direction are shown. The following concepts are illustrated in the figure:

1. The fluid pressure at the level of the bottom of the cylinder is called p_1.
2. The fluid pressure at the level of the top of the cylinder is called p_2.
3. The elevation difference between the top and the bottom of the cylinder is called dz, where $dz = z_2 - z_1$.
4. The pressure change that occurs in the fluid between the level of the bottom and the top of the cylinder is called dp. Therefore, $p_2 = p_1 + dp$.
5. The area of the top and bottom of the cylinder is called A.
6. The volume of the cylinder is the product of the area, A, and the height of the cylinder, dz. That is, $V = A(dz)$.
7. The weight of the fluid within the cylinder is the product of the specific weight of the fluid, γ, and the volume of the cylinder. That is, $w = \gamma V = \gamma A(dz)$. The weight is a force acting on the cylinder in the downward direction through the centroid of the cylindrical volume.
8. The force acting on the bottom of the cylinder due to the fluid pressure p_1 is the product of the pressure and the area, A. That is, $F_1 = p_1 A$. This force acts vertically upward, perpendicular to the bottom of the cylinder.
9. The force acting on the top of the cylinder due to the fluid pressure p_2 is the product of the pressure and the area, A. That is, $F_2 = p_2 A$. This force acts vertically downward, perpendicular to the top of the cylinder. Because $p_2 = p_1 + dp$, another expression for the force F_2 is

$$F_2 = (p_1 + dp)A \qquad\qquad (3\text{–}4)$$

FIGURE 3.6 Forces acting in the vertical direction.

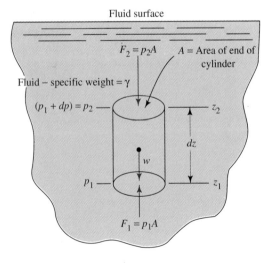

Now we can apply the principle of static equilibrium, which states that the sum of the forces in the vertical direction must be zero. Using upward forces as positive, we get

$$\sum F_v = 0 = F_1 - F_2 - w \tag{3-5}$$

Substituting from Steps 7, 8, and 9 gives

$$p_1 A - (p_1 + dp)A - \gamma(dz)A = 0 \tag{3-6}$$

Notice that the area, A, appears in all terms on the left side of Eq. (3–6). It can be eliminated by dividing all terms by A. The resulting equation is

$$p_1 - p_1 - dp - \gamma(dz) = 0 \tag{3-7}$$

Now the p_1 term can be cancelled out. Solving for dp gives

$$dp = -\gamma(dz) \tag{3-8}$$

Equation (3–8) is the controlling relationship between a change in elevation and a change in pressure. The use of Eq. (3–8), however, depends on the kind of fluid. Remember that the equation was developed for a very small element of the fluid. The process of integration extends Eq. (3–8) to large changes in elevation, as indicated below:

$$\int_{p_1}^{p_2} dp = \int_{z_1}^{z_2} -\gamma(dz) \tag{3-9}$$

To complete the analysis, we must define how the specific weight of the fluid varies with a change in pressure. Eq. (3–9) is developed differently for liquids and for gases.

3.4.1
Liquids

A liquid is considered to be incompressible. Thus, its specific weight, γ, is a constant. This allows γ to be taken outside the integral sign in Eq. (3–9). Then,

$$\int_{p_1}^{p_2} dp = -\gamma \int_{z_1}^{z_2} (dz) \tag{3-10}$$

Completing the integration process and applying the limits gives

$$p_2 - p_1 = -\gamma(z_2 - z_1) \tag{3-11}$$

For convenience, we define $\Delta p = p_2 - p_1$ and $h = z_1 - z_2$. Equation (3–11), then, becomes

$$\Delta p = \gamma h$$

which is identical to Eq. (3–3). The signs for Δp and h can be assigned at the time of use of the formula by recalling that pressure increases as depth in the fluid increases and vice versa.

**3.4.2
Gases**

Because a gas is compressible, its specific weight changes as pressure changes. To complete the integration process called for in Eq. (3–9), you must know the relationship between the change in pressure and the change in specific weight. The relationship is different for different gases, but a complete discussion of those relationships is beyond the scope of this text and requires the study of thermodynamics.

Appendix E describes the properties of air in the *standard atmosphere* as defined by the U.S. National Oceanic and Atmospheric Administration (NOAA).

**3.5
PASCAL'S PARADOX**

In the development of the relationship $\Delta p = \gamma h$, the size of the small volume of fluid does not affect the result. The change in pressure depends only on the change in elevation and the type of fluid, not on the size of the fluid container. Therefore, all the containers shown in Fig. 3.7 would have the same pressure at the bottom, even though they contain vastly different amounts of fluid. This observation is called *Pascal's paradox*.

FIGURE 3.7   Illustration of Pascal's paradox.

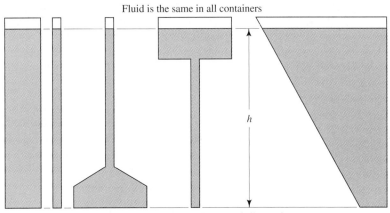

Fluid is the same in all containers

h

Pressure is the same at the bottom of all containers

This phenomenon is useful when a consistently high pressure must be produced on a system of interconnected pipes and tanks. Water systems for cities often include water towers located on high hills, as shown in Fig. 3.8. Besides providing a reserve supply of water, the primary purpose of such tanks is to maintain a sufficiently high pressure in the water system for satisfactory delivery of the water to residential, commercial, and industrial users.

In industrial or laboratory applications, a standpipe containing a static liquid can be used to create a stable pressure on a particular process or system. It is positioned at a high elevation relative to the system and is connected to the system by pipes. Raising or lowering the level of the fluid in the standpipe changes the pressure in the system. Standpipes are sometimes placed on the roofs of buildings to maintain water pressure in local firefighting systems.

FIGURE 3.8 Use of a water tower or standpipe to maintain water system pressure.

3.6 MANOMETERS

This and following sections describe several types of pressure measurement devices. The first is the *manometer,* which uses the relationship between a change in pressure and a change in elevation in a static fluid, $\Delta p = \gamma h$ (see Sections 3.3 and 3.4). Photographs of commercially available manometers are shown in Figs. 3.9, 3.12, and 3.13.

The simplest kind of manometer is the U-tube (Fig. 3.9). One end of the U-tube is connected to the pressure that is to be measured, while the other end is left open to the atmosphere. The tube contains a liquid called the *gage fluid,* which does not mix with the fluid whose pressure is to be measured. Typical gage fluids are water, mercury, and colored light oils.

Under the action of the pressure to be measured, the gage fluid is displaced from its normal position. Since the fluids in the manometer are at rest, the equation $\Delta p = \gamma h$ can be used to write expressions for the changes in pressure that occur throughout the manometer. These expressions can then be combined and solved algebraically for the desired pressure. Because manometers are used in many real situations such as those described in this book, you should learn the following step-by-step procedure:

PROCEDURE FOR WRITING THE EQUATION FOR A MANOMETER

1. Start from a convenient point, usually where the pressure is known, and write this pressure in symbol form (e.g., p_A refers to the pressure at point A).

FIGURE 3.9   U-tube manometer. (Source of photo: Dwyer Instruments, Inc. Michigan City, IN)

(a) Photograph of commercially available model

(b) Sketch showing typical application

2. Using $\Delta p = \gamma h$, write expressions for the changes in pressure that occur from the starting point to the point at which the pressure is to be measured, being careful to include the correct algebraic sign for each term.
3. Equate the expression from Step 2 to the pressure at the desired point.
4. Substitute known values and solve for the desired pressure.

Working several practice problems will help you to apply this procedure correctly. The following problems are written in the programmed instruction format. To work through the program, cover the material below the heading "Programmed Example Problems" and then uncover one panel at a time.

PROGRAMMED EXAMPLE PROBLEMS

☐ **EXAMPLE PROBLEM 3.8**   Using Fig. 3.9, calculate the pressure at point A. Perform Step 1 of the procedure before going to the next panel.

Figure 3.10 is identical to Fig. 3.9(b) except that certain key points have been numbered for use in the problem solution.

The only point for which the pressure is known is the surface of the mercury in the right leg of the manometer, point 1. Now, how can an expression be written for the pressure that exists within the mercury, 0.25 m below this surface at point 2?

The expression is

$$p_1 + \gamma_m(0.25 \text{ m})$$

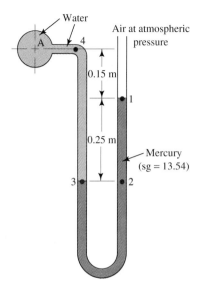

FIGURE 3.10 U-tube manometer.

The term $\gamma_m(0.25 \text{ m})$ is the change in pressure between points 1 and 2 due to a change in elevation, where γ_m is the specific weight of mercury, the gage fluid. This pressure change is added to p_1 because there is an increase in pressure as we descend in a fluid.

So far we have an expression for the pressure at point 2 in the right leg of the manometer. Now write the expression for the pressure at point 3 in the left leg.

This is the expression:

$$p_1 + \gamma_m(0.25 \text{ m})$$

Because points 2 and 3 are on the same level in the same fluid at rest, their pressures are equal.

Continue and write the expression for the pressure at point 4.

$$p_1 + \gamma_m(0.25 \text{ m}) - \gamma_w(0.40 \text{ m})$$

where γ_w is the specific weight of water. Remember, there is a decrease in pressure between points 3 and 4, so this last term must be subtracted from our previous expression.

What must you do to get an expression for the pressure at point A?

Nothing. Because points A and 4 are on the same level, their pressures are equal.

Now perform Step 3 of the procedure.

You should now have

$$p_1 + \gamma_m(0.25 \text{ m}) - \gamma_w(0.40 \text{ m}) = p_A$$

or

$$p_A = p_1 + \gamma_m(0.25 \text{ m}) - \gamma_w(0.40 \text{ m})$$

This is the completed equation for the pressure at point A. Now do Step 4.

Several calculations are required here:

$$p_1 = p_{atm} = 0 \text{ Pa(gage)}$$
$$\gamma_m = (sg)_m(9.81 \text{ kN/m}^3) = (13.54)(9.81 \text{ kN/m}^3) = 132.8 \text{ kN/m}^3$$
$$\gamma_w = 9.81 \text{ kN/m}^3$$

Then, we have

$$p_A = p_1 + \gamma_m(0.25 \text{ m}) - \gamma_w(0.40 \text{ m})$$
$$= 0 \text{ Pa(gage)} + (132.8 \text{ kN/m}^3)(0.25 \text{ m}) - (9.81 \text{ kN/m}^3)(0.40 \text{ m})$$
$$= 0 \text{ Pa(gage)} + 33.20 \text{ kN/m}^2 - 3.92 \text{ kN/m}^2$$
$$p_A = 29.28 \text{ kN/m}^2 = 29.28 \text{ kPa(gage)}$$

Remember to include the units in your calculations. Review this problem to be sure you understand every step before going to the next panel for another problem.

☐ **EXAMPLE PROBLEM 3.9**   Calculate the pressure at point B in Fig. 3.11 if the pressure at A is 22.40 psig. This type of manometer is called a *differential manometer* because it indicates the difference between the pressure at points A and B but not the actual value of either one. Do Step 1 of the procedure to find p_B.

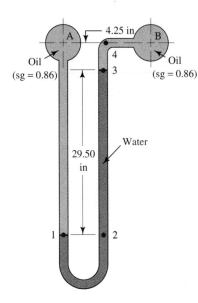

We know the pressure at A, so we start there and call it p_A. Now write an expression for the pressure at point 1.

You should have

$$p_A + \gamma_o(33.75 \text{ in})$$

where γ_o is the specific weight of the oil.
What is the pressure at point 2?

It is the same as the pressure at point 1 because the two points are on the same level. Go on to point 3 in the manometer.

The expression should now look like this:

$$p_A + \gamma_o(33.75 \text{ in}) - \gamma_w(29.5 \text{ in})$$

Now write the expression for the pressure at point 4.

This is the desired expression:

$$p_A + \gamma_o(33.75 \text{ in}) - \gamma_w(29.5 \text{ in}) - \gamma_o(4.25 \text{ in})$$

This is also the expression for the pressure at B since points 4 and B are on the same level. Now do Steps 3 and 4 of the procedure.

Our final expression should be

$$p_A + \gamma_o(33.75 \text{ in}) - \gamma_w(29.5 \text{ in}) - \gamma_o(4.25 \text{ in}) = p_B$$

or

$$p_B = p_A + \gamma_o(33.75 \text{ in}) - \gamma_w(29.5 \text{ in}) - \gamma_o(4.25 \text{ in})$$

The known values are

$$p_A = 22.40 \text{ psig}$$
$$\gamma_o = (sg)_o(62.4 \text{ lb/ft}^3) = (0.86)(62.4 \text{ lb/ft}^3) = 53.7 \text{ lb/ft}^3$$
$$\gamma_w = 62.4 \text{ lb/ft}^3$$

FIGURE 3.11   Differential manometer.

In this case it may help to simplify the expression for p_B before substituting known values. Since two terms are multiplied by γ_o they can be combined as follows:

$$p_B = p_A + \gamma_o(29.5 \text{ in}) - \gamma_w(29.5 \text{ in})$$

Factoring out the common term gives

$$p_B = p_A + (29.5 \text{ in})(\gamma_o - \gamma_w)$$

This looks simpler than the original equation. The difference between p_A and p_B is a function of the *difference* between the specific weights of the two fluids.

The pressure at B, then, is

$$p_B = 22.40 \text{ psig} + (29.5 \text{ in})(53.7 - 62.4) \frac{\text{lb}}{\text{ft}^3} \times \frac{1 \text{ ft}^3}{1728 \text{ in}^3}$$

$$= 22.40 \text{ psig} + \frac{(29.5)(-8.7)\text{lb/in}^2}{1728}$$

$$= 22.40 \text{ psig} - 0.15 \text{ lb/in}^2$$

$$p_B = 22.25 \text{ psig}$$

Notice that using a gage fluid with a specific weight very close to that of the fluid being measured makes the manometer very sensitive. Note that $p_A - p_B = 0.15$ psi. This concludes the programmed instruction.

■

Figure 3.12 shows another type of manometer called the *well-type*. When a pressure is applied to a well-type manometer, the fluid level in the well drops a small amount, while the level in the right leg rises a larger

FIGURE 3.12 Well-type manometer. (Source of photo: Dwyer Instruments, Inc. Michigan City, IN)

amount in proportion to the ratio of the areas of the well and the tube. A scale is placed alongside the tube so that the deflection can be read directly. The scale is calibrated to account for the small drop in the well level.

The *inclined well-type manometer,* shown in Fig. 3.13, has the same features as the well-type but offers a greater sensitivity by placing the scale along the inclined tube. The scale length is increased as a function of the angle of inclination of the tube, θ. For example, if the angle θ in Fig. 3.13(b) is 15°, the ratio of scale length L to manometer deflection h is

$$\frac{h}{L} = \sin \theta$$

or

$$\frac{L}{h} = \frac{1}{\sin \theta} = \frac{1}{\sin 15°} = \frac{1}{0.259} = 3.86$$

FIGURE 3.13   Inclined well-type manometer. (Source of photo: Dwyer Instruments, Inc. Michigan City, IN)

(a)

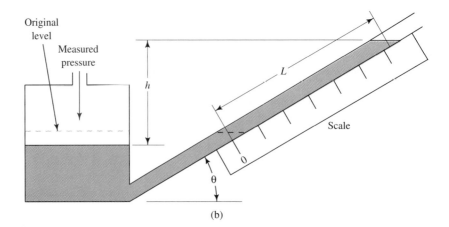

(b)

The scale would be calibrated so that the deflection could be read directly.

3.7 BAROMETERS

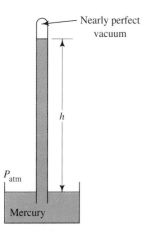

— Nearly perfect vacuum

h

P_{atm}

Mercury

FIGURE 3.14 Barometer.

A device for measuring the atmospheric pressure is called a *barometer*. A simple type is shown in Fig. 3.14. It consists of a long tube closed at one end which is initially filled completely with mercury. The open end is then submerged under the surface of a container of mercury and allowed to come to equilibrium, as shown in Fig. 3.14. A void is produced at the top of the tube which is very nearly a perfect vacuum, containing mercury vapor at a pressure of only 0.17 Pa at 20°C. By starting at this point and writing an equation similar to those for manometers, we get

$$0 + \gamma_m h = p_{atm}$$

or

$$p_{atm} = \gamma_m h \tag{3-12}$$

Since the specific weight of mercury is approximately constant, a change in atmospheric pressure will cause a change in the height of the mercury column. This height is often reported as the barometric pressure. To obtain true atmospheric pressure it is necessary to multiply h by γ_m.

Precision measurement of the atmospheric pressure with a mercury manometer requires that the specific weight of the mercury be adjusted for changes in temperature. But in this book, we will use the values given in Appendix B. In SI units, $\gamma = 132.8 \text{ kN/m}^3$. In U.S. Customary System units, $\gamma = 844.9 \text{ lb/ft}^3$.

The atmospheric pressure varies from time to time, as you hear on weather reports. The atmospheric pressure also varies with altitude. A decrease of approximately 1.0 in of mercury occurs per 1000 ft of increase in altitude. In SI units, the decrease is approximately 85 mm of mercury per 1000 m.

☐ **EXAMPLE PROBLEM 3.10**

A news broadcaster reports that the barometric pressure is 772 mm of mercury. Calculate the atmospheric pressure in kPa(abs).

Solution

In Eq. (3–12),

$$p_{atm} = \gamma_m h$$
$$\gamma_m = 132.8 \text{ kN/m}^3$$
$$h = 0.772 \text{ m}$$

Then we have

$$p_{atm} = (132.8 \text{ kN/m}^3)(0.772 \text{ m}) = 102.5 \text{ kN/m}^2 = 102.5 \text{ kPa(abs)}$$

☐ **EXAMPLE PROBLEM 3.11**

The standard atmospheric pressure is 101.3 kPa. Calculate the height of a mercury column equivalent to this pressure.

Solution   In Eq. (3–12),

$$p_{atm} = \gamma_m h$$

$$h = \frac{p_{atm}}{\gamma_m} = \frac{101.3 \times 10^3 \text{ N}}{\text{m}^2} \times \frac{\text{m}^3}{132.8 \times 10^3 \text{ N}} = 0.763 \text{ m} = 763 \text{ mm}$$

☐ **EXAMPLE PROBLEM 3.12**   A news broadcaster reports that the barometric pressure is 30.40 in of mercury. Calculate the atmospheric pressure in psia.

Solution   In Eq. (3–12),

$$p_{atm} = \gamma_m h$$

$$\gamma_m = 844.9 \text{ lb/ft}^3$$

$$h = 30.40 \text{ in}$$

Then we have

$$p_{atm} = \frac{844.9 \text{ lb}}{\text{ft}^3} \times 30.40 \text{ in} \times \frac{1 \text{ ft}^3}{1728 \text{ in}^3} = 14.9 \text{ lb/in}^2$$

$$p_{atm} = 14.9 \text{ psia}$$

☐ **EXAMPLE PROBLEM 3.13**   The standard atmospheric pressure is 14.7 psia. Calculate the height of a mercury column equivalent to this pressure.

Solution   In Eq. (3–12),

$$p_{atm} = \gamma_m h$$

$$h = \frac{p_{atm}}{\gamma_m} = \frac{14.7 \text{ lb}}{\text{in}^2} \times \frac{\text{ft}^3}{844.9 \text{ lb}} \times \frac{1728 \text{ in}^3}{\text{ft}^3} = 30.06 \text{ in}$$

**3.8
PRESSURE GAGES
AND TRANSDUCERS**

A widely used pressure measuring device is the *Bourdon tube pressure gage** (see Fig. 3.15). The pressure to be measured is applied to the inside of a flattened tube which is normally shaped as a segment of a circle or a spiral. The increased pressure inside the tube causes it to be straightened somewhat. The movement of the end of the tube is transmitted through a linkage which causes a pointer to rotate.

The scale of the gage normally reads zero when the gage is open to atmospheric pressure and is calibrated in pascals (Pa) or other units of pressure above zero. Therefore, this type of gage reads *gage pressure* directly. Some gages are capable of reading pressures below atmospheric.

Figure 3.16 shows a pressure gage using an actuation means called *magnehelic*®.† The pointer is attached to a helix made from a material hav-

---

* Note that two spellings, *gage* and *gauge,* are often used interchangeably.

† Magnehelic® is a registered trade name of Dwyer Instruments, Inc., Michigan City, Indiana.

FIGURE 3.15   Bourdon tube pressure gage. (Source of photo: Ametek/U.S. Gauge, Sellersville, PA)

(a) Front view   (b) Rear view with back of case removed

FIGURE 3.16   Magnehelic® pressure gage. (Source: Dwyer Instruments, Inc., Michigan City, IN)

(a)

(b)   (c)

ing a high magnetic permeability that is supported in sapphire bearings. A leaf spring is driven up and down by the motion of a flexible diaphragm, not shown in the figure. At the end of the spring, the C-shaped element contains a strong magnet placed in close proximity to the outer surface of the helix. As the leaf spring moves up and down, the helix rotates to follow the magnet, moving the pointer. Note that there is no physical contact between the magnet and the helix. Calibration of the gage is accomplished by adjusting the length of the spring at its clamped end.

3.9
PRESSURE
TRANSDUCERS

A *transducer* is an instrument that measures some physical quantity and generates an electrical signal that has a predictable relationship to the measured quantity. The level of the electrical signal then can be recorded, displayed on a meter, or stored in a computer memory for later display or analysis. This section describes several types of pressure transducers.

3.9.1
Strain Gage Pressure
Transducer

Figure 3.17 shows a *strain gage pressure transducer*. The pressure to be measured is introduced through the pressure port and acts on a diaphragm to which foil strain gages are bonded. As the strain gages sense the deformation

FIGURE 3.17 Strain gage pressure transducer and indicator. (Source of photos: Sensotec, Inc. Columbus, OH)

Internal electronic amplifier

Pressure port

Electrical connector

(a) Strain gage pressure transducer

(b) Digital electronic amplifier / indicator

of the diaphragm, their resistance changes. By passing an electrical current through the gages and connecting them into a network, called a *Wheatstone bridge,* a change in electrical voltage is produced. The readout device is typically a digital voltmeter, calibrated in pressure units.

3.9.2
LVDT-Type Pressure Transducer

A *linear variable differential transformer* (LVDT) is composed of a cylindrical electric coil with a movable rod-shaped core. As the core moves along the axis of the coil, a voltage change occurs in relation to the position of the core. This type of transducer is applied to pressure measurement by attaching the core rod to a flexible diaphragm (Fig. 3.18). For gage pressure measurements, one side of the diaphragm is exposed to atmospheric pressure, while the other side is exposed to the pressure to be measured. Changes in pressure cause the diaphragm to move, thus moving the core of the LVDT. The resulting voltage change is recorded or indicated on a meter calibrated in pressure units. Differential pressure measurements can be made by introducing the two pressures to the opposite sides of the diaphragm.

3.9.3
Piezoelectric Pressure Transducers

Certain crystals, such as quartz and barium titanate, exhibit the *piezoelectric effect,* in which the electrical charge across the crystal varies with stress in the crystal. The *piezoelectric pressure transducer* uses this characteristic of the crystal by causing the pressure to exert a force, either directly or indi-

FIGURE 3.18 LVDT-type pressure transducer. (Source of photo: Schaevitz Engineering. Pennsauken, NJ)

Power supply and signal conditioning electronics

LVDT coil

LVDT core

Operating pressure port

Pressure sensing capsule (diaphragm)

Reference pressure port

rectly, on the crystal. A voltage change, which is related to the pressure change, is then produced.

Figure 3.19 shows a commercially available pressure gage that incorporates a piezoelectric pressure transducer. One can indicate pressure or vacuum in any of 18 different display units by simply pressing the *units* button. The gage also incorporates a calibrator signal that indicates a DC milliamp current reading for field calibration of remote transmitters. The *zero* key allows the setting of the reference pressure in the field.

3.9.4
Quartz Resonator Pressure
Transducers

A quartz crystal resonates at a frequency that is dependent on the stress in the crystal. Under increasing tension, the resonant frequency increases. Conversely, the resonant frequency decreases with compression. The

FIGURE 3.19 Digital pressure gage. (Source: Rochester Instrument Systems, Rochester, NY)

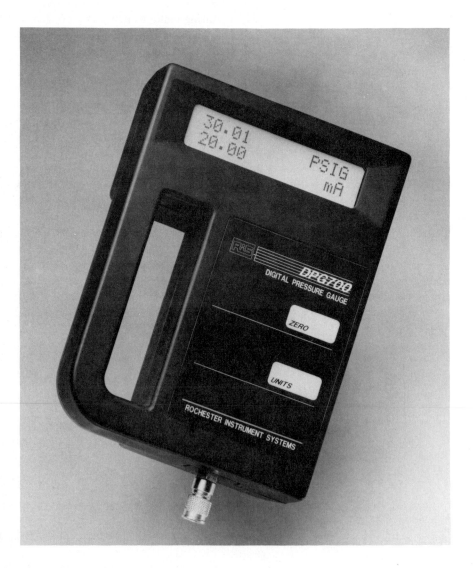

changes in frequency can be measured very precisely by digital electronic systems. Pressure transducers can use this phenomenon by linking bellows, diaphragms, or Bourdon tubes to quartz crystals. Such devices can provide pressure measurement accuracy of 0.01 percent or better.

3.9.5 Solid State Pressure Sensors

Solid state technology permits very small pressure sensors to be made from silicon. Thin film silicon resistors can be used instead of foil strain gages for a Wheatstone bridge–type system. Another type uses two parallel plates whose surfaces are composed of an etched pattern of silicon. Pressure applied to one plate causes it to deflect, changing the air gap between the plates. The resulting change in capacitance can be detected with an oscillator circuit.

3.10 PRESSURE EXPRESSED AS THE HEIGHT OF A COLUMN OF LIQUID

When measuring pressures in some fluid flow systems, such as air flow in heating ducts, the actual magnitude of the pressure reading is often small. Manometers are sometimes used to measure these pressures and their readings are given in units such as *"inches of water"* rather than the conventional units of psi or Pa.

To convert from such units to those needed in calculations, the pressure-elevation relationship must be used. For example, a pressure of 1.0 in of water (1.0 in H_2O) expressed in psi units is,

$$p = \gamma h$$

$$p = \frac{62.4 \text{ lb}}{\text{ft}^3} (1.0 \text{ in } H_2O) \frac{1 \text{ ft}^3}{1728 \text{ in}^3} = 0.0361 \text{ lb/in}^2 = 0.0361 \text{ psi}$$

Then we can use this as a conversion factor,

$$1.0 \text{ in of water} = 0.0361 \text{ psi}$$

Converting this to Pa gives

$$1.0 \text{ in of water} = 249.0 \text{ Pa}$$

Similarly, somewhat higher pressures are measured with a mercury manometer. Using $\gamma = 132.8 \text{ kN/m}^3$ or $\gamma = 844.9 \text{ lb/ft}^3$, we can develop the conversion factors,

$$1.0 \text{ in of mercury} = 0.489 \text{ psi}$$
$$1.0 \text{ mm of mercury} = 0.01926 \text{ psi}$$
$$1.0 \text{ mm of mercury} = 132.8 \text{ Pa}$$

You must remember that temperature of the gage fluid can affect its specific weight and therefore the accuracy of these factors.

REFERENCES

1. Avallone, Eugene A., and Theodore Baumeister III, eds. 1987. *Marks' Standard Handbook for Mechanical Engineers*. 9th ed. New York: McGraw-Hill.

2. Busse, Donald W. 1987 (March). Quartz Transducers for Precision Under Pressure. *Mechanical Engineering Magazine* 109(5):52–56.

3. Walters, Sam. 1987 (March). Inside Pressure Measurement. *Mechanical Engineering Magazine* 109(5):41–47.

4. Worden, Roy D. 1987 (March). Designing a Fused-Quartz Pressure Transducer. *Mechanical Engineering Magazine* 109(5):48–51.

PRACTICE PROBLEMS

Absolute and Gage Pressure

3.1 Write the expression for computing the pressure in a fluid.

3.2 Define *absolute pressure*.

3.3 Define *gage pressure*.

3.4 Define *atmospheric pressure*.

3.5 Write the expression relating gage pressure, absolute pressure, and atmospheric pressure.

State whether statements 3.6 through 3.10 are (or can be) true or false. For those that are false, tell why.

3.6 The value for the absolute pressure will always be greater than that for the gage pressure.

3.7E As long as you stay on the surface of the earth, the atmospheric pressure will be 14.7 psia.

3.8M *The pressure in a certain tank is −55.8 Pa(abs).*

3.9E The pressure in a certain tank is −4.65 psig.

3.10M *The pressure in a certain tank is −150 kPa(gage).*

3.11E If you were to ride in an open-cockpit airplane to an elevation of 4000 ft above sea level, what would the atmospheric pressure be if it conforms to the standard atmosphere?

3.12E The peak of a certain mountain is 13 500 ft above sea level. What is the approximate atmospheric pressure?

3.13 Expressed as a gage pressure, what is the pressure at the surface of a glass of milk?

Problems 3.14 to 3.33 require that you convert the given pressure from gage to absolute pressure or from absolute to gage pressure as indicated. The value of the atmospheric pressure is given.

Problem	Given Pressure		p_{atm}		Express Result As:
3.14M	583	kPa(abs)	103	kPa(abs)	Gage pressure
3.15M	157	kPa(abs)	101	kPa(abs)	Gage pressure
3.16M	30	kPa(abs)	100	kPa(abs)	Gage pressure
3.17M	74	kPa(abs)	97	kPa(abs)	Gage pressure

Problem	Given Pressure		p_{atm}		Express Result As:
3.18M	101	kPa(abs)	104	kPa(abs)	Gage pressure
3.19M	284	kPa(gage)	100	kPa(abs)	Absolute pressure
3.20M	128	kPa(gage)	98.0	kPa(abs)	Absolute pressure
3.21M	4.1	kPa(gage)	101.3	kPa(abs)	Absolute pressure
3.22M	−29.6	kPa(gage)	101.3	kPa(abs)	Absolute pressure
3.23M	−86.0	kPa(gage)	99.0	kPa(abs)	Absolute pressure
3.24E	84.5 psia		14.9 psia		Gage pressure
3.25E	22.8 psia		14.7 psia		Gage pressure
3.26E	4.3 psia		14.6 psia		Gage pressure
3.27E	10.8 psia		14.0 psia		Gage pressure
3.28E	14.7 psia		15.1 psia		Gage pressure
3.29E	41.2 psig		14.5 psia		Absolute pressure
3.30E	18.5 psig		14.2 psia		Absolute pressure
3.31E	0.6 psig		14.7 psia		Absolute pressure
3.32E	−4.3 psig		14.7 psia		Absolute pressure
3.33E	−12.5 psig		14.4 psia		Absolute pressure

Pressure-Elevation Relationship

3.34M *If milk has a specific gravity of 1.08, what is the pressure at the bottom of a milk can 550 mm deep?*

3.35E The pressure in an unknown fluid at a depth of 4.0 ft is measured to be 1.820 psig. Compute the specific gravity of the fluid.

3.36M *The pressure at the bottom of a tank of propyl alcohol at 25°C must be maintained at 52.75 kPa(gage). What depth of alcohol should be maintained?*

3.37E When you dive to a depth of 12.50 ft in seawater, what is the pressure?

3.38E A water storage tank is on the roof of a factory building, and the surface of the water is 50.0 ft above the floor of the factory. If a pipe connects the storage tank to the floor level and the pipe is full of static water, what is the pressure in the pipe at floor level?

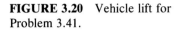

FIGURE 3.20 Vehicle lift for Problem 3.41.

Oil
sg = 0.90

Air

32 in

Lift
cylinder

80 in

A

48 in

3.39M *An open tank contains ethylene glycol at 25°C. Compute the pressure at a depth of 3.0 m.*

3.40M *For the tank of ethylene glycol described in Problem 3.39, compute the pressure at a depth of 12.0 m.*

3.41E Figure 3.20 shows a diagram of the hydraulic sys-tem for a vehicle lift. An air compressor maintains pressure above the oil in the reservoir. What must the air pressure be if the pressure in point A must be at least 180 psig?

3.42E Figure 3.21 shows a clothes washing machine. The pump draws fluid from the tub and delivers it to the

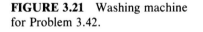

FIGURE 3.21 Washing machine for Problem 3.42.

Tub

Sink

375 mm

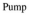

Pump

disposal sink. Compute the pressure at the inlet to the pump when the water is static (no flow). The soapy water solution has a specific gravity of 1.15.

3.43M *An airplane is flying at 10.6 km altitude. In its non-pressurized cargo bay is a container of mercury 325 mm deep. The container is vented to the local atmosphere. What is the absolute pressure at the surface of the mercury and at the bottom of the container? Assume the conditions of the standard atmosphere prevail for pressure. Use sg = 13.54 for the mercury.*

3.44E For the tank shown in Fig. 3.22, determine the reading of the bottom pressure gage in psig if the top of the tank is vented to the atmosphere and the depth of the oil, h, is 28.50 ft.

FIGURE 3.22 Problems 3.44 to 3.47.

3.45E For the tank shown in Fig. 3.22, determine the reading of the bottom pressure gage in psig if the top of the tank is sealed, the top gage reads 50.0 psig, and the depth of the oil, h, is 28.50 ft.

3.46E For the tank shown in Fig. 3.22, determine the reading of the bottom pressure gage in psig if the top of the tank is sealed, the top gage reads −10.8 psig, and the depth of the oil, h, is 6.25 ft.

3.47E For the tank shown in Fig. 3.22, determine the depth of the oil, h, if the reading of the bottom pressure gage is 35.5 psig, the top of the tank is sealed, and the top gage reads 30.0 psig.

3.48M *For the tank in Fig. 3.23, compute the depth of the oil if the depth of the water is 2.80 m and the gage at the bottom of the tank reads 52.3 kPa(gage).*

FIGURE 3.23 Problems 3.48 to 3.50.

3.49M *For the tank in Fig. 3.23, compute the depth of the water if the depth of the oil is 6.90 m and the gage at the bottom of the tank reads 125.3 kPa(gage).*

3.50M *Figure 3.23 represents an oil storage drum which is open to the atmosphere at the top. Some water was accidentally pumped into the tank and settled to the bottom as shown in the figure. Calculate the depth of the water h_2 if the pressure gage at the bottom reads 158 kPa(gage). The total depth h_T = 18.0 m.*

3.51M *A storage tank for sulfuric acid is 1.5 m in diameter and 4.0 m high. If the acid has a specific gravity of 1.80, calculate the pressure at the bottom of the tank. The tank is open to the atmosphere at the top.*

3.52E A storage drum for crude oil (sg = 0.89) is 32 ft deep and open at the top. Calculate the pressure at the bottom.

3.53M *The greatest known depth in the ocean is approximately 11.0 km. Assuming that the specific weight of the water is constant at 10.0 kN/m³, calculate the pressure at this depth.*

3.54M *Figure 3.24 shows a closed tank that contains gasoline floating on water. Calculate the air pressure above the gasoline.*

3.55M *Figure 3.25 shows a closed container holding water and oil. Air at 34 kPa below atmospheric pressure is above the oil. Calculate the pressure at the bottom of the container in kPa(gage).*

3.56M *Determine the pressure at the bottom of the tank in Fig. 3.26.*

FIGURE 3.24   Problem 3.54.

FIGURE 3.26   Problem 3.56.

Manometers

3.57 Describe a simple U-tube manometer.

3.58 Describe a differential U-tube manometer.

3.59 Describe a well-type manometer.

3.60 Describe an inclined well-type manometer.

FIGURE 3.25   Problem 3.55.

3.61 Describe a compound manometer.

3.62M *Water is in the pipe shown in Fig. 3.27. Calculate the pressure at point A in kPa(gage).*

3.63E For the differential manometer shown in Fig. 3.28, calculate the pressure difference between points A and B. The specific gravity of the oil is 0.85.

FIGURE 3.27   Problem 3.62.

3.64E For the manometer shown in Fig. 3.29, calculate
$(p_A - p_B)$.

3.65M *For the manometer shown in Fig. 3.30, calculate*
$(p_A - p_B)$.

3.66M *For the manometer shown in Fig. 3.31, calculate*
$(p_A - p_B)$.

3.67M *For the compound manometer shown in Fig. 3.32,*
calculate the pressure at point A.

FIGURE 3.29   Problem 3.64.

FIGURE 3.28   Problem 3.63.

FIGURE 3.30   Problem 3.65.

FIGURE 3.31   Problem 3.66.

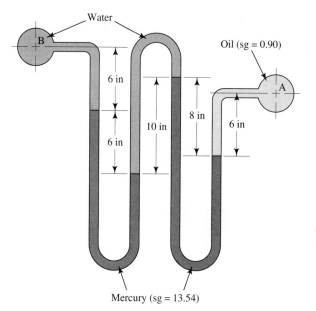

FIGURE 3.33   Problem 3.68.

3.68E For the compound differential manometer in Fig. 3.33, calculate $(p_A - p_B)$.

3.69E Figure 3.34 shows a manometer being used to indicate the difference in pressure between two points in a pipe. Calculate $(p_A - p_B)$.

FIGURE 3.32   Problem 3.67.

FIGURE 3.34   Problem 3.69.

FIGURE 3.35   Problem 3.70.

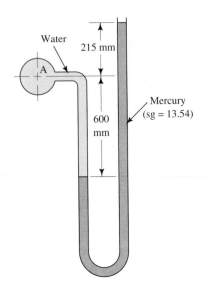

FIGURE 3.37   Problem 3.72.

3.70E For the well-type manometer in Fig. 3.35, calculate p_A.

3.71M *Figure 3.36 shows an inclined well-type manometer in which the distance L indicates the movement of the gage fluid level as the pressure p_A is applied above the well. The gage fluid has a specific gravity of 0.87 and L = 115 mm. Neglecting the drop in fluid level in the well, calculate p_A.*

3.72M *a. Determine the gage pressure at point A in Fig. 3.37.*

   b. If the barometric pressure is 737 mm of mercury, express the pressure at point A in kPa(abs).

Barometers

3.73 What is the function of a barometer?

3.74 Describe the construction of a barometer.

3.75 Why is mercury a convenient fluid to use in a barometer?

3.76 If water were to be used instead of mercury in a barometer, how high would the water column be?

3.77E What is the barometric pressure reading in inches of mercury corresponding to 14.7 psia?

3.78M *What is the barometric pressure reading in millimeters of mercury corresponding to 101.3 kPa(abs)?*

3.79 Why must a barometric pressure reading be corrected for temperature?

FIGURE 3.36   Problem 3.71.

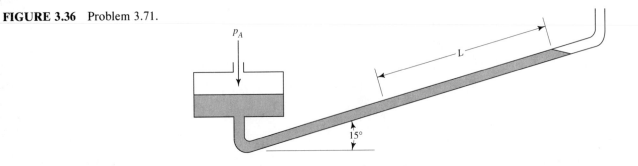

3.80E By how much would the barometric pressure reading decrease from its sea-level value at an elevation of 1250 ft?

3.81C Denver, Colorado, is called the "Mile-High City" because it is located at an elevation of approximately 5200 ft. Assuming that the sea-level pressure is 101.3 kPa(abs), what would be the approximate atmospheric pressure in Denver?

3.82E The barometric pressure is reported to be 28.6 in of mercury. Calculate the atmospheric pressure in psia.

3.83E A barometer indicates the atmospheric pressure to be 30.65 in of mercury. Calculate the atmospheric pressure in psia.

3.84E What would be the reading of a barometer in inches of mercury corresponding to an atmospheric pressure of 14.2 psia?

3.85M *A barometer reads 745 mm of mercury. Calculate the barometric reading in kPa(abs).*

Pressure Gages and Transducers

3.86 Describe a Bourdon tube pressure gage.

3.87 Describe a strain gage–type pressure transducer.

3.88 Describe a quartz crystal pressure transducer that uses the piezoelectric effect.

3.89 Describe a quartz crystal pressure transducer that uses the resonant frequency effect.

4 Forces on Submerged Plane and Curved Areas

4.1
OBJECTIVES

Pressure has been defined by the equation $p = F/A$, where F is a force and A is the area on which the force acts. Figure 4.1 illustrates several cases for which it is necessary to calculate the forces exerted on *plane* (flat) and *curved* areas as a result of fluid pressure. The magnitude and distribution of these forces must be known to design the components of each system.

This chapter presents the methods of analysis used to compute the force exerted on a plane area for three special cases and the general case. Forces on curved surfaces also are analyzed.

After completing this chapter, you should be able to:

1. Compute the force exerted on a plane area by a pressurized gas.
2. Compute the force exerted by any static fluid acting on a horizontal plane area.
3. Compute the resultant force exerted on a rectangular wall by a static liquid.
4. Define the term *center of pressure*.
5. Determine the location of the center of pressure for a liquid exerting a force on a rectangular wall.
6. Compute the resultant force exerted on any submerged plane area by a static liquid.
7. Determine the location of the center of pressure for a liquid exerting a force on any submerged plane area by a static liquid.
8. Show the vector representing the resultant force on any submerged plane area in its proper location and direction.
9. Visualize the distribution of force on a submerged curved surface.
10. Compute the horizontal component of the resultant force on the curved surface.
11. Determine the line of action of the horizontal component.
12. Compute the vertical component of the resultant force on the curved surface.
13. Determine the line of action of the vertical component.
14. Compute the total resultant force on the curved surface.
15. Compute the direction in which the resultant force acts and show its line of action on a sketch of the surface.
16. Include the effect of a pressure head over the liquid on the force on a plane or curved surface.

FIGURE 4.1 Examples of cases where forces on submerged areas must be computed.

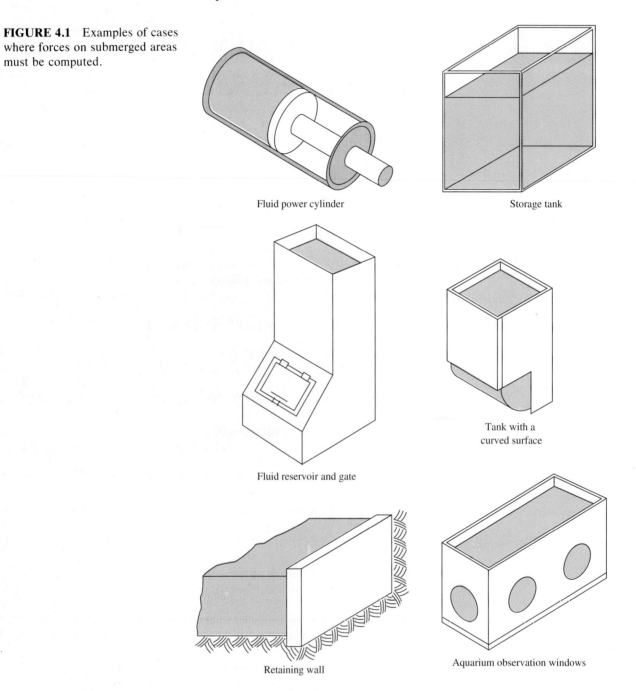

Fluid power cylinder

Storage tank

Fluid reservoir and gate

Tank with a curved surface

Retaining wall

Aquarium observation windows

4.2 GASES UNDER PRESSURE

Figure 4.2 shows a pneumatic cylinder of the type used in automated machinery. The air pressure acts on the piston face, producing a force that causes the linear movement of the rod. The pressure also acts on the end of the cylinder, tending to pull it apart. This is the reason for the four tie rods between the end caps of the cylinder. The distribution of pressure within a gas is very nearly uniform. Therefore, we can calculate the force on the piston and the cylinder ends directly from $F = pA$.

Wear ring

Rod gland

Piston

Rod bushing

Rod seal and
Rod wiper seal

Steel head and caps

Piston seals

Cylinder tube

Tie rods and nuts

Piston rod

FIGURE 4.2 Fluid power cylinder. (Source of photo: Mosier Industries, Brookville, OH)

☐ **EXAMPLE PROBLEM 4.1**

If the cylinder in Fig. 4.2 has an internal diameter of 2 in and operates at a pressure of 300 psig, calculate the force on the ends of the cylinder.

Solution

$$F = pA$$

$$A = \frac{\pi D^2}{4} = \frac{\pi (2 \text{ in})^2}{4} = 3.14 \text{ in}^2$$

$$F = \frac{300 \text{ lb}}{\text{in}^2} \times 3.14 \text{ in}^2 = 942 \text{ lb}$$

Notice that gage pressure was used in the calculation of force instead of absolute pressure. The additional force due to atmospheric pressure acts on both sides of the area and is thus balanced. If the pressure on the outside surface is not atmospheric, then all external forces must be considered to determine a net force on the area.

4.3
HORIZONTAL FLAT
SURFACES UNDER
LIQUIDS

Figure 4.3 shows a cylindrical drum containing oil and water. The pressure in the water at the bottom of the drum is uniform across the entire area since it is a horizontal plane in a fluid at rest. Again, we can simply use $F = pA$ to calculate the force on the bottom.

☐ **EXAMPLE PROBLEM 4.2**

If the drum in Fig. 4.3 is open to the atmosphere at the top, calculate the force on the bottom.

Solution

To use $F = pA$ we must first calculate the pressure at the bottom of the drum p_B and the area of the bottom:

$$p_B = p_{atm} + \gamma_o(2.4 \text{ m}) + \gamma_w(1.5 \text{ m})$$

$$\gamma_o = (sg)_o(9.81 \text{ kN/m}^3) = (0.90)(9.81 \text{ kN/m}^3) = 8.83 \text{ kN/m}^3$$

$$p_B = 0 \text{ Pa(gage)} + (8.83 \text{ kN/m}^3)(2.4 \text{ m}) + (9.81 \text{ kN/m}^3)(1.5 \text{ m})$$
$$= (0 + 21.2 + 14.7) \text{ kPa} = 35.9 \text{ kPa(gage)}$$

$$A = \pi D^2/4 = \pi(3.0 \text{ m})^2/4 = 7.07 \text{ m}^2$$

$$F = p_B A = (35.9 \text{ kN/m}^2)(7.07 \text{ m}^2) = 253.8 \text{ kN}$$

☐ **EXAMPLE PROBLEM 4.3**

Would there be any difference between the force on the bottom of the drum in Fig. 4.3 and that on the bottom of the cone-shaped container in Fig. 4.4?

Solution

The force would be the same because the pressure at the bottom is dependent only upon the depth and specific weight of the fluid in the container. The total weight of fluid has no effect. Recall Pascal's paradox in Section 3.5.

FIGURE 4.3   Cylindrical drum for Example Problem 4.2.

FIGURE 4.4   Cone-shaped container for Example Problem 4.3.

4.4
RECTANGULAR WALLS

The retaining walls shown in Fig. 4.5 are typical examples of rectangular walls exposed to a pressure varying from zero at the surface of the fluid to a maximum at the bottom of the wall. The force due to the fluid pressure tends to overturn the wall or break it at the place where it is fixed to the bottom.

FIGURE 4.5   Rectangular walls.

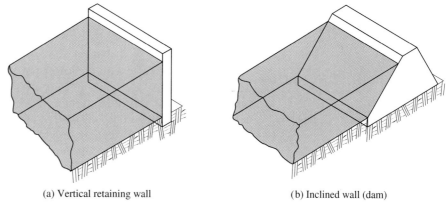

(a) Vertical retaining wall   (b) Inclined wall (dam)

The actual force is distributed over the entire wall, but for the purpose of analysis it is desirable to determine the resultant force and the place where it acts, called the *center of pressure*. That is, if the entire force were concentrated at a single point, where would that point be and what would the magnitude of the force be?

Figure 4.6 shows the pressure distribution on the vertical retaining wall. As indicated by the equation $\Delta p = \gamma h$, the pressure varies linearly (in a straight-line manner) with depth in the fluid. The lengths of the dashed arrows represent the magnitude of the fluid pressure at various points on the wall. Because of this linear variation in pressure, the total resultant force can be calculated by the equation

$$F_R = p_{avg} \times A \qquad\qquad \textbf{(4–1)}$$

where p_{avg} is the average pressure and A is the total area of the wall. But the average pressure is that at the middle of the wall and can be calculated by the equation

$$p_{avg} = \gamma(d/2) \qquad\qquad \textbf{(4–2)}$$

where d is the total depth of the fluid.

FIGURE 4.6   Vertical rectangular wall.

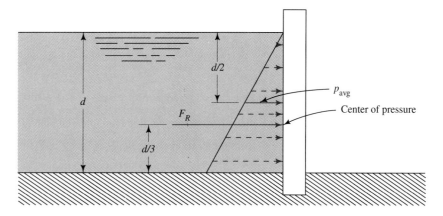

⇨ **RESULTANT FORCE ON A RECTANGULAR WALL**

Therefore, we have

$$F_R = \gamma(d/2)A \qquad (4\text{–}3)$$

The pressure distribution shown in Fig. 4.6 indicates that a greater portion of the force acts on the lower part of the wall than on the upper part. The center of pressure is at the centroid of the pressure distribution triangle, one-third of the distance from the bottom of the wall. The resultant force F_R acts perpendicular to the wall at this point.

The procedure for calculating the magnitude of the resultant force due to fluid pressure and the location of the center of pressure on a rectangular wall such as those shown in Fig. 4.5 is listed below. The procedure applies whether the wall is vertical or inclined.

PROCEDURE FOR COMPUTING THE FORCE ON A RECTANGULAR WALL

1. Calculate the magnitude of the resultant force F_R from

$$F_R = \gamma(d/2)A$$

where

$$\gamma = \text{specific weight of the fluid}$$
$$d = \text{total depth of the fluid}$$
$$A = \text{total area of the wall}$$

2. Locate the center of pressure at a vertical distance of $d/3$ from the bottom of the wall.
3. Show the resultant force acting at the center of pressure perpendicular to the wall.

☐ **EXAMPLE PROBLEM 4.4**   In Fig. 4.6, the fluid is gasoline (sg = 0.68) and the total depth is 12 ft. The wall is 40 ft long. Calculate the magnitude of the resultant force on the wall and the location of the center of pressure.

Solution   Step 1.

$$F_R = \gamma(d/2)A$$
$$\gamma = (0.68)(62.4 \text{ lb/ft}^3) = 42.4 \text{ lb/ft}^3$$
$$A = (12 \text{ ft})(40 \text{ ft}) = 480 \text{ ft}^2$$
$$F_R = \frac{42.4 \text{ lb}}{\text{ft}^3} \times \frac{12 \text{ ft}}{2} \times 480 \text{ ft}^2 = 122,000 \text{ lb}$$

Step 2.   The center of pressure is at a distance of

$$d/3 = 12 \text{ ft}/3 = 4 \text{ ft}$$

from the bottom of the wall.

Step 3.   The force F_R acts perpendicular to the wall at the center of pressure as shown in Fig. 4.6.

☐ **EXAMPLE PROBLEM 4.5**   Figure 4.7 shows a dam, 30.5 m long, which retains 8 m of fresh water and is inclined at an angle, θ, of 60°. Calculate the magnitude of the resultant force on the dam and the location of the center of pressure.

FIGURE 4.7 Inclined rectangular wall.

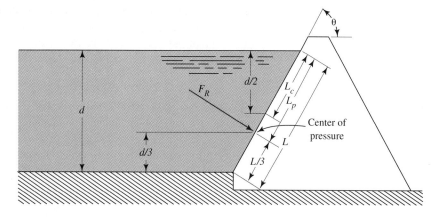

Solution   Step 1.

$$F_R = \gamma(d/2)A$$

To calculate the area of the dam we need the height of its face, called L in Fig. 4.7:

$$\sin \theta = d/L$$
$$L = d/\sin \theta = 8 \text{ m}/\sin 60° = 9.24 \text{ m}$$

Then, the area of the dam is

$$A = (9.24 \text{ m})(30.5 \text{ m}) = 281.8 \text{ m}^2$$

Now we can calculate the resultant force:

$$F_R = \gamma(d/2)A = \frac{9.81 \text{ kN}}{\text{m}^3} \times \frac{8 \text{ m}}{2} \times 281.8 \text{ m}^2$$
$$= 11\ 060 \text{ kN} = 11.06 \text{ MN}$$

Step 2.   The center of pressure is at a vertical distance of

$$d/3 = 8 \text{ m}/3 = 2.67 \text{ m}$$

from the bottom of the dam. Or, measured from the bottom of the dam along the face of the dam, the center of pressure is at:

$$L/3 = 9.24 \text{ m}/3 = 3.08 \text{ m}$$

Measured along the face of the dam we define:

L_p = distance from the free surface of the fluid to the center of pressure
$L_p = L - L/3$
$L_p = 9.24 \text{ m} - 3.08 \text{ m} = 6.16 \text{ m}$

We show F_R acting at the center of pressure perpendicular to the wall.

**4.5
SUBMERGED PLANE
AREAS—GENERAL**

The procedure we will discuss in this section applies to problems dealing with plane areas, either vertical or inclined, that are completely submerged in the fluid. As in previous problems, the procedure will enable us to calculate the magnitude of the resultant force on an area and the location of the center of pressure where we can assume the resultant force to act.

Figure 4.8 shows a tank that has a rectangular window in an inclined wall. The standard dimensions and symbols used in the procedure described later are shown in the figure and defined below:

F_R	Resultant force on the area due to the fluid pressure
—	The *center of pressure* of the area is the point at which the resultant force can be considered to act
—	The *centroid* of the area is the point at which the area would be balanced if suspended from that point. It is equivalent to the center of gravity of a solid body.
θ	Angle of inclination of the area
d_c	Depth of fluid from the free surface to the centroid of the area
L_c	Distance from the level of the free surface of the fluid to the centroid of the area, measured along the angle of inclination of the area
L_p	Distance from the level of the free surface of the fluid to the center of pressure of the area, measured along the angle of inclination of the area
B, H	Dimensions of the area

FIGURE 4.8   Force on submerged plane area.

Centroid

$$I_c = \frac{BH^3}{12}$$

$$A = BH$$

FIGURE 4.9 Properties of a rectangle.

Figure 4.9 shows the location of the centroid of a rectangle. Other shapes are described in Appendix L.

The following procedure will help you to calculate the magnitude of the resultant force on a submerged plane area due to fluid pressure and the location of the center of pressure.

PROCEDURE FOR COMPUTING THE FORCE ON A SUBMERGED PLANE AREA

1. Identify the point where the angle of inclination of the area of interest intersects the level of the free surface of the fluid. This may require the extension of the angled surface. Call this point S.
2. Locate the centroid of the area from its geometry.
3. Determine d_c as the *vertical* distance from the level of the free surface down to the centroid of the area.
4. Determine L_c as the *inclined* distance from the level of the free surface down to the centroid of the area. This is the distance from S to the centroid. Note that d_c and L_c are related by:

$$d_c = L_c \sin \theta$$

5. Calculate the total area A on which the force is to be determined.
6. Calculate the resultant force from

$$F_R = \gamma d_c A \qquad\qquad \textbf{(4–4)}$$

RESULTANT FORCE ON A SUBMERGED PLANE AREA

where γ is the specific weight of the fluid. This equation states that the resultant force is the product of the pressure at the centroid of the area and the total area.

7. Calculate I_c, the moment of inertia of the area about its centroidal axis.
8. Calculate the location of the center of pressure from

$$L_p = L_c + \frac{I_c}{L_c A} \qquad\qquad \textbf{(4–5)}$$

LOCATION OF CENTER OF PRESSURE

Notice that the center of pressure is always below the centroid of an area which is inclined with the horizontal. In some cases it may be of interest to calculate only the difference between L_p and L_c from

$$L_p - L_c = \frac{I_c}{L_c A} \qquad\qquad \textbf{(4–6)}$$

9. Sketch the resultant force F_R acting at the center of pressure, perpendicular to the area.
10. Show the dimension L_p on the sketch in a manner similar to that used in Fig. 4.8.
11. Draw the dimension lines for L_c and L_p from a reference line drawn through point S and perpendicular to the angle of inclination of the area.

We will now use the programmed instruction approach to illustrate the application of this procedure.

PROGRAMMED EXAMPLE PROBLEM

☐ **EXAMPLE PROBLEM 4.6**   The tank shown in Fig. 4.8 contains a lubricating oil with a specific gravity of 0.91. The rectangular gate with the dimensions $B = 4$ ft and $H = 2$ ft is placed in the inclined wall of the tank ($\theta = 60°$). The centroid of the gate is at a depth of 5 ft from the surface of the oil. Calculate (a) the magnitude of the resultant force F_R on the gate and (b) the location of the center of pressure.

Using the procedure described above, perform Steps 1 and 2 before going to the next panel.

Point S is shown in Fig. 4.8.

The area of interest is the rectangular gate sketched as Fig. 4.10. The centroid is at the intersection of the axes of symmetry of the rectangle.

FIGURE 4.10   Rectangular gate for Example Problem 4.6.

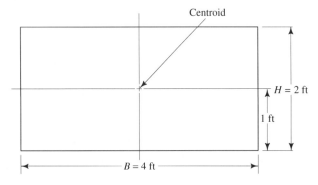

Now for Step 3, what is the distance d_c?

From the problem statement we know that $d_c = 5$ ft, the vertical depth from the free surface of the oil to the centroid of the gate.

Now calculate L_c. See Step 4.

The terms L_c and d_c are related in this case by

$$d_c = L_c \sin \theta$$

Therefore, we have

$$L_c = d_c/\sin \theta = 5 \text{ ft}/\sin 60° = 5.77 \text{ ft}$$

Both d_c and L_c will be needed for later calculations.

Go on to Step 5.

Since the area of the rectangle is BH,

$$A = BH = (4 \text{ ft})(2 \text{ ft}) = 8 \text{ ft}^2$$

Now do Step 6.

In the equation $F_R = \gamma d_c A$ we need the specific weight of the oil:

$$\gamma_o = (sg)_o(62.4 \text{ lb/ft}^3) = (0.91)(62.4 \text{ lb/ft}^3)$$
$$= 56.8 \text{ lb/ft}^3$$

Then we have

$$F_R = \gamma_o d_c A = \frac{56.8 \text{ lb}}{\text{ft}^3} \times 5 \text{ ft} \times 8 \text{ ft}^2 = 2270 \text{ lb}$$

The next steps concern the location of the center of pressure. Go on to Step 7 now.

From Fig. 4.9 we find that for a rectangle,

$$I_c = BH^3/12 = (4 \text{ ft})(2 \text{ ft})^3/12 = 2.67 \text{ ft}^4$$

Now we have all the data necessary to do Step 8.

Since $I_c = 2.67 \text{ ft}^4$, $L_c = 5.77 \text{ ft}$, and $A = 8 \text{ ft}^2$,

$$L_p = L_c + \frac{I_c}{L_c A} = 5.77 \text{ ft} + \frac{2.67 \text{ ft}^4}{(5.77 \text{ ft})(8 \text{ ft}^2)}$$

$$L_p = 5.77 \text{ ft} = 0.058 \text{ ft} = 5.828 \text{ ft}$$

The result is $L_p = 5.828$ ft.

This means that the center of pressure is 0.058 ft (or 0.70 in) below the centroid of the gate.

Steps 9, 10, and 11 are already completed in Fig. 4.8. Be sure you understand how the dimension L_p is drawn from the reference line.

■

4.6
DEVELOPMENT OF THE GENERAL PROCEDURE FOR FORCES ON SUBMERGED PLANE AREAS

Section 4.5 showed the use of the principles for computing the resultant force on a submerged plane area and for finding the location of the center of pressure. Equation (4–4) gives the resultant force, and Eq. (4–6) gives the distance between the centroid of the area of interest and the center of pressure. Figure 4.8 illustrates the various terms. This section shows the development of those relationships.

4.6.1
Resultant Force

The *resultant force* is defined as the summation of the forces on small elements of the area of interest. Figure 4.11 illustrates the concept using the same rectangular window used in Fig. 4.8. Actually, the shape of the area is arbitrary. On any small area dA, there exists a force dF acting perpendicular to the area owing to the fluid pressure, p. But the magnitude of the pressure at a depth h in a static liquid of specific weight γ is $p = \gamma h$. Then, the force is

$$dF = p(dA) = \gamma h(dA) \tag{4–7}$$

FIGURE 4.11 Development of the general procedure for forces on submerged plane areas.

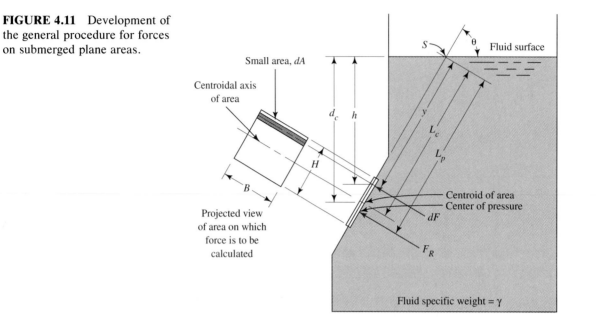

Because the area is inclined at an angle θ it is convenient to work in the plane of the area, using y to denote the position on the area at any depth h. Note that

$$h = y \sin \theta \tag{4–8}$$

where y is measured from the level of the free surface of the fluid along the angle of inclination of the area. Then,

$$dF = \gamma(y \sin \theta)(dA) \tag{4–9}$$

The summation of forces over the entire area is accomplished by the mathematical process of integration.

$$F_R = \int_A dF = \int_A \gamma(y \sin \theta)(dA) = \gamma \sin \theta \int_A y(dA)$$

From mechanics we learn that $\int y(dA)$ is equal to the product of the total area times the distance to the centroid of the area from the reference axis. That is,

$$\int_A y(dA) = L_c A$$

Then, the resultant force F_R is

$$F_R = \gamma \sin \theta (L_c A) \tag{4–10}$$

Now we can substitute $d_c = L_c \sin \theta$, giving

$$F_R = \gamma d_c A \tag{4–11}$$

This is the same form as Eq. (4–4). Because each of the small forces, dF, acted perpendicular to the area, the resultant force also acts perpendicular to the area.

4.6.2
Center of Pressure

The *center of pressure* is that point on an area where the resultant force can be assumed to act to have the same effect as the distributed force over the entire area due to fluid pressure. We can express this effect in terms of the moment of a force with respect to an axis through S perpendicular to the page.

See Fig. 4.11. The moment of each small force, dF, with respect to this axis is

$$dM = dF \cdot y$$

But, $dF = \gamma(y \sin \theta)(dA)$. Then,

$$dM = y[\gamma(y \sin \theta)(dA)] = \gamma \sin \theta(y^2 dA)$$

The moment of all the forces on the entire area is found by integrating over the area. Now, if we assume that the resultant force F_R acts at the center of pressure, its moment with respect to the axis through S is $F_R L_p$. Then,

$$F_R L_p = \int \gamma \sin \theta(y^2 dA) = \gamma \sin \theta \int (y^2 dA)$$

Again, from mechanics, we learn that $\int (y^2 dA)$ is defined as the moment of inertia I of the entire area with respect to the axis from which y is measured. Then,

$$F_R L_p = \gamma \sin \theta(I)$$

Solving for L_p gives

$$L_p = \frac{\gamma \sin \theta(I)}{F_R}$$

Substituting for F_R from Eq. (4–10) gives

$$L_p = \frac{\gamma \sin \theta(I)}{\gamma \sin \theta(L_c A)} = \frac{I}{L_c A} \tag{4–12}$$

A more convenient expression can be developed by using the transfer theorem for moment of inertia from mechanics. That is,

$$I = I_c + AL_c^2$$

where I_c is the moment of inertia of the area of interest with respect to its own centroidal axis, and L_c is the distance from the reference axis to the centroid. Equation (4–12) then becomes

$$L_p = \frac{I}{L_c A} = \frac{I_c + AL_c^2}{L_c A} = \frac{I_c}{L_c A} + L_c \tag{4–13}$$

Rearranging gives the same form as Eq. (4–6):

$$L_p - L_c = \frac{I_c}{L_c A}$$

4.7
PIEZOMETRIC HEAD

In all the problems demonstrated so far, the free surface of the fluid was exposed to the ambient pressure where $p = 0$ (gage). Therefore, our calculations for pressure within the fluid were also gage pressures. It was appropriate to use gage pressures for computing the magnitude of the net force on the areas of interest because the ambient pressure also acts outside the area.

A change is required in our procedure if the pressure above the free surface of the fluid is different from the ambient pressure outside the area. A convenient method uses the concept of *piezometric head*, in which the actual pressure above the fluid, p_a, is converted into an equivalent depth of the fluid, d_a, that would create the same pressure (Fig. 4.12):

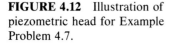   PIEZOMETRIC HEAD

$$d_a = p_a/\gamma \qquad \textbf{(4–14)}$$

This depth is added to any depth d below the free surface to obtain an *equivalent depth*, d_e. That is,

$$d_e = d + d_a \qquad \textbf{(4–15)}$$

Then, d_e can be used in any calculation requiring a depth to compute pressure. For example, in Fig. 4.12, the equivalent depth to the centroid is

$$d_{ce} = d_c + d_a$$

FIGURE 4.12   Illustration of piezometric head for Example Problem 4.7.

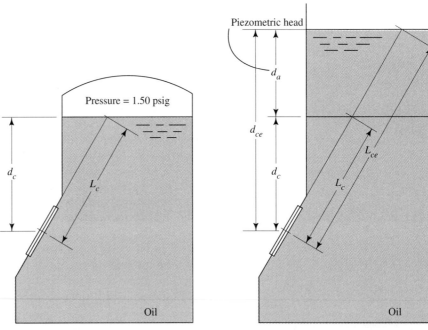

(a) Tank from Fig. 4.8 with pressure above the oil

(b) Tank showing piezometric head equivalent to pressure above the oil

☐ **EXAMPLE PROBLEM 4.7**

Repeat Example Problem 4.6, except consider that the tank shown in Fig. 4.8 is sealed at its top and that there is a pressure of 1.50 psig above the oil.

Solution

Several calculations in the solution to Example Problem 4.6 used the depth to the centroid, d_c, given to be 5.0 ft below the surface of the oil. With the pressure above the oil, we must add the piezometric head d_a from Eq. (4–14). Using $\gamma = 56.8$ lb/ft³, we get

$$d_a = \frac{p_a}{\gamma} = \frac{1.5 \text{ lb}}{\text{in}^2} \frac{144 \text{ in}^2}{\text{ft}^2} \frac{\text{ft}^3}{56.8 \text{ lb}} = 3.80 \text{ ft}$$

Then, the equivalent depth to the centroid is

$$d_{ce} = d_c + d_a = 5.00 \text{ ft} + 3.80 \text{ ft} = 8.80 \text{ ft}$$

The resultant force is then

$$F_R = \gamma d_{ce} A = (56.8 \text{ lb/ft}^3)(8.80 \text{ ft})(8.0 \text{ ft}^2) = 4000 \text{ lb}$$

Compare this with the value of 2270 lb found before for the open tank.

The center of pressure also changes because the distance, L_c, changes to L_{ce}, as shown below.

$$L_{ce} = d_{ce}/\sin \theta = 8.80 \text{ ft}/\sin 60° = 10.16 \text{ ft}$$

$$L_{pe} - L_{ce} = \frac{I_c}{L_{ce} A} = \frac{2.67 \text{ ft}^4}{(10.16 \text{ ft})(8 \text{ ft}^2)} = 0.033 \text{ ft}$$

The corresponding distance from Example Problem 4.6 was 0.058 ft.

**4.8
DISTRIBUTION OF
FORCE ON A
SUBMERGED CURVED
SURFACE**

Figure 4.13 shows a tank holding a liquid with its top surface open to the atmosphere. Part of the left wall is vertical, and the lower portion is a segment of a cylinder. Here we are interested in the force acting on the curved surface due to the fluid pressure.

One way to visualize the total force system involved is to isolate the volume of fluid directly above the surface of interest as a free body and show all the forces acting on it, as shown in Fig. 4.14. Our goal here is to determine the horizontal force F_H and the vertical force F_V exerted on the fluid by the curved surface and their resultant force F_R. The line of action of the resultant force acts through the center of curvature of the curved surface. This is because each of the individual force vectors due to the fluid pressure acts perpendicular to the boundary, which is then along the radius of curvature. Figure 4.14 shows the resulting force vectors.

**4.8.1
Horizontal Component**

The vertical solid wall at the left exerts horizontal forces on the fluid in contact with it in reaction to the forces due to the fluid pressure. This part of the system behaves in the same manner as the vertical walls studied earlier. The resultant force F_1 acting at a distance, $d/3$, from the bottom of the wall.

The force F_{2a} on the right side on the upper part to a depth of d is equal to F_1 in magnitude and acts in the opposite direction. Then, they have no effect on the curved surface.

FIGURE 4.13 Tank with a curved surface containing a static fluid.

Curved surface on which force is to be computed

Tank with a curved surface

By summing forces in the horizontal direction, you can see that F_H must be equal to F_{2b} acting on the lower part of the right side. The area on which F_{2b} acts is the *projection* of the curved surface onto a vertical plane.

The magnitude of F_{2b} and its location can be found using the procedures developed for plane surfaces. That is,

$$F_{2b} = \gamma d_c A \qquad\qquad \textbf{(4–16)}$$

where d_c is the depth to the centroid of the projected area. For the type of surface shown in Fig. 4.14, the projected area is a rectangle. Calling the height of the rectangle s, you can see that $d_c = d + s/2$. Also, the area is sw where w is the width of the curved surface. Then,

$$F_{2b} = F_H = \gamma sw(d + s/2) \qquad\qquad \textbf{(4–17)}$$

The location of F_{2b} is the center of pressure of the projected area. Again, using the principles developed earlier, we get

$$d_p - d_c = I_c/(d_c A)$$

But, for the rectangular projected area,

$$I_c = ws^3/12$$
$$A = sw$$

Then,

$$d_p - d_c = \frac{ws^3}{12(d_c)(sw)} = \frac{s^2}{12d_c} \qquad\qquad \textbf{(4–18)}$$

FIGURE 4.14 Free body diagram of volume of fluid above the curved surface.

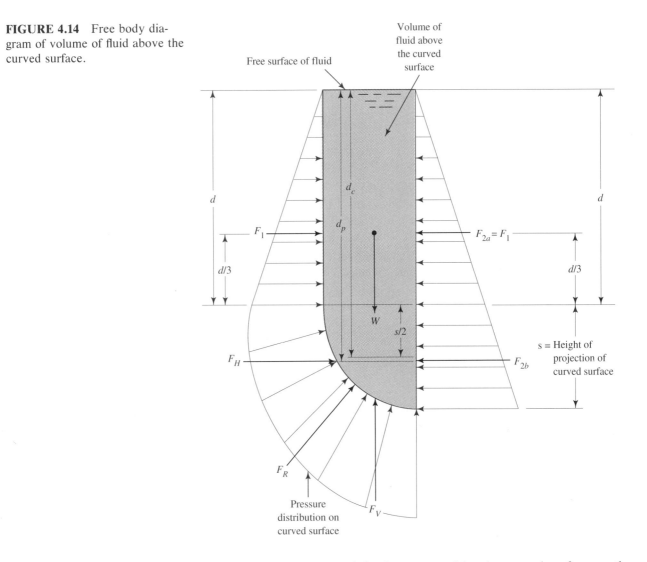

4.8.2
Vertical Component

The vertical component of the force exerted by the curved surface on the fluid can be found by summing forces in the vertical direction. Only the weight of the fluid acts downward, and only the vertical component, F_V, acts upward. Then, the weight and F_V must be equal to each other in magnitude. The weight of the fluid is simply the product of its specific weight times the volume of the isolated body of fluid. The volume is the product of the cross-sectional area of the volume shown in Fig. 4.14 and the length of interest, w. That is,

$$F_V = \gamma(\text{volume}) = \gamma A w \qquad \textbf{(4–19)}$$

4.8.3
Resultant Force

The total resultant force F_R is

$$F_R = \sqrt{F_H^2 + F_V^2} \qquad \textbf{(4–20)}$$

The resultant force acts at an angle, ϕ, relative to the horizontal, found from

$$\phi = \tan^{-1}(F_V/F_H) \qquad\qquad (4-21)$$

4.8.4
Summary of the Procedure
for Computing the Force
on a Submerged
Curved Surface

Given a curved surface submerged beneath a static liquid similar to the configuration shown in Fig. 4.13, the following procedure can be used to compute the magnitude, direction, and location of the resultant force on the surface.

1. Isolate the volume of fluid above the surface.

2. Compute the weight of the isolated volume.

3. The magnitude of the vertical component of the resultant force is equal to the weight of the isolated volume. It acts in line with the centroid of the isolated volume.

4. Draw a projection of the curved surface onto a vertical plane and determine its height, called s.

5. Compute the depth to the centroid of the projected area from

$$d_c = d + s/2$$

where d is the depth of the top of the projected area.

6. Compute the magnitude of the horizontal component of the resultant force from

$$F_H = \gamma sw(d + s/2) = \gamma swd_c$$

7. Compute the depth to the line of action of the horizontal component from

$$d_p = d_c + s^2/(12\,d_c)$$

8. Compute the resultant force from
$$F_R = \sqrt{F_V^2 + F_H^2}$$

9. Compute the angle of inclination of the resultant force relative to the horizontal from

$$\phi = \tan^{-1}(F_V/F_H)$$

10. Show the resultant force acting on the curved surface in such a direction that its line of action passes through the center of curvature of the surface.

☐ **EXAMPLE PROBLEM 4.8**   For the tank shown in Fig. 4.13, the following dimensions apply:

$$d_1 = 3.00 \text{ m}$$
$$d_2 = 4.50 \text{ m}$$
$$w = 2.50 \text{ m}$$
$$\gamma = 9.81 \text{ kN/m}^3 \qquad \text{(water)}$$

Compute the horizontal and vertical components of the resultant force on the curved surface and the resultant force itself. Show these force vectors on a sketch.

Solution Using the steps outlined above:

1. The volume above the curved surface is shown in Fig. 4.15.

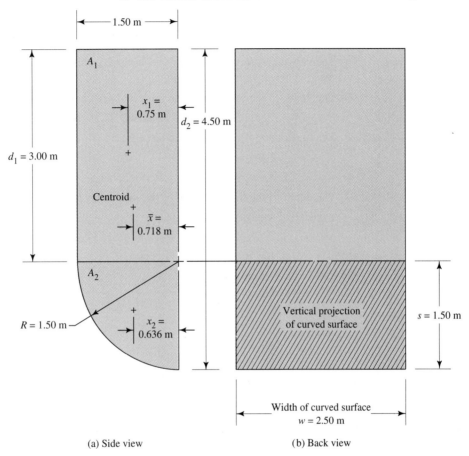

(a) Side view (b) Back view

FIGURE 4.15 Isolated volume above the curved surface for Example Problem 4.8.

2. The weight of the isolated volume is the product of the specific weight of the water times the volume. The volume is the product of the area times the length w.

$$\text{Area} = A_1 + A_2 = d_1 \cdot R + \tfrac{1}{4}(\pi R^2)$$
$$\text{Area} = (3.00 \text{ m})(1.50 \text{ m}) + \tfrac{1}{4}[\pi(1.50 \text{ m})^2] = 4.50 \text{ m}^2 + 1.767 \text{ m}^2$$
$$\text{Area} = 6.267 \text{ m}^2$$
$$\text{Volume} = \text{area} \cdot w = (6.267 \text{ m}^2)(2.50 \text{ m}) = 15.67 \text{ m}^3$$
$$\text{Weight} = \gamma V = (9.81 \text{ kN/m}^3)(15.67 \text{ m}^3) = 153.7 \text{ kN}$$

3. Then, $F_V = 153.7$ kN, acting upward through the centroid of the volume. The location of the centroid is found using the composite area technique. Refer to Fig. 4.15 for the data. Each value should be obvious except x_2, the location of the

centroid of the quadrant. From Appendix L,

$$x_2 = 0.424R = 0.424(1.50 \text{ m}) = 0.636 \text{ m}$$

Then, the location of the centroid for the composite area is

$$\bar{x} = \frac{A_1 x_1 + A_2 x_2}{A_1 + A_2} = \frac{(4.50)(0.75) + (1.767)(0.636)}{4.50 + 1.767} = 0.718 \text{ m}$$

4. The vertical projection of the curved surface is shown in Fig. 4.15. The height, s, equals 1.50 m.

5. The depth to the centroid of the projected area is

$$d_c = d_1 + s/2 = 3.00 \text{ m} + (1.50 \text{ m})/2 = 3.75 \text{ m}$$

6. The magnitude of the horizontal force is

$$F_H = \gamma s w(d_1 + s/2) = \gamma s w d_c$$
$$F_H = (9.81 \text{ kN/m}^3)(1.50 \text{ m})(2.50 \text{ m})(3.75 \text{ m}) = 138.0 \text{ kN}$$

7. The depth to the line of action of the horizontal component is found from

$$d_p = d_c + s^2/(12 d_c)$$
$$d_p = 3.75 \text{ m} + (1.50)^2/[(12)(3.75)] = 3.80 \text{ m}$$

FIGURE 4.16   Results for Example Problem 4.8.

8. The resultant force is computed from

$$F_R = \sqrt{F_V^2 + F_H^2}$$

$$F_R = \sqrt{(153.7 \text{ kN})^2 + (138.0 \text{ kN})^2} = 206.5 \text{ kN}$$

9. The angle of inclination of the resultant force relative to the horizontal is computed from

$$\phi = \tan^{-1}(F_V/F_H)$$

$$\phi = \tan^{-1}(153.7/138.0) = 48.1°$$

10. The horizontal component, the vertical component, and the resultant force are shown in Fig. 4.16. Note that the line of action of F_R is through the center of curvature of the surface. Also, the vertical component is acting through the centroid of the volume of liquid above the surface. The horizontal component is acting through the center of pressure of the projected area at a depth d_p from the level of the free surface of the fluid.

4.9
EFFECT OF A PRESSURE ABOVE THE FLUID SURFACE

In the preceding discussion of force on a submerged curved surface, the magnitude of the force was directly dependent on the depth of the static fluid above the surface of interest. If an additional pressure exists above the fluid or if the fluid itself is pressurized, the effect is to add to the actual depth a depth of fluid d_a equivalent to p/γ. This is the same procedure, called *piezometric head*, used in Section 4.7. The new equivalent depth is used to compute both the vertical and horizontal forces.

4.10
FORCES ON A CURVED SURFACE WITH FLUID BELOW IT

To this point, problems have considered curved surfaces supporting a fluid above. An important concept presented for such problems was that the vertical force on the curved surface was equal to the weight of the fluid above the surface.

Now, consider the type of curved surface shown in Fig. 4.17, in which the fluid is restrained below the surface. Fluid pressure on such a surface causes forces that tend to push it upward and to the right. The surface and its connections then would have to exert reaction forces downward and to the left on the contained fluid.

The pressure in the fluid at any pont is dependent on the depth of fluid to that point from the level of the free surface. This situation is equivalent to having the curved surface supporting a volume of liquid *above* it, except for the direction of the force vectors. Figure 4.18 shows that we can visualize an imaginary volume of fluid extending from the surface of interest to the level of the free surface or to the piezometric line if the fluid is under an additional pressure. Then, as before, the horizontal component of the force exerted by the curved surface on the fluid is the force on the projection of the curved surface on a vertical plane. The vertical component is equal to the weight of the imaginary volume of fluid above the surface.

FIGURE 4.17 Curved surface restraining a fluid below it.

FIGURE 4.18 Forces exerted by curved surface on the fluid.

FIGURE 4.19 Semicylindrical gate.

Oil
sg = 0.90

1.75 m

1.40-m diameter

4.11
FORCES ON CURVED
SURFACES WITH FLUID
ABOVE AND BELOW

Figure 4.19 shows a semicylindrical gate projecting into a tank containing an oil. The force due to fluid pressure would have a horizontal component acting to the right on the gate. This force acts on the projection of the surface on a vertical plane and is computed in the same manner as used in Section 4.8.

FIGURE 4.20 Volumes used to compute net vertical force on the gate.

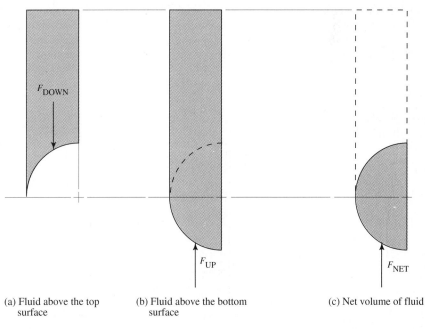

F_{DOWN}

F_{UP}

F_{NET}

(a) Fluid above the top
    surface

(b) Fluid above the bottom
    surface

(c) Net volume of fluid

In the vertical direction, the force on the top of the gate would act downward and would equal the weight of the oil above the gate. However, there is also a force acting upward on the bottom surface of the gate equal to the total weight of the fluid, both real and imaginary, above that surface. The net vertical force is the difference between the two forces, equal to the weight of the semicylindrical volume of fluid displaced by the gate itself (see Fig. 4.20).

PRACTICE PROBLEMS

Forces Due to Gas Pressure

4.1E Figure 4.21 shows a vacuum tank which has a flat circular observation window in one end. If the pressure in the tank is 0.12 psia when the barometer reads 30.5 in of mercury, calculate the total force on the window.

4.2E The flat left end of the tank shown in Fig. 4.21 is secured with a bolted flange. If the inside diameter of the tank is 30 in and the internal pressure is raised to +14.4 psig, calculate the total force that must be resisted by the bolts in the flange.

4.3E An exhaust system for a room creates a partial vacuum in the room of 1.20 in of water relative to the atmospheric pressure outside the room. Compute the net force exerted on a 36-by-80-in door to this room.

4.4E A piece of 14-in Schedule 40 pipe is used as a pressure vessel by capping its ends. Compute the force on the caps if the pressure in the pipe is raised to 325 psig. See Appendix F for the dimensions of the pipe.

4.5M *A pressure relief valve is designed so that the gas pressure in the tank acts on a piston with a diameter of 30 mm. How much spring force must be applied to the outside of the piston to hold the valve closed under a pressure of 3.50 MPa?*

4.6M *A gas-powered cannon shoots projectiles by introducing nitrogen gas at 20.5 MPa into a cylinder having an inside diameter of 50 mm. Compute the force exerted on the projectile.*

4.7M *The egress hatch of a manned spacecraft is designed so that the internal pressure in the cabin applies a force to help maintain the seal. If the internal pressure is 34.4 kPa(abs) and the external pressure is a perfect vacuum, calculate the force on a square hatch 800 mm on a side.*

Forces on Horizontal Flat Surfaces Under Liquids

4.8E A tank containing liquid ammonia at 77°F has a flat horizontal bottom. A rectangular door, 24 in by 18 in, is installed in the bottom to provide access for cleaning. Compute the force on the door if the depth of ammonia is 12.3 ft.

FIGURE 4.21   Tank for Problems 4.1 and 4.2.

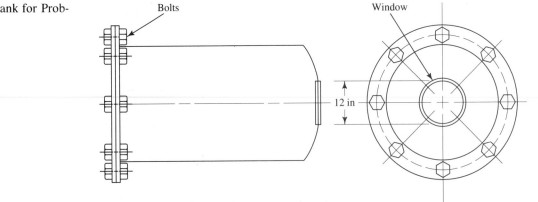

4.9E The bottom of a laboratory vat has a hole in it to allow the liquid mercury to pour out. The hole is sealed by a rubber stopper pushed in the hole and held by friction. What force tends to push the 0.75-in-diameter stopper out of the hole if the depth of the mercury is 28.0 in?

4.10M *A simple shower for remote locations is designed with a cylindrical tank 500 mm in diameter and 1.800 m high as shown in Fig. 4.22. The water flows through a flapper valve in the bottom through a 75-mm-diameter opening. The flapper must be pushed upward in order to open the valve. How much force is required to open the valve?*

4.11M *Calculate the total force on the bottom of the closed tank shown in Fig. 4.23 if the air pressure is 52 kPa(gage).*

4.12M *If the length of the tank in Fig. 4.24 is 1.2 m, calculate the total force on the bottom of the tank.*

4.13M *An observation port in a small submarine is located in a horizontal surface of the sub. The shape of the port is shown in Fig. 4.25. Compute the total force acting on the port when the pressure inside the sub is 100 kPa(abs) and the sub is operating at a depth of 175 m in sea water.*

Forces on Rectangular Walls

4.14E A rectangular gate is installed in a vertical wall of a reservoir, as shown in Fig. 4.26. Compute the magnitude of the resultant force on the gate and the location of the center of pressure.

4.15E A vat has a sloped side, as shown in Fig. 4.27. Compute the resultant force on this side if the vat contains 15.5 ft of glycerine. Also, compute the location of the center of pressure, and show on a sketch the resultant force.

4.16E The wall shown in Fig. 4.28 is 20 ft long. (a) Calculate the total force on the wall due to water pressure and locate the center of pressure; and (b) calculate the moment due to this force at the base of the wall.

4.17M *If the wall in Fig. 4.29 is 4 m long, calculate the total force on the wall due to the oil pressure. Also determine the location of the center of pressure and show the resultant force on the wall.*

Forces on Submerged Plane Areas

For each of the cases shown in Figs. 4.30 through 4.41, compute the magnitude of the resultant force on the indicated area and the location of the center of pressure. Show the resultant force on the area and clearly dimension its location. See p. 104 for Problems 4.18 through 4.29.

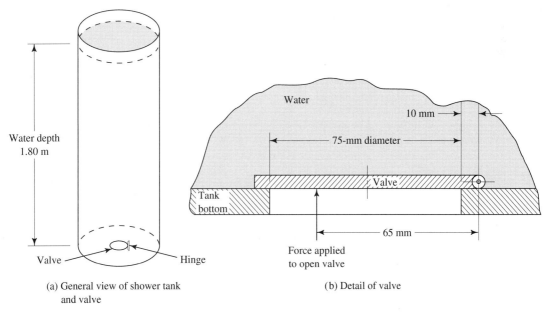

(a) General view of shower tank and valve

(b) Detail of valve

FIGURE 4.22 Shower tank and valve for Problem 4.10.

FIGURE 4.23 Problem 4.11.

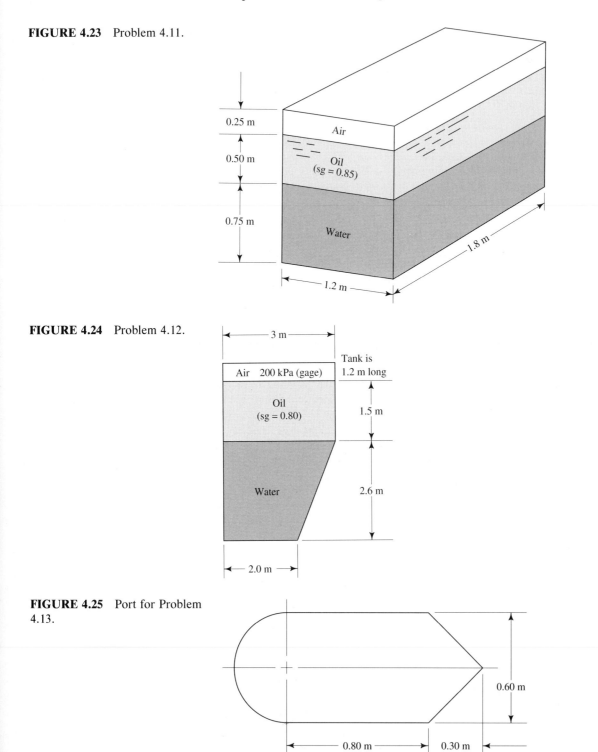

0.25 m

Air

0.50 m

Oil
(sg = 0.85)

0.75 m

Water

1.8 m

1.2 m

FIGURE 4.24 Problem 4.12.

3 m

Air   200 kPa (gage)

Tank is
1.2 m long

Oil
(sg = 0.80)

1.5 m

Water

2.6 m

2.0 m

FIGURE 4.25 Port for Problem
4.13.

0.60 m

0.80 m

0.30 m

FIGURE 4.26 Gate in a reservoir wall for Problem 4.14.

FIGURE 4.27 Vat for Problem 4.15.

FIGURE 4.28 Problem 4.16.

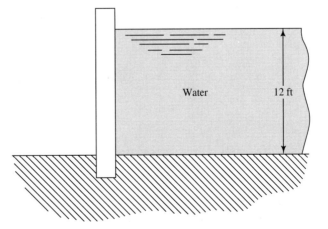

FIGURE 4.29   Problem 4.17.

FIGURE 4.30   Problem 4.18.

Reservoir for a
hydraulic system.
Length = 3.50 ft.
Compute force on
side AB.

FIGURE 4.31   Problems 4.19
and 4.43.

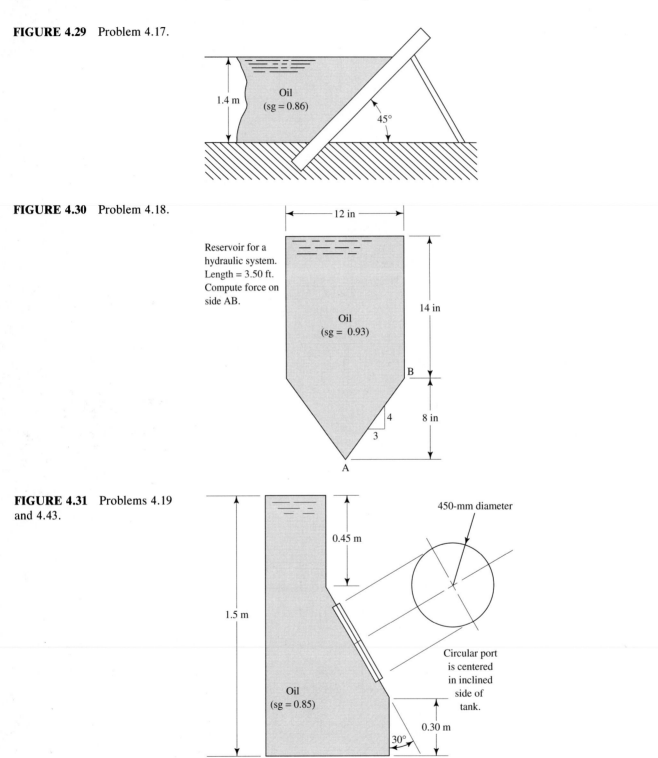

FIGURE 4.32 Problems 4.20, 4.36, 4.37, and 4.44.

FIGURE 4.33 Problem 4.21.

FIGURE 4.34   Problem 4.22.

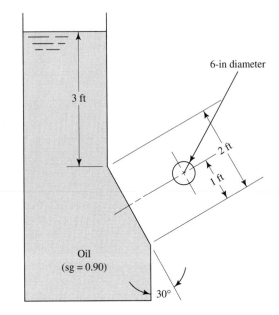

4.18E Refer to Fig. 4.30.

4.19M *Refer to Fig. 4.31.*

4.20M *Refer to Fig. 4.32.*

4.21E Refer to Fig. 4.33.

4.22E Refer to Fig. 4.34.

4.23M *Refer to Fig. 4.35.*

4.24E Refer to Fig. 4.36.

4.25M *Refer to Fig. 4.37.*

4.26E Refer to Fig. 4.38.

4.27M *Refer to Fig. 4.39.*

4.28E Refer to Fig. 4.40.

4.29E Refer to Fig. 4.41.

4.30M *Figure 4.42 shows a gasoline tank filled into the filler pipe. The gasoline has a specific gravity of*

FIGURE 4.35   Problems 4.23, 4.38, and 4.39.

FIGURE 4.36 Problem 4.24.

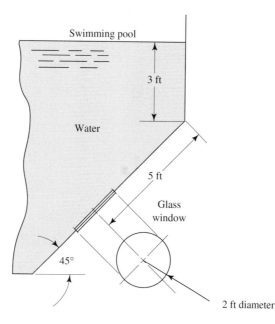

0.67. Calculate the total force on each flat end of the tank and determine the location of the center of pressure.

4.31M *If the tank in Fig. 4.42 is filled just to the bottom of the filler pipe with gasoline (sg = 0.67), calculate the magnitude and location of the resultant force on the flat end.*

4.32M *If the tank in Fig. 4.42 is only half full of gasoline*

(sg = 0.67), calculate the magnitude and location of the resultant force on the flat end.

4.33E For the water tank shown in Fig. 4.43, compute the magnitude and location of the total force on the vertical back wall.

4.34E For the water tank shown in Fig. 4.43, compute the magnitude and location of the total force on each vertical end wall.

FIGURE 4.37 Problem 4.25.

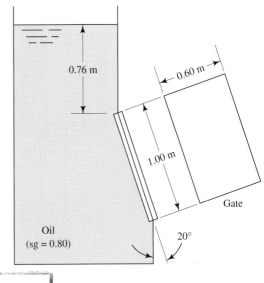

FIGURE 4.38   Problems 4.26
and 4.45.

4.35E For the water tank shown in Fig. 4.43, compute the magnitude and location of the total force on the inclined wall.

4.36M *For the orange-drink tank shown in Fig. 4.32, compute the magnitude and location of the total force on each vertical end wall. The tank is 3.0 m long.*

4.37M *For the orange-drink tank shown in Fig. 4.32, compute the magnitude and location of total force on the vertical back wall. The tank is 3.0 m long.*

4.38M *For the oil tank shown in Fig. 4.35, compute the magnitude and location of the total force on each vertical end wall. The tank is 1.2 m long.*

FIGURE 4.39   Problem 4.27.

FIGURE 4.40   Problems 4.28 and 4.46.

10 in

26 in

10 in

Ethylene glycol
sg = 1.10

Tank is symmetrical

8.0 in

20-in radius

30°

4.39M *For the oil tank shown in Fig. 4.35, compute the magnitude and location of the total force on the vertical back wall. The tank is 1.2 m long.*

4.40E Figure 4.44 shows a rectangular gate holding water behind it. If the water is 6.00 ft deep, compute the

magnitude and location of the resultant force on the gate. Then, compute the forces on the hinge at the top and on the stop at the bottom.

4.41M *Figure 4.45 shows a gate hinged at its bottom and held by a simple support at its top. The gate sepa-*

FIGURE 4.41   Problem 4.29.

18 in

6 in

20 in

30 in

Water

50°

FIGURE 4.42   Problems 4.30 to 4.32.

rates two fluids. Compute the net force on the gate due to the fluid on each side. Then, compute the force on the hinge and on the support.

4.42E Figure 4.46 shows a tank of water with a circular pipe connected to its bottom. A circular gate seals the pipe opening to prohibit flow. To drain the tank, a winch is used to pull the gate open. Compute the amount of force that the winch cable must exert to open the gate.

Piezometric Head

4.43M *Repeat Problem 4.19M (Fig. 4.31), except that the tank is now sealed at the top with a pressure of 13.8 kPa above the oil.*

FIGURE 4.43   Problems 4.33 to 4.35.

FIGURE 4.44 Problem 4.40.

Hinge

4.00 ft

Water

Stop

Rectangular gate, 1.25 ft wide

4.44M *Repeat Problem 4.20M (Fig. 4.32), except that the tank is now sealed at the top with a pressure of 25.0 kPa above the fluid.*

4.45E Repeat Problem 4.26E (Fig. 4.38), except that the tank is now sealed at the top with a pressure of 2.50 psig above the fluid.

4.46E Repeat Problem 4.28E (Fig. 4.40), except that the

tank is now sealed at the top with a pressure of 4.0 psig above the fluid.

Forces on Curved Surfaces

General note for Problems 4.47 through 4.54. For each problem, one curved surface is shown restraining a body of

FIGURE 4.45 Problem 4.41.

Support

Gate, 0.60 m wide

2.80 m

2.50 m

2.0 m

Water

Oil
sg = 0.90

Hinge

FIGURE 4.46   Problem 4.42.

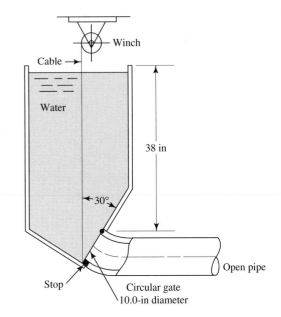

static fluid. Compute the magnitude of the horizontal component of the force, and compute the vertical component of the force exerted by the fluid on that surface. Then, compute the magnitude of the resultant force and its direction. Show the resultant force acting on the curved surface. In each case the surface of interest is a portion of a cylinder with the length of the surface given in the problem statement.

4.47M *Use Fig. 4.47. The surface is 2.00 m long.*

FIGURE 4.47   Problems 4.47 and 4.55.

4.48M *Use Fig. 4.48. The surface is 2.50 m long.*

FIGURE 4.48   Problems 4.48 and 4.56.

4.49E Use Fig. 4.49. The surface is 5.00 ft long.

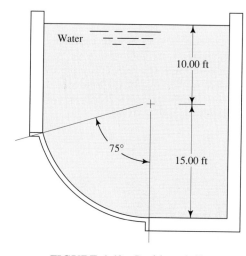

FIGURE 4.49 Problem 4.49.

4.50E Use Fig. 4.50. The surface is 3.50 ft long.

FIGURE 4.50 Problem 4.50.

4.51M *Use Fig. 4.51. The surface is 4.00 m long.*

FIGURE 4.51. Problem 4.51.

4.52M *Use Fig. 4.52. The surface is 1.50 m long.*

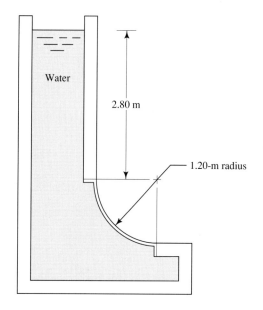

FIGURE 4.52 Problem 4.52.

4.53M *Use Fig. 4.53. The surface is 1.50 m long.*

FIGURE 4.53   Problem 4.53.

4.54E Use Fig. 4.54. The surface is 60 in long.

FIGURE 4.54   Problem 4.54.

4.55M *Repeat Problem 4.47 using Fig. 4.47, except that there is now 7.50 kPa air pressure above the fluid.*

4.56M *Repeat Problem 4.48 using Fig. 4.48, except that there is now 4.65 kPa air pressure above the fluid.*

4.57E Figure 4.55 shows a solid cylinder sitting on the bottom of a tank holding a static volume of water. Compute the force exerted by the cylinder on the bottom of the tank for the following data: $D = 6.00$ in; $L = 10.00$ in; $\gamma_c = 0.284$ lb/in^3 (steel); $\gamma_f = 62.4$ lb/ft^3 (water); $h = 30$ in.

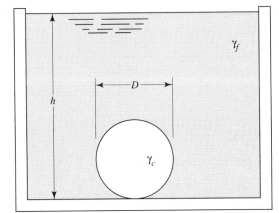

Cylinder length = L

FIGURE 4.55   Problems 4.57 to 4.63.

4.58E Repeat Problem 4.57, except use $\gamma_c = 0.100$ lb/in^3 (aluminum).

4.59E Repeat Problem 4.57, except use $\gamma_c = 30.00$ lb/ft^3 (wood).

4.60 For the situation described in Problem 4.57, specify the required relationship between the specific weight of the cylinder and that of the fluid so that no force at all is exerted on the bottom of the tank.

4.61E Repeat Problem 4.57 for a depth of $h = 10.00$ in.

4.62E Repeat Problem 4.57 for a depth of $h = 5.00$ in.

4.63E For the situation described in Problem 4.57, compute the force exerted on the bottom of the tank for varying depths of fluid from $h = 30$ in to $h = 0$. Use any convenient increments of change in depth that will produce a well-defined curve of force versus depth.

COMPUTER PROGRAMMING ASSIGNMENTS

1. Write a program to solve Problem 4.41 with any combination of data for the variables in Fig. 4.45, including the depth on either side of the gate and the specific gravity of the fluids.

2. Write a program to solve Problem 4.42 (Fig. 4.46) with any combination of data, including the size of the gate, the depth of the fluid, the specific gravity of the fluid, and the angle of inclination of the gate.

3. Write a program to solve curved surface problems of the type shown in Fig. 4.47 through 4.51 for any combination of variables, including the depth of the fluid, the angular size of the curved segment, the specific gravity of the fluid, and the radius of the surface.

4. Write a program to solve Problem 4.57 with any combination of data for the variables shown in Fig. 4.55.

5. For Program 1 above, cause the depth h to vary over some specified range, giving output for each value.

■ ■ ■ ■   5 Buoyancy and Stability

**5.1
OBJECTIVES**

Buoyancy is the tendency of a fluid to exert a supporting force on a body placed in the fluid. *Stability* refers to the ability of a body to return to its original position after being tilted about a horizontal axis.

Figure 5.1 shows several devices for which an analysis of buoyancy and stability should be completed. The buoy (a) and the ship (e) must be designed to float stably. The instrument package (b) would tend to float if not restrained by the anchoring cable. The diving bell (c) must be supported by the crane on the ship to keep it from sinking. However, the submarine (d) has the ability to hover at any depth, dive deeper into the water, or rise to the surface and float.

After completing this chapter, you should be able to:

1. Write the equation for the buoyant force.
2. Analyze the case of bodies floating on a fluid.
3. Use the principle of static equilibrium to solve for the forces involved in buoyancy problems.
4. Define the conditions that must be met for a body to be stable when completely submerged in a fluid.
5. Define the conditions that must be met for a body to be stable when floating on a fluid.
6. Define the term *metacenter* and compute its location.

**5.2
BUOYANCY**

A body in a fluid, whether floating or submerged, is buoyed up by a force equal to the weight of the fluid displaced. The buoyant force acts vertically upward through the centroid of the displaced volume and can be defined mathematically by Archimedes' principle as stated below.

▷ **BUOYANT FORCE**

$$F_b = \gamma_f V_d \tag{5-1}$$

where

$$F_b = \text{buoyant force}$$
$$\gamma_f = \text{specific weight of the fluid}$$
$$V_d = \text{displaced volume of the fluid}$$

When a body is floating freely, it displaces a sufficient volume of fluid to just balance its own weight.

The analysis of problems dealing with buoyancy requires the application of the equation of static equilibrium in the vertical direction, $\Sigma F_v = 0$,

115

FIGURE 5.1 Examples of buoyancy problem types.

assuming the object is at rest in the fluid. The following procedure is recommended for all problems, whether they involve floating or submerged bodies:

PROCEDURE FOR SOLVING BUOYANCY PROBLEMS

1. Determine the objective of the problem solution. Are you to find a force, weight, volume, or specific weight?
2. Draw a free body diagram of the object in the fluid. Show all forces that act on the free body in the vertical direction, including the weight of the body, the buoyant force, and all external forces. If the direction of some force is not known, assume the most probable direction and show it on the free body.
3. Write the equation of static equilibrium in the vertical direction, $\Sigma F_v = 0$, assuming the positive direction to be upward.
4. Solve for the desired force, weight, volume, or specific weight, remembering the following concepts:
   a. The buoyant force is calculated from $F_b = \gamma_f V_d$.
   b. The weight of a solid object is the product of its total volume and its specific weight; that is, $w = \gamma V$.
   c. An object with an average specific weight less than that of the fluid will tend to float because $w < F_b$ with the object submerged.
   d. An object with an average specific weight greater than that of the fluid will tend to sink because $w > F_b$ with the object submerged.
   e. *Neutral buoyancy* occurs when a body stays in a given position wherever it is submerged in the fluid. An object whose average specific weight is equal to that of the fluid would be neutrally buoyant.

PROGRAMMED EXAMPLE PROBLEMS

☐ **EXAMPLE PROBLEM 5.1**   A cube, 0.50 m on a side, is made of bronze having a specific weight of 86.9 kN/m³. Determine the magnitude and direction of the force required to hold the cube in equilibrium completely submerged (a) in water and (b) in mercury. The specific gravity of mercury is 13.54.

Consider part (a) first. Imagine the cube of bronze submerged in water. Now do Step 1 of the procedure.

Assuming that the bronze cube will not stay in equilibrium by itself, some external force is required. The objective is to find the magnitude of this force and the direction in which it would act—that is, up or down.

Now do Step 2 of the procedure before looking at the next panel.

The free body is simply the cube itself. There are three forces acting on the cube in the vertical direction, as shown in Fig. 5.2: the weight of the cube w, acting downward through its center of gravity; the buoyant force F_b, acting upward through the centroid of the displaced volume; and the externally applied supporting force F_e.

FIGURE 5.2 Free body diagram of cube.

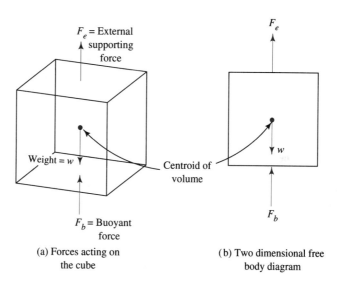

(a) Forces acting on the cube

(b) Two dimensional free body diagram

Part (a) of Fig. 5.2 shows the cube as a three-dimensional object with the three forces acting along a vertical line through the centroid of the volume. This is the preferred visualization of the free body diagram. However, for most problems it is suitable to use a simplified two-dimensional sketch as shown in part (b).

How do we know to draw the force F_e in the upward direction?

We really do not know for certain. However, experience should indicate that without an external force the solid bronze cube would tend to sink in water. Therefore, an upward force seems to be required to hold the cube in equilibrium. If our choice is wrong, the final result will indicate that to us.

Now, assuming that the forces are as shown in Fig. 5.2, go on to Step 3.

The equation should look like this. (Assume that positive forces act upward.)

$$\Sigma F_v = 0$$
$$F_b + F_e - w = 0 \qquad \text{(5–2)}$$

As a part of Step 4, solve this equation algebraically for the desired term.

You should now have

$$F_e = w - F_b \qquad \text{(5–3)}$$

since the objective is to find the external force.

How do we calculate the weight of the cube w?

Item b under Step 4 of the procedure indicates that $w = \gamma_B V$, where γ_B is the specific weight of the bronze cube and V is its total volume. For the cube, since each side is 0.50 m, we have

$$V = (0.50 \text{ m})^3 = 0.125 \text{ m}^3$$

and

$$w = \gamma_B\ V = (86.9 \text{ kN/m}^3)(0.125 \text{ m}^3) = 10.86 \text{ kN}$$

There is another unknown on the right side of Eq. (5–3). How do we calculate F_b?

Check item a under Step 4 of the procedure if you have forgotten.

$$F_b = \gamma_f V_d$$

In this case γ_f is the specific weight of the water (9.81 kN/m³), and the displaced volume V_d is equal to the total volume of the cube that we already know to be 0.125 m³. Then, we have

$$F_b = \gamma_f V_d = (9.81 \text{ kN/m}^3)(0.125 \text{ m}^3) = 1.23 \text{ kN}$$

Now we can complete our solution for F_e.

The solution is

$$F_e = w - F_b = 10.86 \text{ kN} - 1.23 \text{ kN} = 9.63 \text{ kN}$$

Notice that the result is positive. This means that our assumed direction for F_e was correct. Then the solution to the problem is that an upward force of 9.63 kN is required to hold the block of bronze in equilibrium under water.

What about part (b) of the problem, where the cube is submerged in mercury? Our objective is the same as before—to determine the magnitude and direction of the force required to hold the cube in equilibrium.

Now do Step 2 of the procedure.

FIGURE 5.3 Two possible free body diagrams.

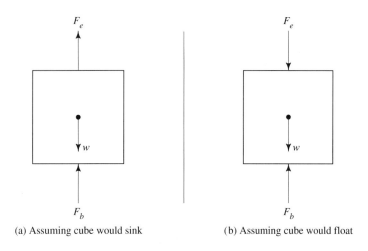

(a) Assuming cube would sink   (b) Assuming cube would float

Either of two free body diagrams is correct as shown in Fig. 5.3, depending on the assumed direction for the external force F_e. The solution for the two diagrams will be carried out simultaneously so you can check your work regardless of which diagram looks like yours and to demonstrate that either approach will yield the correct answer.

Now do Step 3 of the procedure.

These are the correct equations of equilibrium. Notice the differences and relate them to the figures.

$$F_b + F_e - w = 0 \quad | \quad F_b - F_e - w = 0$$

Now solve for F_e.

You should now have

$$F_e = w - F_b \quad | \quad F_e = F_b - w$$

Since the magnitude of w and F_b are the same for each equation, they can now be calculated.

As in part (a) of the problem, the weight of the cube is

$$w = \gamma_B V = (86.9 \text{ kN/m}^3)(0.125 \text{ m}^3) = 10.86 \text{ kN}$$

For the buoyant force F_b, you should have

$$F_b = \gamma_m V = (\text{sg})_m (9.81 \text{ kN/m}^3)(V)$$

where the subscript m refers to mercury. We then have

$$F_b = (13.54)(9.81 \text{ kN/m}^3)(0.125 \text{ m}^3) = 16.60 \text{ kN}$$

Now go on with the solution for F_e.

The correct answers are

$$F_e = w - F_b$$
$$= 10.86 \text{ kN} - 16.60 \text{ kN}$$
$$= -5.74 \text{ kN}$$

$$F_e = F_b - w$$
$$= 16.60 \text{ kN} - 10.86 \text{ kN}$$
$$= +5.74 \text{ kN}$$

Notice that both solutions yield the same numerical value, but they have opposite signs. The negative sign for the solution on the left means that the assumed direction for F_e in Fig. 5.3(a) was wrong. Therefore, both approaches give the same result. The required external force is a downward force of 5.74 kN.

How could you have reasoned from the start that a downward force would be required?

Items c and d of Step 4 of the procedure suggest that the specific weight of the cube and the fluid be compared. In this case:

For the bronze cube, $\gamma_B = 86.9 \text{ kN/m}^3$

For the fluid (mercury); $\gamma_m = (13.54)(9.81 \text{ kN/m}^3)$
$$= 132.8 \text{ kN/m}^3$$

Since the specific weight of the cube is less than that of the mercury, it would tend to float without an external force. Therefore, a downward force, as pictured in Fig. 5.3(b), would be required to hold it in equilibrium under the surface of the mercury.

This example problem is concluded.

☐ **EXAMPLE PROBLEM 5.2**   A certain solid metal object has such an irregular shape that it is difficult to calculate its volume by geometry. We can use the principle of buoyancy to calculate its volume.

First, the weight of the object is determined in the normal manner to be 60 lb. Then, using a setup similar to that in Fig. 5.4, we find its apparent weight while submerged in water to be 46.5 lb. Using these data and the procedure for analyzing buoyancy problems, we can find the volume of the object.

FIGURE 5.4   Metal object suspended in a fluid.

Balance beam

Total weight = 46.5 lb

Water

FIGURE 5.5 Free body diagram.

Now draw the free body diagram of the object while it is suspended in the water.

The free body diagram of the object while it is suspended in the water should look like Fig. 5.5. In this figure, what are the two forces F_e and w?

From the problem statement we should know that w = 60 lb, the weight of the object in air, and F_e = 46.5 lb, the supporting force exerted by the balance shown in Fig. 5.4.

Now do Step 3 of the procedure.

Using $\Sigma\, F_v = 0$, we get

$$F_b + F_e - w = 0$$

Our objective is to find the total volume V of the object. How can we get V from this equation?

We use this equation:

$$F_b = \gamma_f V$$

where γ_f is the specific weight of the water, 62.4 lb/ft^3.

Substitute this into the preceding equation and solve for V.

You should now have

$$F_b + F_e - w = 0$$
$$\gamma_f V + F_e - w = 0$$
$$\gamma_f V = w - F_e$$
$$V = \frac{w - F_e}{\gamma_f}$$

Now we can put in the known values and calculate V.

The result is V = 0.216 ft^3. This is how it is done:

$$V = \frac{w - F_e}{\gamma_f} = (60 - 46.5)\text{lb}\left(\frac{\text{ft}^3}{62.4\ \text{lb}}\right) = \frac{13.5\ \text{ft}^3}{62.4} = 0.216\ \text{ft}^3$$

Now that the volume of the object is known, the specific weight of the material can be found.

$$\gamma = \frac{w}{V} = \frac{60\ \text{lb}}{0.216\ \text{ft}^3} = 278\ \text{lb/ft}^3$$

This is approximately the specific weight of a titanium alloy.

The next two problems are worked out in detail and should serve to check your ability to solve buoyancy problems. After reading the problem statement, you should complete the solution yourself before reading the panel on which a correct solution is given. Be sure to read the problem carefully and use the proper units in your calculations. Although there is more than one way to solve some problems, it is possible to get the correct answer by the wrong method. If your method is different from that given, be sure yours is based on sound principles before assuming it is correct.

☐ **EXAMPLE PROBLEM 5.3**   A cube, 80 mm on a side, is made of a rigid foam material and floats in water with 60 mm below the surface. Calculate the magnitude and direction of the force required to hold it completely submerged in glycerine, which has a specific gravity of 1.26.

Complete the solution before looking at the next panel.

Solution   First calculate the weight of the cube, then the force required to hold the cube submerged in glycerine. Use the free body diagrams in Fig. 5.6: (a) cube floating on water and (b) cube submerged in glycerine.

FIGURE 5.6   Free body diagrams.

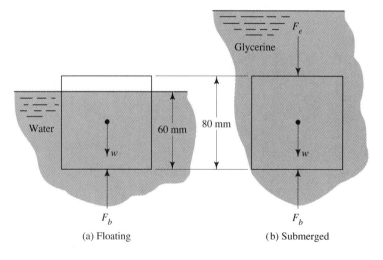

From Fig. 5.6(a), we have

$$\sum F_v = 0$$

$$F_b - w = 0$$

$$w = F_b = \gamma_f V_d$$

$$V_d = (80 \text{ mm})(80 \text{ mm})(60 \text{ mm}) = 384 \times 10^3 \text{ mm}^3$$
(submerged volume of cube)

$$w = \left(\frac{9.81 \times 10^3 \text{ N}}{\text{m}^3}\right)(384 \times 10^3 \text{ mm}^3)\left(\frac{1 \text{ m}^3}{(10^3 \text{ mm})^3}\right)$$

$$= 3.77 \text{ N}$$

From Fig. 5.6(b), we have

$$\sum F_v = 0$$

$$F_b - F_e - w = 0$$

$$F_e = F_b - w = \gamma_f V_d - 3.77 \text{ N}$$

$$V_d = (80 \text{ mm})^3 = 512 \times 10^3 \text{ mm}^3$$

(total volume of cube)

$$\gamma_f = (1.26)(9.81 \text{ kN/m}^3) = 12.36 \text{ kN/m}^3$$

$$F_e = \gamma_f V_d - 3.77 \text{ N}$$

$$= \left(\frac{12.36 \times 10^3 \text{ N}}{\text{m}^3}\right)(512 \times 10^3 \text{ mm}^3)\left(\frac{1 \text{ m}^3}{(10^3 \text{ mm})^3}\right) - 3.77 \text{ N}$$

$$= 6.33 \text{ N} - 3.77 \text{ N} = 2.56 \text{ N}$$

A downward force of 2.56 N is required to hold the cube submerged in glycerine.

☐ **EXAMPLE PROBLEM 5.4**   A brass cube, 6 in on a side, weighs 67 lb. We want to hold this cube in equilibrium under water by attaching a light foam buoy to it. If the foam weighs 4.5 lb/ft³, what is the minimum required volume of the buoy?

Complete the solution yourself before looking at the next panel.

Solution   Calculate the minimum volume of foam to hold the brass cube in equilibrium.

Notice that the foam and brass in Fig. 5.7 are considered as parts of a single system and that there is a buoyant force on each. The subscript F refers to the foam, and the subscript B refers to the brass. No external force is required.

FIGURE 5.7 Free body diagram for brass and foam together.

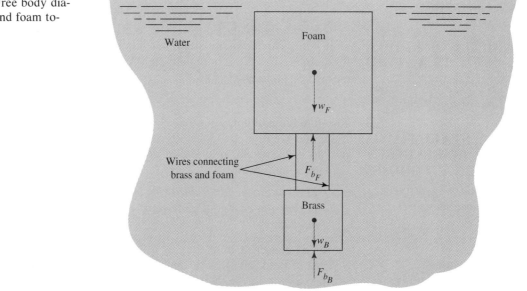

The equilibrium equation is

$$\sum F_v = 0$$

$$0 = F_{b_B} + F_{b_F} - w_B - w_F \tag{5-4}$$

$$w_B = 67 \text{ lb} \quad \text{(given)}$$

$$F_{b_B} = \gamma_f V_{d_B} = \left(\frac{62.4 \text{ lb}}{\text{ft}^3}\right)(6 \text{ in})^3 \left(\frac{\text{ft}^3}{1728 \text{ in}^3}\right) = 7.8 \text{ lb}$$

$$w_F = \gamma_F V_F$$

$$F_{b_F} = \gamma_f V_F$$

Substitute these quantities into Eq. (5–4):

$$F_{b_B} + F_{b_F} - w_B - w_F = 0$$

$$7.8 \text{ lb} + \gamma_f V_F - 67 \text{ lb} - \gamma_F V_F = 0$$

Solve for V_F, using $\gamma_f = 62.4$ lb/ft^3 and $\gamma_F = 4.5$ lb/ft^3:

$$\gamma_f V_F - \gamma_F V_F = 67 \text{ lb} - 7.8 \text{ lb} = 59.2 \text{ lb}$$

$$V_F(\gamma_f - \gamma_F) = 59.2 \text{ lb}$$

$$V_F = \frac{59.2 \text{ lb}}{\gamma_f - \gamma_F} = \frac{59.2 \text{ lb ft}^3}{(62.4 - 4.5) \text{ lb}}$$

$$= 1.02 \text{ ft}^3$$

This means that if 1.02 ft^3 of foam were attached to the brass cube, the combination would be in equilibrium in water without any external force. It would be neutrally buoyant.

This completes the programmed example problems.

■

5.3
STABILITY OF BODIES IN A FLUID

A body in a fluid is considered stable if it will return to its original position after being rotated a small amount about a horizontal axis. The conditions for stability are different depending on whether the body is completely submerged or floating.

5.4
STABILITY OF COMPLETELY SUBMERGED BODIES

Two familiar examples of bodies completely submerged in a fluid are submarines and weather balloons. It is important for these kinds of objects to remain in a specific orientation despite the action of currents, winds, or maneuvering forces.

⇨ CONDITION OF STABILITY FOR
SUBMERGED BODIES

The condition for stability of bodies completely submerged in a fluid is that the center of gravity of the body must be below the center of buoyancy.

The center of buoyancy of a body is at the centroid of the displaced volume of fluid, and it is through this point that the buoyant force acts in a vertical direction. The weight of the body acts vertically downward through the center of gravity.

FIGURE 5.8 Grumman-BEN FRANKLIN, undersea research vessel. (Source of photo: Grumman Aerospace Corp. Ocean Systems Department, Bethpage, NY)

The undersea research vehicle shown in Fig. 5.8 has a stable configuration due to its shape and the location of equipment within the hull. Figure 5.9 shows the approximate cross-sectional shape of the vessel. The circular section is a hollow cylinder providing a manned cabin and housing for delicate instruments and life-support systems. The rectangular section at the bottom contains heavy batteries and other durable equipment. With this

FIGURE 5.9 Stability of submerged body.

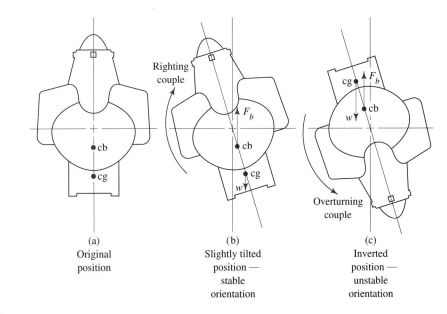

(a)	(b)	(c)
Original position	Slightly tilted position — stable orientation	Inverted position — unstable orientation

distribution of weight and volume, the center of gravity (cg) and the center of buoyancy (cb) are located approximately as shown in Fig. 5.9(a).

Figure 5.9(b) shows the action of the buoyant force and the weight to produce a couple that tends to rotate the vessel back to its original position after being rotated slightly. Thus, the body is stable. Contrast this with Fig. 5.9(c), which shows what would happen if the configuration were upside down from that in Fig. 5.9(a). When this body is rotated a small amount, the weight and the buoyant force produce a couple that tends to overturn it. Therefore, this orientation of the body is unstable.

If the center of gravity and center of buoyancy of a body coincide, as with a solid object, the weight and buoyant force act through the same point, producing no couple. In this case the body would have neutral stability and would remain in any orientation in which it is placed.

5.5
STABILITY OF
FLOATING BODIES

The condition for stability of floating bodies is different from that for completely submerged bodies; the reason is illustrated in Fig. 5.10, which shows the approximate cross section of a ship's hull. In part (a) of the figure, the floating body is at its equilibrium orientation and the center of gravity (cg) is above the center of buoyancy (cb). A vertical line through these points will be called the *vertical axis* of the body. Figure 5.10(b) shows that if the body is rotated slightly, the center of buoyancy shifts to a new position because the geometry of the displaced volume has changed. The buoyant force and the weight now produce a righting couple that tends to return the body to its original orientation. Thus, the body is stable.

FIGURE 5.10   Method of finding the metacenter.

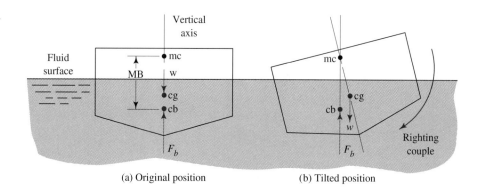

(a) Original position   (b) Tilted position

In order to state the condition for stability of a floating body, we must define a new term, *metacenter*. The metacenter (mc) is defined as the intersection of the vertical axis of a body when in its equilibrium position and a vertical line through the new position of the center of buoyancy when the body is rotated slightly. This is illustrated in Fig. 5.10(b).

▷ CONDITION OF STABILITY FOR
FLOATING BODIES

A floating body is stable if its center of gravity is below the meta-center.

It is possible to determine analytically if a floating body is stable by calculating the location of its metacenter. The distance to the *metacenter* from the center of *buoyancy* is called MB and is calculated from

$$\text{MB} = I/V_d \qquad \qquad \textbf{(5–5)}$$

In this equation, V_d is the displaced volume of fluid and I is the *least* moment of inertia of a horizontal section of the body taken at the surface of the fluid. *If the distance MB places the metacenter above the center of gravity, the body is stable.*

PROCEDURE FOR EVALUATING THE STABILITY OF FLOATING BODIES

1. Determine the position of the floating body, using the principles of buoyancy.
2. Locate the center of buoyancy, cb; and compute the distance from some reference axis to cb, called y_{cb}. Usually, the bottom of the object is taken as the reference axis.
3. Locate the center of gravity, cg; and compute y_{cg} measured from the same reference axis.
4. Determine the shape of the area at the fluid surface, and compute the *smallest* moment of inertia, I, for that shape.
5. Compute the displaced volume, V_d.
6. Compute $\text{MB} = I/V_d$.
7. Compute $y_{mc} = y_{cb} + \text{MB}$.
8. If $y_{mc} > y_{cg}$, the body is stable.
9. If $y_{mc} < y_{cg}$, the body is unstable.

PROGRAMMED EXAMPLE PROBLEMS

☐ **EXAMPLE PROBLEM 5.5**   Figure 5.11(a) shows a flatboat hull that, when fully loaded, weighs 150 kN. Parts (b), (c), and (d) show the top, front, and side views of the boat. Note the location of the center of gravity, cg. We would like to determine if the boat is stable in fresh water.

Find out if the boat will float.

This is done by finding how far the boat will sink into the water, using the principles of buoyancy stated in Section 5.2. Complete that calculation before going to the next panel.

The depth of submergence or draft of the boat is 1.06 m, as shown in Fig. 5.12, found by the method described below.

Equation of equilibrium: $\sum F_v = 0 = F_b - w$

$$w = F_b$$

(a) Loaded flatboat

(b) Top view and horizontal
cross section

(c) Front view and vertical
cross section

(d) Side view

FIGURE 5.11   Shape of hull.

FIGURE 5.12   Free body diagram.

Submerged volume: $V_d = B \times L \times X$

Buoyant force: $F_b = \gamma_f V_d = \gamma_f \times B \times L \times X$

Then, we have

$$w = F_b = \gamma_f \times B \times L \times X$$

$$X = \frac{w}{B \times L \times \gamma_f} = \frac{150 \text{ kN}}{(2.4 \text{ m})(6.0 \text{ m})} \times \frac{\text{m}^3}{(9.81 \text{ kN})} = 1.06 \text{ m}$$

It floats with 1.06 m submerged. Where is the center of buoyancy?

It is at the center of the displaced volume of water. In this case, as shown in Fig. 5.13, it is on the vertical axis of the boat at a distance of 0.53 m from the bottom. That is half of the draft, X. Then $y_{cb} = 0.53$ m.

FIGURE 5.13   Location of center of buoyancy and center of gravity.

Since the center of gravity is above the center of buoyancy, we must locate the metacenter to determine if the boat is stable. Using Eq. (5–5), calculate the distance MB and show it on the sketch.

The result is MB = 0.45 m, as shown in Fig. 5.14. Here is how it is done.

$$MB = I/V_d$$

$$V_d = L \times B \times X = (6.0 \text{ m})(2.4 \text{ m})(1.06 \text{ m}) = 15.26 \text{ m}^3$$

The moment of inertia, I, is determined about the axis X-X in Fig. 5.11(b), since this would yield the smallest value for I:

$$I = \frac{LB^3}{12} = \frac{(6.0 \text{ m})(2.4 \text{ m})^3}{12} = 6.91 \text{ m}^4$$

FIGURE 5.14   Location of
metacenter.

Then, the distance from the center of buoyancy to the metacenter is

$$MB = I/V_d = 6.91 \text{ m}^4/15.26 \text{ m}^3 = 0.45 \text{ m}$$

and

$$y_{mc} = y_{cb} + MB = 0.53 \text{ m} + 0.45 \text{ m} = 0.98 \text{ m}$$

Is the boat stable?

Yes, it is. Since the metacenter is above the center of gravity, as shown in Fig.
5.14, the boat is stable. That is, $y_{mc} > y_{cg}$.

Now, read the next panel for another problem.

□ **EXAMPLE PROBLEM 5.6**   A solid cylinder is 3.0 ft in diameter, is 6.0 ft high, and weighs 1550 lb. If the cylinder
is placed in oil (sg = 0.90), with its axis vertical, would it be stable?

The complete solution is shown in the next panel. Do this problem and then
look at the solution.

Solution   Position of cylinder in oil (Fig. 5.15):

$$V_d = \text{submerged volume} = AX = \frac{\pi D^2}{4}(X)$$

Equilibrium equation: $\sum F_v = 0$

$$w = F_b = \gamma_o V_d = \gamma_o \frac{\pi D^2}{4}(X)$$

$$X = \frac{4w}{\pi D^2 \gamma_o} = \frac{(4)(1550 \text{ lb}) \text{ ft}^3}{(\pi)(3.0 \text{ ft})^2(0.90)(62.4 \text{ lb})} = 3.90 \text{ ft}$$

The center of buoyancy, cb, is at a distance $X/2$ from the bottom of the cylinder:

$$y_{cb} = X/2 = 3.90 \text{ ft}/2 = 1.95 \text{ ft}$$

FIGURE 5.15 Complete solution for Example Problem 5.6.

The center of gravity, cg, is at $H/2 = 3$ ft from the bottom of the cylinder, assuming the material of the cylinder is of uniform specific weight. The position of the metacenter, mc, using Eq. (5–5), is

$$MB = I/V_d$$

$$I = \frac{\pi D^4}{64} = \frac{\pi (3 \text{ ft})^4}{64} = 3.98 \text{ ft}^4$$

$$V_d = AX = \frac{\pi D^2}{4}(X) = \frac{\pi (3 \text{ ft})^2}{4}(3.90 \text{ ft}) = 27.6 \text{ ft}^3$$

$$MB = I/V_d = 3.98 \text{ ft}^4/27.6 \text{ ft}^3 = 0.144 \text{ ft}$$

$$y_{mc} = y_{cb} + MB = 1.95 \text{ ft} + 0.14 \text{ ft} = 2.09 \text{ ft}$$

Since this places the metacenter below the center of gravity, as shown in Fig. 5.15, the cylinder is not stable. That is, $y_{mc} < y_{cg}$.

This completes the programmed instruction.

■

The conditions for stability of bodies in a fluid can be summarized as follows.

- *Completely submerged bodies are stable if the center of gravity is below the center of buoyancy.*
- *Floating bodies are stable if the center of gravity is below the metacenter.*

5.6
DEGREE OF STABILITY

Although the limiting case of stability has been stated as any design for which the metacenter is above the center of gravity, some objects can be more stable than others. One measure of relative stability is called the *metacentric height* defined as the distance to the metacenter from the center of gravity.

Refer to Fig. 5.16. The metacentric height is labeled MG. Using the procedures discussed in this chapter, we can compute MG from

$$MG = y_{mc} - y_{cg} \qquad \qquad (5\text{--}6)$$

Reference 1 states that small seagoing vessels should have a minimum value of MG of 1.5 ft (0.46 m). Large ships should have MG > 3.5 ft (1.07 m). But, the metacentric height should not be too large because the ship may then have the uncomfortable rocking motions that cause seasickness.

FIGURE 5.16   Degree of stability as indicated by metacentric height and righting arm.

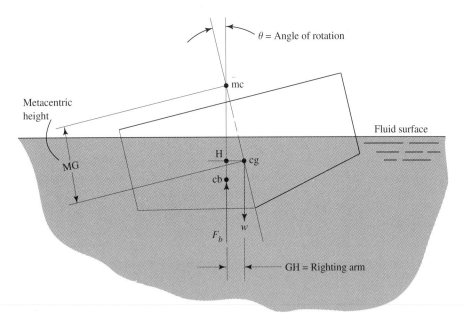

□ **EXAMPLE PROBLEM 5.7**   Compute the metacentric height for the flatboat hull described in Example Problem 5.5.

Solution   From the results of Example Problem 5.5,

$$y_{mc} = 0.98 \text{ m from the bottom of the hull}$$
$$y_{cg} = 0.80 \text{ m}$$

Then, the metacentric height is

$$MG = y_{mc} - y_{cg} = 0.98 \text{ m} - 0.80 \text{ m} = 0.18 \text{ m}$$

5.6.1
Static Stability Curve

Another measure of the stability of a floating object is the amount of offset between the line of action of the weight of the object acting through the center of gravity and that of the buoyant force acting through the center of buoyancy. Earlier, in Fig. 5.10, it was shown that the product of one of these forces and the amount of the offset produces the righting couple that causes the object to return to its original position and thus to be stable.

Figure 5.16 shows a sketch of a boat hull in a rotated position with the weight and the buoyant force shown. A horizontal line drawn through the center of gravity intersects the line of action of the buoyant force at point *H*. The horizontal distance, GH, is called the *righting arm* and is a measure of the magnitude of the righting couple. The distance GH varies as the angle of rotation varies, and Fig. 5.17 shows a characteristic plot of the righting arm versus the angle of rotation for a ship. Such a plot is called a *static stability curve*. As long as the value of GH remains positive, the ship is stable. Conversely, when GH becomes negative, the ship is unstable and it will overturn.

FIGURE 5.17 Static stability curve for a floating body.

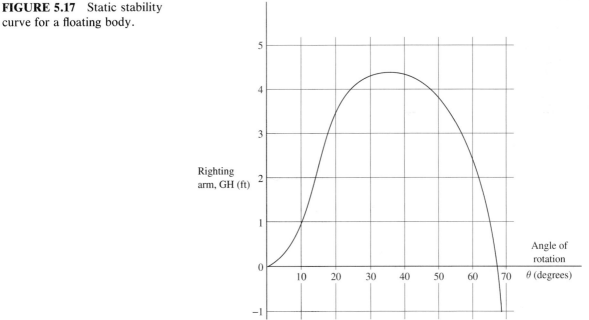

REFERENCE
1. Avallone, Eugene A., and Theodore Baumeister III, eds. 1987. *Marks' Standard Handbook for Mechanical Engineers.* 9th ed. New York: McGraw-Hill.

PRACTICE PROBLEMS

Buoyancy

5.1M *The instrument package shown in Fig. 5.18 weighs 258 N. Calculate the tension in the cable if the package is completely submerged in seawater having a specific weight of 10.05 kN/m³.*

5.2M *A 1.0-m-diameter hollow sphere weighing 200 N is attached to a solid concrete block weighing 4.1 kN. If the concrete has a specific weight of 23.6 kN/m³, will the two objects together float or sink in water?*

5.3M *A certain standard steel pipe has an outside diameter of 168 mm, and a 1-m length of the pipe weighs 277 N. Would the pipe float or sink in glycerine (sg = 1.26) if its ends are sealed?*

5.4E A cylindrical float has a 10-in diameter and is 12 in long. What should be the specific weight of the float material if it is to have nine-tenths of its volume below the surface of a fluid with a specific gravity of 1.10?

5.5M *A buoy is a solid cylinder 0.3 m in diameter and 1.2 m long. It is made of a material with a specific weight of 7.9 kN/m³. If it floats upright, how much of its length is above the water?*

5.6M *A float to be used as a level indicator is being designed to float in oil, which has a specific gravity of*

0.90. It is to be a cube, 100 mm on a side, and is to have 75 mm submerged in the oil. Calculate the required specific weight of the float material.

5.7M *A concrete block with a specific weight of 23.6 kN/m³ is suspended by a rope in a solution with a specific gravity of 1.15. What is the volume of the concrete block if the tension in the rope is 2.67 kN?*

5.8E Figure 5.19 shows a pump partially submerged in oil (sg = 0.90) and supported by springs. If the total weight of the pump is 14.6 lb and the submerged volume is 40 in³, calculate the supporting force exerted by the springs.

5.9M *A steel cube 100 mm on a side weighs 80 N. We want to hold the cube in equilibrium under water by attaching a light foam buoy to it. If the foam weighs 470 N/m³, what is the minimum required volume of the buoy?*

5.10E A cylindrical drum is 2 ft in diameter, is 3 ft long, and weighs 30 lb when empty. Aluminum weights are to be placed inside the drum in order to make it neutrally buoyant in fresh water. What volume of aluminum will be required if it weighs 0.100 lb/in³?

5.11E If the aluminum weights described in Problem 5.10

FIGURE 5.18   Problem 5.1.

FIGURE 5.19   Problem 5.8.

are placed outside the drum, what volume will be required?

5.12 Figure 5.20 shows a cube floating in a fluid. Derive an expression relating the submerged depth X, the specific weight of the cube, and the specific weight of the fluid.

5.13E A hydrometer is a device for indicating the specific gravity of liquids. Figure 5.21 shows the design for a hydrometer in which the bottom part is a hollow cylinder with a 1.00-in diameter, and the top is a tube with a 0.25-in diameter. The empty hydrometer weighs 0.020 lb. What weight of steel balls should be added to make the hydrometer float in the position shown in fresh water? (Note that this is for a specific gravity of 1.00.)

FIGURE 5.21   Hydrometer for Problems 5.13, 5.14, and 5.15.

FIGURE 5.20   Problems 5.12 and 5.60.

5.14E For the hydrometer designed in Problem 5.13, what will be the specific gravity of the fluid in which the hydrometer would float at the top mark?

5.15E For the hydrometer designed in Problem 5.13, what will be the specific gravity of the fluid in which the hydrometer would float at the bottom mark?

5.16E A buoy is to support a cone-shaped instrument package, as shown in Fig. 5.22. The buoy is made from a uniform material having a specific weight of 8.00 lb/ft³. At least 1.50 ft of the buoy must be above the surface of the seawater for safety and visibility. Calculate the maximum allowable weight of the instrument package.

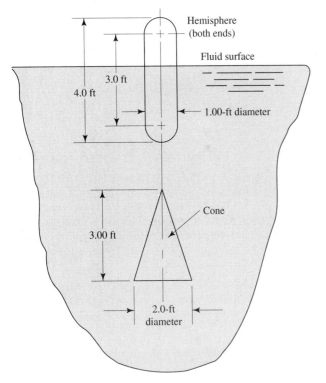

FIGURE 5.22   Problems 5.16, 5.23, and 5.24.

5.17E A cube has side dimensions of 18.00 in. It is made of steel having a specific weight of 491 lb/ft³. What force is required to hold it in equilibrium under fresh water?

5.18E A cube has side dimensions of 18.00 in. It is made of steel having a specific weight of 491 lb/ft³. What force is required to hold it in equilibrium under mercury?

5.19M *A ship has a mass of 292 Mg. Compute the volume of sea water it will displace when floating.*

5.20M *An iceberg has a specific weight of 8.72 kN/m³. What portion of its volume is above the surface when in seawater?*

5.21M *A cylindrical log has a diameter of 450 mm and a length of 6.75 m. When the log is floating in fresh water with its long axis horizontal, 110 mm of its diameter is above the surface. What is the specific weight of the wood in the log?*

5.22M *The cylinder shown in Fig. 5.23 is made from a uniform material. What is its specific weight?*

FIGURE 5.23   Problems 5.22, 5.24, and 5.52.

5.23M *If the cylinder from Problem 5.22 is placed in fresh water at 95°C, how much of its height would be above the surface?*

5.24M *A brass weight is to be attached to the bottom of the cylinder described in Problems 5.22 and 5.23, so that the cylinder will be completely submerged and neutrally buoyant in water at 95°C. The brass is*

FIGURE 5.24   Problems 5.24 and 5.25.

Water surface

Water at 95°C

Brass plate
$\gamma = 84.0$ kN/m³

$t = ?$

to be a cylinder with the same diameter as the original cylinder shown in Fig. 5.24. What is the required thickness of the brass?

5.25M *For the cylinder with the added brass (described in Problem 5.24), what will happen if the water were cooled to 15°C?*

5.26M *For the composite cylinder shown in Fig. 5.25, what thickness of brass is required to cause the cylinder to float in the position shown in carbon tetrachloride at 25°C?*

5.27M *A vessel for a special experiment has a hollow cylinder for its upper part and a solid hemisphere for its lower part, as shown in Fig. 5.26. What must be the total weight of the vessel if it is to sit upright, submerged to a depth of 0.75 m, in a fluid having a specific gravity of 1.16?*

5.28M *A light foam cup similar to a disposable coffee cup has a weight of 0.05 N, a uniform diameter of 82.0 mm, and a length of 150 mm. How much of its height would be submerged if placed in water?*

5.29M *A light foam cup similar to a disposable coffee cup has a weight of 0.05 N. A steel bar is placed inside the cup. The bar has a specific weight of 76.8 kN/*

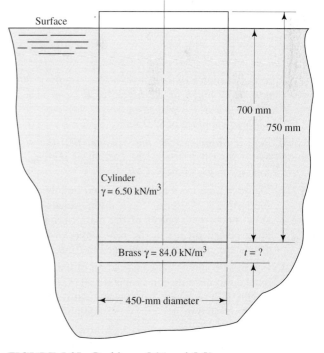

Surface

700 mm

750 mm

Cylinder
$\gamma = 6.50$ kN/m³

Brass $\gamma = 84.0$ kN/m³

$t = ?$

450-mm diameter

FIGURE 5.25   Problems 5.26 and 5.53.

FIGURE 5.26   Problems 5.27 and 5.48.

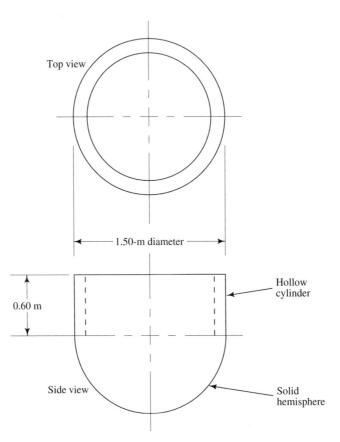

Top view

1.50-m diameter

0.60 m

Hollow cylinder

Side view

Solid hemisphere

m^3, a diameter of 38.0 mm, and a length of 80.0 mm. How much of the height of the cup will be submerged if it is placed in water? The cup has a uniform diameter of 82.0 mm.

5.30M *Repeat Problem 5.29, but consider that the steel bar is fastened outside the bottom of the cup instead of being placed inside.*

5.31E Figure 5.27 shows a raft made of four hollow drums supporting a platform. Each drum weighs 30 lb. How much total weight of the platform and anything placed on it can the raft support when the drums are completely submerged in fresh water?

5.32E Figure 5.28 shows the construction of the platform for the raft described in Problem 5.31. Compute its weight if it is made from wood with a specific weight of 40.0 lb/ft³.

5.33E For the raft shown in Fig. 5.27, how much of the drums will be submerged when only the platform is being supported? Refer to Problems 5.31 and 5.32 for data.

36 in

21-in diameter

FIGURE 5.27   Problems 5.31, 5.33, and 5.34.

FIGURE 5.28 Raft construction
for Problems 5.32 and 5.34.

5.34E For the raft and platform shown in Figs. 5.27 and 5.28 and described in Problems 5.31 and 5.32, what extra weight will cause all the drums and the platform itself to be submerged? Assume that no air is trapped beneath the platform.

5.35E A float in an ocean harbor is made from a uniform foam having a specific weight of 12.00 lb/ft³. It is made in the shape of a rectangular solid, 18.00 in square and 48.00 in long. A concrete (specific weight = 150 lb/ft³) block weighing 600 lb in air is attached to the float by a cable. The length of the cable is adjusted so that 14.00 in of the height of the float is above the surface with the long axis vertical. Compute the tension in the cable.

5.36E Describe how the situation described in Problem 5.35 will change if the water level rises by 18 in during high tide.

5.37E A cube, 6.00 in on a side, is made from aluminum having a specific weight of 0.100 lb/in³. If the cube is suspended on a wire with half its volume in water and the other half in oil (sg = 0.85), what is the tension in the string?

5.38E Figure 4.55 (Chapter 4) shows a solid cylinder sitting on the bottom of a tank holding a static volume of fluid. Compute the force exerted by the cylinder on the bottom of the tank for the following data: $D = 6.00$ in; $L = 10.00$ in; $\gamma_c = 0.284$ lb/in³ (steel); $\gamma_f = 62.4$ lb/ft³; $h = 30.00$ in.

Stability

5.39M *A cylindrical block of wood is 1.00 m in diameter and 1.00 m long and has a specific weight of 8.00 kN/m³. Will it float in a stable manner in water with its axis vertical?*

FIGURE 5.29   Problem 5.41.

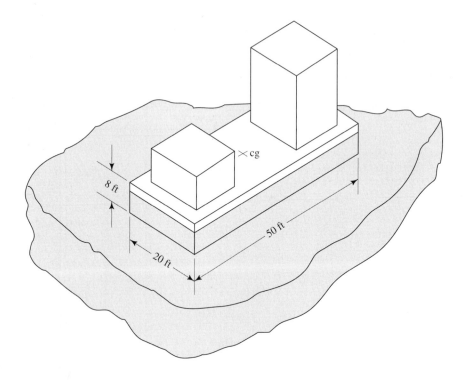

5.40E A container for an emergency beacon is a rectangular shape 30.0 in wide, 40.0 in long, and 22.0 in high. Its center of gravity is 10.50 in above its base. The container weighs 250 lb. Will the box be stable with the 30 × 40 in side parallel to the surface in plain water?

5.41E The large platform shown in Fig. 5.29 carries equipment and supplies to offshore installations. The total weight of the system is 450 000 lb, and its center of gravity is even with the top of the platform, 8.00 ft from the bottom. Will the platform be stable in seawater in the position shown?

5.42E Will the cylindrical float described in Problem 5.4 be stable if placed in the fluid with its axis vertical?

5.43M *Will the buoy described in Problem 5.5 be stable if placed in the water with its axis vertical?*

5.44M *Will the float described in Problem 5.6 be stable if placed in the oil with its top surface horizontal?*

5.45E A closed, hollow, empty drum has a diameter of 24.0 in, a length of 48.0 in, and a weight of 70.0 lb. Will it float stably if placed upright in water?

5.46E Figure 5.30 shows a river scow used to carry bulk materials. Assume that the scow's center of gravity

FIGURE 5.30   Problems 5.46 and 5.47.

is at its centroid and that it floats with 8.00 ft submerged. Determine the minimum width that will ensure stability in seawater.

5.47E Repeat Problem 5.46, except assume that crushed coal is added to the scow so that the scow is submerged to a depth of 16.0 ft and its center of gravity is raised to 13.50 ft from the bottom. Determine the minimum width for stability.

5.48M *For the vessel shown in Fig. 5.26 and described in Problem 5.27, assume that it floats with just the entire hemisphere submerged and that its center of gravity is 0.65 m from the top. Is it stable in the position shown?*

5.49M *For the foam cup described in Problem 5.28, will it float stably if placed in the water with its axis vertical?*

5.50M *Referring to Problem 5.29, assume that the steel bar is placed inside the cup with its long axis vertical. Will the cup float stably?*

5.51M *Referring to Problem 5.30, assume that the steel bar is fastened to the bottom of the cup with the long axis of the bar horizontal. Will the cup float stably?*

5.52M *Will the cylinder shown in Fig. 5.23 and described in Problem 5.22 be stable in the position shown?*

5.53M *Will the cylinder together with the brass plate shown in Fig. 5.25 and described in Problem 5.26 be stable in the position shown?*

5.54E A proposed design for a part of a seawall consists of a rectangular solid weighing 3840 lb with dimensions of 8.00 ft by 4.00 ft by 2.00 ft. The 8.00-ft side

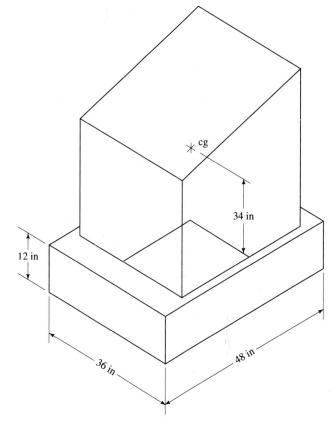

FIGURE 5.31   Problem 5.55.

FIGURE 5.32   Problem 5.59.

FIGURE 5.33   Problem 5.61.

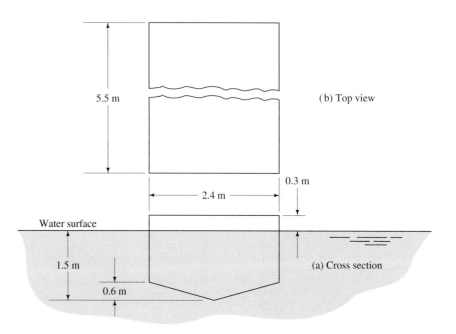

5.5 m

2.4 m

0.3 m

(b) Top view

Water surface

1.5 m

0.6 m

(a) Cross section

is to be vertical. Will this object float stably in sea-water?

5.55E A platform is being designed to support some water pollution equipment. As shown in Fig. 5.31, its base is 36.00 in wide, 48.00 in long, and 12.00 in high. The entire system weighs 130 lb, and its center of gravity is 34.0 in above the top surface of the plat-form. Is the proposed system stable when floating in seawater?

5.56E A block of wood with a specific weight of 32 lb/ft^3 is 6 by 6 by 12 in. If it is placed in oil (sg = 0.90) with the 6 by 12-inch surface parallel to the surface, would it be stable?

5.57E A barge is 60 ft long, 20 ft wide, and 8 ft deep. When empty, it weighs 210 000 lb, and its center of gravity is 1.5 ft above the bottom. Is it stable when floating in water?

5.58E If the barge in Problem 5.57 is loaded with 240 000 lb of loose coal having an average density of 45 lb/ft^3, how much of the barge would be below the water? Is it stable?

5.59M *A piece of cork having a specific weight of 2.36 kN/m^3 is shaped as shown in Fig. 5.32. (a) To what depth will it sink in turpentine (sg = 0.87) if placed in the orientation shown? (b) Is it stable in this position?*

5.60M *Figure 5.20 shows a cube floating in a fluid. (a) Derive an expression for the depth of submergence X that would ensure that the cube is stable in the position shown. (b) Using the expression derived in*

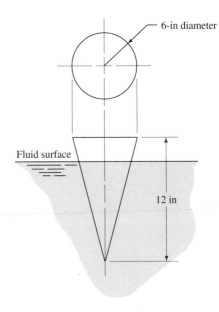

6-in diameter

Fluid surface

12 in

FIGURE 5.34   Problem 5.62.

(a), determine the required distance X for a cube 75 mm on a side.

5.61M *A boat has the cross section shown in Fig. 5.33(a). Its geometry at the water line is shown in the top view, Fig. 5.33(b). The hull is solid. Is the boat stable?*

5.62E (a) If the cone shown in Fig. 5.34 is made of pine wood with a specific weight of 30 lb/ft^3, will it be stable in the position shown floating in water? (b) Will it be stable if it is made of teak wood with a specific weight of 55 lb/ft^3?

COMPUTER PROGRAMMING ASSIGNMENTS

1. Write a program to evaluate the stability of a circular cylinder placed in a fluid with its axis vertical. Call for input data for diameter, length, weight (or specific weight) of the cylinder; location of the center of gravity; and specific weight of the fluid. Solve for the position of the cylinder when it is floating, the location of the center of buoyancy, and the metacenter. Compare the location of the metacenter with the center of gravity to evaluate stability.

2. For any cylinder of a uniform density, floating in any fluid and containing a specified volume, vary the diameter from a small value to larger values in selected increments. Then, compute the required height of the cylinder to obtain the specified volume. Finally, evaluate the stability of the cylinder if it is placed in the fluid with its axis vertical.

3. For the results found in Programming Assignment 2, compute the metacentric height (as described in Section 5.6). Plot the metacentric height versus the diameter of the cylinder.

4. Write a program to evaluate the stability of a rectangular block placed in a fluid in a specified orientation. Call for input data for length, width, height, weight (or specific weight) of the block; location of the center of gravity; and specific weight of the fluid. Solve for the position of the block when it is floating, the location of the center of buoyancy, and the metacenter. Compare the location of the metacenter with the center of gravity to evaluate stability.

5. Write a program to determine the stability of a rectangular block with a given length and height as the width varies. Call for input data for length, height, weight (or specific weight), and fluid specific weight. Vary the width in selected increments from small values to larger values, and compute the range of widths for which the metacentric height is positive, that is, for which the design would be stable. Plot a graph of metacentric height versus width.

6 Flow of Fluids and Bernoulli's Equation

6.1 OBJECTIVES

The material in Chapters 6 through 12 is concerned mainly with the flow of fluids in closed circular pipes and tubes and with the devices used to control the flow. We will discuss applications involving fluid power systems, fluid distribution systems, pumps, turbines, valves, elbows, and other fittings. The techniques of analysis for many different situations are presented first. Then, in Chapters 11 and 12, these techniques are combined in the solution of more complex pipe line problems.

When dealing with flow in pipes and tubes, we will assume that the fluid completely fills the available flow area unless otherwise stated. The analysis of open channel flow is presented in Chapter 13.

Most of the problems concerned with the flow of fluids in pipes and tubes involve the prediction of the conditions at one section in a system when the conditions at some other section are known. This is illustrated in Fig. 6.1, which shows a portion of a fluid distribution system with the flow going from section 1 at the bottom to section 2 at the top. At any section in such a system, we are usually concerned with the pressure of the fluid, the velocity of flow, and the elevation of the section. Remember, *elevation* is the term used to define the vertical distance from some reference level to a point of interest and is called z. When dealing with pipes and tubes, we measure elevation to the centerline of the section of interest.

After completing this chapter, you should be able to:

1. Define *volume flow rate* and its units.
2. Define *weight flow rate* and its units.
3. Define *mass flow rate* and its units.
4. Define *steady flow*.
5. Define the *principle of continuity*.
6. Write the continuity equation, and use it to relate the mass, area, and velocity of flow between two points in a fluid flow system.
7. Describe four types of commercially available pipe and tubing: steel pipe, ductile iron pipe, steel tubing, and copper tubing.
8. Specify the desired size of pipe or tubing to carry a given flow rate of fluid at a specified velocity.
9. State recommended velocities of flow and typical volume flow rates for various types of systems.
10. State the principle of conservation of energy as it applies to fluid flow systems.

145

FIGURE 6.1   Portion of a fluid distribution system.

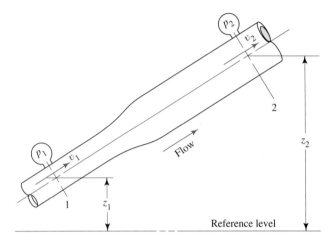

11. Define *potential energy, kinetic energy,* and *flow energy* as they relate to fluid flow systems.

12. Apply the principle of conservation of energy to develop *Bernoulli's equation.*

13. Define the terms *pressure head, elevation head, velocity head,* and *total head.*

14. State the restrictions on the use of Bernoulli's equation.

15. Apply Bernoulli's equation to fluid flow systems.

16. Define *Torricelli's theorem.*

17. Compute the flow rate of fluid under a falling head.

6.2
FLUID FLOW RATE

The quantity of fluid flowing in a system per unit time can be expressed by three different terms as defined below.

Q   The *volume flow rate* is the volume of fluid flowing past a section per unit time.

W   The *weight flow rate* is the weight of fluid flowing past a section per unit time.

M   The *mass flow rate* is the mass of fluid flowing past a section per unit time.

The most fundamental of these three terms is the volume flow rate Q, which is calculated from

  VOLUME FLOW RATE

$$Q = Av \tag{6–1}$$

where A is the area of the section and v is the average velocity of flow. The units of Q can be derived as follows, using SI units for illustration:

$$Q = Av = m^2 \times m/s = m^3/s$$

The weight flow rate W is related to Q by

WEIGHT FLOW RATE

$$W = \gamma Q \qquad (6\text{–}2)$$

where γ is the specific weight of the fluid. The units of W are then

$$W = \gamma Q = N/m^3 \times m^3/s = N/s$$

The mass flow rate M is related to Q by

MASS FLOW RATE

$$M = \rho Q \qquad (6\text{–}3)$$

where ρ is the density of the fluid. The units of M are then

$$M = \rho Q = kg/m^3 \times m^3/s = kg/s$$

TABLE 6.1 Flow rates

Symbol	Name	Definition	SI Units	U.S. Customary System Units
Q	Volume flow rate	$Q = Av$	m^3/s	ft^3/s
W	Weight flow rate	$W = \gamma Q$	N/s	lb/s
		$W = \gamma Av$		
M	Mass flow rate	$M = \rho Q$	kg/s	$slugs/s$
		$M = \rho Av$		

Table 6.1 summarizes these three types of fluid flow rates and gives the standard units in both the SI system and the U.S. Customary System. Because both the cubic meter per second and cubic foot per second are very large flow rates, other units are frequently used, such as liters per minute (L/min)* and gallons per minute (gal/min or gpm; this text will use gal/min). Useful conversions are

$$1.0 \text{ L/min} = 16.67 \times 10^{-6} \text{ m}^3/s$$
$$1.0 \text{ m}^3/s = 60\,000 \text{ L/min}$$
$$1.0 \text{ gal/min} = 3.785 \text{ L/min}$$
$$1.0 \text{ gal/min} = 6.309 \times 10^{-5} \text{ m}^3/s$$
$$1.0 \text{ ft}^3/s = 449 \text{ gal/min}$$

Table 6.2 lists typical volume flow rates for different kinds of systems.

---

* The international symbol for liter is the lowercase "l," which can easily be confused with the numeral "1." The American National Metric Council, on the basis of a recommendation from the Department of Commerce, has therefore suggested the symbol "L" for United States use.

TABLE 6.2   Typical volume flow rates

Flow rate (L/min)		Flow rate (gal/min)
10–100	Industrial oil hydraulic systems	3–30
100–600	Hydraulic systems for mobile equipment	30–150
40–4500	Centrifugal pumps in chemical processes	10–1200
15–125	Reciprocating pumps handling heavy fluids and slurries	4–33
200–4000	Flood control and drainage pumps	50–1000
40–15 000	Centrifugal pumps handling mine wastes	10–4000
1800–9500	Centrifugal fire fighting pumps	500–2500

☐ **EXAMPLE PROBLEM 6.1**   Convert a flow rate of 30 gal/min to ft³/s.

Solution   The flow rate is

$$Q = 30 \text{ gal/min} \left(\frac{1.0 \text{ ft}^3/\text{s}}{449 \text{ gal/min}}\right) = 6.68 \times 10^{-2} \text{ ft}^3/\text{s}$$

☐ **EXAMPLE PROBLEM 6.2**   Convert a flow rate of 600 L/min to m³/s.

Solution

$$Q = 600 \text{ L/min} \left(\frac{1.0 \text{ m}^3/\text{s}}{60\ 000 \text{ L/min}}\right) = 0.010 \text{ m}^3/\text{s}$$

☐ **EXAMPLE PROBLEM 6.3**   Convert a flow rate of 30 gal/min to L/min.

Solution

$$Q = 30 \text{ gal/min} \left(\frac{3.785 \text{ L/min}}{1.0 \text{ gal/min}}\right) = 113.6 \text{ L/min}$$

**6.3
CONTINUITY
EQUATION**

The method of calculating the velocity of flow of a fluid in a closed pipe system depends on the principle of continuity. Consider the pipe in Fig. 6.1. A fluid is flowing from section 1 to section 2 at a constant rate. That is, the quantity of fluid flowing past any section in a given amount of time is constant. This is referred to as *steady flow*. Now if there is no fluid added, stored, or removed between section 1 and section 2, then the mass of fluid flowing past section 2 in a given amount of time must be the same as that flowing past section 1. This can be expressed in terms of the mass flow rate as

$$M_1 = M_2$$

or, since $M = \rho A v$, we have

$$\rho_1 A_1 v_1 = \rho_2 A_2 v_2 \qquad (6\text{–}4)$$

 **CONTINUITY EQUATION FOR ANY
FLUID**

Equation (6–4) is a mathematical statement of the principle of continuity and is called the *continuity equation*. It is used to relate the fluid density, flow area, and velocity of flow at two sections of a system in which there is steady flow. It is valid for all fluids, whether gas or liquid.

If the fluid in the pipe in Fig. 6.1 is a liquid which can be considered incompressible, then the terms ρ_1 and ρ_2 in Eq. (6–4) are equal. The equation then becomes

 CONTINUITY EQUATION FOR LIQUIDS

$$A_1 v_1 = A_2 v_2 \qquad (6\text{–}5)$$

or, since $Q = Av$, we have

$$Q_1 = Q_2$$

Equation (6–5) is the continuity equation as applied to liquids; it states that for steady flow the volume flow rate is the same at any section. It can also be used for gases at low velocity, i.e., less than 100 m/s, with little error.

☐ **EXAMPLE PROBLEM 6.4**   In Fig. 6.1 the inside diameters of the pipe at sections 1 and 2 are 50 mm and 100 mm, respectively. Water at 70°C is flowing with an average velocity of 8 m/s at section 1. Calculate the following:

(a) Velocity at section 2
(b) Volume flow rate
(c) Weight flow rate
(d) Mass flow rate

Solution   (a) Velocity at section 2
From Eq. (6–5) we have

$$A_1 v_1 = A_2 v_2$$

$$v_2 = v_1 \left(\frac{A_1}{A_2} \right)$$

$$A_1 = \frac{\pi D_1^2}{4} = \frac{\pi (50 \text{ mm})^2}{4} = 1963 \text{ mm}^2$$

$$A_2 = \frac{\pi D_2^2}{4} = \frac{\pi (100 \text{ mm})^2}{4} = 7854 \text{ mm}^2$$

Then the velocity at section 2 is

$$v_2 = v_1 \left(\frac{A_1}{A_2} \right) = \frac{8.0 \text{ m}}{\text{s}} \times \frac{1963 \text{ mm}^2}{7854 \text{ mm}^2} = 2.0 \text{ m/s}$$

Notice that for steady flow of a liquid, as the flow area increases, the velocity decreases. This is independent of pressure and elevation.

(b) Volume flow rate, Q
From Table 6.1, $Q = Av$. Because of the principle of continuity we could use the conditions either at section 1 or at section 2 to calculate Q. At section 1 we have

$$Q = A_1 v_1 = 1963 \text{ mm}^2 \times \frac{8.0 \text{ m}}{\text{s}} \times \frac{1 \text{ m}^2}{(10^3 \text{ mm})^2} = 0.0157 \text{ m}^3/\text{s}$$

(c) Weight flow rate, W

From Table 6.1, $W = \gamma Q$. At 70°C the specific weight of water is 9.59 kN/m³. Then the weight flow rate is

$$W = \gamma Q = \frac{9.59 \text{ kN}}{\text{m}^3} \times \frac{0.0157 \text{ m}^3}{\text{s}} = 0.151 \text{ kN/s}$$

(d) Mass flow rate, M

From Table 6.1, $M = \rho Q$. At 70°C the density of water is 978 kg/m³. Then the mass flow rate is

$$M = \rho Q = \frac{978 \text{ kg}}{\text{m}^3} \times \frac{0.0157 \text{ m}^3}{\text{s}} = 15.36 \text{ kg/s}$$

☐ **EXAMPLE PROBLEM 6.5**   At one section in an air distribution system, air at 14.7 psia and 100°F has an average velocity of 1200 ft/min and the duct is 12 in square. At another section, the duct is round with a diameter of 18 in, and the velocity is measured to be 900 ft/min. Calculate the density of the air in the round section and the weight flow rate of air in pounds per hour. At 14.7 psia and 100°F, the density of air is 2.20×10^{-3} slugs/ft³ and the specific weight is 7.09×10^{-2} lb/ft³.

Solution   According to the continuity equation for gases, Eq. (6–4), we have

$$\rho_1 A_1 v_1 = \rho_2 A_2 v_2$$

Then, we can calculate the area of the two sections and solve for ρ_2.

$$\rho_2 = \rho_1 \left(\frac{A_1}{A_2}\right)\left(\frac{v_1}{v_2}\right)$$

$$A_1 = (12 \text{ in})(12 \text{ in}) = 144 \text{ in}^2$$

$$A_2 = \frac{\pi D_2^2}{4} = \frac{\pi (18 \text{ in})^2}{4} = 254 \text{ in}^2$$

Then, the density of the air in the round section is

$$\rho_2 = (2.20 \times 10^{-3} \text{ slugs/ft}^3)\left(\frac{144 \text{ in}^2}{254 \text{ in}^2}\right)\left(\frac{1200 \text{ ft/min}}{900 \text{ ft/min}}\right)$$

$$\rho_2 = 1.66 \times 10^{-3} \text{ slugs/ft}^3$$

The weight flow rate can be found at section 1 from $W = \gamma_1 A_1 v_1$. Then, the weight flow rate is

$$W = \gamma_1 A_1 v_1$$

$$W = (7.09 \times 10^{-2} \text{ lb/ft}^3)(144 \text{ in}^2)\left(\frac{2100 \text{ ft}}{\text{min}}\right)\left(\frac{1 \text{ ft}^2}{144 \text{ in}^2}\right)\left(\frac{60 \text{ min}}{\text{h}}\right)$$

$$W = 5100 \text{ lb/h}$$

6.4 COMMERCIALLY AVAILABLE PIPE AND TUBING   The actual outside and inside diameters of standard commercially available pipe and tubing may be quite different from the nominal size given. We will describe several widely used types of standard pipe and tubing in this section. Data are given in the appendices for outside diameter, inside diameter, wall thickness, and flow area for some of these types. The nominal sizes for commercially available pipe still refer to an "inch" size even though the transition to the SI system is an international trend. Since the nominal size is used solely to designate a certain pipe or tube, the standard conventional size will be used in this book. However, as you can see in Appendices F, G, H, and I, the dimensions are listed in millimeters (mm) for outside diameter, inside diameter, and wall thickness. The flow area is listed in square meters (m²) to help you maintain consistent units in calculations. Data are also given in the U.S. Customary System.

6.4.1
Steel Pipe

General-purpose pipe lines are often constructed of steel pipe. Standard pipe sizes are designated by the nominal size and schedule number. Schedule numbers are related to the permissible operating pressure of the pipe and to the allowable stress of the steel in the pipe. The range of schedule numbers is from 10 to 160, with the higher numbers indicating a heavier wall thickness. Since all schedules of pipe of a given nominal size have the same outside diameter, the higher schedules have a smaller inside diameter. The most complete series of steel pipe available are Schedules 40 and 80. Data for these two schedules are given in SI units and in U.S. Customary System units in Appendix F. Refer to ANSI/ASME Standard 31.1-1992 *Power Piping* for a method of computing the minimum acceptable wall thickness for pipes. See Reference 1.

6.4.2
Steel Tubing

Standard steel tubing is used in hydraulic systems, condensers, heat exchangers, engine fuel systems, and industrial fluid processing systems. Sizes are designated by outside diameter and wall thickness. Standard sizes from ⅛ in to 2 in for several wall thickness gauges are tabulated in Appendix G. Other wall thicknesses are available.

6.4.3
Copper Tubing

Household plumbing, refrigerant lines, and compressed air lines often use copper tubing manufactured as Type K or Type L. Type K has the greater wall thickness and is recommended for underground service. Type L is suitable for general-purpose interior plumbing. The nominal size of copper tubing is ⅛ in less than the actual outside diameter of the tube. Data for wall thickness, inside diameter, and flow area are given in Appendix H for Type K tubing.

6.4.4
Ductile Iron Pipe

Water, gas, and sewage lines are often made of ductile iron pipe because of its strength, ductility, and relative ease of handling. It has replaced cast iron in many applications. Standard fittings are supplied with the pipe for convenient installation above or below ground. Several classes of ductile iron pipe are available for use in systems with a range of pressures. Appendix I lists the dimensions of Class 150 pipe for 150 psi (1.03 MPa) service in nominal sizes from 3 to 24 inches. Actual inside and outside diameters are larger than the nominal sizes.

6.4.5
Other Types of
Pipe and Tubing

Brass pipe is used for corrosive fluids, as is stainless steel. Other materials used are aluminum, lead, tin, vitrified clay, concrete, and many types of plastics, such as polyethylene, nylon, and polyvinyl chloride.

6.4.6
Areas of Standard
Size Circles

When problems in this book identify a particular size and type of pipe or tube, look in the corresponding appendix table for the necessary diameters and flow areas. When actual diameters are given, you may compute the area from $A = \pi D^2/4$. For convenience, Appendix J gives the areas in ft^2 for circles from 0.25 to 24 in. SI metric diameters are also listed from 6 to 600 mm with areas given in m^2.

6.5
RECOMMENDED
VELOCITY OF FLOW IN
PIPE AND TUBING

Factors affecting the selection of a satisfactory velocity of flow in fluid systems are numerous. Some of the important ones are the type of fluid, the length of the flow system, the type of pipe or tube, the pressure drop that can be tolerated, the devices (such as pumps, valves, etc.) that may be connected to the pipe or tube, temperature, pressure, and noise.

When we discussed the continuity equation in Section 6.3, we learned that the velocity of flow increases as the area of the flow path decreases. Therefore, smaller tubes will cause high velocities, and larger tubes will provide low velocities. Later we will explain that energy losses and the corresponding pressure drop increase dramatically as the flow velocity increases. For this reason it is desirable to keep velocities low. But, since larger pipes and tubes are more costly, some limits are necessary.

A reasonable flow velocity for fluid distribution systems is approximately 3.0 m/s (about 10 ft/s). This would apply to water, oil, and other common liquids in pipes beyond the outlet of pumps. Proper pump performance requires lower velocities at the inlet to pumps, approximately 1.0 m/s (about 3 ft/s). Table 6.3 lists general guidelines for flow velocities in oil hydraulic fluid power systems. Chapter 15 presents a more extensive discussion of recommended velocity and pipe sizes.

TABLE 6.3 Recommended flow velocities in fluid power systems

	SI Units	U.S. Customary System Units
Pump inlets: (Suction lines)	0.6 to 2.2 m/s	2.0 to 7.0 ft/s
Pump outlets: (Discharge or pressure lines)	2.4 to 7.5 m/s	8.0 to 25.0 ft/s
Return lines:	0.6 to 2.6 m/s	2.0 to 8.5 ft/s

□ **EXAMPLE PROBLEM 6.6**

Determine the maximum allowable volume flow rate in L/min that can be carried through a standard steel tube with an outside diameter of 1¼ in and a 0.065 in wall thickness if the maximum velocity is to be 3.0 m/s.

Solution   Using the definition of volume flow rate, we have

$$Q = Av$$
$$A = 6.356 \times 10^{-4} \text{ m}^2 \qquad \text{(from Appendix G)}$$

Then we find the flow rate:

$$Q = (6.356 \times 10^{-4} \text{ m}^2)(3.0 \text{ m/s}) = 1.907 \times 10^{-3} \text{ m}^3/\text{s}$$

Converting to L/min, we have

$$Q = 1.907 \times 10^{-3} \text{ m}^3/\text{s} \left(\frac{60\ 000 \text{ L/min}}{1.0 \text{ m}^3/\text{s}}\right) = 114 \text{ L/min}$$

□ **EXAMPLE PROBLEM 6.7**   Determine the required size standard Schedule 40 steel pipe to carry 3200 L/min of water with a maximum velocity of 6.0 m/s.

Solution   Since Q and v are known, the required area can be found from

$$Q = Av$$
$$A = Q/v$$

First, we must convert the volume flow rate to the units of m³/s:

$$Q = 3200 \text{ L/min} \left(\frac{1.0 \text{ m}^3/\text{s}}{60\ 000 \text{ L/min}}\right) = 0.0533 \text{ m}^3/\text{s}$$

Then, we have

$$A = \frac{Q}{v} = \frac{0.0533 \text{ m}^3/\text{s}}{6.0 \text{ m/s}} = 0.008\ 88 \text{ m}^2 = 8.88 \times 10^{-3} \text{ m}^2$$

This must be interpreted as the *minimum* allowable area since any smaller area would produce a velocity higher than 6.0 m/s. Therefore, we must look in Appendix F for a standard pipe with a flow area just larger than 8.88×10^{-3} m². A standard 5-in Schedule 40 steel pipe, with a flow area of 1.291×10^{-2} m², is required. The actual velocity of flow when this pipe carries 0.0533 m³/s of water is

$$v = \frac{Q}{A} = \frac{0.0533 \text{ m}^3/\text{s}}{1.291 \times 10^{-2} \text{ m}^2} = 4.13 \text{ m/s}$$

If the next smaller pipe (a 4-in Schedule 40 pipe) were used, the velocity would have been

$$v = \frac{Q}{A} = \frac{0.0533 \text{ m}^3/\text{s}}{8.213 \times 10^{-3} \text{ m}^2} = 6.49 \text{ m/s} \qquad \text{(too high)}$$

6.6
FLOW IN
NONCIRCULAR
SECTIONS

The continuity equation applies equally well to flow in noncircular sections as it does in circular pipes and tubes. In the formula for volume flow rate, $Q = Av$, the area, A, is the *net flow area* and v is the average velocity of flow when carrying the given volume flow rate Q.

☐ **EXAMPLE PROBLEM 6.8**   Figure 6.2 shows a heat exchanger used to transfer heat from the fluid flowing inside
the inner tube to that flowing in the space between the outside of the tube and the
inside of the square shell that surrounds the tube. Such a device is often called a *shell
and tube heat exchanger*. Compute the volume flow rate in gal/min that would
produce a velocity of 8.0 ft/s both inside the tube and in the shell.

FIGURE 6.2   Shell and tube heat exchanger for Example Problem 6.8.

Solution   We use the formula for volume flow rate, $Q = Av$, for each part.
  (a) Inside the ½-in Type K copper tube:
      From Appendix H, we can read:

$$OD = 0.625 \text{ in}$$
$$ID = 0.527 \text{ in}$$
$$\text{Wall thickness} = 0.049 \text{ in}$$
$$A_t = 1.515 \times 10^{-3} \text{ ft}^2$$

Then, the volume flow rate inside the tube is

$$Q_t = A_t v = (1.515 \times 10^{-3} \text{ ft}^2)(8.0 \text{ ft/s}) = 0.01212 \text{ ft}^3\text{/s}$$

Converting to gal/min gives

$$Q_t = 0.01212 \text{ ft}^3\text{/s} \, \frac{449 \text{ gal/min}}{1.0 \text{ ft}^3\text{/s}} = 5.44 \text{ gal/min}$$

  (b) In the shell:
      The net flow area is the difference between the area *inside* the square shell and
      the *outside* of the tube. Then,

$$A_s = S^2 - \pi OD^2/4$$
$$= (0.80 \text{ in})^2 - \pi(0.625 \text{ in})^2/4 = 0.3332 \text{ in}^2$$

Converting to ft^2 gives

$$A_s = 0.3332 \text{ in}^2 \frac{1.0 \text{ ft}^2}{144 \text{ in}^2} = 2.314 \times 10^{-3} \text{ ft}^2$$

The required volume flow rate is then

$$Q_s = A_s v = (2.314 \times 10^{-3} \text{ ft}^2)(8.0 \text{ ft/s}) = 0.01851 \text{ ft}^3/\text{s}$$

$$Q_s = 0.01851 \text{ ft}^3/\text{s} \frac{449 \text{ gal/min}}{1.0 \text{ ft}^3/\text{s}} = 8.31 \text{ gal/min}$$

The ratio of the flow in the shell to the flow in the tube is

$$\text{Ratio} = Q_s/Q_t = 8.31/5.44 = 1.53$$

**6.7
CONSERVATION
OF ENERGY—
BERNOULLI'S
EQUATION**

The analysis of a pipe line problem such as that illustrated in Fig. 6.1 accounts for all the energy within the system. In physics you learned that energy can be neither created nor destroyed, but it can be transformed from one form to another. This is a statement of the law of *conservation of energy.*

There are three forms of energy which are always considered when analyzing a pipe flow problem. Consider an element of fluid, as shown in Fig. 6.3, that may be inside a pipe in a flow system. It would be located at a certain elevation z, have a certain velocity v, and have a pressure p. The element of fluid would possess the following forms of energy:

1. *Potential Energy.* Due to its elevation, the potential energy of the element relative to some reference level is

$$\text{PE} = wz \qquad (6\text{--}6)$$

where w is the weight of the element.

2. *Kinetic Energy.* Due to its velocity, the kinetic energy of the element is

$$\text{KE} = wv^2/2g \qquad (6\text{--}7)$$

3. *Flow Energy.* Sometimes called *pressure energy* or *flow work,* this represents the amount of work necessary to move the element of fluid across a certain section against the pressure p. Flow energy is abbreviated FE and is calculated from

$$\text{FE} = wp/\gamma \qquad (6\text{--}8)$$

FIGURE 6.3 Element of a fluid in a pipe.

Equation (6–8) can be derived as follows. Figure 6.4 shows the element of fluid in the pipe being moved across a section. The force on the element is pA, where p is the pressure at the section and A is the area of the section. In moving the element across the section, the force moves a distance L equal to the length of the element. Therefore, the work done is

$$\text{Work} = pAL = pV$$

where V is the volume of the element. The weight of the element w is

$$w = \gamma V$$

where γ is the specific weight of the fluid. Then, the volume of the element is

$$V = w/\gamma$$

and we have

$$\text{Work} = pV = pw/\gamma$$

which is called flow energy in Eq. (6–8).

FIGURE 6.4   Flow energy.

Element of fluid

The total amount of energy of these three forms possessed by the element of fluid would be the sum, called E:

$$E = \text{FE} + \text{PE} + \text{KE}$$
$$= wp/\gamma + wz + wv^2/2g$$

Each of these terms is expressed in units of energy which are newton-meters (N·m) in the SI unit system or foot-pounds (ft-lb) in the U.S. Customary System.

Now consider the element of fluid in Fig. 6.5, which moves from a section 1 to a section 2. The values for p, z, and v will be different at the two sections. At section 1, the total energy is

$$E_1 = \frac{wp_1}{\gamma} + wz_1 + \frac{wv_1^2}{2g}$$

At section 2, the total energy is

$$E_2 = \frac{wp_2}{\gamma} + wz_2 + \frac{wv_2^2}{2g}$$

FIGURE 6.5 Fluid elements used in Bernoulli's equation.

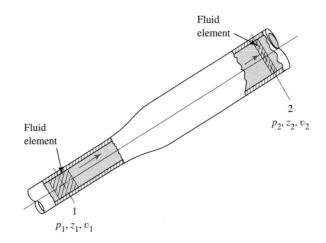

Fluid element

Fluid element

2
p_2, z_2, v_2

1
p_1, z_1, v_1

If no energy is added to the fluid or lost between sections 1 and 2, then the principle of conservation of energy requires that

$$E_1 = E_2$$

$$\frac{wp_1}{\gamma} + wz_1 + \frac{wv_1^2}{2g} = \frac{wp_2}{\gamma} + wz_2 + \frac{wv_2^2}{2g}$$

The weight of the element w is common to all terms and can be divided out. The equation then becomes

BERNOULLI'S EQUATION

$$\frac{p_1}{\gamma} + z_1 + \frac{v_1^2}{2g} = \frac{p_2}{\gamma} + z_2 + \frac{v_2^2}{2g} \qquad (6-9)$$

This is referred to as *Bernoulli's equation.*

**6.8
INTERPRETATION
OF BERNOULLI'S
EQUATION**

Each term of Bernoulli's Eq. (6–9) resulted from dividing an expression for energy by the weight of an element of the fluid. Therefore, it is appropriate to refer to the resulting forms as *energy possessed by the fluid per unit weight of fluid flowing in the system.* The units for each term would be newton-meters per newton (N·m/N) in the SI system and pounds-feet per pound (lb·ft/lb) in the U.S. Customary System.

But the weight unit, newton (N) or pound (lb) can be cancelled, leaving only a unit of height, meter (m) or foot (ft). Therefore, the terms in Bernoulli's equation are often expressed as "head," referring to a height above a reference level. The term p/γ is called *pressure head*; z is called *elevation head*; and $v^2/2g$ is called *velocity head*. The sum of these three is called the *total head.* Because each term represents a height, a diagram similar to that shown in Fig. 6.6 is helpful to visualize the relationship among the three types of energy. Notice that because of the assumption that no energy is lost or added, the total head remains at a constant level. Then, the relative height of each head term varies as predicted by Bernoulli's equation.

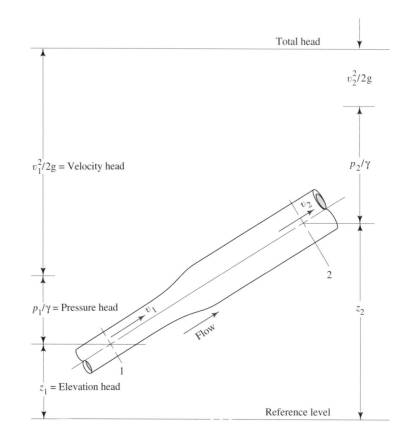

In Fig. 6.6 you should see that the velocity head at section 2 will be less than that at section 1. This can be shown by the continuity equation,

$$A_1v_1 = A_2v_2$$
$$v_2 = v_1 \, (A_1/A_2)$$

Because $A_1 < A_2$, v_2 must be less than v_1. And because the velocity is squared in the velocity head term, $v_2^2/2g$ is much less than $v_1^2/2g$.

Typically, when the section size expands as it does in Fig. 6.6, the pressure head increases because the velocity head decreases. That is how Fig. 6.6 is constructed. However, the actual change is also affected by the change in the elevation head.

In summary, Bernoulli's equation accounts for the changes in elevation head, pressure head, and velocity head between two points in a fluid flow system. It is assumed that there are no energy losses or additions between the two points so the total head remains constant.

When writing Bernoulli's equation, it is essential that the pressures at the two reference points both be expressed as absolute pressures or both as gage pressures. That is, they must both have the same reference pressure. In most problems it will be convenient to use gage pressure because parts of the fluid system exposed to the atmosphere will then have zero pressure. Also, most pressures are measured by a gage relative to the local atmospheric pressure.

**6.9
RESTRICTIONS ON
BERNOULLI'S
EQUATION**

Although Bernoulli's equation is applicable to a larger number of practical problems, there are several limitations that must be understood in order to apply it properly.

1. It is valid only for incompressible fluids since the specific weight of the fluid is assumed to be the same at the two sections of interest.
2. There can be no mechanical devices between the two sections of interest that would add energy to or remove energy from the system, since the equation states that the total energy in the fluid is constant.
3. There can be no heat transferred into or out of the fluid.
4. There can be no energy lost due to friction.

In reality no system satisfies all these restrictions. However, there are many systems for which only a negligible error will result when Bernoulli's equation is used. Also, the use of this equation may allow a fast estimate of a result when that is all that is required. In Chapter 7, limitations 2 and 4 will be eliminated by expanding Bernoulli's equation into the *general energy equation.*

**6.10
APPLICATIONS OF
BERNOULLI'S
EQUATION**

We will present several programmed example problems to illustrate the use of Bernoulli's equation. Although it is not possible to cover all types of problems with a certain solution method, the general approach to problems of fluid flow is described here.

PROCEDURE FOR APPLYING BERNOULLI'S EQUATION

1. Decide which items are known and what is to be found.
2. Decide which two sections in the system will be used when writing Bernoulli's equation. One section is chosen for which much data is known. The second is usually the section at which something is to be calculated.
3. Write Bernoulli's equation for the two selected sections in the system. It is important that the equation is written *in the direction of flow*. That is, the flow must proceed *from* the section on the left side of the equation *to* that on the right side.
4. Simplify the equation, if possible, by cancelling terms that are zero or those that are equal on both sides of the equation.
5. Solve the equation algebraically for the desired term.
6. Substitute known quantities and calculate the result, being careful that consistent units are used throughout the calculation.

PROGRAMMED EXAMPLE PROBLEM

☐ **EXAMPLE PROBLEM 6.9**

In Fig. 6.6, water at 10°C is flowing from section 1 to section 2. At section 1, which is 25 mm in diameter, the gage pressure is 345 kPa and the velocity of flow is 3.0 m/s.

Section 2, which is 50 mm in diameter, is 2.0 m above section 1. Assuming there are no energy losses in the system, calculate the pressure p_2.

List the items that are known from the problem statement before looking at the next panel.

$D_1 = 25$ mm $v_1 = 3.0$ m/s $z_2 - z_1 = 2.0$ m

$D_2 = 50$ mm $p_1 = 345$ kPa(gage)

The pressure p_2 is to be found. In other words, we are asked to calculate the pressure at section 2, which is different from the pressure at section 1 because there is a change in elevation and flow area between the two sections.

We are going to use Bernoulli's equation to solve the problem. Which two sections should be used when writing the equation?

In this case, sections 1 and 2 are the obvious choices. At section 1, we know p_1, v_1, and z_1. The unknown pressure p_2 is at section 2.

Now write Bernoulli's equation. [See Eq. (6–9).]

It should look like this:

$$\frac{p_1}{\gamma} + z_1 + \frac{v_1^2}{2g} = \frac{p_2}{\gamma} + z_2 + \frac{v_2^2}{2g}$$

The three terms on the left refer to section 1, while the three on the right refer to section 2.

Solve for p_2 in terms of the other variables.

The algebraic solution for p_2 could look like this:

$$\frac{p_1}{\gamma} + z_1 + \frac{v_1^2}{2g} = \frac{p_2}{\gamma} + z_2 + \frac{v_2^2}{2g}$$

$$\frac{p_2}{\gamma} = \frac{p_1}{\gamma} + z_1 + \frac{v_1^2}{2g} - z_2 - \frac{v_2^2}{2g}$$

$$p_2 = \gamma \left(\frac{p_1}{\gamma} + z_1 + \frac{v_1^2}{2g} - z_2 - \frac{v_2^2}{2g} \right)$$

This is correct. However, it is convenient to group the elevation heads and velocity heads together. Also, since $\gamma(p_1/\gamma) = p_1$, the final solution for p_2 should be

$$p_2 = p_1 + \gamma \left(z_1 - z_2 + \frac{v_1^2 - v_2^2}{2g} \right) \qquad (6\text{–}10)$$

Are the values of all the terms on the right side of this equation known?

Everything was given except γ, v_2, and g. Of course, $g = 9.81$ m/s². Since water at 10°C is flowing in the system, $\gamma = 9.81$ kN/m³. How can v_2 be determined?

The continuity equation is used:

$$A_1 v_1 = A_2 v_2$$
$$v_2 = v_1(A_1/A_2)$$

Calculate v_2 now.

You should have $v_2 = 0.75$ m/s. This is found from

$$A_1 = \pi D_1^2/4 = \pi(25 \text{ mm})^2/4 = 491 \text{ mm}^2$$
$$A_2 = \pi D_2^2/4 = \pi(50 \text{ mm})^2/4 = 1963 \text{ mm}^2$$
$$v_2 = v_1(A_1/A_2) = 3.0 \text{ m/s}(491 \text{ mm}^2/1963 \text{ mm}^2) = 0.75 \text{ m/s}$$

Now substitute the known values into Eq. (6–10).

$$p_2 = 345 \text{ kPa} + \frac{9.81 \text{ kN}}{\text{m}^3}\left(-2.0 \text{ m} + \frac{(3.0 \text{ m/s})^2 - (0.75 \text{ m/s})^2}{2(9.81 \text{ m/s}^2)}\right)$$

Notice that $z_1 - z_2 = -2.0$ m. Neither z_1 nor z_2 is known, but it is known that z_2 is 2.0 m greater than z_1. Therefore, the difference $z_1 - z_2$ must be negative. Now complete the calculation for p_2.

The final answer is $p_2 = 329.6$ kPa. This is 15.4 kPa less than p_1. The details of the solution are

$$p_2 = 345 \text{ kPa} + \frac{9.81 \text{ kN}}{\text{m}^3}\left(-2.0 \text{ m} + \frac{(9.0 - 0.563)\text{m}^2/\text{s}^2}{2(9.81)\text{m/s}^2}\right)$$

$$p_2 = 345 \text{ kPa} + \frac{9.81 \text{ kN}}{\text{m}^3}(-2.0 \text{ m} + 0.43 \text{ m})$$

$$p_2 = 345 \text{ kPa} - 15.4 \text{ kN/m}^2 = 345 \text{ kPa} - 15.4 \text{ kPa}$$

$$p_2 = 329.6 \text{ kPa}$$

The pressure p_2 is a gage pressure because it was computed relative to p_1, which was also a gage pressure. In later problem solutions, we will assume the pressures to be gage unless otherwise stated.

■

6.10.1

Tanks, Reservoirs, and Nozzles Exposed to the Atmosphere

Figure 6.7 shows a fluid flow system in which a siphon draws fluid from a tank or reservoir and delivers it through a nozzle at the end of the pipe. Note that the surface of the tank (point A) and the free stream of fluid exiting the nozzle (section F) are not confined by solid boundaries and are exposed to the prevailing atmosphere. Therefore, the pressure at those sections is zero gage pressure. When such points are used as reference points in Bernoulli's equation, the pressure head terms will be zero and can be cancelled.

The tank from which the fluid is being drawn can be assumed to be quite large compared to the size of the flow area inside the pipe. Now,

FIGURE 6.7 Siphon for Example Problem 6.10.

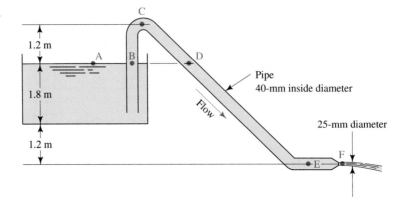

because $v = Q/A$, the velocity at the surface of such a tank will be very small. Furthermore, when we use the velocity to compute the velocity head, $v^2/2g$, we *square* the velocity. The process of squaring a small number produces an even smaller number. For these reasons, we consider the velocity head at the surface of a tank or reservoir to be very nearly zero and we cancel it from the equation.

6.10.2
When Both Reference Points Are in the Same Pipe

Also notice in Fig. 6.7 that several points of interest (points B, C, D, and E) are inside the pipe that has a uniform flow area. Under the conditions of steady flow assumed for these problems, the velocity will be the same throughout the pipe. Then, if the two points of reference for Bernoulli's equation were both inside the pipe, the velocity head terms on both sides of the equation will be equal and can be cancelled.

6.10.3
When Elevations Are Equal at Both Reference Points

If the elevations at the two reference points used in Bernoulli's equation are equal, those terms can be cancelled from the equation.

The four observations made in sections 6.10.1–3 enable the simplification of Bernoulli's equation and make the algebraic manipulations easier. Example Problem 6.10 uses these observations.

PROGRAMMED EXAMPLE PROBLEM

☐ **EXAMPLE PROBLEM 6.10**

Figure 6.7 shows a siphon that is used to draw water from a swimming pool. The pipe that makes up the siphon has an inside diameter of 40 mm and terminates with a 25-mm diameter nozzle. Assuming that there are no energy losses in the system, calculate the volume flow rate through the siphon and the pressure at points B, C, D, and E.

The first step in this problem solution is to calculate the volume flow rate Q, using Bernoulli's equation. The two most convenient points to use for this calculation are A and F. What is known about point A?

Point A is the free surface of the water in the pool. Therefore, $p_A = 0$ Pa. Also, since the surface area of the pool is very large, the velocity of the water at the surface is very nearly zero. Therefore, we will assume $v_A = 0$.

What do we know about point F?

Point F is in the free stream of water outside the nozzle. Since this stream is exposed to atmospheric pressure, the pressure $p_F = 0$ Pa. We also know that point F is 3.0 m below point A.

Now write Bernoulli's equation for points A and F.

You should have

$$\frac{p_A}{\gamma} + z_A + \frac{v_A^2}{2g} = \frac{p_F}{\gamma} + z_F + \frac{v_F^2}{2g}$$

Taking into account the information in the previous two panels, how can we simplify this equation?

Since $p_A = 0$ Pa, $p_F = 0$ Pa, and v_A is approximately zero, we can cancel them from the equation. What remains is this:

$$z_A = z_F + \frac{v_F^2}{2g}$$

The objective is to calculate the volume flow rate, which depends on the velocity v_F. Solve for v_F now.

You should have

$$v_F = \sqrt{(z_A - z_F)2g}$$

What is $z_A - z_F$?

From Fig. 6.7 we see that $z_A - z_F = 3.0$ m. Notice that the difference is positive since z_A is greater than z_F. We can now solve for the value of v_F.

The result is

$$v_F = \sqrt{(3.0 \text{ m})(2)(9.81 \text{ m/s}^2)} = \sqrt{58.9} \text{ m/s} = 7.67 \text{ m/s}$$

Now, how can Q be calculated?

Using the continuity equation $Q = Av$, compute the volume flow rate.

The result is

$$Q = A_F v_F$$

$$v_F = 7.67 \text{ m/s}$$

$$A_F = \pi(25 \text{ mm})^2/4 = 491 \text{ mm}^2$$

$$Q = 491 \text{ mm}^2\left(\frac{7.67 \text{ m}}{\text{s}}\right)\left(\frac{1 \text{ m}^2}{10^6 \text{ mm}^2}\right) = 3.77 \times 10^{-3} \text{ m}^3/\text{s}$$

The first part of the problem is now complete. Now use Bernoulli's equation to determine p_B. What two points should be used?

Points A and B are the best. As shown in the previous panels, using point A allows the equation to be simplified greatly. And since we are looking for p_B, we must choose point B.

Write Bernoulli's equation for points A and B, simplify it as before and solve for p_B.

Here is one possible solution procedure:

$$\frac{p_A}{\gamma} + z_A + \frac{v_A^2}{2g} = \frac{p_B}{\gamma} + z_B + \frac{v_B^2}{2g}$$

Since $p_A = 0$ Pa and $v_A = 0$, we have

$$z_A = \frac{p_B}{\gamma} + z_B + \frac{v_B^2}{2g}$$

$$p_B = \gamma[(z_A - z_B) - v_B^2/2g] \tag{6–11}$$

What is $z_A - z_B$?

It is zero. Since the two points are on the same level, their elevations are the same. Can you find v_B?

We can calculate v_B by using the continuity equation:

$$Q = A_B v_B$$

$$v_B = Q/A_B$$

The area of a 40-mm diameter pipe can be found in Appendix J. Complete the calculation for v_B.

The result is

$$v_B = Q/A_B$$

$$Q = 3.77 \times 10^{-3} \text{ m}^3/\text{s}$$

$$A_B = 1.257 \times 10^{-3} \text{ m}^2$$

$$v_B = \frac{3.77 \times 10^{-3} \text{ m}^3}{\text{s}} \times \frac{1}{1.257 \times 10^{-3} \text{ m}^2} = 3.00 \text{ m/s}$$

We now have all the data we need to calculate p_B from Eq. (6–11).

The pressure at point B is

$$p_B = \gamma[(z_A - z_B) - v_B^2/2g]$$

$$\frac{v_B^2}{2g} = \frac{(3.00)^2 \text{ m}^2}{\text{s}^2} \times \frac{\text{s}^2}{(2)(9.81) \text{ m}} = 0.459 \text{ m}$$

$$p_B = (9.81 \text{ kN/m}^3)(0 - 0.459 \text{ m})$$

$$p_B = -4.50 \text{ kN/m}^2$$

$$p_B = -4.50 \text{ kPa}$$

The negative sign indicates that p_B is 4.50 kPa below atmospheric pressure. Notice that when we deal with fluids in motion, the concept that points on the same level have the same pressure does *not* apply as it does with fluids at rest.

The next three panels present the solutions for the pressures p_C, p_D, and p_E, which can be found in a manner very similar to that used for p_B. Complete the solution for p_C yourself before looking at the next panel.

The answer is $p_C = -16.27$ kPa. We use Bernoulli's equation:

$$\frac{p_A}{\gamma} + z_A + \frac{v_A^2}{2g} = \frac{p_C}{\gamma} + z_C + \frac{v_C^2}{2g}$$

Since $p_A = 0$ and $v_A = 0$, the pressure at point C is

$$z_A = \frac{p_C}{\gamma} + z_C + \frac{v_C^2}{2g}$$

$$p_C = \gamma[(z_A - z_C) - v_C^2/2g]$$

$$z_A - z_C = -1.2 \text{ m} \quad \text{(negative, since } z_C \text{ is greater than } z_A)$$

$$v_C = v_B = 3.00 \text{ m/s} \quad \text{(since } A_C = A_B)$$

$$\frac{v_C^2}{2g} = \frac{v_B^2}{2g} = 0.459 \text{ m}$$

$$p_C = (9.81 \text{ kN/m}^3)(-1.2 \text{ m} - 0.459 \text{ m})$$

$$p_C = -16.27 \text{ kN/m}^2$$

$$p_C = -16.27 \text{ kPa}$$

Complete the calculation for p_D before looking at the next panel.

The answer is $p_D = -4.50$ kPa. This is the same as p_B because the elevation and the velocity at points B and D are equal. Solution by Bernoulli's equation would prove this.

Now find p_E.

The pressure at point E is 24.93 kPa. We use Bernoulli's equation:

$$\frac{p_A}{\gamma} + z_A + \frac{v_A^2}{2g} = \frac{p_E}{\gamma} + z_E + \frac{v_E^2}{2g}$$

Since $p_A = 0$ and $v_A = 0$, we have

$$z_A = \frac{p_E}{\gamma} + z_E + \frac{v_E^2}{2g}$$

$$p_E = \gamma[(z_A - z_E) - v_E^2/2g]$$

$$z_A - z_E = +3.0 \text{ m}$$

$$v_E = v_B = 3.00 \text{ m/s}$$

$$\frac{v_E^2}{2g} = \frac{v_B^2}{2g} = 0.459 \text{ m}$$

$$p_E = (9.81 \text{ kN/m}^3)(3.0 \text{ m} - 0.459 \text{ m})$$

$$p_E = 24.93 \text{ kN/m}^2$$

$$p_E = 24.93 \text{ kPa}$$

SUMMARY OF THE RESULTS OF EXAMPLE PROBLEM 6.10

1. The velocity of flow from the nozzle, and therefore the volume flow rate delivered by the siphon, depends on the elevation difference between the free surface of the fluid and the outlet of the nozzle.

2. The pressure at point B is below atmospheric pressure even though it is on the same level as point A, which is exposed to the atmosphere. In Eq. (6–11), Bernoulli's equation shows that the pressure head at B is decreased by the amount of the velocity head. That is, some of the energy is converted to kinetic energy, resulting in a lower pressure at B.

3. The velocity of flow is the same at all points where the pipe size is the same, when steady flow exists.

4. The pressure at point C is the lowest in the system because point C is at the highest elevation.

5. The pressure at point D is the same as that at point B, because both are on the same elevation and the velocity head at both points is the same.

6. The pressure at point E is the highest in the system because point E is at the lowest elevation.

■

6.10.4
Venturi Meters and Other
Closed Systems with
Unknown Velocities

Figure 6.8 shows a device called a *venturi meter* that can be used to measure the velocity of flow in a fluid flow system. A more complete description of the venturi meter is given in Chapter 14. However, the analysis of such a device is based on the application of Bernoulli's equation. The reduced diameter section at B causes the velocity of flow to increase there with a corresponding decrease in the pressure. It will be shown that the velocity of flow is dependent on the *difference* in pressure between points A and B. Therefore, a differential manometer as shown is convenient to use.

FIGURE 6.8   Venturi meter system for Example Problem 6.11.

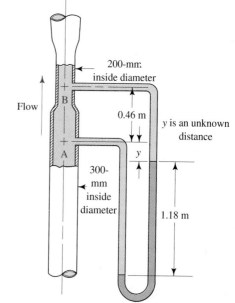

We will also show in the solution to the following problem that we must combine the continuity equation with Bernoulli's equation to solve for the desired velocity of flow.

☐ **EXAMPLE PROBLEM 6.11**   The venturi meter shown in Fig. 6.8 carries water at 60°C. The specific gravity of the gage fluid in the manometer is 1.25. Calculate the velocity of flow at section A and the volume flow rate of water.

Solution   The problem solution will be shown in the steps outlined at the beginning of this section but the programmed technique will not be used.

1. Decide what items are known and what is to be found. The elevation difference between points A and B is known. The manometer allows the determination of the difference in pressure between points A and B. The sizes of the sections at A and B are known.

   The velocity is not known at any point in the system and the velocity at point A was specifically requested.

2. Decide on sections of interest. Points A and B are the obvious choices.
3. Write Bernoulli's equation between points A and B.

$$\frac{p_A}{\gamma} + z_A + \frac{v_A^2}{2g} = \frac{p_B}{\gamma} + z_B + \frac{v_B^2}{2g}$$

The specific weight γ is for water at 60°C, which is 9.65 kN/m³.
4. Simplify the equation, if possible, by eliminating terms that are zero or terms that are equal on both sides of the equation. No simplification can be done here.
5. Solve the equation algebraically for the desired term. This step will require significant effort. First, note that *both* of the velocities are unknown. But we can find the difference in pressures between A and B and the elevation difference is known. Therefore, it is convenient to bring both pressure terms and both elevation terms onto the left side of the equation in the form of differences. Then the two velocity terms can be moved to the right side. The result is

$$\frac{p_A - p_B}{\gamma} + (z_A - z_B) = \frac{v_B^2 - v_A^2}{2g} \qquad (6\text{--}12)$$

6. Calculate the result. Several steps are required. The elevation difference is

$$z_A - z_B = -0.46 \text{ m} \qquad (6\text{--}13)$$

The value is negative because B is higher than A. This value will be used in Eq. (6–12) later.

The pressure head difference term can be evaluated by writing the equation for the manometer. We will use γ_g for the specific weight of the gage fluid where

$$\gamma_g = 1.25(\gamma_w \text{ at } 4°\text{C}) = 1.25(9.81 \text{ kN/m}^3) = 12.26 \text{ kN/m}^3$$

A new problem occurs here because the data in Fig. 6.8 do not include the vertical distance from point A to the level of the gage fluid in the right leg of the manometer. We will show that this problem will be eliminated by simply calling this unknown distance y or any other variable name.

Now we can write the manometer equation starting at A.

$$p_A + \gamma(y) + \gamma(1.18 \text{ m}) - \gamma_g(1.18 \text{ m}) - \gamma(y) - \gamma(0.46 \text{ m}) = p_B$$

Note that the two terms containing the unknown y variable can be cancelled out. Solving for the pressure difference $p_A - p_B$,

$$p_A - p_B = \gamma(0.46 \text{ m} - 1.18 \text{ m}) + \gamma_g(1.18 \text{ m})$$
$$p_A - p_B = \gamma(-0.72 \text{ m} + \gamma_g(1.18 \text{ m})$$

But notice in Eq. 6–12 above, we really need $(p_A - p_B)/\gamma$. If we divide both sides of the above equation by γ we get the desired term.

$$\frac{p_A - p_B}{\gamma} = -0.72 \text{ m} + \frac{\gamma_g(1.18 \text{ m})}{\gamma} = -0.72 \text{ m} + \frac{12.26 \text{ kN/m}^3 (1.18 \text{ m})}{9.65 \text{ kN/m}^3}$$

$$(p_A - p_B)/\gamma = -0.72 \text{ m} + 1.50 \text{ m} = 0.78 \text{ m} \qquad (6\text{--}14)$$

The entire left side of Eq. 6–12 has now been evaluated. But note that there are still *two* unknowns on the right side, v_A and v_B. We can eliminate one unknown by finding another independent equation that relates those two variables. A convenient equation is the *continuity equation*.

$$A_A v_A = A_B v_B$$

Solving for v_B in terms of v_A,

$$v_B = v_A(A_A/A_B)$$

The areas for the 200 mm and 300 mm diameter sections can be found in Appendix J. Then,

$$v_B = v_A \, (7.069 \times 10^{-2}/3.142 \times 10^{-2}) = 2.25 \, v_A$$

But we need v_B^2.

$$v_B^2 = 5.06 \, v_A^2$$

Then,

$$v_B^2 - v_A^2 = 5.06 \, v_A^2 - v_A^2 = 4.06 \, v_A^2 \qquad \textbf{(6–15)}$$

We can now take this result, the elevation head difference [Eq. (6–13)] and the pressure head difference [Eq. (6–14)] back into Eq. (6–12). We can complete the solution now. Equation (6–12) becomes

$$0.78 \text{ m} - 0.46 \text{ m} = 4.06 \, v_A^2/2g$$

Solving for v_A gives

$$v_A = \sqrt{\frac{2g(0.32 \text{ m})}{4.06}} = \sqrt{\frac{2(9.81 \text{ m/s}^2) \, (0.32 \text{ m})}{4.06}}$$

$$v_A = 1.24 \text{ m/s}$$

The problem statement also asked for the volume flow rate which can be computed from

$$Q = A_A v_A = (7.069 \times 10^{-2} \text{ m}^2)(1.24 \text{ m/s}) = 8.77 \times 10^{-2} \text{ m}^3/\text{s}$$

This example problem is completed.

6.11
TORRICELLI'S
THEOREM

In the siphon analyzed in Example Problem 6.10, it was observed that the velocity of flow from the siphon depends on the elevation difference between the free surface of the fluid and the outlet of the siphon. A classic application of this observation is shown in Fig. 6.9. Fluid is flowing from the side of a tank through a smooth, rounded nozzle. To determine the velocity of flow from the nozzle, write Bernoulli's equation between a reference point on the fluid surface and a point in the jet issuing from the nozzle:

$$\frac{p_1}{\gamma} + z_1 + \frac{v_1^2}{2g} = \frac{p_2}{\gamma} + z_2 + \frac{v_2^2}{2g}$$

FIGURE 6.9   Flow from a
tank.

But $p_1 = p_2 = 0$, and v_1 is approximately zero. Then, solving for v_2 gives

$$v_2 = \sqrt{2g(z_1 - z_2)}$$

Letting $h = (z_1 - z_2)$, we have

$$v_2 = \sqrt{2gh} \qquad\qquad (6\text{--}16)$$

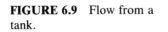 TORRICELLI'S THEOREM

Equation (6–16) is called *Torricelli's theorem* in honor of Evangelista Torricelli, who discovered it in approximately 1645.

□ **EXAMPLE PROBLEM 6.12**   For the tank shown in Fig. 6.9, compute the velocity of flow from the nozzle for a fluid depth h of 3.00 m.

Solution   This is a direct application of Torricelli's theorem:

$$v_2 = \sqrt{2gh} = \sqrt{(2)(9.81 \text{ m/s}^2)(3.0 \text{ m})} = 7.67 \text{ m/s}$$

□ **EXAMPLE PROBLEM 6.13**   For the tank shown in Fig. 6.9, compute the velocity of flow from the nozzle and the volume flow rate for a range of depths from 3.0 m to 0.50 m in steps of 0.50 m. The diameter of the jet at the nozzle is 50 mm.

Solution   The same procedure used in Example Problem 6.12 can be used to determine the velocity at any depth. So, at $h = 3.0$ m, $v_2 = 7.67$ m/s. The volume flow rate is computed by multiplying this velocity by the area of the jet:

$$A_j = 1.963 \times 10^{-3} \text{ m}^2 \quad \text{from Appendix J.}$$

Then,

$$Q = A_j v_2 = (1.963 \times 10^{-3} \text{ m}^2)(7.67 \text{ m/s}) = 1.51 \times 10^{-2} \text{ m}^3/2$$

Using the same procedure, we compute the following data:

Depth, h (m)	v_2 (m/s)	Q (m³/s)
3.0	7.67	1.51×10^{-2}
2.5	7.00	1.38×10^{-2}
2.0	6.26	1.23×10^{-2}
1.5	5.42	1.07×10^{-2}
1.0	4.43	0.87×10^{-2}
0.5	3.13	0.61×10^{-2}

Figure 6.10 is a plot of velocity and volume flow rate versus depth.

FIGURE 6.10 Jet velocity and volume flow rate vs. fluid depth.

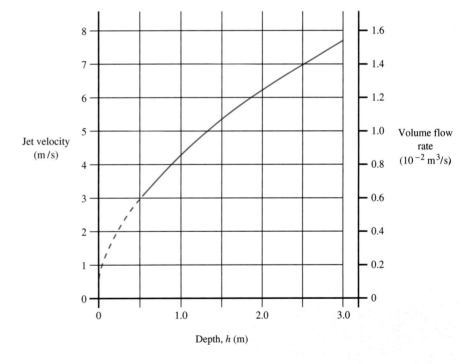

Another interesting application of Torricelli's theorem is shown in Fig. 6.11, in which a jet of fluid is shooting upward. If no energy losses occur, the jet will reach a height equal to the elevation of the free surface of the fluid in the tank. And, of course, at this height the velocity in the stream is zero. This can be demonstrated

FIGURE 6.11   Vertical jet.

using Bernoulli's equation. First, obtain an expression for the velocity of the jet at point 2:

$$\frac{p_1}{\gamma} + z_1 + \frac{v_1^2}{2g} = \frac{p_2}{\gamma} + z_2 + \frac{v_2^2}{2g}$$

This is an identical situation to that encountered in the initial development of Torricelli's theorem. Then, as in Eq. (6–16),

$$v_2 = \sqrt{2gh}$$

Now, write Bernoulli's equation between point 2 and point 3 at the level of the free surface of the fluid but in the fluid stream.

$$\frac{p_2}{\gamma} + z_2 + \frac{v_2^2}{2g} = \frac{p_3}{\gamma} + z_3 + \frac{v_3^2}{2g}$$

But, $p_2 = p_3 = 0$. Then, solving for v_3, we have

$$v_3 = \sqrt{v_2^2 + 2g(z_2 - z_3)}$$

From Eq. (6–16), $v_2^2 = 2gh$. Also, $(z_2 - z_3) = -h$. Then,

$$v_3 = \sqrt{2gh + 2g(-h)} = 0$$

This result verifies that the stream just reaches the height of the free surface of the fluid in the tank.

To make a jet go higher (as with some decorative fountains, for example), a greater pressure can be developed above the fluid in the reservoir, or a pump can be used to develop a higher pressure.

☐ **EXAMPLE PROBLEM 6.14**

Using a system similar to that shown in Fig. 6.12, compute the required air pressure above the water to cause the jet to rise 40.0 ft from the nozzle. The depth $h = 6.0$ ft.

Solution

First, use Bernoulli's equation to obtain an expression for the velocity of flow from the nozzle as a function of the air pressure.

$$\frac{p_1}{\gamma} + z_1 + \frac{v_1^2}{2g} = \frac{p_2}{\gamma} + z_2 + \frac{v_2^2}{2g}$$

Here, we can see $v_1 = 0$ and $p_2 = 0$. Solving for v_2 gives

$$v_2 = \sqrt{2g[(p_1/\gamma) + (z_1 - z_2)]}$$

As before, letting $h = (z_1 - z_2)$, we have

$$v_2 = \sqrt{2g[(p_1/\gamma) + h]} \qquad\qquad (6\text{–}17)$$

This is similar to Torricelli's theorem. It was shown above that for $v = \sqrt{2gh}$, the jet rises to a height h. By analogy, the pressurized system would cause the jet to rise to a height of $[(p_1/\gamma) + h]$. Then, in this problem, if we want a height of 40.0 ft and $h = 6.0$ ft,

$$p_1/\gamma = 40.0 \text{ ft} - 6.0 \text{ ft} = 34.0 \text{ ft}$$

FIGURE 6.12 Pressurized tank delivering a vertical jet. Also for Problems 6.93 and 6.94.

And,

$$p_1 = \gamma(34.0 \text{ ft})$$
$$p_1 = (62.4 \text{ lb/ft}^3)(34.0 \text{ ft})(1 \text{ ft}^2)/(144 \text{ in}^2)$$
$$p_1 = 14.73 \text{ psig}$$

In Chapter 4, we defined the pressure head, p/γ, in such applications as the *piezometric head*. Then the *total head* above the nozzle is $p_1/\gamma + h$.

6.12
FLOW DUE TO A
FALLING HEAD

As stated earlier, most problems considered in this book are for situations in which the flow rate is constant. However, in Section 6.11, it was shown that the flow rate depends on the pressure head available to cause the flow. The results of Example Problem 6.13, plotted in Fig. 6.10, show that the velocity and volume flow rate issuing from an orifice in a tank decrease in a nonlinear manner as the fluid flows from the tank and the depth of the fluid decreases.

In this section, we will develop a method for computing the time required to empty a tank, considering the variation of velocity as the depth decreases. Figure 6.13 shows a tank with a smooth, well-rounded nozzle in

FIGURE 6.13 Flow from a tank with falling head. Also for Problems 6.95–6.106.

the bottom through which fluid is discharging. For a given depth of fluid h, Torricelli's theorem tells us that the velocity of flow in the jet is

$$v_j = \sqrt{2gh}$$

The volume flow rate through the nozzle is $Q = A_j v_j$ in such units as cubic meters per second (m³/s) or cubic feet per second (ft³/s). In a small amount of time, dt, the volume of fluid flowing through the nozzle is

$$\text{Volume flowing} = Q(dt) = A_j v_j (dt) \qquad \textbf{(6–18)}$$

Meanwhile, because fluid is leaving the tank, the fluid level is decreasing. During the small time increment, dt, the fluid level drops a small distance, dh. Then, the volume of fluid removed from the tank is

$$\text{Volume removed} = -A_t dh \qquad \textbf{(6–19)}$$

These two volumes must be equal. Then,

$$A_j v_j (dt) = -A_t dh \qquad \textbf{(6–20)}$$

Solving for the time dt, we have

$$dt = \frac{-(A_t/A_j)}{v_j} \, dh \qquad \textbf{(6–21)}$$

From Torricelli's theorem, we can substitute $v_j = \sqrt{2gh}$. Then,

$$dt = \frac{-(A_t/A_j)}{\sqrt{2gh}}\, dh \qquad (6-22)$$

Rewriting to separate the terms involving h gives

$$dt = \frac{-(A_t/A_j)}{\sqrt{2g}}\, h^{-1/2}dh \qquad (6-23)$$

The time required for the fluid level to fall from one depth, h_1, to another depth, h_2, can be found by integrating Eq. (6–23):

$$\int_{t_1}^{t_2} dt = \frac{-(A_t/A_j)}{\sqrt{2g}} \int_{h_1}^{h_2} h^{-1/2}dh \qquad (6-24)$$

$$t_2 - t_1 = \frac{-(A_t/A_j)}{\sqrt{2g}} \frac{(h_2^{1/2} - h_1^{1/2})}{\tfrac{1}{2}} \qquad (6-25)$$

We can reverse the two terms involving h and remove the minus sign. At the same time, clearing the $\tfrac{1}{2}$ from the denominator, we get

⇨ **TIME REQUIRED TO DRAIN A TANK**

$$t_2 - t_1 = \frac{2(A_t/A_j)}{\sqrt{2g}} (h_1^{1/2} - h_2^{1/2}) \qquad (6-26)$$

Equation (6–26) can be used to compute the time required to drain a tank from h_1 to h_2.

☐ **EXAMPLE PROBLEM 6.15**   For the tank shown in Fig. 6.13, find the time required to drain the tank from a level of 3.0 m to 0.50 m. The tank has a diameter of 1.50 m, and the nozzle has a diameter of 50 mm.

Solution   To use Eq. (6–26), the required areas are

$$A_t = \pi(1.50 \text{ m})^2/4 = 1.767 \text{ m}^2$$
$$A_j = \pi(0.05 \text{ m})^2/4 = 0.001963 \text{ m}^2$$

The ratio of these two areas is required:

$$\frac{A_t}{A_j} = \frac{1.767 \text{ m}^2}{0.001963 \text{ m}^2} = 900$$

Now in Eq. (6–26),

$$t_2 - t_1 = \frac{2(A_t/A_j)}{\sqrt{2g}} (h_1^{1/2} - h_2^{1/2})$$

$$t_2 - t_1 = \frac{2(900)}{\sqrt{2(9.81 \text{ m/s}^2)}} [(3.0 \text{ m})^{1/2} - (0.5 \text{ m})^{1/2}]$$

$$t_2 - t_1 = 417 \text{ s}$$

This is equivalent to 6 min and 57 s.

6.12.1
Draining a
Pressurized Tank

If the tank in Fig. 6.13 is sealed with a pressure above the fluid, the piezo-metric head, p/γ, should be added to the actual fluid depth before completing the calculations called for in Eq. (6–25).

6.12.2
Effect of the
Type of Nozzle

The development of Eq. (6–26) assumes that the diameter of the jet of fluid flowing from the nozzle is the same as the diameter of the nozzle itself. This is very nearly true for the well-rounded nozzles depicted in Figs. 6.9, 6.11, and 6.13. However, if the nozzle is sharper, the minimum diameter of the jet is significantly smaller than the diameter of the opening. For example, Fig. 6.14 shows the flow from a tank through a sharp-edged orifice. The proper area to use for A_j in Eq. (6–26) is that at the smallest diameter. This point, called the *vena contracta*, occurs slightly outside the orifice. For the sharp-edged orifice, $A_j = 0.62A_o$ is a good approximation.

FIGURE 6.14   Flow through a sharp-edged orifice.

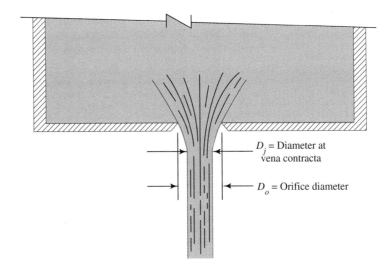

D_j = Diameter at vena contracta

D_o = Orifice diameter

REFERENCE

1. American Society of Mechanical Engineers. 1992. *ANSI/ASME Standard B31.1-1992 Power Piping*. New York: Author.

PRACTICE PROBLEMS

Conversion Factors

6.1 Convert a volume flow rate of 3.0 gal/min to m³/s.

6.2 Convert 459 gal/min to m³/s.

6.3 Convert 8720 gal/min to m³/s.

6.4 Convert 84.3 gal/min to m³/s.

6.5 Convert a volume flow rate of 125 L/min to m³/s.

6.6 Convert 4500 L/min to m³/s.

6.7 Convert 15 000 L/min to m³/s.

6.8 Convert 459 gal/min to L/min.

6.9 Convert 8720 gal/min to L/min.

6.10 Convert 23.5 cm³/s to m³/s.

6.11 Convert 0.296 cm³/s to m³/s.

6.12 Convert 0.105 m³/s to L/min.

6.13 Convert 3.58×10^{-3} m³/s to L/min.

6.14 Convert 5.26×10^{-6} m³/s to L/min.

6.15 Convert 459 gal/min to ft³/s.

6.16 Convert 20 gal/min to ft³/s.

6.17 Convert 2500 gal/min to ft³/s.

6.18 Convert 2.50 gal/min to ft³/s.

6.19 Convert 1.25 ft³/s to gal/min.

6.20 Convert 0.060 ft³/s to gal/min.

6.21 Convert 7.50 ft³/s to gal/min.

6.22 Convert 0.008 ft³/s to gal/min.

6.23 Table 6.2 lists the range of typical volume flow rates for centrifugal fire-fighting pumps to be 500 to 2500 gal/min. Express this range in the units of ft³/s and m³/s.

6.24 Table 6.2 lists the range of typical volume flow rates for industrial oil hydraulic systems pumps to be 3 to 30 gal/min. Express this range in the units of ft³/s and m³/s.

6.25 A certain deep well pump for a residence is rated to deliver 745 gal/h of water. Express this flow rate in ft³/s.

6.26 A small pump delivers 0.85 gal/h of liquid fertilizer. Express this flow rate in ft³/s.

6.27 A small metering pump delivers 11.4 gal of a water treatment chemical per 24 h. Express this flow rate in ft³/s.

6.28 A small metering pump delivers 19.5 mL/min of water to dilute a waste stream. Express this flow rate in m³/s.

General Note: In the following problems you may be required to refer to the appendix for fluid properties, dimensions of pipe and tubing, or conversion factors. Assume that there are no energy losses in all problems. Unless otherwise stated, the pipe sizes given are actual inside diameters.

Fluid Flow Rates

6.29M *Water at 10°C is flowing at 0.075 m³/s. Calculate the weight flow rate and the mass flow rate.*

6.30M *Oil for a hydraulic system (sg = 0.90) is flowing at 2.35×10^{-3} m³/s. Calculate the weight flow rate and the mass flow rate.*

6.31M *A liquid refrigerant (sg = 1.08) is flowing at a weight flow rate of 28.5 N/h. Calculate the volume flow rate and the mass flow rate.*

6.32M *After the refrigerant from Problem 6.31 flashes into a vapor, its specific weight is 12.50 N/m³. If the weight flow rate remains at 28.5 N/h, compute the volume flow rate.*

6.33C A fan delivers 640 ft³/min (CFM) of air. If the density of the air is 1.20 kg/m³, compute the mass flow rate in slugs/s and the weight flow rate in lb/h.

6.34E A large blower for a furnace delivers 45 700 ft³/min (CFM) of air having a specific weight of 0.075 lb/ft³. Calculate the weight flow rate and mass flow rate.

6.35E A furnace requires 1200 lb/h of air for efficient combustion. If the air has a specific weight of 0.062 lb/ft³, compute the required volume flow rate.

6.36E If a pump removes 1.65 gal/min of water from a tank, how long will it take to empty the tank if it contains 7425 lb of water?

Continuity Equation

6.37E Calculate the diameter of a pipe that would carry 75.0 ft³/s of a liquid at an average velocity of 10.0 ft/s.

6.38E If the velocity of a liquid is 1.65 ft/s in a 12-in diameter pipe, what is the velocity in a 3-in diameter jet exiting from a nozzle attached to the pipe?

6.39M *When 2000 L/min of water flow through a 300-mm diameter pipe that later reduces to a 150-mm diameter pipe, calculate the average velocity of flow in each pipe.*

6.40M *Water flows at 1.20 m/s in a 150-mm diameter pipe. Calculate the velocity of flow in a 300-mm pipe connected to it.*

6.41M *A 150-mm diameter pipe carries 0.072 m³/s of water. The pipe branches into two pipes, as shown in Fig. 6.15. If the velocity in the 50-mm pipe is 12.0 m/s, what is the velocity in the 100-mm pipe?*

6.42E A standard Schedule 40 steel pipe is to be selected to carry 10 gal/min of water with a maximum velocity of 1.0 ft/s. What size pipe should be used?

6.43E If water at 180°F is flowing with a velocity of 4.50 ft/s in a standard 6-in Schedule 40 pipe, calculate the weight flow rate in lb/h.

FIGURE 6.15   Problem 6.41.

Flow

Reducing tee

100-mm inside diameter

150-mm inside diameter

50-mm inside diameter

6.44M *A standard 1-in OD (outside diameter) steel tube (0.065-in wall thickness) is carrying 19.7 L/min of oil. Calculate the velocity of flow.*

6.45E Table 6.3 shows that the recommended velocity of flow in the discharge line of an oil hydraulic system is in the range of 8.0 to 25.0 ft/s. If the pump delivers 30 gal/min of oil, specify the smallest and largest suitable sizes of steel tubing.

6.46E Repeat Problem 6.45, except specify suitable sizes for the suction lines to maintain the velocity between 2.0 and 7.0 ft/s for 30 gal/min of flow.

6.47M *Table 6.2 shows the typical volume flow rate for centrifugal firefighting pumps is in the range of 1800 to 9500 L/min. Specify the smallest suitable size of Schedule 40 steel pipe for each flow rate that will maintain the maximum velocity of flow at 2.0 m/s.*

6.48M *Repeat Problem 6.47, but use Schedule 80 pipe.*

6.49M *Compute the resulting velocity of flow if 400 L/min of fluid flow through a 2-in Schedule 40 pipe.*

6.50M Repeat Problem 6.49 for a 2-in Schedule 80 pipe.

6.51E Compute the resulting velocity of flow if 400 gal/min of fluid flow through a 4-in Schedule 40 pipe.

6.52E Repeat Problem 6.51 for a 4-in Schedule 80 pipe.

6.53M *From the list of standard steel tubing in Appendix G, select the smallest size that would carry 2.80 L/min of oil with a maximum velocity of 0.30 m/s.*

6.54E A standard 6-in Schedule 40 steel pipe is carrying 95 gal/min of water. The pipe then branches into two standard 3-in pipes. If the flow divides evenly between the branches, calculate the velocity of flow in all three pipes.

6.55E A shell and tube heat exchanger is made of two standard steel tubes, as shown in Fig. 6.16. Each tube has a wall thickness of 0.049 in. Calculate the required ratio of the volume flow rate in the shell to that in the tube if the average velocity of flow is to be the same in each.

6.56M *Figure 6.17 shows a heat exchanger in which each of two 6-in Schedule 40 pipes carries 450 L/min of water. The pipes are inside a rectangular duct whose inside dimensions are 200 mm by 400 mm. Compute the velocity of flow in the pipes. Then, compute the required volume flow rate of water in the duct to obtain the same average velocity.*

6.57E Figure 6.18 shows the cross section of a shell-and-tube heat exchanger. Compute the volume flow rate required in each small pipe and in the shell to obtain an average velocity of flow of 25 ft/s in all parts.

6.58M *A venturi meter is a device that uses a constriction in a flow system to measure the velocity of flow. Figure 6.19 illustrates one type of design. If the main pipe section is a standard 4-in Type K copper tube, compute the volume flow rate when the velocity there is 3.0 m/s. Then, for that volume flow*

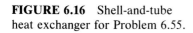

FIGURE 6.16 Shell-and-tube
heat exchanger for Problem 6.55.

FIGURE 6.17 Problem 6.56.

FIGURE 6.18 Problem 6.57.

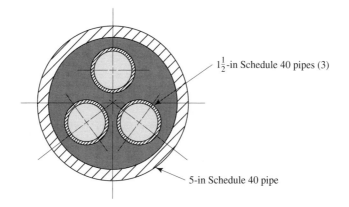

FIGURE 6.19   Venturi meter for Problem 6.58.

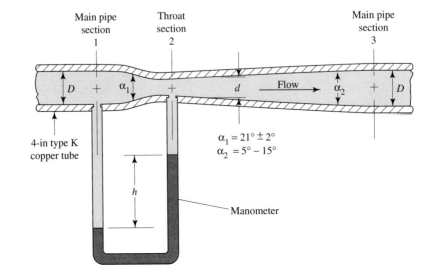

rate, specify the required size of the throat section that would make the velocity there at least 15.0 m/s.

6.59E   A flow nozzle, shown in Fig. 6.20, is used to measure the velocity of flow. If the nozzle is installed inside a 14-in Schedule 40 pipe and the nozzle diameter is 2.75 in, compute the velocity of flow at section 1 and the throat of the nozzle when 7.50 ft³/s of water flow through the system.

Bernoulli's Equation

6.60M   *Gasoline (sg = 0.67) is flowing at 0.11 m³/s in the pipe shown in Fig. 6.21. If the pressure before the reduction is 415 kPa, calculate the pressure in the 75-mm diameter pipe.*

6.61M   *Water at 10°C is flowing from point A to point B through the pipe shown in Fig. 6.22 at the rate of 0.37 m³/s. If the pressure at A is 66.2 kPa, calculate the pressure at B.*

FIGURE 6.20.   Nozzle meter for Problem 6.59.

FIGURE 6.21   Problem 6.60.

FIGURE 6.22   Problem 6.61.

FIGURE 6.23   Problem 6.62.

6.62M *Calculate the volume flow rate of water at 5°C through the system shown in Fig. 6.23.*

6.63E Calculate the pressure required in the pipe just ahead of the nozzle in Fig. 6.24 to produce a jet velocity of 75 ft/s. The fluid is water at 180°F.

FIGURE 6.24   Problem 6.63.

FIGURE 6.25 Problem 6.65.

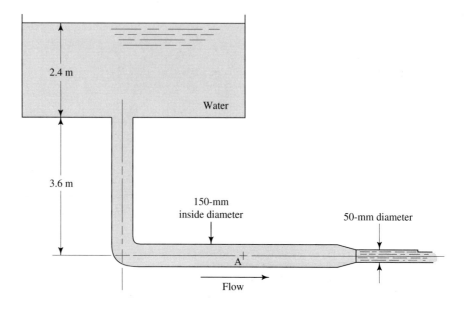

6.64E Kerosene with a specific weight of 50.0 lb/ft³ is flowing at 10 gal/min from a standard 1-in Schedule 40 steel pipe to a standard 2-in Schedule 40 steel pipe. Calculate the difference in pressure in the two pipes.

6.65M *For the system shown in Fig. 6.25, calculate (a) the volume flow rate of water from the nozzle and (b) the pressure at point A.*

6.66M *For the system shown in Fig. 6.26, calculate (a) the volume flow rate of oil from the nozzle and (b) the pressures at A and B.*

6.67E For the tank shown in Fig. 6.27, calculate the volume flow rate of water from the nozzle. The tank

is sealed with a pressure of 20 psig above the water. The depth, *h*, is 8 ft.

6.68E Calculate the pressure of the air in the sealed tank shown in Fig. 6.27 that would cause the velocity of flow to be 20 ft/s from the nozzle. The depth, *h*, is 10 ft.

6.69M *For the siphon in Fig. 6.28, calculate (a) the volume flow rate of water through the nozzle and (b) the pressure at points A and B. The distance X = 4.6 m and Y = 0.90 m.*

6.70M *For the siphon in Fig. 6.28, calculate the distance X required to obtain a volume flow rate of 7.1 × 10⁻³ m³/s.*

FIGURE 6.26 Problem 6.66.

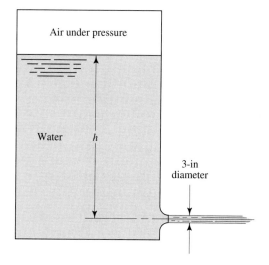

FIGURE 6.27   Problems 6.67 and 6.68.

FIGURE 6.29   Problems 6.72 and 6.83

the volume flow rate of oil from the tank and (b) the pressures at points A, B, C, and D.

6.73E   For the pipe reducer shown in Fig. 6.30, the pressure at A is 50.0 psig and the pressure at B is 42.0

6.71M   *For the siphon in Fig. 6.28, assume that the volume flow rate is 5.6 × 10⁻³ m³/s. Determine the maximum allowable distance Y if the minimum allowable pressure in the system is −18 kPa (gage).*

6.72M   *For the siphon shown in Fig. 6.29, calculate (a)*

FIGURE 6.28   Problems 6.69, 6.70, and 6.71

FIGURE 6.30   Problems 6.73
and 6.84

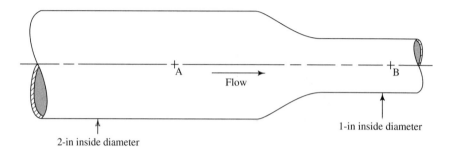

2-in inside diameter

1-in inside diameter

psig. Calculate the velocity of flow of water at point B.

6.74E In the enlargement shown in Fig. 6.31, the pressure at A is 25.6 psig and the pressure at B is 28.2 psig. Calculate the volume flow rate of oil (sg = 0.90).

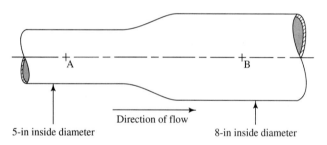

5-in inside diameter

Direction of flow

8-in inside diameter

FIGURE 6.31   Problem 6.74.

6.75M *Fig. 6.32 shows a manometer being used to indicate the pressure difference between two points in a pipe system. Calculate the volume flow rate of water in the system if the manometer deflection, h,*

is 250 mm. (This arrangement is called a venturi meter, which is often used for flow measurement.)

6.76M *For the venturi meter shown in Fig. 6.32, calculate the manometer deflection, h, if the velocity of flow of water in the 25-mm diameter section is 10 m/s.*

6.77M *Oil with a specific weight of 8.64 kN/m³ flows from A to B through the system shown in Fig. 6.33. Calculate the volume flow rate of oil.*

6.78M *The venturi meter shown in Fig. 6.34 carries oil (sg = 0.90). The specific gravity of the gage fluid in the manometer is 1.40. Calculate the volume flow rate of oil.*

6.79E Oil with a specific gravity of 0.90 is flowing downward through the venturi meter shown in Fig. 6.35. If the manometer deflection h is 28 in, calculate the volume flow rate of oil.

6.80E Oil with a specific gravity of 0.90 is flowing downward through the venturi meter shown in Fig. 6.35. If the velocity of flow in the 2-in diameter section is 10.0 ft/s, calculate the deflection, h, of the manometer.

6.81E Gasoline (sg = 0.67) is flowing at 4.0 ft³/s in the

FIGURE 6.32   Problems 6.75
and 6.76.

Direction of flow

50-mm
diameter

25-mm
diameter

Mercury
(sg = 13.54)

h

50-mm
inside diameter

Flow

100-mm
inside diameter

B

A

600 mm

150
mm

200
mm

Water

FIGURE 6.33   Problem 6.77.

75-mm
inside diameter

Flow

B

A

0.25 m

200-
mm
inside
diameter

0.60 m

FIGURE 6.34   Problem 6.78.

FIGURE 6.35   Problems 6.79
and 6.80.

4-in
inside diameter

A

2-in
inside diameter

B

Flow

h

Mercury
(sg = 13.54)

FIGURE 6.36   Problem 6.81.

pipe shown in Fig. 6.36. If the pressure before the reduction is 60 psig, calculate the pressure in the 3-in diameter pipe.

6.82E Oil with a specific weight of 55.0 lb/ft³ flows from A to B through the system shown in Fig. 6.37. Calculate the volume flow rate of the oil.

FIGURE 6.37   Problem 6.82.

6.83M *Draw a plot of elevation head, pressure head, velocity head, and total head for the siphon system shown in Fig. 6.29 and analyzed in Problem 6.70.*

6.84E Draw a plot of elevation head, pressure head, velocity head, and total head for the system shown in Fig. 6.30 and analyzed in Problem 6.73.

6.85E Fig. 6.38 shows a system in which water flows from a tank through a pipe system having several sizes and elevations. For points A through G, compute the elevation head, the pressure head, the velocity head, and the total head. Plot these values on a sketch similar to that shown in Fig. 6.6.

6.86M *Fig. 6.39 shows a venturi meter with a U-tube manometer to measure the velocity of flow. When no flow occurs, the mercury column is balanced and its top is 300 mm below the throat. Compute the volume flow rate through the meter that will cause the mercury to flow into the throat. Note that for a given manometer deflection, h, the left side will move down h/2 while the right side would rise h/2.*

6.87E For the tank shown in Fig. 6.40, compute the velocity of flow from the outlet nozzle at varying depths from 10.0 ft to 2.0 ft in 2.0-ft increments. Then, use increments of 0.5 ft to zero. Plot the velocity versus depth.

6.88E What depth of fluid above the outlet nozzle is required to deliver 200 gal/min of water from the tank shown in Fig. 6.40? The nozzle has a 3.00-in diameter.

Torricelli's Theorem

6.89 Derive Torricelli's theorem for the velocity of flow from a tank through an orifice opening into the atmosphere under a given depth of fluid.

6.90E Solve Problem 6.88 using the direct application of Torricelli's theorem.

6.91M *To what height will the jet of fluid rise for the conditions shown in Fig. 6.41?*

6.92E To what height will the jet of water rise for the conditions shown in Fig. 6.42?

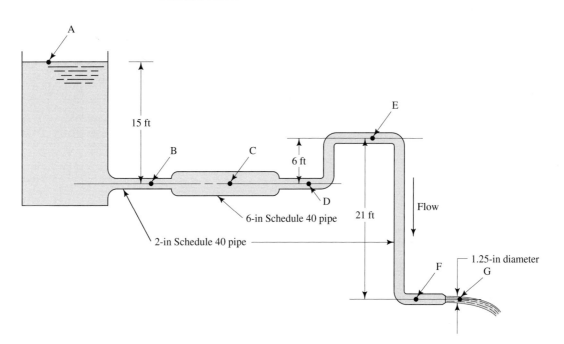

FIGURE 6.38 Flow system for Problem 6.85.

FIGURE 6.39 Venturi meter for Problem 6.86.

FIGURE 6.40   Tank for Problems 6.87–6.88.

FIGURE 6.42   Problem 6.92.

6.93E What pressure is required above the fluid in Fig. 6.12 to cause the jet to rise to 28.0 ft? The fluid depth is 4.50 ft.

6.94M *What pressure is required above the fluid in Fig. 6.12 to cause the jet to rise to 9.50 m? The fluid depth is 1.50 m.*

Flow Due to Falling Head

6.95M *Compute the time required to empty the tank shown in Fig. 6.13 if the original depth is 2.68 m. The tank diameter is 3.00 m, and the orifice diameter is 150 mm.*

6.96M *Compute the time required to empty the tank shown in Fig. 6.13 if the original depth is 55 mm. The tank diameter is 300 mm, and the orifice diameter is 20 mm.*

6.97E Compute the time required to empty the tank shown in Fig. 6.13 if the original depth is 15 ft. The tank diameter is 12.0 ft, and the orifice diameter is 6.00 in.

6.98E Compute the time required to empty the tank shown in Fig. 6.13 if the original depth is 18.5 in. The tank diameter is 22.0 in, and the orifice diameter is 0.50 in.

6.99M *Compute the time required to reduce the depth in the tank shown in Fig. 6.13 by 1.50 m if the origi-*

FIGURE 6.41   Problem 6.91.

nal depth is 2.68 m. The tank diameter is 2.25 m, and the orifice diameter is 50 mm.

6.100M *Compute the time required to reduce the depth in the tank shown in Fig. 6.13 by 225 mm if the original depth is 1.38 mm. The tank diameter is 1.25 m, and the orifice diameter is 25 mm.*

6.101E Compute the time required to reduce the depth in the tank shown in Fig. 6.13 by 12.5 in if the original depth is 38 in. The tank diameter is 6.25 ft, and the orifice diameter is 0.625 in.

6.102E Compute the time required to reduce the depth in the tank shown in Fig. 6.13 by 21.0 ft if the original

depth is 23.0 ft. The tank diameter is 46.5 ft, and the orifice diameter is 8.75 in.

6.103E Repeat Problem 6.97 if the tank is sealed and a pressure of 5.0 psig is above the water in the tank.

6.104E Repeat Problem 6.101 if the tank is sealed and a pressure of 2.8 psig is above the water in the tank.

6.105M *Repeat Problem 6.96 if the tank is sealed and a pressure of 20 kPa(gage) is above the water in the tank.*

6.106M *Repeat Problem 6.100 if the tank is sealed and a pressure of 35 kPa(gage) is above the water in the tank.*

COMPUTER PROGRAMMING ASSIGNMENTS

1. Write a program to compute the values of the pressure head, the velocity head, the elevation head, and the total head for given values for pressure, velocity, and elevation.

2. Enhance the program in Assignment 1 by causing it to plot side-by-side in several combinations the various head components, in order to compare one with another.

3. In the program in Assignment 1, include the ability to compute the velocity of flow from given data for volume flow rate and pipe size.

4. Write a program to compute, using Eq. (6–26), the time required to decrease the fluid level in a tank between two values for any combination of tank size and nozzle diameter.

5. Add the ability to pressurize the system to the program in Assignment 2.

6. Write a program to compute the velocity of flow from an orifice, using Torricelli's theorem for any depth of fluid and any amount of pressure above the fluid.

7 General Energy Equation

7.1 OBJECTIVES

Some of the restrictions placed on the use of Bernoulli's equation when it was developed in Chapter 6 can be eliminated by expanding the equation to what is called the *general energy equation*. After completing this chapter, you should be able to:

1. Identify the conditions under which energy losses occur in fluid flow systems.
2. Identify the means by which energy can be added to a fluid flow system.
3. Identify the means by which energy can be removed from a fluid flow system.
4. Expand Bernoulli's equation to form the general energy equation by considering energy losses, energy additions, and energy removals.
5. Apply the general energy equation to a variety of practical problems.
6. Compute the power added to a fluid by pumps.
7. Define the *efficiency of pumps.*
8. Compute the power required to drive pumps.
9. Compute the power delivered by a fluid to a fluid motor.
10. Define the *efficiency of fluid motors.*
11. Compute the power output from a fluid motor.

7.2 ENERGY LOSSES AND ADDITIONS

When we developed Bernoulli's equation in Chapter 6, we placed four restrictions on its use, namely:

1. It is valid only for incompressible fluids.
2. There can be no mechanical devices between the two sections of interest.
3. There can be no heat transferred into or out of the fluid.
4. There can be no energy lost due to friction.

In several example problems in Chapter 6 the assumption that these restrictions are satisfied would produce a negligible error in the result. However, for a flow system such as that shown in Fig. 7.1, there are definitely some energy losses and additions between the two sections of interest. For systems such as this, Bernoulli's equation is not valid.

Although details of how to calculate the magnitude of energy losses and additions will be presented later, the general conditions under which they would occur are described in this section.

FIGURE 7.1 Typical pipeline installation, showing a pump, valves, tees, and other fittings. (Source of photo: Ingersoll-Rand Co., Phillipsburg, NJ)

7.2.1
Mechanical Devices

With regard to their effect on a flow system, mechanical devices can be classified according to whether the device delivers energy to the fluid or the fluid delivers energy to the device.

A pump is a common example of a mechanical device that adds energy to a fluid. An electric motor or some other prime power device drives a rotating shaft in the pump. The pump then takes this kinetic energy and delivers it to the fluid, resulting in increased fluid pressure and fluid flow.

Many configurations are used in pump designs. The system pictured in Fig. 7.1 contains a centrifugal pump mounted inline with the process piping. Figures 7.2 and 7.3 show two types of fluid power pumps capable of produc-

FIGURE 7.2 Gear pump. (Source of photo: Danfoss Fluid Power, a division of Danfoss, Inc., Racine, WI; source of drawing: *Machine Design Magazine*)

(a) Cutaway   (b) Sketch of flow path

FIGURE 7.3 Piston pump. (Source of photo: Danfoss Fluid Power, a division of Danfoss, Inc., Racine, WI; source of drawing: *Machine Design Magazine*)

(a) Cutaway   (b) Sketch of flow path

ing very high pressures in the range from 1500 to 5000 psi (10.3 to 34.5 MPa). Chapter 15 gives an extensive discussion of these and several other styles of pumps along with their selection and application.

Fluid motors, turbines, rotary actuators, and linear actuators are examples of devices that take energy from a fluid and deliver it in the form of work, causing the rotation of a shaft or the linear movement of a piston.

Many fluid motors have the same basic configurations as the pumps shown in Figs. 7.2 and 7.3. The major difference between a pump and a fluid motor is that, when acting as a motor, the fluid drives the rotating elements of the device. The reverse is true for pumps. For some designs, such as the gear-on-gear type in Fig. 7.2, a pump could act as a motor by forcing a flow of fluid through the device. In other types, a change in the valve arrangement or in the configuration of the rotating elements would be required.

The hydraulic motor shown in Fig. 7.4 is often used as a drive for the wheels of construction equipment and trucks, and for rotating components of material transfer systems, conveyors, agricultural equipment, special machines, and automation equipment. The design incorporates a stationary internal gear with a special shape. The rotating component is like an external gear, sometimes called a *gerotor,* that has one fewer teeth than the internal gear. The external gear rotates in a circular orbit around the center of the internal gear. High pressure fluid entering the cavity between the two gears acts on the rotor and develops a torque that rotates the output shaft. The magnitude of the output torque depends on the pressure difference between the input and output sides of the rotating gear. The speed of rotation is a function of the displacement of the motor (volume per revolution) and the volume flow rate of fluid through the motor. Fig. 7.5 shows a photograph of a cutaway model of a fluid power cylinder or linear actuator.

7.2.2
Fluid Friction

A fluid in motion offers frictional resistance to flow. Part of the energy in the system is converted into *thermal energy* (heat), which is dissipated through the walls of the pipe in which the fluid is flowing. The magnitude of the energy loss is dependent on the properties of the fluid, the flow velocity, the pipe size, the smoothness of the pipe wall, and the length of the pipe. We will develop methods of calculating this frictional energy loss in later chapters.

Stationary
internal
gear

Rotor

(b) Rotor and internal gear

Rotor

Output shaft

(a) Cutaway

FIGURE 7.4   Hydraulic motor. (Source: Danfoss Fluid Power, a division of
Danfoss, Inc., Racine, WI)

Wear ring

Rod gland

Piston

Rod bushing

Rod seal and
Rod wiper seal

Steel head and caps   Piston seals

Cylinder tube   Tie rods and nuts   Piston rod

FIGURE 7.5   Fluid power cylinder. (Source of photo: Mosier Industries,
(Brookville, OH)

7.2.3
Valves and Fittings

Elements that control the direction or flow rate of a fluid in a system typically set up local turbulence in the fluid, causing energy to be dissipated as heat. Whenever there is a restriction, a change in flow velocity, or a change in the direction of flow, these energy losses occur. In a large system the magnitude of losses due to valves and fittings is usually small compared with frictional losses in the pipes. Therefore, such losses are referred to as *minor losses.*

7.3
NOMENCLATURE OF
ENERGY LOSSES
AND ADDITIONS

We will account for energy losses and additions in a system in terms of energy per unit weight of fluid flowing in the system. This is also known as "head," as described in Chapter 6. As an abbreviation for head we will use the symbol h for energy losses and additions. Specifically, we will use the following terms throughout the next several chapters:

h_A = *Energy added* to the fluid with a mechanical device such as a pump. This is often referred to as the *total head* on the pump.

h_R = *Energy removed* from the fluid by a mechanical device such as a fluid motor.

h_L = *Energy losses* from the system due to friction in pipes or minor losses due to valves and fittings.

We will not consider the effects of heat transferred into or out of the fluid at this time because they are negligible in the types of problems with which we are dealing.

The magnitude of energy losses produced by many kinds of valves and fittings is directly proportional to the velocity head of the fluid. This can be expressed mathematically as

$$h_L = K(v^2/2g)$$

The term K is the *resistance coefficient*, usually found experimentally. Chapter 10 gives more details about minor losses.

7.4
GENERAL ENERGY
EQUATION

The general energy equation as used in this text is an expansion of Bernoulli's equation, which makes it possible to solve problems in which energy losses and additions occur. The logical interpretation of the energy equation can be seen in Fig. 7.6, which represents a flow system. The terms E_1' and E_2' denote the energy possessed by the fluid per unit weight at sections 1 and 2, respectively. The energy additions, removals, and losses, h_A, h_R, and h_L, are shown. For such a system the expression of the principle of conservation of energy is

$$E_1' + h_A - h_R - h_L = E_2' \qquad (7\text{–}1)$$

The energy possessed by the fluid per unit weight is

$$E' = \frac{p}{\gamma} + z + \frac{v^2}{2g} \qquad (7\text{–}2)$$

FIGURE 7.6   Fluid flow system illustrating the general energy equation.

Equation (7–1) then becomes

GENERAL ENERGY EQUATION

$$\frac{p_1}{\gamma} + z_1 + \frac{v_1^2}{2g} + h_A - h_R - h_L = \frac{p_2}{\gamma} + z_2 + \frac{v_2^2}{2g} \tag{7–3}$$

This is the form of the energy equation that we will use most often in this book. As with Bernoulli's equation, each term in Eq. (7–3) represents a quantity of energy per unit weight of fluid flowing in the system. Typical SI units are N·m/N, or meters. U.S. Customary System units are lb-ft/lb or ft.

    It is essential that the general energy equation be written *in the direction of flow*, that is, *from* the reference point on the left side of the equation *to* that on the right side. Algebraic signs are critical because the left side of Equation (7–3) states that an element of fluid having a certain amount of energy per unit weight at section 1 may have energy added ($+h_A$), energy removed ($-h_R$), or energy lost ($-h_L$) from it before it reaches section 2. There it contains a different amount of energy per unit weight as indicated by the terms on the right side of the equation.

    For example, in Fig. 7.6, reference points are shown to be points 1 and 2 with the pressure head, elevation head, and velocity head indicated at each point. After the fluid leaves point 1 it enters the pump where energy is added. A prime mover such as an electric motor drives the pump and the impeller of the pump transfers the energy to the fluid ($+h_A$). Then the fluid flows through a piping system composed of a valve, elbows, and lengths of pipe in which energy is dissipated from the fluid and is lost ($-h_L$). Before reaching point 2, the fluid flows through a fluid motor which removes some of the energy to drive an external device $(-h_R)$. The general energy equation accounts for all of these energies.

In a particular problem, it is possible that not all of the terms in the general energy equation will be required. For example, if there is no mechanical device between the sections of interest, the terms h_A and h_R will be zero and can be left out of the equation. If energy losses are so small that they can be neglected, the term h_L can be left out. If both of these conditions exist, it can be seen that Eq. (7–3) reduces to Bernoulli's equation.

PROGRAMMED EXAMPLE PROBLEMS

☐ **EXAMPLE PROBLEM 7.1** Water flows from a large reservoir at the rate of 1.20 ft³/s through a pipe system as shown in Fig. 7.7. Calculate the total amount of energy lost from the system because of the valve, the elbows, the pipe entrance, and fluid friction.

FIGURE 7.7 Pipe system for Example Problem 7.1.

Using an approach similar to that used with Bernoulli's equation, select the two sections of interest and write the general energy equation before looking at the next panel.

The sections at which we know the most information about pressure, velocity, and elevation are the surface of the reservoir and the free stream of fluid at the exit of the pipe. Call these section 1 and section 2, respectively. Then, the complete general energy equation [Eq. (7–3)] is

$$\frac{p_1}{\gamma} + z_1 + \frac{v_1^2}{2g} + h_A - h_R - h_L = \frac{p_2}{\gamma} + z_2 + \frac{v_2^2}{2g}$$

The value of some of these terms is zero. Determine which are zero and simplify the energy equation accordingly.

The value of the following terms is zero:

$$p_1 = 0 \qquad \text{Surface of reservoir exposed to the atmosphere}$$
$$p_2 = 0 \qquad \text{Free stream of fluid exposed to the atmosphere}$$
$$v_1 = 0 \qquad \text{(Approximately) Surface area of reservoir is large}$$
$$h_A = h_R = 0 \qquad \text{No mechanical device in the system}$$

Then, the energy equation becomes

$$z_1 - h_L = z_2 + v_2^2/2g$$

Since we are looking for the total energy lost from the system, solve this equation for h_L.

You should have

$$h_L = (z_1 - z_2) - v_2^2/2g$$

Now evaluate the terms on the right side of the equation to determine h_L in the units of lb-ft/lb.

The answer is $h_L = 15.75$ lb-ft/lb. Here is how it is done:

$$z_1 - z_2 = \ + 25 \text{ ft}$$
$$v_2 = Q/A_2$$

Since Q was given as 1.20 ft³/s and the area of a 3-in diameter jet is 0.0491 ft², we have

$$v_2 = \frac{Q}{A_2} = \frac{1.20 \text{ ft}^3}{\text{s}} \times \frac{1}{0.0491 \text{ ft}^2} = 24.4 \text{ ft/s}$$

$$\frac{v_2^2}{2g} = \frac{(24.4)^2 \text{ ft}^2}{\text{s}^2} \times \frac{\text{s}^2}{(2)(32.2) \text{ ft}} = 9.25 \text{ ft}$$

Then the total amount of energy lost from the system is

$$h_L = (z_1 - z_2) - v_2^2/2g = 25 \text{ ft} - 9.25 \text{ ft}$$
$$h_L = 15.75 \text{ ft, or } 15.75 \text{ lb-ft/lb}$$

☐ **EXAMPLE PROBLEM 7.2**   The volume flow rate through the pump shown in Fig. 7.8 is 0.014 m³/s. The fluid being pumped is oil with a specific gravity of 0.86. Calculate the energy delivered by the pump to the oil per unit weight of oil flowing in the system. Neglect any energy losses in the system.

   Using the sections where the pressure gages are located as the sections of interest, write the energy equation for the system, including only the necessary terms.

FIGURE 7.8 Pump system for Example Problem 7.2.

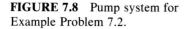

FIGURE 7.8 Pump system for
Example Problem 7.2.

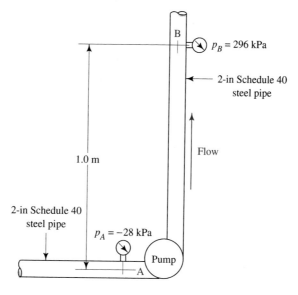

You should have

$$\frac{p_A}{\gamma} + z_A + \frac{v_A^2}{2g} + h_A = \frac{p_B}{\gamma} + z_B + \frac{v_B^2}{2g}$$

Notice that the terms h_R and h_L have been left out of the general energy equation.

The objective of the problem is to calculate the energy added to the oil by the pump. Solve for h_A before looking at the next panel.

One correct solution is

$$h_A = \frac{p_B - p_A}{\gamma} + (z_B - z_A) + \frac{v_B^2 - v_A^2}{2g} \qquad (7-4)$$

Notice that similar terms have been grouped. This will be convenient when performing the calculations.

What is the value of the term $(v_B^2 - v_A^2)/2g$?

It is zero. Notice that the pipe size is the same at section A and at section B. The volume flow rate at each point is also the same. Then, since $v = Q/A$, we can conclude that $v_A = v_B$. Now evaluate $z_B - z_A$.

You should have $z_B - z_A = 1.0$ m. Now, in the remaining term, $(p_B - p_A)/\gamma$, what is the value of γ?

Remember that the specific weight of the fluid being pumped must be used. In this case, the specific weight of the oil is

$$\gamma = (sg)(\gamma_w) = (0.86)(9.81 \text{ kN/m}^3) = 8.44 \text{ kN/m}^3$$

Now complete the evaluation of $(p_B - p_A)/\gamma$.

Since $p_B = 296$ kPa and $p_A = -28$ kPa, we have

$$\frac{p_B - p_A}{\gamma} = \frac{[296 - (-28)] \text{ kN}}{\text{m}^2} \times \frac{\text{m}^3}{8.44 \text{ kN}} = 38.4 \text{ m}$$

We can now calculate h_A from Eq. (7–4).

The energy added to the system is

$$h_A = 38.4 \text{ m} + 1.0 \text{ m} + 0 \text{ m} = 39.4 \text{ m, or } 39.4 \text{ N·m/N}$$

That is, the pump delivers 39.4 N·m of energy to each newton of oil flowing through it.

This completes the programmed instruction.

■

7.5
POWER REQUIRED
BY PUMPS

Power is defined as the rate of doing work. In fluid mechanics we can modify this statement and consider that power is the rate at which energy is being transferred. The unit for power in the SI system is the watt (W), which is equivalent to 1.0 N·m/s.

In Example Problem 7.2 we found that the pump was delivering 39.4 N·m of energy to each newton of oil as it flowed through the pump. In order to calculate the power delivered to the oil, we must determine how many newtons of oil are flowing through the pump in a given amount of time. This is called the *weight flow rate W*, which we defined in Chapter 6, and is expressed in units of N/s. Power is calculated by multiplying the energy transferred per newton of fluid by the weight flow rate. This is

$$P_A = h_A W$$

But since $W = \gamma Q$, we can also write

▷ POWER ADDED TO A FLUID BY
A PUMP

$$P_A = h_A \gamma Q \qquad\qquad (7\text{–}5)$$

where P_A denotes power added to the fluid, γ is the specific weight of the fluid flowing through the pump, and Q is the volume flow rate of the fluid.

Using the data of Example Problem 7.2, we can find the power delivered by the pump to the oil as follows:

$$P_A = h_A \gamma Q$$

We know from Example Problem 7.2 that

$$h_A = 39.4 \text{ N} \cdot \text{m/N}$$
$$\gamma = 8.44 \text{ kN/m}^3$$
$$Q = 0.014 \text{ m}^3/\text{s}$$

Substituting these values into Eq. (7–5), we get

$$P_A = \frac{39.4 \text{ N} \cdot \text{m}}{\text{N}} \times \frac{8.44 \times 10^3 \text{ N}}{\text{m}^3} \times \frac{0.014 \text{ m}^3}{\text{s}} = 4660 \text{ N} \cdot \text{m/s}$$

Expressing the result in watts gives

$$P_A = 4660 \text{ W} = 4.66 \text{ kW}$$

7.5.1
Power in the U.S.
Customary System

The unit for power in the U.S. Customary System is lb-ft/s. Since it is common practice to refer to power in horsepower (hp), the conversion factor required is

$$1 \text{ hp} = 550 \text{ lb-ft/s}$$

In Eq. (7–5) the energy added, h_A, is expressed in feet of the fluid flowing in the system. Then, expressing the specific weight of the fluid in lb/ft^3 and the volume flow rate in ft^3/s would yield the weight flow rate γQ in lb/s. Finally, in the power equation $P_A = h_A \gamma Q$, power would be expressed in lb-ft/s.

To convert these units to the SI system we use the factors

$$1 \text{ lb-ft/s} = 1.356 \text{ W}$$
$$1 \text{ hp} = 745.7 \text{ W}$$

7.5.2
Mechanical Efficiency
of Pumps

The term *efficiency* is used to denote the ratio of the power delivered by the pump to the fluid to the power supplied to the pump. Because of energy losses due to mechanical friction in pump components, fluid friction in the pump, and excessive fluid turbulence in the pump, not all of the input power is delivered to the fluid. Then, using the symbol e_M for mechanical efficiency, we have

▷ PUMP EFFICIENCY

$$e_M = \frac{\text{Power delivered to fluid}}{\text{Power put into pump}} = \frac{P_A}{P_I} \qquad (7\text{–}6)$$

The value of e_M will always be less than 1.0.

Continuing with the data of Example Problem 7.2, we could calculate the power input to the pump if e_M is known. For commercially available pumps the value of e_M is published as a part of the performance data. If we assume that for the pump in this problem the efficiency is 82 percent, then

$$P_I = P_A/e_M = 4.66/0.82 = 5.68 \text{ kW}$$

The value of the mechanical efficiency of pumps depends not only on the design of the pump but also on the conditions under which it is operating, particularly the total head and the flow rate. For pumps used in hydraulic systems, such as those shown in Figs. 7.2 and 7.3, efficiencies range from about 70 to 90 percent. For centrifugal pumps used primarily to transfer or circulate liquids, the efficiencies range from about 50 to 85 percent.

The following programmed example problem illustrates a possible setup for measuring pump efficiency.

PROGRAMMED EXAMPLE PROBLEM

□ **EXAMPLE PROBLEM 7.3**   For the pump test arrangement shown in Fig. 7.9, determine the mechanical efficiency of the pump if the power input is measured to be 3.85 hp when pumping 500 gal/min of oil (γ = 56.0 lb/ft³).

FIGURE 7.9 Pump test system for Example Problem 7.3.

To begin, write the energy equation for this system.

Using the points identified as 1 and 2 in Fig. 7.9, we have

$$\frac{p_1}{\gamma} + z_1 + \frac{v_1^2}{2g} + h_A = \frac{p_2}{\gamma} + z_2 + \frac{v_2^2}{2g}$$

Because we must find the power delivered by the pump to the fluid, we should now solve for h_A.

This equation is used:

$$h_A = \frac{p_2 - p_1}{\gamma} + (z_2 - z_1) + \frac{v_2^2 - v_1^2}{2g} \qquad (7\text{--}7)$$

It is convenient to solve for each term individually and then combine the results. The manometer enables us to calculate $(p_2 - p_1)/\gamma$ because it measures the pressure difference. Using the procedure outlined in Chapter 3, write the manometer equation between points 1 and 2.

Starting at point 1, we have

$$p_1 + \gamma_o y + \gamma_m(20.4 \text{ in}) - \gamma_o(20.4 \text{ in}) - \gamma_o y = p_2$$

where y is the unknown distance from point 1 to the top of the mercury column in the left leg of the manometer. The terms involving y cancel out. Also, in this equation γ_o is the specific weight of the oil and γ_m is the specific weight of the mercury gage fluid.

The desired result for use in Eq. (7–7) is $(p_2 - p_1)/\gamma_o$. Solve for this now and compute the result.

The correct solution is $(p_2 - p_1)/\gamma_o = 24.0$ ft. Here is one way to do it:

$$\gamma_m = (13.54)(\gamma_w) = (13.54)(62.4 \text{ lb/ft}^3) = 844.9 \text{ lb/ft}^3$$

$$p_2 = p_1 + \gamma_m(20.4 \text{ in}) - \gamma_o(20.4 \text{ in})$$

$$p_2 - p_1 = \gamma_m(20.4 \text{ in}) - \gamma_o(20.4 \text{ in})$$

$$\frac{p_2 - p_1}{\gamma_o} = \frac{\gamma_m(20.4 \text{ in})}{\gamma_o} - 20.4 \text{ in} = \left(\frac{\gamma_m}{\gamma_o} - 1\right)20.4 \text{ in}$$

$$= \left(\frac{844.9 \text{ lb/ft}^3}{56.0 \text{ lb/ft}^3} - 1\right)20.4 \text{ in} = (15.1 - 1)(20.4 \text{ in})$$

$$= (14.1)(20.4 \text{ in})\left(\frac{1 \text{ ft}}{12 \text{ in}}\right) = 24.0 \text{ ft}$$

The next term in Eq. (7–7) is $z_2 - z_1$. What is its value?

It is zero. Both points are on the same elevation. Now find $(v_2^2 - v_1^2)/2g$.

You should have $(v_2^2 - v_1^2)/2g = 1.99$ ft.

$$Q = 500 \text{ gal/min} \left(\frac{1 \text{ ft}^3/\text{s}}{449 \text{ gal/min}}\right) = 1.11 \text{ ft}^3/\text{s}$$

Using $A_1 = 0.2006 \text{ ft}^2$ and $A_2 = 0.0884 \text{ ft}^2$ from Appendix F, we get

$$v_1 = \frac{Q}{A_1} = \frac{1.11 \text{ ft}^3}{\text{s}} \times \frac{1}{0.2006 \text{ ft}^2} = 5.55 \text{ ft/s}$$

$$v_2 = \frac{Q}{A_2} = \frac{1.11 \text{ ft}^3}{\text{s}} \times \frac{1}{0.0884 \text{ ft}^2} = 12.6 \text{ ft/s}$$

$$\frac{v_2^2 - v_1^2}{2g} = \frac{(12.6)^2 - (5.55)^2}{(2)(32.2)} \frac{\text{ft}^2}{\text{s}^2} \frac{\text{s}^2}{\text{ft}} = 1.99 \text{ ft}$$

Now place these results into Eq. (7–7) and solve for h_A.

Solving for h_A, we get

$$h_A = 24.0 \text{ ft} + 0 + 1.99 \text{ ft} = 25.99 \text{ ft}$$

We can now calculate the power delivered to the oil, P_A.

The result is $P_A = 2.95$ hp.

$$P_A = h_A \gamma Q = 25.99 \text{ ft} \left(\frac{56.0 \text{ lb}}{\text{ft}^3}\right)\left(\frac{1.11 \text{ ft}^3}{\text{s}}\right)$$

$$= 1620 \text{ lb-ft/s} \left(\frac{1 \text{ hp}}{550 \text{ lb-ft/s}}\right) = 2.95 \text{ hp}$$

The final step is to calculate e_M, the mechanical efficiency of the pump.

From Eq. (7–6) we get

$$e_M = P_A/P_I = 2.95/3.85 = 0.77$$

Expressed as a percentage, the pump is 77 percent efficient at the stated conditions. This completes the programmed instruction.

■

7.6
POWER DELIVERED
TO FLUID MOTORS

The energy delivered by the fluid to a mechanical device such as a fluid motor or turbine is denoted in the general energy equation by the term h_R. This is a measure of the energy delivered by each unit weight of fluid as it passes through the device. We find the power delivered by multiplying h_R by the weight flow rate W:

POWER REMOVED FROM A FLUID
BY A MOTOR

$$P_R = h_R W = h_R \gamma Q \qquad (7\text{–}8)$$

where P_R is the power delivered by the fluid to the fluid motor.

7.6.1
Mechanical Efficiency
of Fluid Motors

As was described for pumps, energy losses in a fluid motor are produced by mechanical and fluid friction. Therefore, not all the power delivered to the motor is ultimately converted to power output from the device. Mechanical efficiency is then defined as

MOTOR EFFICIENCY

$$e_M = \frac{\text{Power output from motor}}{\text{Power delivered by fluid}} = \frac{P_O}{P_R} \qquad (7\text{–}9)$$

Here again, the value of e_M is always less than 1.0.
Refer to Section 7.5 for power units.

PROGRAMMED EXAMPLE PROBLEM

☐ **EXAMPLE PROBLEM 7.4**

Water at 10°C is flowing at a rate of 115 L/min through the fluid motor shown in Fig. 7.10. The pressure at A is 700 kPa and the pressure at B is 125 kPa. It is estimated that due to friction in the piping there is an energy loss of 4.0 N·m/N of water

FIGURE 7.10 Fluid motor for Example Problem 7.4.

flowing. (a) Calculate the power delivered to the fluid motor by the water. (b) If the mechanical efficiency of the fluid motor is 85 percent, calculate the power output.

Start the solution by writing the energy equation.

Choosing points A and B as our reference points, we get

$$\frac{p_A}{\gamma} + z_A + \frac{v_A^2}{2g} - h_R - h_L = \frac{p_B}{\gamma} + z_B + \frac{v_B^2}{2g}$$

The value of h_R is needed to determine the power output. Solve the energy equation for this term.

Compare this equation with your result:

$$h_R = \frac{p_A - p_B}{\gamma} + (z_A - z_B) + \frac{v_A^2 - v_B^2}{2g} - h_L \qquad \textbf{(7–10)}$$

Before looking at the next panel, solve for the value of each term in this equation using the unit of N·m/N or m.

The correct results are

1. $\dfrac{p_A - p_B}{\gamma} = \dfrac{(700 - 125)(10^3)\text{N}}{\text{m}^2} \times \dfrac{\text{m}^3}{9.81 \times 10^3 \text{ N}} = 58.6 \text{ m}$

2. $z_A - z_B = 1.8 \text{ m}$

3. Solving for $(v_A^2 - v_B^2)/2g$,

$$Q = 115 \text{ L/min} \times \frac{1.0 \text{ m}^3/\text{s}}{60\,000 \text{ L/min}} = 1.92 \times 10^{-3} \text{ m}^3/\text{s}$$

$$v_A = \frac{Q}{A_A} = \frac{1.92 \times 10^{-3} \text{ m}^3}{\text{s}} \times \frac{1}{4.909 \times 10^{-4} \text{ m}^2} = 3.91 \text{ m/s}$$

$$v_B = \frac{Q}{A_B} = \frac{1.92 \times 10^{-3} \text{ m}^3}{\text{s}} \times \frac{1}{4.418 \times 10^{-3} \text{ m}^2} = 0.43 \text{ m/s}$$

$$\frac{v_A^2 - v_B^2}{2g} = \frac{(3.91)^2 - (0.43)^2}{(2)(9.81)} \frac{\text{m}^2}{\text{s}^2} \frac{\text{s}^2}{\text{m}} = 0.77 \text{ m}$$

4. $h_L = 4.0$ m

Complete the solution of Eq. (7–10) for h_R now.

The energy delivered by the water to the turbine is

$$h_R = (58.6 + 1.8 + 0.77 - 4.0) \text{ m} = 57.2 \text{ m}$$

To complete part (a) of the problem, calculate P_R.

Substituting the known values into Eq. (7–8), we get

$$P_R = h_R \gamma Q$$

$$= 57.2 \text{ m} \times \frac{9.81 \times 10^3 \text{ N}}{\text{m}^3} \times \frac{1.92 \times 10^{-3} \text{ m}^3}{\text{s}} = 1080 \text{ N} \cdot \text{m/s}$$

$$= 1.08 \text{ kW}$$

This is the power delivered to the fluid motor by the water. How much useful power can be expected to be put out by the motor?

Because the efficiency of the motor is 85 percent, we get 0.92 kW of power out. Using Eq. (7–9), $e_M = P_O/P_R$, we get

$$P_O = e_M P_R$$
$$= (0.85)(1.08 \text{ kW})$$
$$= 0.92 \text{ kW}$$

This completes the programmed example problem.

∎

PRACTICE PROBLEMS

It may be necessary to refer to the appendix for data concerning the dimensions of pipes or the properties of fluids. Assume there are no energy losses unless stated otherwise.

7.1E A horizontal pipe carries oil with a specific gravity of 0.83. If two pressure gages along the pipe read 74.6 psig and 62.2 psig, respectively, calculate the energy loss between the two gages.

7.2E Water at 40°F is flowing downward through the pipe shown in Fig. 7.11. At point A the velocity is 10 ft/s

and the pressure is 60 psig. The energy loss between points A and B is 25 lb-ft/lb. Calculate the pressure at point B.

7.3M *Find the volume flow rate of water exiting from the tank shown in Fig. 7.12. The tank is sealed with a pressure of 140 kPa above the water. There is an energy loss of 2.0 N·m/N as the water flows through the nozzle.*

7.4M *A long 6-in Schedule 40 steel pipe discharges 0.085*

FIGURE 7.11   Problem 7.2.

FIGURE 7.12   Problem 7.3.

m^3/s of water from a reservoir into the atmosphere as shown in Fig. 7.13. Calculate the energy loss in the pipe.

7.5E Figure 7.14 shows a setup to determine the energy loss due to a certain piece of apparatus. The inlet is through a 2-in Schedule 40 pipe and the outlet is a 4-in Schedule 40 pipe. Calculate the energy loss between points A and B if water is flowing upward at 0.20 ft^3/s. The gage fluid is mercury (sg = 13.54).

7.6E A test setup to determine the energy loss as water flows through a valve is shown in Fig. 7.15. Calcu-

late the energy loss if 0.10 ft^3/s of water at 40°F is flowing. Also, calculate the *resistance* coefficient *K* if the energy loss is expressed as $K(v^2/2g)$.

7.7M *The setup shown in Fig. 7.16 is being used to measure the energy loss across a valve. The velocity of flow of the oil is 1.2 m/s. Calculate the value of K if the energy loss is expressed as $K(v^2/2g)$.*

7.8M *A pump is being used to transfer water from an open tank to one that has air at 500 kPa above the water, as shown in Fig. 7.17. If 2250 L/min are being pumped, calculate the power delivered by the pump to the water. Assume that the level of the surface in each tank is the same.*

FIGURE 7.13   Problem 7.4.

FIGURE 7.14   Problem 7.5.

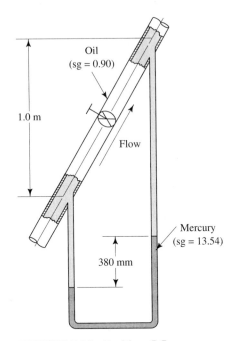

FIGURE 7.16   Problem 7.7.

7.9M *In Problem 7.8 (Fig. 7.17), if the left-hand tank were also sealed and air pressure above the water is 68 kPa, calculate the pump power.*

7.10E A commercially available sump pump is capable of delivering 2800 gal/h of water through a vertical lift of 20 ft. The inlet to the pump is just below the water surface and the discharge is to the atmosphere through a 1¼-in Schedule 40 pipe. (a) Calcu-

late the power delivered by the pump to the water. (b) If the pump draws 0.5 hp, calculate its efficiency.

7.11E A submersible deep-well pump delivers 745 gal/h of water through a 1-in Schedule 40 pipe when operating in the system sketched in Fig. 7.18. An energy

FIGURE 7.15   Problem 7.6.

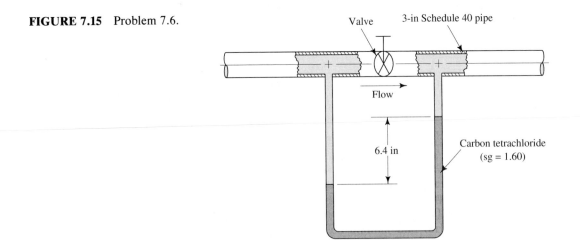

FIGURE 7.17 Problems 7.8 and 7.9.

FIGURE 7.18 Problem 7.11.

FIGURE 7.19 Problem 7.13.

loss of 10.5 lb-ft/lb occurs in the piping system. (a) Calculate the power delivered by the pump to the water. (b) If the pump draws 1 hp, calculate its efficiency.

7.12M *In a pump test the suction pressure at the pump inlet is 30 kPa below atmospheric pressure. The discharge pressure at a point 750 mm above the inlet is 520 kPa. Both pipes are 75 mm in diameter. If the volume flow rate of water is 75 L/min, calculate the power delivered by the pump to the water.*

7.13M *The pump shown in Fig. 7.19 is delivering hydraulic oil with a specific gravity of 0.85 at a rate of 75 L/min. The pressure at A is −20 kPa while the pressure at B is 275 kPa. The energy loss in the system is 2.5 times the velocity head in the discharge pipe. Calculate the power delivered by the pump to the oil.*

7.14E The pump in Fig. 7.20 delivers water from the lower to the upper reservoir at the rate of 2.0 ft³/s. The energy loss between the suction pipe inlet and the pump is 6 lb-ft/lb and between the pump outlet and the upper reservoir is 12 lb-ft/lb. Both pipes are 6-in Schedule 40 steel pipe. Calculate (a) the pressure at the pump inlet, (b) the pressure at the pump outlet, (c) the total head on the pump, and (d) the power delivered by the pump to the water.

7.15E Repeat Problem 7.14, but assume that the level of the lower reservoir is 10 ft above the pump instead of below it. All other data remain the same.

7.16M *Figure 7.21 shows a pump delivering 840 L/min of crude oil (sg = 0.85) from an underground storage drum to the first stage of a processing system. (a) If the total energy loss in the system is 4.2 N·m/N of oil flowing, calculate the power delivered by the pump. (b) If the energy loss in the suction pipe is 1.4 N·m/N of oil flowing, calculate the pressure at the pump inlet.*

7.17M *Figure 7.22 shows a submersible pump being used to circulate 60 L/min of a water-based coolant (sg = 0.95) to the cutter of a milling machine. The outlet is through a ¾-in Schedule 40 steel pipe. Assuming a total energy loss due to the piping of*

FIGURE 7.20   Problems 7.14 and 7.15.

FIGURE 7.21   Problem 7.16.

FIGURE 7.22   Problem 7.17.

3.0 N·m/N, calculate the total head developed by the pump and the power delivered to the coolant.

7.18M *Figure 7.23 shows a small pump in an automatic washer discharging into a laundry sink. The washer tub is 525 mm in diameter and 250 mm deep. The average head above the pump is 375 mm as shown. The discharge hose has an inside diameter of 18 mm. The energy loss in the hose system is 0.22 N·m/N. If the pump empties the tub in 90 s, calculate the average total head on the pump.*

7.19E The water being pumped in the system shown in Fig. 7.24 discharges into a tank which is being

weighed. It is found that 556 lb of water are collected in 10 s. If the pressure at A is 2.0 psi below atmospheric pressure, calculate the horsepower delivered by the pump to the water.

7.20E A manufacturer's rating for a gear pump states that 0.85 hp is required to drive the pump when it is pumping 9.1 gal/min of oil (sg = 0.90) with a total head of 257 ft. Calculate the mechanical efficiency of the pump.

7.21M *The specifications for an automobile fuel pump state that it should pump 1 L of gasoline in 40 s with a suction pressure of 150 mm of mercury vacuum*

FIGURE 7.23   Problem 7.18.

FIGURE 7.24   Problem 7.19.

and a discharge pressure of 30 kPa. Assuming that the pump efficiency is 60 percent, calculate the power drawn from the engine. See Fig. 7.25. The suction and discharge lines are the same size.

7.22E Figure 7.26 shows the arrangement of a circuit for a hydraulic system. The pump draws oil with a specific gravity of 0.90 from a reservoir and delivers it to the hydraulic cylinder. The cylinder has an inside diameter of 5.0 in, and in 15 s the piston must travel 20 in while exerting a force of 11 000 lb. It is estimated that there are energy losses of 11.5 lb-ft/lb in the suction pipe and 35.0 lb-ft/lb in the discharge pipe. Both pipes are ⅜-in Schedule 80 steel pipes. Calculate:

a. The volume flow rate through the pump.
b. The pressure at the cylinder.
c. The pressure at the outlet of the pump.
d. The pressure at the inlet to the pump.
e. The power delivered to the oil by the pump.

7.23M *Calculate the power delivered to the hydraulic motor in Fig. 7.27 if the pressure at A is 6.8 MPa and the pressure at B is 3.4 MPa. The motor inlet is a 1-in steel tube (0.065-in wall thickness) and the outlet is a 2-in steel tube (0.065-in wall thickness). The fluid is oil (sg = 0.90) and the velocity of flow is 1.5 m/s at point B.*

7.24E Water flows through the turbine shown in Fig. 7.28 at a rate of 3400 gal/min when the pressure at A is 21.4 psig and the pressure at B is −5 psig. The friction energy loss between A and B is twice the velocity head in the 12-in pipe. Determine the power delivered by the water to the turbine.

7.25M *Calculate the power delivered by the oil to the fluid motor shown in Fig. 7.29 if the volume flow rate is 0.25 m³/s. There is an energy loss of 1.4 N·m/N in the piping system. If the motor has an efficiency of 75 percent, calculate the power output.*

7.26E What hp must the pump shown in Fig. 7.30 deliver to a fluid having a specific weight of 60.0 lb/ft³ if energy losses of 3.40 lb-ft/lb occur between points 1 and 2? The pump delivers 40 gal/min of fluid.

7.27E If the pump in Problem 7.26 operates with an efficiency of 75 percent, what is the power input to the pump?

7.28M *The system shown in Fig. 7.31 delivers 600 L/min of water. The outlet is directly into the atmosphere. Determine the energy losses in the system.*

7.29M *Kerosene (sg = 0.823) flows at 0.060 m³/s in the pipe shown in Fig. 7.32. Compute the pressure at B if the total energy loss in the system is 4.60 N·m/N.*

7.30E Water at 60°F flows from a large reservoir through a turbine at the rate of 1000 gal/min in the system shown in Fig. 7.33. If the turbine removes 37 hp from the fluid, calculate the energy losses in the system.

FIGURE 7.25   Automobile fuel pump for Problem 7.21.

FIGURE 7.26   Problem 7.22.

7.31E Figure 7.34 shows a portion of a fire protection system in which a pump draws 1500 gal/min of water at 50°F from a reservoir and delivers it to a point B. The energy loss between the reservoir and point A at the inlet to the pump is 0.65 lb-ft/lb. Specify the

FIGURE 7.27   Problem 7.23.

FIGURE 7.28   Problem 7.24.

FIGURE 7.29   Problem 7.25.

required depth h to maintain at least 5.0 psig pressure at point A.

7.32E For the conditions of Problem 7.31, and if we assume that the pressure at point A is 5.0 psig, calculate the power delivered by the pump to the water

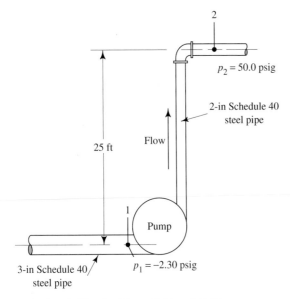

FIGURE 7.30   Problems 7.26 and 7.27.

FIGURE 7.31   Problem 7.28.

to maintain a pressure of 85 psig at point B. Energy losses between the pump and point B total 28.0 lb-ft/lb.

7.33M *In Fig. 7.35, kerosene at 25°C is flowing at 500 L/min from the lower tank to the upper tank through 2-in Type K copper tubing and a valve. If the pressure above the fluid is 15.0 psig, how much energy loss occurs in the system?*

7.34M *For the system shown in Fig. 7.35 and analyzed in Problem 7.33, assume that the energy loss is proportional to the velocity head in the tubing. Compute the pressure in the tank required to cause a flow of 1000 L/min.*

General Data for Problems 7.35E Through 7.40E

Figure 7.36 shows a diagram of a fluid power system for a hydraulic press used to extrude rubber parts. The following data are known:

1. The fluid is oil (sg = 0.93).
2. Volume flow rate is 175 gal/min.
3. Power input to the pump is 28.4 hp.
4. Pump efficiency is 80 percent.
5. Energy loss from point 1 to 2 is 2.80 lb-ft/lb.
6. Energy loss from point 3 to 4 is 28.50 lb-ft/lb.
7. Energy loss from point 5 to 6 is 3.50 lb-ft/lb.

7.35E Compute the power removed from the fluid by the press.

7.36E Compute the pressure at point 2 at the pump inlet.

7.37E Compute the pressure at point 3 at the pump outlet.

FIGURE 7.32   Problem 7.29.

20 m

B

3-in Schedule 40   3 m

FIGURE 7.33   Problem 7.30.

165 ft

Gate valve

Turbine

Flow

8-in Schedule 40
steel pipe

FIGURE 7.34   Problems 7.31
and 7.32.

B

8-in Schedule 40 steel pipe

25 ft

Flow

h

Pump

A

10-in Schedule 40 steel pipe

FIGURE 7.35 Problems 7.33 and 7.34.

7.38E Compute the pressure at point 4 at the press inlet.

7.39E Compute the pressure at point 5 at the press outlet.

7.40E Compare the velocity of flow in the suction line, in the discharge line, and in the return line of the system, using the recommendations listed in Table 6.3.

Are they satisfactory? If not, specify suitable pipe sizes.

7.41E The portable, pressurized fuel can shown in Fig. 7.37 is used to deliver fuel to a race car during a pit stop. What pressure must be above the fuel to de-

FIGURE 7.36 Problems 7.35–7.40.

FIGURE 7.37 Problem 7.41.

FIGURE 7.38 Problems 7.42 and 7.43.

1$\frac{1}{2}$-in Schedule 80 steel pipe

2$\frac{1}{2}$-in Schedule 80 steel pipe

Motor

Flow

38.5 in

Mercury
sg = 13.54

FIGURE 7.39   Problems 7.44 and 7.45.

liver 40 gal in 8.0 s? The specific gravity of the fuel is 0.76. An energy loss of 4.75 lb-ft/lb occurs at the nozzle.

7.42E Professor Crocker is building a cabin on a hillside and has proposed the water system shown in Fig. 7.38. The distribution tank in the cabin maintains a pressure of 30.0 psig above the water. There is an energy loss of 15.5 lb-ft/lb in the piping. When the pump is delivering 40 gal/min of water, compute the horsepower delivered by the pump to the water.

7.43E If Professor Crocker's pump, described in Problem 7.42, has an efficiency of 72 percent, what size motor is required to drive the pump?

7.44E The test setup in Fig. 7.39 measures the pressure difference between the inlet and outlet of the fluid motor. The flow rate of hydraulic oil (sg = 0.90) is 135 gal/min. Compute the power removed from the fluid by the motor.

7.45E If the fluid motor in Problem 7.44 has an efficiency of 78 percent, how much power is delivered by the motor?

8 Reynolds Number, Laminar Flow, and Turbulent Flow

8.1
OBJECTIVES

To compute the amount of energy loss due to friction in a fluid flow system, it is necessary to characterize the nature of the flow. Slow, smooth flow is called *laminar flow*, while rapid, chaotic flow is called *turbulent flow*. The methods used to compute the energy loss are different for each type of flow.

The type of flow can be predicted by computing a dimensionless number, the *Reynolds number*, relating the important variables describing the flow: velocity, size of the flow path, fluid density, and viscosity.

After completing this chapter, you should be able to:

1. Describe the appearance of laminar flow and turbulent flow.
2. State the relationship used to compute the Reynolds number.
3. Identify the limiting values of the Reynolds number by which you can predict whether flow is laminar or turbulent.
4. Compute the Reynolds number for the flow of fluids in round pipes and tubes.
5. Define the term *hydraulic radius* as it applies to the description of the size of noncircular flow paths.
6. Compute the Reynolds number for flow in noncircular flow paths.
7. Describe the velocity profile for laminar and turbulent flow.
8. Describe the laminar boundary layer as it occurs in turbulent flow.

8.2
LAMINAR FLOW AND TURBULENT FLOW

When analyzing a fluid in a flow stream, it is important to be able to determine the character of the flow. Under some conditions the fluid will appear to flow in layers in a smooth and regular manner. You can observe this by opening a water faucet slowly until the flow is smooth and steady. This type of flow is called *laminar flow*. If the water faucet is opened wider, allowing the velocity of flow to increase, a point would be reached when the flow is no longer smooth and regular. The water in the stream would appear to be moving in a rather chaotic manner. The flow would then be termed *turbulent flow*.

Figure 8.1 shows one way of visualizing laminar flow in a circular pipe. Concentric rings of fluid are flowing in a straight, smooth path. There is little or no mixing of the fluid across the "boundaries" of each layer as the flow proceeds through the pipe. In reality, however, an infinite number of layers makes up the flow.

Another way to visualize laminar flow is depicted in Fig. 8.2, which shows a transparent fluid, such as water, flowing in a clear glass tube. A

FIGURE 8.1   Illustration of laminar flow in a circular pipe.

particle or a stream of another fluid injected into the flow stream will proceed in a straight line and will not become mixed with the bulk of the fluid. The dark line in the figure is a colored fluid, such as a dye, that remains intact as long as the flow remains laminar.

Laminar flow also occurs in open channel flow, as shown in Fig. 8.3. In an open channel, laminar flow is called *tranquil*, a very descriptive term. Tranquil flow over a wall would appear as a smooth sheet of fluid, often seen in decorative fountains.

In contrast to laminar flow, turbulent flow appears chaotic and rough with much intermixing of the fluid. A dye stream introduced into turbulent flow would immediately dissipate throughout the primary fluid in the system, as shown in Fig. 8.4.

FIGURE 8.2   Dye stream in laminar flow.

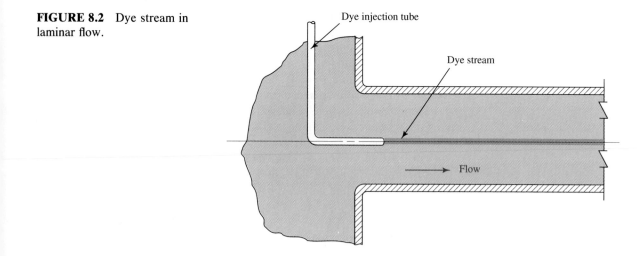

Dye injection tube

Dye stream

Flow

FIGURE 8.3 Tranquil (laminar) flow over a wall.

Indeed, an important reason for creating turbulent flow is to promote mixing in such applications as:

1. Blending two or more fluids.
2. Hastening chemical reactions.
3. Increasing heat transfer into or out of a fluid.

8.3 REYNOLDS NUMBER

The behavior of a fluid, particularly with regard to energy losses, is quite dependent on whether the flow is laminar or turbulent, as will be demonstrated in Chapter 9. For this reason we want to have a means of predicting

FIGURE 8.4 Dye stream mixing with turbulent flow.

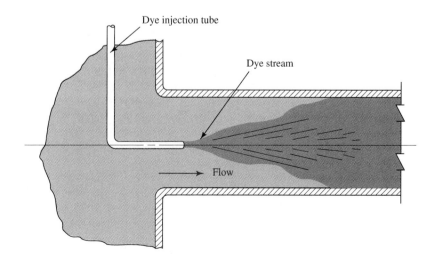

the type of flow without actually observing it. Indeed, direct observation is impossible for fluids in opaque pipes. It can be shown experimentally and verified analytically that the character of flow in a round pipe depends on four variables: fluid density ρ, fluid viscosity μ, pipe diameter D, and average velocity of flow v. Osborne Reynolds was the first to demonstrate that laminar or turbulent flow can be predicted if the magnitude of a dimensionless number, now called the Reynolds number (N_R), is known. Equation (8–1) shows the basic definition of the Reynolds number.

⇨ **REYNOLDS NUMBER—CIRCULAR SECTIONS**

$$N_R = \frac{vD\rho}{\mu} = \frac{vD}{\nu} \qquad (8\text{–}1)$$

These two forms of the equation are equivalent because $\nu = \mu/\rho$ as discussed in Chapter 2.

We can demonstrate that the Reynolds number is dimensionless by substituting standard SI units into Eq. (8–1):

$$N_R = \frac{vD\rho}{\mu} = v \times D \times \rho \times \frac{1}{\mu}$$

$$N_R = \frac{m}{s} \times m \times \frac{kg}{m^3} \times \frac{m \cdot s}{kg}$$

Because all units can be cancelled, N_R is dimensionless. However, it is essential that all terms in the equation be in consistent units in order to obtain the correct numerical value for N_R. In U.S. Customary System units, v should be in feet per second, D in feet, ρ in slugs/ft^3, μ in pound second per foot squared (lb·s/ft^2), and ν in foot squared per second (ft^2/s).

Reynolds number is one of several dimensionless numbers useful in the study of fluid mechanics and heat transfer. The process called *dimensional analysis* can be used to determine dimensionless numbers (see Reference 2). Reynolds number is the ratio of the inertia force on an element of fluid to the viscous force. The inertia force is developed from Newton's second law of motion, $F = ma$. As discussed in Chapter 2, the viscous force is related to the product of the shear stress times area.

Flows having large Reynolds numbers, typically because of high velocity and/or low viscosity, tend to be turbulent. Those fluids having high viscosity and/or moving at low velocities will have low Reynolds numbers and will tend to be laminar. The following section gives some quantitative data with which to predict whether a given flow system will be laminar or turbulent.

The formula for Reynolds number takes a different form for noncircular cross sections, open channels, and for the flow of fluid around immersed bodies. These situations are discussed elsewhere in this book.

8.4
CRITICAL REYNOLDS NUMBERS

For practical applications in pipe flow we find that if the Reynolds number for the flow is less than 2000, the flow will be laminar. Also, if the Reynolds number is greater than 4000, the flow can be assumed to be turbulent. In the range of Reynolds numbers between 2000 and 4000, it is impossible to predict which type of flow exists; therefore this range is called the *critical region*. Typical applications involve flows that are well within the laminar flow range or well within the turbulent flow range, so the existence of this region of uncertainty does not cause great difficulty. If the flow in a system is found to be in the critical region, the usual practice is to change the flow rate or pipe diameter to cause the flow to be definitely laminar or turbulent. More precise analysis is then possible.

By carefully minimizing external disturbances, it is possible to maintain laminar flow for Reynolds numbers as high as 50 000. However, when N_R is greater than about 4000, a minor disturbance of the flow stream will cause the flow to suddenly change from laminar to turbulent. For this reason, and because we are dealing with practical applications in this book, we will assume the following:

$$\text{If } N_R < 2000, \text{ the flow is laminar.}$$

$$\text{If } N_R > 4000, \text{ the flow is turbulent.}$$

☐ **EXAMPLE PROBLEM 8.1**   Determine whether the flow is laminar or turbulent if glycerine at 25°C flows in a pipe with a 150-mm inside diameter. The average velocity of flow is 3.6 m/s.

Solution   We must first evaluate the Reynolds number using Eq. (8–1):

$$N_R = vD\rho/\mu$$

$$v = 3.6 \text{ m/s}$$

$$D = 0.15 \text{ m}$$

$$\rho = 1258 \text{ kg/m}^3 \quad \text{(from Appendix B)}$$

$$\mu = 9.60 \times 10^{-1} \text{ Pa} \cdot \text{s} \quad \text{(from Appendix B)}$$

Then we have

$$N_R = \frac{(3.6)(0.15)(1258)}{9.60 \times 10^{-1}} = 708$$

Because $N_R = 708$, which is less than 2000, the flow is laminar. Notice that each term was converted to consistent SI units before N_R was evaluated.

_____

☐ **EXAMPLE PROBLEM 8.2**   Determine if the flow is laminar or turbulent if water at 70°C flows in a 1-in Type K copper tube with a flow rate of 285 L/min.

Solution   Evaluate the Reynolds number, using Eq. (8–1):

$$N_R = \frac{vD\rho}{\mu} = \frac{vD}{\nu}$$

For a 1-in Type K copper tube, $D = 0.02527$ m and $A = 5.017 \times 10^{-4}$ m^2 (from Appendix H). Then we have

$$v = \frac{Q}{A} = \frac{285 \text{ L/min}}{5.017 \times 10^{-4} \text{ m}^2} \times \frac{1 \text{ m}^3/\text{s}}{60\ 000 \text{ L/min}} = 9.47 \text{ m/s}$$

$$\nu = 4.11 \times 10^{-7} \text{ m}^2/\text{s} \quad \text{(from Appendix A)}$$

$$N_R = \frac{(9.47)(0.02527)}{4.11 \times 10^{-7}} = 5.82 \times 10^5$$

Because the Reynolds number is greater than 4000, the flow is turbulent.

□ **EXAMPLE PROBLEM 8.3**   Determine the range of average velocity of flow for which the flow would be in the critical region if SAE 10 oil at 60°F is flowing in a 2-in Schedule 40 steel pipe. The oil has a specific gravity of 0.89.

Solution   The flow would be in the critical region if $2000 < N_R < 4000$. First, we use the Reynolds number and solve for velocity:

$$N_R = \frac{\nu D \rho}{\mu}$$

$$\nu = \frac{N_R \mu}{D \rho} \tag{8-2}$$

Then we find the values for μ, D, and ρ:

$$D = 0.1723 \text{ ft} \quad \text{(from Appendix F)}$$

$$\mu = 2.10 \times 10^{-3} \text{ lb-s/ft}^2 \quad \text{(from Appendix D)}$$

$$\rho = (\text{sg})(1.94 \text{ slugs/ft}^3) = (0.89)(1.94 \text{ slugs/ft}^3) = 1.73 \text{ slugs/ft}^3$$

Substituting these values into Eq. (8-2), we get

$$\nu = \frac{N_R(2.10 \times 10^{-3})}{(0.1723)(1.73)} = (7.05 \times 10^{-3})N_R$$

For $N_R = 2000$, we have

$$\nu = (7.05 \times 10^{-3})(2 \times 10^3) = 14.1 \text{ ft/s}$$

For $N_R = 4000$, we have

$$\nu = (7.05 \times 10^{-3})(4 \times 10^3) = 28.2 \text{ ft/s}$$

Therefore, if $14.1 < \nu < 28.2$ ft/s, the flow will be in the critical region.

**8.5
VELOCITY PROFILES**   Unless otherwise stated, we assume that the term *velocity* indicates the average velocity of flow found from the continuity equation, $\nu = Q/A$. However, in some cases we must determine the fluid velocity at a point within the flow stream. The magnitude of velocity is by no means uniform across a particular section of a pipe, and the manner in which the velocity varies with position is dependent on the type of flow that exists, as shown in Fig. 8.5. We observed in Chapter 2 (Section 2.2) that the velocity of a fluid in contact with a stationary solid boundary is zero. The maximum velocity for any type of flow occurs at the center of the pipe. The reason for the different shapes of the velocity profiles is that, because of the rather chaotic motion and violent mixing of fluid molecules in turbulent flow, there is a transfer of momentum

FIGURE 8.5 Velocity profiles for pipe flow.

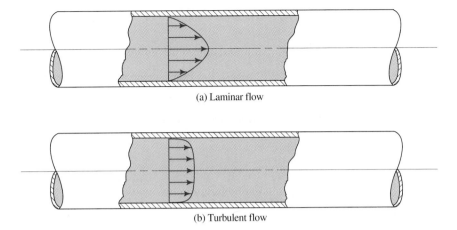

(a) Laminar flow

(b) Turbulent flow

between molecules resulting in a more uniform velocity distribution than in the case of laminar flow. Because laminar flow is essentially made up of layers of fluid, the momentum transfer between molecules is less and the velocity profile becomes parabolic.

Note in Fig. 8.5(b) that, even though the flow as a whole is turbulent, there exists a thin layer of fluid near the pipe wall where the velocity is quite small and in which the flow is actually laminar. This is known as the *boundary layer*. The actual thickness of the boundary layer and the velocity distribution within it are very important in analyzing heat transfer into the fluid or in determining the drag on bodies submerged in a fluid.

Because of the regularity of the velocity profile in laminar flow, we can define an equation for the local velocity at any point within the flow path. If we call the local velocity U at a radius r, the maximum radius r_o, and the average velocity v,

$$U = 2v[1 - (r/r_o)^2] \qquad (8-3)$$

☐ **EXAMPLE PROBLEM 8.4** Using the data from Example Problem 8.1, compute points on the velocity profile from the pipe wall to the middle of the pipe at increments of 15 mm. Plot the data for U versus r. Also, show the average velocity on the plot.

Solution From Example Problem 8.1, we find the Reynolds number to be 708, which indicates laminar flow. The average velocity of flow was given to be 3.60 m/s. Then, we compute r_o:

$$r_o = D/2 = 150/2 = 75 \text{ mm}$$

At $r = 75$ mm $= r_o$ at the pipe wall, $r/r_o = 1$ and $U = 0$ from Eq. (8–3). This is consistent with the observation that the velocity of a fluid at a solid boundary is equal to the velocity of that boundary.

At $r = 60$ mm,

$$U = 2(3.6 \text{ m/s})[1 - (60/75)^2] = 2.59 \text{ m/s}$$

If we use a similar technique, we can compute the following values:

r (mm)	r/r_o	U (m/s)	
75	1.0	0	(at the pipe wall)
60	0.8	2.59	
45	0.6	4.61	
30	0.4	6.05	
15	0.2	6.91	
0	0.0	7.20	(middle of the pipe)

Notice that the local velocity at the middle of the pipe is 2.0 times the average velocity. Figure 8.6 shows the plot of U versus r.

FIGURE 8.6 Results of Example Problems 8.4 and 8.5. Velocity profile for laminar flow.

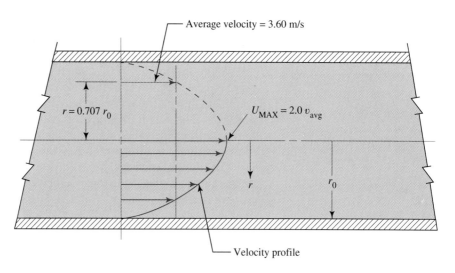

Average velocity = 3.60 m/s

$r = 0.707\, r_0$

$U_{MAX} = 2.0\, v_{avg}$

Velocity profile

□ **EXAMPLE PROBLEM 8.5**   Compute the radius at which the local velocity U would equal the average velocity v for laminar flow.

Solution   In Eq. (8–3), for the condition that $U = v$, we can first divide by U to obtain

$$1 = 2[1 - (r/r_o)^2]$$

Now, solving for r gives

$$r = \sqrt{0.5}\, r_o = 0.707 r_o \qquad (8\text{–}4)$$

For the data from Example Problem 8.1, the local velocity is equal to the average velocity 3.6 m/s at

$$r = 0.707(75 \text{ mm}) = 53.0 \text{ mm}$$

Computing the local velocity for points within turbulent flow requires knowledge of the *friction factor, f,* which is developed in Chapter 9. Further discussion of the velocity profile for turbulent flow will be presented there.

**8.6
HYDRAULIC RADIUS
FOR NONCIRCULAR
CROSS SECTIONS**

All equations and examples up to this point for computing the Reynolds number have been for fluid flowing in a full circular pipe. In such cases, the characteristic dimension of the flow path is the inside diameter. However, many practical fluid mechanics problems involve flow in noncircular cross sections.

Noncircular cross sections can be either closed conduits running full or open channels, one surface of which is exposed to the local atmosphere. One type is quite different from the other. Here, we will work with only the sections running full, because the analysis of energy losses presented in the next chapter is quite similar for either circular pipes or closed, noncircular cross sections running full. Open channel flow is discussed in Chapter 13, and a revised form of the Reynolds number will be presented there.

Examples of typical closed, noncircular cross sections are shown in Fig. 8.7. The sections shown could represent (a) a shell-and-tube heat exchanger, (b) and (c) air distribution ducts, and (d) a flow path inside a machine.

FIGURE 8.7 Examples of closed noncircular cross sections.

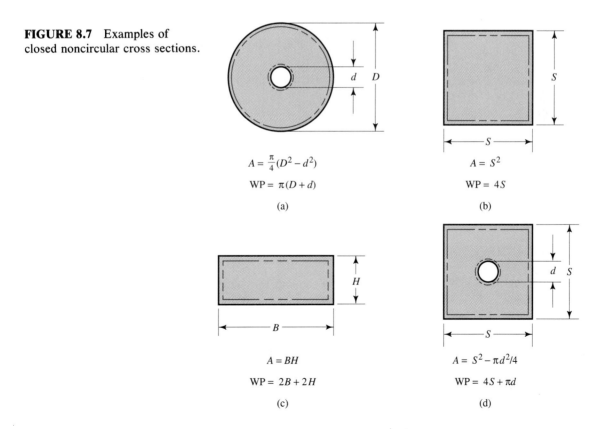

$$A = \frac{\pi}{4}(D^2 - d^2)$$

$$WP = \pi(D + d)$$

(a)

$$A = S^2$$

$$WP = 4S$$

(b)

$$A = BH$$

$$WP = 2B + 2H$$

(c)

$$A = S^2 - \pi d^2/4$$

$$WP = 4S + \pi d$$

(d)

The characteristic dimension of noncircular cross sections is called the *hydraulic radius, R,* defined as the ratio of the net cross-sectional area of a flow stream to the wetter perimeter of the section. That is,

HYDRAULIC RADIUS

$$R = \frac{A}{WP} = \frac{\text{area}}{\text{wetted perimeter}} \quad (8\text{--}5)$$

The unit for R is the meter in the SI unit system. In the U.S. Customary System, R is expressed in feet.

In the calculation of the hydraulic radius, the net cross-sectional area should be evident from the geometry of the section. The *wetted perimeter* is defined as the sum of the length of the boundaries of the section actually in contact with (that is, wetted by) the fluid. Expressions for the area A and the wetted perimeter WP are given in Fig. 8.7 for those sections illustrated. In each case, the fluid flows in the shaded portion of the section. A dashed line is shown adjacent to the boundaries that make up the wetted perimeter.

☐ **EXAMPLE PROBLEM 8.6**   Determine the hydraulic radius of the section shown in Fig. 8.7(d) if the inside dimension of each side of the square is 250 mm and the outside diameter of the tube is 150 mm.

Solution   The net flow area is the difference between the area of the square and the area of the circle:

$$A = S^2 - \pi d^2/4 = (250)^2 - \pi(150)^2/4 = 44\ 829 \text{ mm}^2$$

The wetted perimeter is the sum of the four sides of the square and the circumference of the circle:

$$WP = 4S + \pi d = 4(250) + \pi(150) = 1471 \text{ mm}$$

Then, the hydraulic radius R is

$$R = \frac{A}{WP} = \frac{44\ 829 \text{ mm}^2}{1471 \text{ mm}} = 30.5 \text{ mm} = 0.0305 \text{ m}$$

**8.7
REYNOLDS NUMBER
FOR CLOSED
NONCIRCULAR CROSS
SECTIONS**

When the fluid completely fills the available cross-sectional area and is under pressure, the average velocity of flow is determined by using the volume flow rate and the net flow area in the familiar continuity equation. That is,

$$v = Q/A$$

Note that the area is the same as that used to compute the hydraulic radius.

The Reynolds number for flow in noncircular sections is computed in a very similar manner to that used for circular pipes and tubes. The only alteration to Eq. (8–1) is the replacement of the diameter D with $4R$, four times the hydraulic radius. The result is

REYNOLDS NUMBER—
NONCIRCULAR SECTIONS

$$N_R = \frac{v(4R)\rho}{\mu} = \frac{v(4R)}{\nu} \quad (8\text{--}6)$$

The validity of this substitution can be demonstrated by calculating the hydraulic radius for a circular pipe:

$$R = \frac{A}{WP} = \frac{\pi D^2/4}{\pi D} = \frac{D}{4}$$

Then,

$$D = 4R$$

Therefore, $4R$ is equivalent to D for the circular pipe. Thus, by analogy, the use of $4R$ as the characteristic dimension for noncircular cross sections is appropriate. This approach will give reasonable results as long as the cross section has an *aspect ratio* not much different from that of the circular cross section. In this context, aspect ratio is the ratio of the width of the section to its height. So, for a circular section, the aspect ratio is 1.0. In Fig. 8.7, all the examples shown have reasonable aspect ratios.

An example of a shape that has an unacceptable aspect ratio is a rectangle for which the width is more than four times the height. (See Reference 2.) For such shapes, the hydraulic radius is approximately one-half the height. (See Reference 1.) Some annular shapes, similar to that shown in Fig. 8.7(a), would have high aspect ratios if the space between the two pipes was small. However, general data are not readily available for what constitutes a "small" space or for how to determine the hydraulic radius. Performance testing of such sections is recommended.

☐ **EXAMPLE PROBLEM 8.7**   Compute the Reynolds number for the flow of ethylene glycol at 25°C through the section shown in Fig. 8.7(d). The volume flow rate is 0.16 m³/s. Use the dimensions given in Example Problem 8.6.

Solution   The result for the hydraulic radius for the section from Example Problem 8.6 can be used: $R = 0.0305$ m. Now the Reynolds number can be computed from Eq. (8–6). We can use $\mu = 1.62 \times 10^{-2}$ Pa · s and $\rho = 1100$ kg/m³ (from Appendix B). The area must be converted to m².

$$A = (44\ 829\ \text{mm}^2)(1\ \text{m}^2/10^6\ \text{mm}^2) = 0.0448\ \text{m}^2$$

The average velocity of flow is

$$v = \frac{Q}{A} = \frac{0.16\ \text{m}^3/\text{s}}{0.0448\ \text{m}^2} = 3.57\ \text{m/s}$$

The Reynolds number can now be calculated:

$$N_R = \frac{v(4R)\rho}{\mu} = \frac{(3.57)(4)(0.0305)(1100)}{1.62 \times 10^{-2}}$$

$$N_R = 2.96 \times 10^4$$

REFERENCES

1. Crane Co. 1988. *Flow of Fluids Through Valves, Fittings, and Pipe*. Technical Paper No. 410. Joliet, IL: Author.

2. Fox, Robert W., and Alan T. McDonald. 1992. *Introduction to Fluid Mechanics*. 4th ed. New York: John Wiley & Sons, Inc.

PRACTICE PROBLEMS

The following problems require the use of the reference data listed below:

- Appendices A–C: Properties of liquids
- Appendix D: Dynamic viscosity of fluids
- Appendices F–J: Dimensions of pipe and tubing
- Appendix K: Conversion factors
- Appendix L: Properties of areas

8.1E A 4-in diameter pipe carries 0.20 ft³/s of glycerine (sg = 1.26) at 100°F. Is the flow laminar or turbulent?

8.2C Calculate the minimum velocity of flow in ft/s and m/s of water at 160°F in a 2-in diameter pipe for which the flow is turbulent.

8.3M *Calculate the maximum volume flow rate of fuel oil at 45°C at which the flow will remain laminar in a 100-mm-diameter pipe. For the fuel oil, use sg = 0.895 and dynamic viscosity = 4.0×10^{-2} Pa·s.*

8.4E Calculate the Reynolds number for the flow of each of the following fluids in a 2-in Schedule 40 steel pipe if the volume flow rate is 0.25 ft³/s: (a) water at 60°F, (b) acetone at 77°F, (c) castor oil at 77°F, and (d) SAE 10 oil at 210°F (sg = 0.87).

8.5M *Determine the smallest Type K copper tube size that will carry 4 L/min of the following fluids while maintaining laminar flow: (a) water at 40°C, (b) gasoline (sg = 0.68) at 25°C, (c) ethyl alcohol (sg = 0.79) at 0°C, and (d) heavy fuel oil at 25°C.*

8.6M *In an existing installation, SAE 10 oil (sg = 0.89) must be carried in a 3-in Schedule 40 steel pipe at the rate of 850 L/min. Efficient operation of a certain process requires that the Reynolds number of the flow be approximately 5×10^4. To what temperature must the oil be heated to accomplish this?*

8.7E From the data in Appendix C, we can see that automotive hydraulic oil and the medium machine tool hydraulic oil have nearly the same kinematic viscosity at 212°F. However, because of their different viscosity index, their viscosities at 104°F are quite different. Calculate the Reynolds number for the flow of each oil at each temperature in a 5-in Schedule 80 steel pipe at 10 ft/s velocity. Are the flows laminar or turbulent?

8.8M *Compute the Reynolds number for the flow of 325 L/min of water at 10°C in a standard 2-in steel tube with a wall thickness of 0.065 in. Is the flow laminar or turbulent?*

8.9M *Benzene (sg = 0.86) at 60°C is flowing at 25 L/min in a 1-in Schedule 80 steel pipe. Is the flow laminar or turbulent?*

8.10M *Hot water at 80°C is flowing to a dishwasher at a rate of 15.0 L/min through a ½-in Type K copper tube. Is the flow laminar or turbulent?*

8.11E A major water main is an 18-in ductile iron pipe. Compute the Reynolds number if the pipe carries 16.5 ft³/s of water at 50°F.

8.12E An engine crankcase contains SAE 10 motor oil (sg = 0.88). The oil is distributed to other parts of the engine by an oil pump through an ⅛-in steel tube with a wall thickness of 0.032 in. The ease with which the oil is pumped is obviously affected by its viscosity. Compute the Reynolds number for the flow of 0.40 gal/h of the oil at 40°F.

8.13E Repeat Problem 8.12 for an oil temperature of 160°F.

8.14E At approximately what volume flow rate will propyl alcohol at 77°F become turbulent when flowing in a 3-in Type K copper tube?

8.15M *SAE 30 oil (sg = 0.89) is flowing at 45 L/min through a ⅞-in steel tube with a wall thickness of 0.065 in. If the oil is at 110°C, is the flow laminar or turbulent?*

8.16M *Repeat Problem 8.15 for an oil temperature of 0°C.*

8.17M *Repeat Problem 8.15, except the tube is a 2-in size with a wall thickness of 0.065.*

8.18M *Repeat Problem 8.17 for an oil temperature of 0°C.*

8.19C The lubrication system for a punch press delivers 1.65 gal/min of a light lubricating oil (see Appendix

C) through 5/16-in steel tubes having a wall thickness of 0.049 in. Shortly after the press is started, the oil temperature is 104°F. Compute the Reynolds number for the oil flow.

8.20C After the press has run for some time, the lubricating oil described in Problem 8.19 heats to 212°F. Compute the Reynolds number for the oil flow at this temperature. Discuss the possible operating difficulty as the oil heats up.

8.21E A system is being designed to carry 500 gal/min of ethylene glycol at 77°F. Chapter 6 recommended that the maximum velocity should be approximately 10.0 ft/s. Specify the smallest standard Schedule 40 steel pipe to meet this condition. Then, for the selected pipe, compute the Reynolds number for the flow.

8.22E The range of Reynolds numbers between 2000 and 4000 is described as the *critical region*, because it is not possible to predict whether the flow is laminar or turbulent. One should avoid operation of fluid flow systems in this range. Compute the range of volume flow rates in gal/min of water at 60°F for which the flow would be in the critical region in a 3/4-in Type K copper tube.

8.23E The water line described in Problem 8.22 was a cold water distribution line. At another point in the system, the same size tube delivers water at 180°F. Compute the range of volume flow rates for which the flow would be in the critical region.

8.24C In a dairy, milk at 100°F is reported to have a kinematic viscosity of 1.30 centistokes. Compute the Reynolds number for the flow of the milk at 45 gal/min through a 1¼-in steel tube with a wall thickness of 0.065 in.

8.25C In a soft-drink bottling plant, the concentrated syrup used to make the drink has a kinematic viscosity of 17.0 centistokes at 80°F. Compute the Reynolds number for the flow of 215 L/min of the syrup through a 1-in Type K copper tube.

8.26C A certain jet fuel has a kinematic viscosity of 1.20 centistokes. If the fuel is being delivered to the engine at 200 L/min through a 1-in steel tube with a wall thickness of 0.065 in, compute the Reynolds number for the flow.

8.27M *A small velocity probe is to be inserted through a pipe wall. If we measure from the outside of the 6-in Schedule 80 pipe, how far (in mm) should the probe*

be inserted to sense the average velocity if the flow in the pipe is laminar?

8.28M *If the accuracy of positioning the probe described in Problem 8.27 is plus or minus 5.0 mm, compute the possible error in measuring the average velocity.*

8.29M *An alternative scheme for using the velocity probe described in Problem 8.27 is to place it in the middle of the pipe, where the velocity is expected to be 2.0 times the average velocity. Compute the amount of insertion required to center the probe. Then, if the accuracy of placement is again plus or minus 5.0 mm, compute the possible error in measuring the average velocity.*

8.30M *An existing fixture inserts the velocity probe described in Problem 8.27 exactly 60.0 mm from the outside surface of the pipe. If the probe reads 2.48 m/s, compute the actual average velocity of flow, assuming the flow is laminar. Then, check to see if the flow actually is laminar if the fluid is a heavy fuel oil with a kinematic viscosity of 850 centistokes.*

Noncircular Cross Sections

8.31M *Air with a specific weight of 12.5 N/m³ and a dynamic viscosity of 2.0 × 10⁻⁵ Pa · s flows through the shaded portion of the duct shown in Fig. 8.8 at the rate of 150 m³/h. Calculate the Reynolds number of the flow.*

FIGURE 8.8 Problem 8.31.

8.32E Carbon dioxide with a specific weight of 0.114 lb/ft³ and a dynamic viscosity of 3.34 × 10⁻⁷ lb-s/ft² flows in the shaded portion of the duct shown in Fig. 8.9. If the volume flow rate is 200 ft³/min, calculate the Reynolds number of the flow.

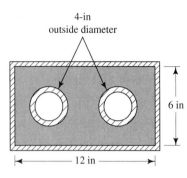

4-in
outside diameter

6 in

12 in

FIGURE 8.9   Problem 8.32.

8.33E Water at 90°F flows in the space between a 6-in Schedule 40 steel pipe and a square duct with inside dimensions of 10.0 in. The duct is similar to that shown in Fig. 8.7(d). Compute the Reynolds number if the volume flow rate is 4.00 ft³/sec.

8.34M *Refer to the shell-and-tube heat exchanger shown in Fig. 6.16. Both tubes are standard steel tubes with 0.049-in wall thicknesses. The inside tube carries 4.75 gal/min of water at 95°C, while the shell carries 15.0 gal/min of ethylene glycol at 25°C to carry heat away from the water. Compute the Reynolds number for the flow in both the tube and the shell.*

8.35M *Refer to Fig. 6.17, which shows two 6-in Schedule 40 pipes inside a rectangular duct. Each pipe carries 450 L/min of water at 20°C. Compute the Reynolds number for the flow of water. Then, for benzene (sg = 0.862) at 70°C flowing inside the duct, compute the volume flow rate required to produce the same Reynolds number.*

8.36E Refer to Fig. 6.18, which shows three pipes inside a larger pipe. The inside pipes carry water at 200°F, while the large pipe carries water at 60°F. The average velocity of flow is 25.0 ft/s in each pipe; compute the Reynolds number for each.

8.37M *Water at 10°C is flowing in the shell shown in Fig. 8.10 at the rate of 850 L/min. The shell is a 2-in Type K copper tube, while the tubes are ³/₈-in Type K copper tubes. Compute the Reynolds number for the flow.*

FIGURE 8.10   Problem 8.37.

8.38E Figure 8.11 shows the cross section of a heat exchanger used to cool a bank of electronic devices. Ethylene glycol at 77°F flows in the shaded area. Compute the volume flow rate required to produce a Reynolds number of 1500.

8.39M *Figure 8.12 shows a liquid-to-air heat exchanger in which air flows at 50 m³/h inside a rectangular passage and around a set of five vertical tubes. Each tube is a ¹/₂-in steel tube with a wall thickness of*

FIGURE 8.11   Problem 8.38.

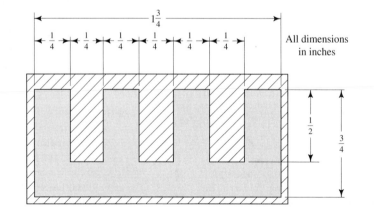

$1\frac{3}{4}$

$\frac{1}{4}$ $\frac{1}{4}$ $\frac{1}{4}$ $\frac{1}{4}$ $\frac{1}{4}$ $\frac{1}{4}$

All dimensions
in inches

$\frac{1}{2}$ $\frac{3}{4}$

FIGURE 8.12 Problem 8.39.

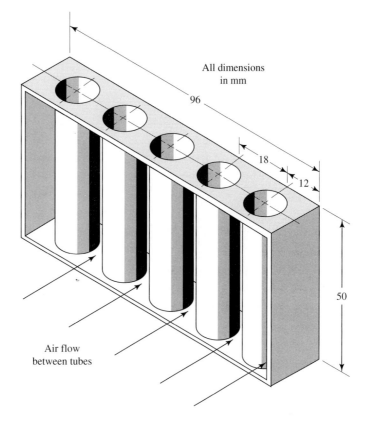

All dimensions
in mm

96

18

12

50

Air flow
between tubes

0.049 in. The air has a density of 1.15 kg/m³ and a
dynamic viscosity of 1.63 × 10⁻⁵ Pa·s. Compute
the Reynolds number for the air flow.

8.40M *Glycerine (sg = 1.26) at 40°C flows in the portion of
the duct outside the square tubes shown in Fig.
8.13. Calculate the Reynolds number for a flow rate
of 0.10 m³/s.*

Both 150 mm square outside

300 mm

450 mm

FIGURE 8.13 Problems 8.40 and 8.41.

8.41M *Each of the square tubes shown in Fig. 8.13 carries
0.75 m³/s of water at 90°C. The thickness of the
walls of the tubes is 2.77 mm. Compute the Reynolds number of the flow of water.*

8.42E A heat sink for an electronic circuit is made by
machining a pocket into a block of aluminum and
then covering it with a flat plate to provide a passage for cooling water as shown in Fig. 8.14. Compute the Reynolds number if the flow of water at
50°F is 78.0 gal/min.

8.43E Figure 8.15 shows the cross section of a cooling
passage for an odd-shaped device. Compute the
volume flow rate of water at 50°F that would produce a Reynolds number of 5 × 10⁵.

8.44E Figure 8.16 shows the cross section of a flow path
machined from a casting using a ¾-in-diameter
milling cutter. Considering all the fillets, compute
the hydraulic radius for the passage, and then compute the volume flow rate of acetone at 77°F required to produce a Reynolds number for the flow
of 2.6 × 10⁴.

FIGURE 8.14   Problem 8.42.

FIGURE 8.15   Problem 8.43.

FIGURE 8.16   Problem 8.44.

FIGURE 8.17 Problem 8.45.

8.0 mm typical

2.0 mm typical

8.45M *The blade of a gas turbine engine contains internal cooling passages, as shown in Fig. 8.17. Compute the volume flow rate of air required to produce an average velocity of flow in each passage of 25.0 m/s. The air flow distributes evenly to all six passages. Then, compute the Reynolds number if the air has a density of 1.20 kg/m³ and a dynamic viscosity of 1.50×10^{-5} Pa · s.*

9 Energy Losses Due to Friction

9.1
OBJECTIVES

As a fluid flows through a pipe, tube, or other conduit, energy loss occurs because of internal friction within the fluid. As indicated by the general energy equation, these energy losses result in the decrease in pressure between two points in the flow system. It is very important to be able to compute the magnitude of such energy losses. After completing this chapter, you should be able to:

1. State *Darcy's equation* for computing energy loss due to friction.
2. Define the *friction factor*, sometimes called the *coefficient of friction*.
3. State the *Hagen-Poiseuille equation* for computing energy loss due to friction in laminar flow.
4. Compute the equivalent friction factor for laminar flow for use in Darcy's equation.
5. Select design values for the pipe wall roughness.
6. Compute the relative roughness.
7. Determine the friction factor for turbulent flow using Moody's diagram.
8. Compute the magnitude of the energy loss for either laminar flow or turbulent flow in round pipes and tubes and use the energy loss in the general energy equation.
9. Use formulas for computing the magnitude of the friction factor.
10. Determine the friction factor and energy losses for flow in noncircular cross sections.
11. Use the Hazen-Williams formula to compute energy loss due to friction for the flow of water in pipes.

9.2
DARCY'S EQUATION

In the general energy equation

$$\frac{p_1}{\gamma} + z_1 + \frac{v_1^2}{2g} + h_A - h_R - h_L = \frac{p_2}{\gamma} + z_2 + \frac{v_2^2}{2g}$$

the term h_L is defined as the energy loss from the system. One component of the energy loss is due to friction in the flowing fluid. Friction is proportional to the velocity head of the flow and to the ratio of the length to the diameter of the flow stream, for the case of flow in pipes and tubes. This is expressed mathematically as Darcy's equation:

▷ DARCY'S EQUATION
FOR ENERGY LOSS

$$h_L = f \times \frac{L}{D} \times \frac{v^2}{2g}$$

(9–1)

where h_L = energy loss due to friction (N·m/N, m, lb-ft/lb, or ft)

L = length of flow stream (m or ft)

D = pipe diameter (m or ft)

v = average velocity of flow (m/s or ft/s)

f = friction factor (dimensionless)

Darcy's equation can be used to calculate the energy loss due to friction in long straight sections of round pipe for both laminar and turbulent flow. The difference between the two is in the evaluation of the dimensionless friction factor f.

9.3
FRICTION LOSS IN LAMINAR FLOW

When laminar flow exists, the fluid seems to flow as several layers, one on another. Because of the viscosity of the fluid, a shear stress is created between the layers of fluid. Energy is lost from the fluid by the action of overcoming the frictional forces produced by the shear stress. Since laminar flow is so regular and orderly, we can derive a relationship between the energy loss and the measurable parameters of the flow system. This relationship is known as the *Hagen-Poiseuille equation*:

▷ **HAGEN-POISEUILLE EQUATION**

$$h_L = \frac{32\mu L v}{\gamma D^2} \qquad (9\text{--}2)$$

The parameters involved are the fluid properties of viscosity and specific weight, the geometrical features of length and pipe diameter, and the dynamics of the flow characterized by the average velocity. The Hagen-Poiseuille equation has been verified experimentally many times. You should observe from Eq. (9–2) that the energy loss in laminar flow is independent of the condition of the pipe surface. Viscous friction losses within the fluid govern the magnitude of the energy loss.

The Hagen-Poiseuille equation is valid only for laminar flow ($N_R <$ 2000). However, we stated earlier that Darcy's equation, Eq. (9–1), could also be used to calculate friction loss for laminar flow. If the two relationships for h_L are set equal to each other, we can solve for the value of the friction factor:

$$f \times \frac{L}{D} \times \frac{v^2}{2g} = \frac{32\mu L v}{\gamma D^2}$$

$$f = \frac{32\mu L v}{\gamma D^2} \times \frac{D 2g}{L v^2} = \frac{64\mu g}{v D \gamma}$$

Since $\rho = \gamma/g$, we get

$$f = \frac{64\mu}{v D \rho}$$

The Reynolds number is defined as $N_R = v D \rho / \mu$. Then we have

▷ **FRICTION FACTOR FOR LAMINAR FLOW**

$$f = \frac{64}{N_R} \qquad (9\text{--}3)$$

In summary, the energy loss due to friction in *laminar flow* can be calculated either from the Hagen-Poiseuille equation:

$$h_L = \frac{32\,\mu L v}{\gamma D^2}$$

or from Darcy's equation:

$$h_L = f \times \frac{L}{D} \times \frac{v^2}{2g}$$

where $f = 64/N_R$.

□ **EXAMPLE PROBLEM 9.1**   Determine the energy loss if glycerine at 25°C flows 30 m through a 150-mm diameter pipe with an average velocity of 4.0 m/s.

Solution   First, we must determine whether the flow is laminar or turbulent by evaluating the Reynolds number:

$$N_R = \frac{v D \rho}{\mu}$$

From Appendix B, we find that for glycerine at 25°C

$$\rho = 1258 \text{ kg/m}^3$$

$$\mu = 9.60 \times 10^{-1} \text{ Pa} \cdot \text{s}$$

Then, we have

$$N_R = \frac{(4.0)(0.15)(1258)}{9.60 \times 10^{-1}} = 786$$

Because $N_R < 2000$, the flow is laminar.

Using Darcy's equation, we get

$$h_L = f \times \frac{L}{D} \times \frac{v^2}{2g}$$

$$f = \frac{64}{N_R} = \frac{64}{786} = 0.081$$

$$h_L = 0.081 \times \frac{30}{0.15} \times \frac{(4.0)^2}{2(9.81)} \text{ m} = 13.2 \text{ m}$$

Notice that each term in each equation is expressed in the units of the SI unit system. Therefore, the resulting units for h_L are m or $\text{N} \cdot \text{m/N}$. This means that 13.2 $\text{N} \cdot \text{m}$ of energy is lost by each newton of the glycerine as it flows along the 30 m of pipe.

9.4
FRICTION LOSS IN
TURBULENT FLOW

For turbulent flow of fluids in circular pipes it is most convenient to use Darcy's equation to calculate the energy loss due to friction. We cannot determine the friction factor f by a simple calculation as we did for laminar flow because turbulent flow does not conform to regular predictable motions. It is rather chaotic and is constantly varying. For these reasons we must rely on experimental data to determine the value of f.

FIGURE 9.1 Pipe wall roughness (exaggerated).

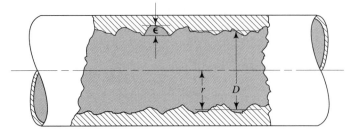

Tests have shown that the dimensionless number f is dependent on two other dimensionless numbers, the Reynolds number and the relative roughness of the pipe. The relative roughness is the ratio of the pipe diameter D to the average pipe wall roughness ϵ (Greek letter epsilon). Figure 9.1 illustrates pipe wall roughness (exaggerated) as the height of the peaks of the surface irregularities. The condition of the pipe surface is very much dependent on the pipe material and the method of manufacture. For commercially available pipe and tubing, the design value of the wall roughness ϵ has been determined as shown in Table 9.1. *These are only average values for new, clean pipe. Some variation should be expected. After a pipe has been in service for a time, the roughness could change due to the formation of deposits on the wall or due to corrosion.*

TABLE 9.1 Pipe roughness—Design values

Material	Roughness, ϵ (m)	Roughness, ϵ (ft)
Glass, plastic	Smooth	Smooth
Copper, brass, lead (tubing)	1.5×10^{-6}	5×10^{-6}
Cast iron—uncoated	2.4×10^{-4}	8×10^{-4}
Cast iron—asphalt coated	1.2×10^{-4}	4×10^{-4}
Commercial steel or welded steel	4.6×10^{-5}	1.5×10^{-4}
Wrought iron	4.6×10^{-5}	1.5×10^{-4}
Riveted steel	1.8×10^{-3}	6×10^{-3}
Concrete	1.2×10^{-3}	4×10^{-3}

One of the most widely used methods for evaluating the friction factor employs the Moody diagram shown in Fig. 9.2. The diagram shows the friction factor f plotted versus the Reynolds number N_R, with a series of parametric curves related to the relative roughness D/ϵ. These curves were generated from experimental data by L. F. Moody (see Reference 2).

Both f and N_R are plotted on logarithmic scales because of the broad range of values encountered. At the left end of the chart, for Reynolds numbers less than 2000, the straight line shows the relationship $f = 64/N_R$ for laminar flow. For $2000 < N_R < 4000$, no curves are drawn since this is the critical zone between laminar and turbulent flow and it is not possible to

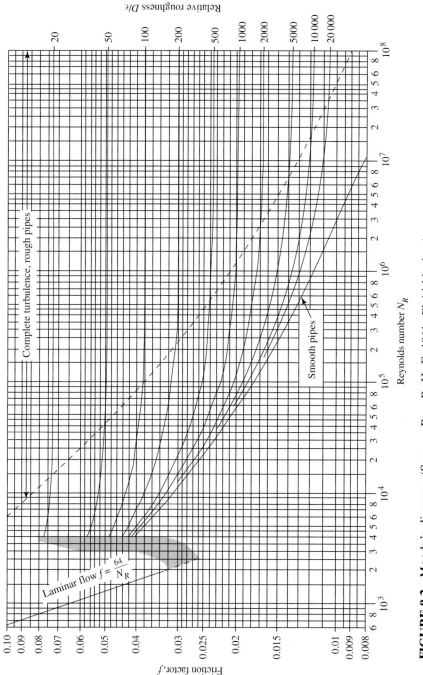

FIGURE 9.2 Moody's diagram. (Source: Pao, R. H. F. 1961. *Fluid Mechanics.* New York: John Wiley & Sons, p. 284.)

predict the type of flow. Beyond $N_R = 4000$, the family of curves for different values of D/ϵ are plotted. Several important observations can be made from these curves:

1. For a given Reynolds number of flow, as the relative roughness D/ϵ is increased, the friction factor f decreases.
2. For a given relative roughness D/ϵ, the friction factor f decreases with increasing Reynolds number until the zone of complete turbulence is reached.
3. Within the zone of complete turbulence, the Reynolds number has no effect on the friction factor.
4. As the relative roughness D/ϵ increases, the value of the Reynolds number at which the zone of complete turbulence begins also increases.

Figure 9.3 is a simplified sketch of Moody's diagram in which the various zones are identified. The *laminar zone* at the left has already been discussed. At the right of the dashed line downward across the diagram is the *zone of complete turbulence*. The lowest possible friction factor for turbulent flow is indicated by the *smooth pipes* line.

Between the smooth pipes line and the line marking the start of the complete turbulence zone is the *transition zone*. Here, the various D/ϵ lines are curved, and care must be exercised to evaluate the friction factor prop-

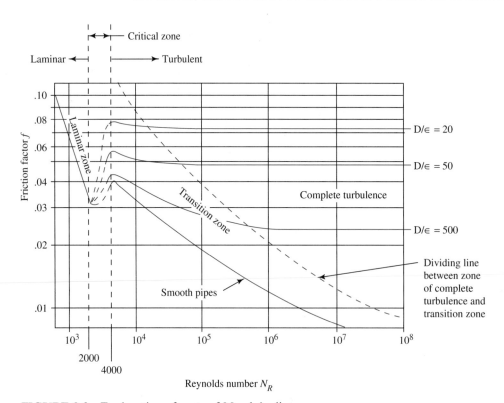

FIGURE 9.3   Explanation of parts of Moody's diagram.

erly. You can see, for example, that the value of the friction factor for a relative roughness of 500 decreases from 0.0420 at N_R = 4000 to 0.0240 at N_R = 6.0 × 10⁵, where the zone of complete turbulence starts.

As stated in Chapter 8, the critical zone between the Reynolds numbers of 2000 and 4000 is to be avoided if possible, because within this range the type of flow cannot be predicted. The dashed curves show how the friction factor could change according to the value of the relative roughness. For low values of D/ϵ (indicating large pipe wall roughness), the increase in friction factor is great as the flow changes from laminar to turbulent. For example, for flow in a pipe with D/ϵ = 20, the friction factor would increase from 0.032 for N_R = 2000 at the end of the laminar range to approximately 0.077 at N_R = 4000 at the beginning of the turbulent range, an increase of 240 percent. Moreover, the value of the Reynolds number where this would occur cannot be predicted. Because the energy loss is directly proportional to the friction factor, changes of such magnitude are significant.

It should be noted that because relative roughness is defined as D/ϵ, a high relative roughness indicates a low value of ϵ, that is, a smooth pipe. In fact, the curve labeled *smooth pipes* is used for materials such as glass which have such a low roughness that D/ϵ would be an extremely large number. Some texts and references use other conventions for reporting relative roughness, such as ϵ/D, ϵ/r, or r/ϵ, where r is the pipe radius. We feel that the convention used in this book makes calculations and interpolations easier.

9.4.1
Use of the Moody Diagram

The Moody diagram is used to help determine the value of the friction factor f for turbulent flow. The value of the Reynolds number and the relative roughness must be known. Therefore, the basic data required are the pipe inside diameter, the pipe material, the flow velocity, and the kind of fluid and its temperature, from which the viscosity can be found. The following example problems illustrate the procedure for finding f.

□ **EXAMPLE PROBLEM 9.2**

Determine the friction factor f if water at 160°F is flowing at 30.0 ft/s in an uncoated cast iron pipe having an inside diameter of 1 in.

Solution

The Reynolds number must first be evaluated to determine whether the flow is laminar or turbulent:

$$N_R = \frac{vD}{\nu}$$

But D = 1 in = 0.0833 ft, and ν = 4.38 × 10⁻⁶ ft²/s. We now have

$$N_R = \frac{(30.0)(0.0833)}{4.38 \times 10^{-6}} = 5.70 \times 10^5$$

Thus, the flow is turbulent. Now the relative roughness must be evaluated. From Table 9.1 we find ϵ = 8 × 10⁻⁴ ft. Then, the relative roughness is

$$\frac{D}{\epsilon} = \frac{0.0833 \text{ ft}}{8 \times 10^{-4} \text{ ft}} = 1.04 \times 10^2 = 104$$

Notice that in order for D/ϵ to be a dimensionless ratio, both D and ϵ must be in the same units.

The final steps in the procedure are

1. Locate the Reynolds number on the abscissa of the Moody diagram:

$$N_R = 5.70 \times 10^5$$

2. Project vertically until the curve for $D/\epsilon = 104$ is reached. Since 104 is so close to 100, that curve can be used.
3. Project horizontally to the left, and read $f = 0.038$.

□ **EXAMPLE PROBLEM 9.3**   If the flow rate of water in Problem 9.2 was 0.45 ft/s with all other conditions being the same, determine the friction factor f.

Solution

$$N_R = \frac{vD}{\nu} = \frac{(0.45)(0.0833)}{4.38 \times 10^{-6}} = 8.55 \times 10^3$$

$$\frac{D}{\epsilon} = \frac{0.0833}{8 \times 10^{-4}} = 104$$

Then, from Fig. 9.2, $f = 0.044$. Notice that this is on the curved portion of the D/ϵ curve and that there is a significant increase in the friction factor over that in Example Problem 9.2.

□ **EXAMPLE PROBLEM 9.4**   Determine the friction factor f if ethyl alcohol at 25°C is flowing at 5.3 m/s in a standard 1½-in Schedule 80 steel pipe.

Solution   Evaluating the Reynolds number, we use the equation

$$N_R = \frac{vD\rho}{\mu}$$

From Appendix B, $\rho = 787$ kg/m^3 and $\mu = 1.00 \times 10^{-3}$ Pa·s. Also, for a 1½-in Schedule 80 pipe, $D = 0.0381$ m. Then we have

$$N_R = \frac{(5.3)(0.0381)(787)}{1.00 \times 10^{-3}} = 1.59 \times 10^5$$

Thus, the flow is turbulent. For a steel pipe, $\epsilon = 4.6 \times 10^{-5}$ m, so the relative roughness is

$$\frac{D}{\epsilon} = \frac{0.0381 \text{ m}}{4.6 \times 10^{-5} \text{ m}} = 828$$

From Fig. 9.2, $f = 0.0225$. You must interpolate on both N_R and D/ϵ to determine this value, and you should expect some variation. However, you should be able to read the value of the friction factor f within ±0.0005 in this portion of the graph.

The following is a programmed example problem illustrating a typical fluid piping situation. The energy loss due to friction must be calculated as a part of the solution.

PROGRAMMED EXAMPLE PROBLEM

☐ **EXAMPLE PROBLEM 9.5**   In a chemical processing plant, benzene at 50°C (sg = 0.86) must be delivered to point B with a pressure of 550 kPa. A pump is located at point A 21 m below point B, and the two points are connected by 240 m of plastic pipe having an inside diameter of 50 mm. If the volume flow rate is 110 L/min, calculate the required pressure at the outlet of the pump.

Write the energy equation between points A and B.

The relation is

$$\frac{p_A}{\gamma} + z_A + \frac{v_A^2}{2g} - h_L = \frac{p_B}{\gamma} + z_B + \frac{v_B^2}{2g}$$

The term h_L is required because there is an energy loss due to friction between points A and B. Point A is at the outlet of the pump, and the objective of the problem is to calculate p_A. Solve algebraically for p_A now.

The equation is

$$p_A = p_B + \gamma \left[(z_B - z_A) + \frac{v_B^2 - v_A^2}{2g} + h_L \right] \qquad \textbf{(9–4)}$$

What is the value of $z_B - z_A$?

We find that $z_B - z_A = + 21$ m because point B is higher than point A. Now, what is $(v_B^2 - v_A^2)/2g$?

It is zero. Since points A and B are both in the same size pipe under steady flow, the velocities are equal.

This brings us to h_L, the energy loss due to friction between A and B. What is the first step?

The evaluation of the Reynolds number is the first step. The type of flow, laminar or turbulent, must be determined. Complete the calculation of the Reynolds number before looking at the next panel.

The correct value is $N_R = 9.54 \times 10^4$. Here is how it is found:

$$N_R = vD\rho/\mu$$

For a 50-mm pipe, $D = 0.050$ m and $A = 1.963 \times 10^{-3}$ m². Then, we have

$$Q = (110 \text{ L/min}) \left(\frac{1 \text{ m}^3/\text{s}}{60\ 000 \text{ L/min}} \right) = 1.83 \times 10^{-3} \text{ m}^3/\text{s}$$

$$v = \frac{Q}{A} = \frac{1.83 \times 10^{-3} \text{ m}^3/\text{s}}{1.963 \times 10^{-3} \text{ m}^2} = 0.932 \text{ m/s}$$

For benzene at 50°C with a specific gravity of 0.86, we find that

$$\rho = (0.86)(1000 \text{ kg/m}^3) = 860 \text{ kg/m}^3$$

$$\mu = 4.2 \times 10^{-4} \text{ Pa·s} \quad \text{(from Appendix D)}$$

Then we have

$$N_R = \frac{(0.932)(0.050)(860)}{4.2 \times 10^{-4}} = 9.54 \times 10^4$$

Therefore, the flow is turbulent. What relationship should be used to calculate h_L?

For turbulent flow, Darcy's equation should be used:

$$h_L = f \times \frac{L}{D} \times \frac{v^2}{2g}$$

In order to use the Moody diagram to find the value of f, is the value of the relative roughness D/ϵ needed?

Not in this case. Because the pipe is plastic, the inner surface is smooth and we can use the curve in the Moody diagram labeled *smooth pipes*. Evaluate f now.

The result is $f = 0.018$. Now the calculation of h_L can be completed.

The correct value is $h_L = 3.83$ m.

$$h_L = f \times \frac{L}{D} \times \frac{v^2}{2g} = 0.018 \times \frac{240}{0.050} \times \frac{(0.932)^2}{2(9.81)} \text{ m}$$

$$h_L = 3.83 \text{ m}$$

Going back to Eq. (9–4), you can now calculate p_A.

You should have $p_A = 759$ kPa.

$$p_A = p_B + \gamma \left[(z_B - z_A) + \frac{v_B^2 - v_A^2}{2g} + h_L \right]$$

$$p_A = 550 \text{ kPa} + \frac{(0.86)(9.81 \text{ kN})}{\text{m}^3} (21 \text{ m} + 0 + 3.83 \text{ m})$$

$$p_A = 550 \text{ kPa} + 209 \text{ kN/m}^2 = 550 \text{ kPa} + 209 \text{ kPa}$$

$$p_A = 759 \text{ kPa}$$

∎

9.5 EQUATIONS FOR FRICTION FACTOR

The Moody diagram in Fig. 9.2 is a convenient and sufficiently accurate means of determining the value of the friction factor when solving problems by manual calculations. However, if the calculations are to be automated for solution on a computer or a programmable calculator, it is necessary to have equations for the friction factor.

The equations used in the work by Moody (1944) cover three different zones of the diagram.* In the *laminar flow zone*, for values of Reynolds number below 2000, f can be found from Eq. (9–3):

FRICTION FACTOR FOR LAMINAR FLOW

$$f = 64/N_R$$

This relationship, developed in Section 9.2, plots in the Moody diagram as a straight line at the left side of the chart.

Of course, for Reynolds numbers from 2000 to 4000, the flow is in the critical range and it is impossible to predict the value of f.

Above the Reynolds number of 4000, the flow is generally referred to as *turbulent*. However, there are essentially two zones of interest here. Toward the right side of the diagram, the flow is in the *complete turbulence zone*. It can be seen that the value of f is not dependent on the Reynolds number but only on the relative roughness D/ϵ. In this range the following formula applies:

$$\frac{1}{\sqrt{f}} = 2 \log_{10}(3.7\ D/\epsilon) \qquad \textbf{(9–5)}$$

The boundary of this zone is the dashed line running generally from the upper left to the lower right of the Moody diagram. The equation of this line is

$$\frac{1}{\sqrt{f}} = \frac{N_R}{200(D/\epsilon)} \qquad \textbf{(9–6)}$$

The third zone of the Moody diagram, called the *transition zone*, lies between the zone of complete turbulence and the line labeled *smooth pipes*. The "smooth pipes" line has the equation

$$\frac{1}{\sqrt{f}} = 2 \log_{10}\left(\frac{N_R \sqrt{f}}{2.51}\right) \qquad \textbf{(9–7)}$$

Being smooth, these pipes present no surface irregularities to the flow, so the friction factor is a function only of the Reynolds number. Pipes made of glass and plastic are essentially smooth. Also, large pipes made of drawn brass or copper have a value of relative roughness which brings them close to the smooth pipes line.

In the transition zone, the friction factor is a function of both the Reynolds number and the relative roughness. C. F. Colebrook developed the relation for the friction factor in this zone:

$$\frac{1}{\sqrt{f}} = -2 \log_{10}\left(\frac{1}{3.7(D/\epsilon)} + \frac{2.51}{N_R\sqrt{f}}\right) \qquad \textbf{(9–8)}$$

---

*   Earlier work in developing the equations was done by several researchers, most notably C. F. Colebrook, L. Prandtl, H. Rouse, T. van Karman, and J. Nikuradse, whose papers are listed in the bibliography of Moody's ASME paper (see Reference 2).

Equation (9–8) approaches the equation for complete turbulence, Eq. (9–5), for large Reynolds numbers as the second term inside the parentheses becomes very small. Also, for very large values of D/ϵ, the first term becomes small and the equation reduces to that for smooth pipes.

Since Eq. (9–8) requires a trial-and-error solution procedure, it is inconvenient for automated calculation of the friction factor.

9.5.1
Explicit Equation
for Friction Factor

The following equation, which allows the direct calculation of the value of the friction factor, was developed by P. K. Swamee and A. K. Jain and is reported in Reference 4 (1976).

⟳ FRICTION FACTOR FOR
TURBULENT FLOW

$$f = \frac{0.25}{\left[\log \left(\dfrac{1}{3.7\,(D/\epsilon)} + \dfrac{5.74}{N_R^{0.9}} \right) \right]^2} \tag{9–9}$$

Equation (9–9) produces values for f that are within ± 1.0 percent of those from the Colebrook equation (9–8) within the range of relative roughness, D/ϵ, from 100 to 1×10^6 and for Reynolds numbers from 5×10^3 to 1×10^8. This is virtually the entire turbulent zone of the Moody diagram.

Summary
To calculate the value of the friction factor, f, when the Reynolds number and relative roughness are known, use Eq. (9–3) for laminar flow and Eq. (9–9) for turbulent flow.

☐ **EXAMPLE PROBLEM 9.6**   Compute the value for the friction factor if the Reynolds number for the flow is 1×10^5 and the relative roughness is 2000.

Solution   Because this is in the turbulent zone, Eq. (9–9) is used.

$$f = \frac{0.25}{\left[\log \left(\dfrac{1}{3.7(2000)} + \dfrac{5.74}{(1 \times 10^5)^{0.9}} \right) \right]^2}$$

$$f = 0.0204$$

This value compares closely with the value read from Fig. 9.2

9.6
FRICTION LOSS
IN NONCIRCULAR
CROSS SECTIONS

Darcy's equation for friction loss can be used for noncircular cross sections if the geometry is represented by the hydraulic radius instead of the pipe diameter, as is used for circular sections. The definition of hydraulic radius and the method of computing its value were presented in Chapter 8.

After computing the hydraulic radius, we can compute the Reynolds number by substituting $4R$ for the diameter D. The equations then become

⟳ REYNOLDS NUMBER FOR
NONCIRCULAR SECTIONS

$$N_R = \frac{v(4R)\rho}{\mu} = \frac{v(4R)}{\nu} \tag{9–10}$$

A similar substitution can be made for computing friction loss. In Darcy's equation, replacing D with $4R$ gives

⇨ **DARCY'S EQUATION FOR NONCIRCULAR SECTIONS**

$$h_L = f \frac{L}{4R} \frac{v^2}{2g} \qquad (9\text{–}11)$$

The relative roughness D/ϵ becomes $4R/\epsilon$. The friction factor can be found from the Moody diagram.

☐ **EXAMPLE PROBLEM 9.7**   Determine the pressure drop for a 50-m length of a duct with the cross section shown in Fig. 9.4. Ethylene glycol at 25°C is flowing at the rate of 0.16 m³/s. The inside dimension of the square is 250 mm, and the outside diameter of the tube is 150 mm. Use $\epsilon = 3 \times 10^{-5}$ m, somewhat smoother than commercial steel pipe.

FIGURE 9.4   Cross section for duct for Example Problem 9.7.

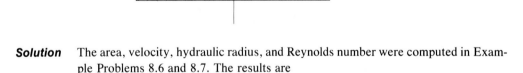

150-mm diameter

250 mm

Solution   The area, velocity, hydraulic radius, and Reynolds number were computed in Example Problems 8.6 and 8.7. The results are

$$A = 0.0448 \text{ m}^2$$
$$v = 3.57 \text{ m/s}$$
$$R = 0.0305 \text{ m}$$
$$N_R = 2.96 \times 10^4$$

The flow is turbulent, and Darcy's equation can be used to calculate the energy loss between two points 50 m apart. In order to determine the friction factor, we must first find the relative roughness:

$$4R/\epsilon = (4)(0.0305)/(3 \times 10^{-5}) = 4067$$

From the Moody diagram, $f = 0.0245$. Then, we have

$$h_L = f \times \frac{L}{4R} \times \frac{v^2}{2g} = 0.0245 \times \frac{50}{(4)(0.0305)} \times \frac{(3.57)^2}{(2)(9.81)} \text{ m}$$
$$h_L = 6.52 \text{ m}$$

If the duct is horizontal,

$$h_L = \Delta p / \gamma$$

$$\Delta p = \gamma h_L$$

where Δp is the pressure drop caused by the energy loss. Then, we have

$$\Delta p = \frac{10.79 \text{ kN}}{\text{m}^3} \times 6.52 \text{ m} = 70.4 \text{ kPa}$$

9.7
VELOCITY PROFILE
FOR TURBULENT
FLOW

In Chapter 8 we showed that the velocity profile for laminar flow was a parabolic shape for which the local velocity at any point could be computed simply from Eq. (8–4). The only variables were the average velocity of flow and the radius of the pipe. Figure 8.6 showed a detailed plot of local velocity versus position within the fluid.

The velocity profile for turbulent flow is far different from the parabolic distribution for laminar flow. As shown in Fig. 9.5, the fluid velocity near the wall of the pipe changes rapidly from zero at the wall to a nearly uniform velocity distribution throughout the bulk of the cross section. The actual shape of the velocity profile varies with the friction factor, f, which in turn varies with the Reynolds number and the relative roughness of the pipe. The governing equation (from Reference 1) is

$$U = v[1 + 1.43\sqrt{f} + 2.15\sqrt{f} \ \log_{10}(1 - r/r_o)] \qquad \textbf{(9–12)}$$

An alternate form of this equation can be developed by defining the distance from the wall of the pipe as $y = r_o - r$. Then, the argument of the logarithm term becomes

$$1 - \frac{r}{r_o} = \frac{r_o - r}{r_o} = \frac{y}{r_o}$$

FIGURE 9.5   General form of velocity profile for turbulent flow.

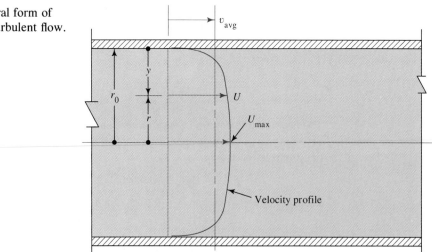

Equation (9–12) is then

$$U = v[1 + 1.43\sqrt{f} + 2.15\sqrt{f}\,\log_{10}(y/r_o)] \qquad (9\text{–}13)$$

When evaluating Eq. (9–12) or (9–13), recall that the logarithm of zero is undefined. You may allow r to *approach* r_o, but not equal it. Similarly, y can only approach zero.

The maximum velocity occurs at the center of the pipe ($r = 0$ or $y = r_o$), and its value can be computed from

$$U_{max} = v(1 + 1.43\sqrt{f}) \qquad (9\text{–}14)$$

Figure 9.6 compares the velocity profiles for laminar flow and for turbulent flow at a variety of Reynolds numbers.

FIGURE 9.6   Velocity profiles in laminar and turbulent flow in a smooth pipe. (Source: Miller, R. W. 1983. *Flow Measurement Engineering Handbook*. New York: McGraw-Hill.)

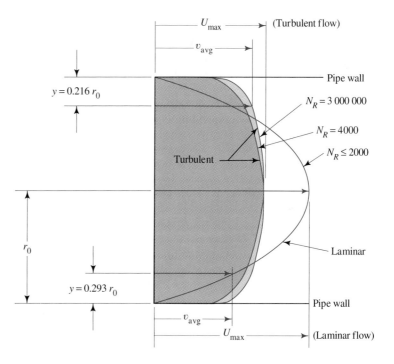

☐ **EXAMPLE PROBLEM 9.8**   For the data from Example Problem 9.5, compute the expected maximum velocity of flow, and compute several points on the velocity profile. Plot the velocity versus the distance from the pipe wall.

Solution   From Example Problem 9.5, the following data can be found:

$$D = 50 \text{ mm} = \text{pipe inside diameter}$$
$$v = 0.932 \text{ m/s} = \text{average velocity}$$
$$f = 0.018 = \text{friction factor}$$
$$N_R = 9.54 \times 10^4 \qquad \text{(turbulent)}$$

Now, from Eq. (9–14), we see that the maximum velocity of flow is

$$U_{max} = v(1 + 1.43\sqrt{f}) = (0.932 \text{ m/s})(1 + 1.43\sqrt{0.018})$$

$$U_{max} = 1.111 \text{ m/s at the center of the pipe}$$

Equation (9–13) can be used to determine the points on the velocity profile. We know that the velocity equals zero at the pipe wall ($y = 0$). Also, the rate of change of velocity with position is greater near the wall than near the center of the pipe. Therefore, increments of 0.5 mm will be used from $y = 0.5$ to $y = 2.5$ mm. Then, increments of 2.5 mm will be used up to $y = 10$ mm. Finally, increments of 5.0 mm will provide sufficient definition of the profile near the center of the pipe. At $y = 1.0$ mm, and $r_o = 25$ mm,

$$U = v[1 + 1.43\sqrt{f} + 2.15\sqrt{f} \log_{10}(y/r_o)]$$

$$U = (0.932 \text{ m/s})[1 + 1.43\sqrt{0.018} + 2.15\sqrt{0.018} \log_{10}(1/25)]$$

$$U = 0.735 \text{ m/s}$$

Using similar calculations, we can compute the following values:

y (mm)	y/r_o	U (m/s)	
0.5	0.02	0.654	
1.0	0.04	0.735	
1.5	0.06	0.782	
2.0	0.08	0.816	
2.5	0.10	0.842	
5.0	0.20	0.923	
7.5	0.30	0.970	
10.0	0.40	1.004	
15.0	0.60	1.051	
20.0	0.80	1.085	
25.0	1.00	1.111	(U_{max} at center of pipe)

Figure 9.7 is the plot of y versus velocity in the form in which the velocity profile is normally shown.

**9.8
HAZEN-WILLIAMS
FORMULA FOR
WATER FLOW**

The Darcy equation presented in this chapter for calculating energy loss due to friction is applicable for any newtonian fluid. An alternate approach is convenient for the special case of the flow of water in pipe line systems.

The *Hazen-Williams formula* is one of the most popular formulas for the design and analysis of water systems. Its use is limited to the flow of water in pipes larger than 2.0 in and smaller than 6.0 ft in diameter. The velocity of flow should not exceed 10.0 ft/s. Also, it has been developed for

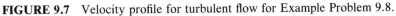

FIGURE 9.7 Velocity profile for turbulent flow for Example Problem 9.8.

water at 60°F. Use at temperatures much lower or higher would result in some error.

The Hazen-Williams formula is unit-specific. In the U.S. Customary unit system it takes the form

$$v = 1.32\, C_h\, R^{0.63}\, s^{0.54} \qquad\qquad (9\text{–}15)$$

where v = average velocity of flow (ft/s)
C_h = Hazen-Williams coefficient (dimensionless)
R = Hydraulic radius of flow conduit (ft)
s = Ratio of h_L/L: energy loss/length of conduit (ft/ft)

The use of the hydraulic radius in the formula allows its application to noncircular sections as well as circular pipes. Recall that $R = D/4$ for circular pipes.

The coefficient C_h is dependent only on the condition of the surface of the pipe or conduit. Table 9.2 gives typical values. Note that some are described as pipe in new, clean condition while the design value accounts for the accumulation of deposits that develop on the inside surfaces of pipe after time, even when clean water flows through them. Smoother pipes have higher values of C_h as compared with rougher pipes.

TABLE 9.2 Hazen-Williams coefficient, C_h

Type of Pipe	Average for New, Clean Pipe	Design Value
Steel, ductile iron, or cast iron with centrifugally applied cement or bituminous lining	150	140
Plastic, copper, brass, glass	140	130
Steel or cast iron	130	100
Concrete	120	100
Corrugated steel	60	60

The Hazen-Williams formula for SI units is:

⇨ **HAZEN-WILLIAMS FORMULA**
SI UNITS

$$v = 0.85 \, C_h \, R^{0.63} \, s^{0.54} \tag{9–16}$$

where v = average velocity of flow (m/s)
             C_h = Hazen-Williams coefficient (dimensionless)
             R = hydraulic radius of flow conduit (m)
             s = ratio of h_L/L: energy loss/length of conduit (m/m)

As before, the volume flow rate can be computed from $Q = Av$.

□ **EXAMPLE PROBLEM 9.9**   For what velocity of flow of water in a new, clean 6-in Schedule 40 steel pipe would an energy loss of 20 ft of head occur over a length of 1000 ft? Compute the volume flow rate at that velocity.

Solution   We can use Equation (9–15).

$$s = h_L/L = (20 \text{ ft})/(1000 \text{ ft}) = 0.02$$
$$R = D/4 = (0.5054 \text{ ft})/4 = 0.126 \text{ ft}$$
$$C_h = 130$$

Then,

$$v = 1.32 \, C_h \, R^{0.63} \, s^{0.54}$$
$$v = (1.32)(130)(0.126)^{0.63}(0.02)^{0.54} = 5.64 \text{ ft/s}$$
$$Q = Av = (0.2006 \text{ ft}^2)(5.64 \text{ ft/s}) = 1.13 \text{ ft}^3/\text{s}$$

Note that the velocity and the volume flow rate are both directly proportional to the value of C_h. If the pipe of Example Problem 9.9 degrades to a value of $C_h = 100$ after use, the allowable volume flow rate to limit the energy loss to the same value of 20 ft per 1000 ft of pipe length would be

$$v = (5.64 \text{ ft/s}) \, (100/130) = 4.34 \text{ ft/s}$$
$$Q = (1.13 \text{ ft}^3/\text{s}) \, (100/130) = 0.869 \text{ ft}^3/\text{s}$$

9.9
OTHER FORMS OF THE
HAZEN-WILLIAMS
FORMULA

Equations (9–15) and (9–16) allow the direct computation of the velocity of flow for a given type and size of flow conduit when the energy loss per unit length is known or specified. The volume flow rate can be simply calculated by using $Q = Av$. Other types of calculations that are often desired are:

1. To determine the required size of pipe to carry a given flow rate while limiting the energy loss to some specified value.
2. To determine the energy loss for a given flow rate through a given type and size of pipe of a known length.

Table 9.3 shows several forms of the Hazen-Williams formula that facilitate such calculations.

TABLE 9.3   Alternate forms of the Hazen-Williams formula

U.S. Customary Units	SI Units
$v = 1.32 \, C_h \, R^{0.63} \, s^{0.54}$	$v = 0.85 \, C_h \, R^{0.63} \, s^{0.54}$
$Q = 1.32 \, A \, C_h \, R^{0.63} \, s^{0.54}$	$Q = 0.85 \, A \, C_h \, R^{0.63} \, s^{0.54}$
$h_L = L \left[\dfrac{Q}{1.32 \, A \, C_h \, R^{0.63}} \right]^{1.852}$	$h_L = L \left[\dfrac{Q}{0.85 \, A \, C_h \, R^{0.63}} \right]^{1.852}$
$D = \left[\dfrac{2.31 \, Q}{C_h \, s^{0.54}} \right]^{0.380}$	$D = \left[\dfrac{3.59 \, Q}{C_h \, s^{0.54}} \right]^{0.380}$
Note: Units must be consistent:	
v in ft/s	v in m/s
Q in ft³/s	Q in m³/s
A in ft²	A in m²
$h_L, L, R,$ and D in ft	$h_L, L, R,$ and D in m
s in ft/ft (dimensionless)	s in m/m (dimensionless)

9.10
NOMOGRAPH FOR
SOLVING
HAZEN-WILLIAMS
FORMULA

The nomograph shown in Figure 9.8 allows the solution of the Hazen-Williams formula to be done by simply aligning known quantities with a straight edge and reading the desired unknowns at the intersection of the straight edge with the appropriate vertical axis. *Note that this nomograph is constructed for the value of the Hazen-Williams coefficient of $C_h = 100$. If the actual pipe condition warrants the use of a different value of C_h, the following formulas can be used to adjust the results. The subscript "100" refers to the value read from the nomograph for $C_h = 100$. The subscript "c" refers to the value for the given C_h.*

$$v_c = v_{100}(C_h/100) \qquad \text{[velocity]} \qquad\qquad \textbf{(9–17)}$$

$$Q_c = Q_{100}(C_h/100) \qquad \text{[volume flow rate]} \qquad \textbf{(9–18)}$$

$$D_c = D_{100}(100/C_h)^{0.38} \qquad \text{[pipe diameter]} \qquad \textbf{(9–19)}$$

$$s_c = s_{100}(100/C_h)^{1.85} \qquad \text{[head loss/length]} \qquad \textbf{(9–20)}$$

FIGURE 9.8 Nomograph for the solution of the Hazen-Williams formula for $C_h = 100$.

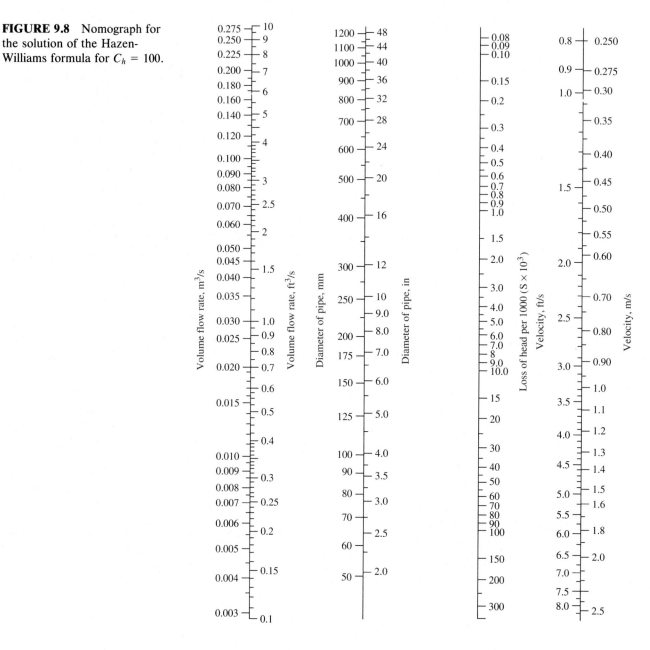

One frequent use of a nomograph like that in Fig. 9.8 is to determine the required size of pipe to carry a given flow rate while limiting the energy loss to some specified value. Thus it is a convenient design tool.

□ **EXAMPLE PROBLEM 9.10**   Specify the required size of Schedule 40 steel pipe to carry 1.20 ft³/s of water with no more than 4.0 ft of head loss over a 1000 ft length of pipe. Use the design value for C_h.

Solution   Table 9.2 suggests $C_h = 100$. Now, using Fig. 9.8, we can place a straight edge from $Q = 1.20$ ft³/s on the volume flow rate line to the value of $s = (4.0$ ft$)/(1000$ ft$)$ on the energy loss line. The straight edge then intersects the pipe size line at approximately 9.7 in. The next larger standard pipe size listed in Appendix F is the nominal 10-in pipe with an inside diameter of 10.02 in.

Returning to the chart in Fig. 9.8 and slightly realigning $Q = 1.20$ ft³/s with $D = 10.02$ in, we can read an average velocity of $v = 2.25$ ft/s. This is relatively low for a water distribution system and the pipe is quite large. If the pipeline is long, the cost for piping would be excessively large.

If we allow the velocity of flow to increase to approximately 6.0 ft/s for the same volume flow rate, we can use the chart to show that a 6-in pipe could be used with a head loss of approximately 37 ft per 1000 ft of pipe. The lower cost of the pipe compared with the 10-in pipe would have to be compared with the higher energy cost required to overcome the additional head loss.

REFERENCES

1. Miller, R. W. 1983. *Flow Measurement Engineering Handbook*. New York: McGraw-Hill.

2. Moody, L. F. 1944. Friction Factors for Pipe Flow. *Transactions of the ASME* 66(8):671–684. New York: American Society of Mechanical Engineers.

3. Steel, E. W., and Terence J. McGhee. 1979. *Water Supply and Sewerage*. New York: McGraw-Hill.

4. Swamee, P. K., and A. K. Jain. 1976. "Explicit Equations for Pipe-flow Problems." *Journal of the Hydraulics Division*, 102(HY5):657–664. New York: American Society of Civil Engineers.

PRACTICE PROBLEMS

9.1M *Crude oil is flowing vertically downward through 60 m of 1-in Schedule 80 steel pipe at a velocity of 0.64 m/s. The oil has a specific gravity of 0.86 and is at 0°C. Calculate the pressure difference between the top and bottom of the pipe.*

9.2M *Water at 75°C is flowing in a ½-in Type K copper tube at a rate of 12.9 L/min. Calculate the pressure difference between two points 45 m apart if the tube is horizontal.*

9.3E Fuel oil is flowing in a 4-in Schedule 40 steel pipe at the maximum rate for which the flow is laminar. If the oil has a specific gravity of 0.895 and a dynamic viscosity of 8.3×10^{-4} lb-s/ft², calculate the energy loss per 100 ft of pipe.

9.4E A 3-in Schedule 40 steel pipe is 5000 ft long and carries a lubricating oil between two points A and B such that the Reynolds number is 800. Point B is 20 ft higher than A. The oil has a specific gravity of 0.90 and a dynamic viscosity of 4×10^{-4} lb-s/ft². If

the pressure at A is 50 psig, calculate the pressure at B.

9.5M *Benzene at 60°C is flowing in a 1-in Schedule 80 steel pipe at the rate of 20 L/min. The specific weight of the benzene is 8.62 kN/m³. Calculate the pressure difference between two points 100 m apart if the pipe is horizontal.*

9.6M *As a test to determine the wall roughness of an existing pipe installation, water at 10°C is pumped through it at the rate of 225 L/min. The pipe is standard 1½-in commercial steel tubing with a wall thickness of 0.083 in. Pressure gages located 30 m apart in a horizontal run of the pipe read 1035 kPa and 669 kPa. Determine the pipe wall roughness.*

9.7E Water at 80°F flows from a storage tank through 550 ft of 6-in Schedule 40 steel pipe, as shown in Fig. 9.9. Taking the energy loss due to friction into account, calculate the required head h above the pipe inlet to produce a volume flow rate of 2.50 ft³/s.

FIGURE 9.9   Problem 9.7.

6-in Schedule 40 steel pipe

550 ft

9.8E A water main is an 18-in diameter concrete pressure pipe. Calculate the pressure drop over a 1-mi length due to pipe wall friction if the pipe carries 15.0 ft³/s of water at 50°F.

9.9E Figure 9.10 shows a portion of a fire protection system in which a pump draws water at 60°F from a reservoir and delivers it to a point B at the flow rate of 1500 gal/min.
  a. Calculate the required height *h* of the water level in the tank in order to maintain 5.0 psig pressure at point A.
  b. Assuming that the pressure at A is 5.0 psig, calculate the power delivered by the pump to the water in order to maintain the pressure at point B at 85 psig.

Include energy losses due to friction but neglect any other energy losses.

9.10E A submersible deep-well pump delivers 745 gal/h of water at 60°F through a 1-in Schedule 40 steel pipe when operating on the system shown in Fig. 9.11. If the total length of pipe is 140 ft, calculate the power delivered by the pump to the water.

9.11E On a farm, water at 60°F is delivered from a pressurized storage tank to an animal watering trough through 300 ft of 1½-in Schedule 40 steel pipe as shown in Fig. 9.12. Calculate the required air pressure above the water in the tank to produce 75 gal/min of flow.

9.12M *Figure 9.13 shows a system for delivering lawn fertilizer in liquid form. The nozzle on the end of the*

FIGURE 9.10   Problem 9.9.

B

Flow

2600 ft long
8-in Schedule
40 steel pipe

25 ft

Flow

Pump

h

A

45 ft long
10-in Schedule 40 steel pipe

Storage tank

Air
40 psig

Vent

Flow

Well casing

120 ft

Well level

Pump

FIGURE 9.11   Problem 9.10.

hose requires 140 kPa of pressure to operate effectively. The hose is smooth plastic with an inside diameter of 25 mm. The fertilizer solution has a specific gravity of 1.10 and a dynamic viscosity of 2.0×10^{-3} Pa·s. If the length of hose is 85 m, determine (a) the power delivered by the pump to the solution and (b) the pressure at the outlet of the pump. Neglect the energy losses on the suction side of the pump. The flow rate is 95 L/min.

9.13M *A pipeline transporting crude oil (sg = 0.93) at 1200 L/min is made of 6-in Schedule 80 steel pipe. Pumping stations are spaced 3.2 km apart. If the oil is at 10°C, calculate (a) the pressure drop between stations and (b) the power required to maintain the same pressure at the inlet of each pump.*

9.14M *For the pipeline described in Problem 9.13, consider that the oil is to be heated to 100°C to decrease its viscosity.*

a. How does this affect the pump power requirement?

FIGURE 9.12   Problem 9.11.

Air
p = ?

300 ft

3 ft

Flow

FIGURE 9.13   Problems 9.12 and 9.14.

Vent

Pump

10 m

1.2 m

1.5 m

FIGURE 9.14   Problem 9.15.

b. At what distance apart could the pumps be placed with the same pressure drop as that from Problem 9.13?

9.15M *Water at 10°C flows at the rate of 900 L/min from the reservoir and through the pipe shown in Fig. 9.14. Compute the pressure at point B, considering the energy loss due to friction, but neglecting other losses.*

9.16E For the system shown in Fig. 9.15, compute the power delivered by the pump to the water to pump 50 gal/min of water at 60°F to the tank. The air in the tank is at 40 psig. Consider the friction loss in the 225-ft-long discharge pipe, but neglect other losses.

9.17E Fuel oil (sg = 0.94) is being delivered to a furnace at a rate of 60 gal/min through a 1½-in Schedule 40 steel pipe. Compute the pressure difference be-

FIGURE 9.15   Problem 9.16.

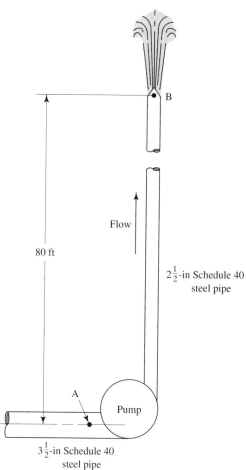

FIGURE 9.16   Problem 9.18.

80 ft

$2\frac{1}{2}$-in Schedule 40
steel pipe

Flow

B

A

Pump

$3\frac{1}{2}$-in Schedule 40
steel pipe

tween two points 40.0 ft apart if the pipe is horizontal and the oil is at 85°F.

9.18E   Figure 9.16 shows a system used to spray polluted water into the air to increase the water's oxygen content and to cause volatile solvents in the water to vaporize. The pressure at point B just ahead of the nozzle must be 25.0 psig for proper nozzle performance. The pressure at point A (the pump inlet) is −3.50 psig. The volume flow rate is 0.50 ft³/s. The dynamic viscosity of the fluid is 4.0×10^{-5} lb·s/ft². The specific gravity of the fluid is 1.026. Compute the power delivered by the pump to the fluid, considering friction energy loss in the discharge line.

9.19E   In a chemical processing system, the flow of glycerine at 60°F (sg = 1.24) in a copper tube must remain laminar with a Reynolds number approximately equal to but not exceeding 300. Specify the smallest standard Type K copper tube that will carry a flow rate of 0.90 ft³/s. Then, for a flow of 0.90 ft³/s in the tube you specified, compute the pressure drop between two points 55.0 ft apart if the pipe is horizontal.

9.20E   Water at 60°F is being pumped from a stream to a reservoir whose surface is 210 ft above the pump. See Figure 9.17. The pipe from the pump to the reservoir is an 8-in Schedule 40 steel pipe 2500 ft long. If 4.00 ft³/s is being pumped, compute the pressure at the outlet of the pump. Consider the

Flow

Reservoir

210 ft

Pump

Stream

FIGURE 9.17   Problems 9.20 and 9.21.

friction loss in the discharge line, but neglect other losses.

9.21E For the pump described in Problem 9.20, if the pressure at the pump inlet is −2.36 psig, compute the power delivered by the pump to the water.

9.22E Gasoline at 50°F flows from A to B through 3200 ft of standard 10-in Schedule 40 steel pipe at the rate of 4.25 ft³/s. Point B is 85 ft above point A and the pressure at B must be 40.0 psig. Considering the friction loss in the pipe, compute the required pressure at A.

9.23E Figure 9.18 shows a pump recirculating 300 gal/min of heavy machine tool lubricating oil at 104°F to test the oil's stability. The total length of 4-in pipe is 25.0 ft, and the total length of 3-in pipe is 75.0 ft. Compute the power delivered by the pump to the oil.

9.24E Linseed oil at 25°C flows at 3.65 m/s in a ¾-in Type K copper tube. Compute the pressure difference between two points in the tube 17.5 m apart if the first point is 1.88 m above the second point.

9.25M *Glycerine at 25°C flows through a straight copper tube (3-in type K) at a flow rate of 180 L/min. Compute the pressure difference between two points 25.8 m apart if the first point is 0.68 m below the second point.*

NOTE: For Problems 9.26 through 9.36, use the equations for friction factor from Section 9.5 to compute the friction factor.

9.26M *Water at 75°C flows in a ½-in Type K copper tube at a rate of 12.9 L/min.*

9.27M *Benzene (sg = 0.88) at 60°C flows in a 1-in Schedule 80 steel pipe at the rate of 20 L/min.*

9.28E Water at 80°F flows in a 6-in ductile pipe at a rate of 2.50 ft³/s. Assume the roughness $\epsilon = 2.0 \times 10^{-4}$ ft.

9.29E Water at 50°F flows at 15.0 ft³/s in a concrete pipe with an inside diameter of 18.0 in.

9.30E Water at 60°F flows at 1500 gal/min in a 10-in Schedule 40 steel pipe.

9.31M *A liquid fertilizer solution (sg = 1.10) with a dynamic viscosity of 2.0×10^{-3} Pa·s flows at 95 L/min through a 25-mm diameter smooth plastic hose.*

9.32M *Crude oil (sg = 0.93) at 100°C flows at a rate of 1200 L/min in a 6-in Schedule 80 steel pipe.*

9.33M *Water at 65°C flows in a 1½-in Schedule 40 steel pipe at a rate of 10 m/s.*

9.34M *Propyl alcohol flows in a 3-in Type K copper tube at 25°C at a rate of 0.026 m³/s.*

9.35E Water at 70°F flows in a 12-in diameter concrete pipe at 3.0 ft³/s.

FIGURE 9.18   Problem 9.23.

6 ft

22 ft

Flow

15 ft

Discharge line
3-in Schedule 40
steel pipe

Suction line
4-in Schedule 40
steel pipe

Pump

9.36E Heavy fuel oil at 77°F flows in a 6-in Schedule 40 steel pipe at 12 ft/s.

Noncircular Cross Sections

9.37E For the system described in Problem 8.33, compute the pressure difference between two points 30.0 ft apart if the duct is horizontal. Use $\epsilon = 8.5 \times 10^{-5}$ ft.

9.38M *For the shell-and-tube heat exchanger described in Problem 8.34, compute the pressure difference for both fluids between two points 5.25 m apart if the heat exchanger is horizontal.*

9.39M *For the system described in Problem 8.35, compute the pressure drop for both fluids between two points 3.80 m apart if the duct is horizontal. Use the roughness for steel pipe for all surfaces.*

9.40E For the system described in Problem 8.36, compute the pressure difference in both the small pipes and the large pipe between two points 50.0 ft apart if the pipes are horizontal. Use the roughness for steel pipe for all surfaces.

9.41M *For the shell-and-tube heat exchanger described in Problem 8.37, compute the pressure drop for the flow of water in the shell. Use the roughness for copper for all surfaces. The length is 3.60 m.*

9.42E For the heat exchanger described in Problem 8.38, compute the pressure drop for a length of 57 in.

9.43M *For the glycerine described in Problem 8.40, compute the pressure drop for a horizontal duct 22.6 m long. All surfaces are copper.*

9.44M *For the flow of water in the square tubes described in Problem 8.41, compute the pressure drop over a length of 22.6 m. All surfaces are copper, and the duct is horizontal.*

9.45E If the heat sink described in Problem 8.42 is 105 in long, compute the pressure drop for the water. Use $\epsilon = 2.5 \times 10^{-5}$ ft for the aluminum.

9.46E Compute the energy loss for the flow of water in the cooling passage described in Problem 8.43 if its total length is 45 in.

9.47E In Fig. 9.19, ethylene glycol (sg = 1.10) at 77°F flows around the tubes and inside the rectangular passage. Calculate the volume flow rate of ethylene glycol in gal/min required for the flow to have a Reynolds number of 8000. Then, compute the energy loss over a length of 128 in. All surfaces are brass.

1.50 in

$\frac{3}{8}$ O.D.
0.049 wall thickness brass tubes

2.25 in

FIGURE 9.19 Problem 9.47.

9.48M *Figure 9.20 shows a duct in which methyl alcohol at 25°C flows at the rate of 3000 L/min. Compute the energy loss over a 2.25-m length of the duct. All surfaces are smooth plastic.*

9.49E A furnace heat exchanger has a cross section like that shown in Fig. 9.21. The air flows around the three thin passages in which hot gases flow. The air is at 140°F and has a density of 2.06×10^{-3} slugs/ft³ and a dynamic viscosity of 4.14×10^{-7} lb·s/ft². Compute the Reynolds number for the flow if the velocity is 20 ft/sec.

9.50E Figure 9.22 shows a system in which methyl alcohol at 77°F flows outside the three tubes while ethyl alcohol at 0°F flows inside the tubes. Compute the volume flow rate of each fluid required to produce a Reynolds number of 3.5×10^4 in all parts of the system. Then, compute the pressure difference between two points 10.5 ft apart if the system is horizontal. All surfaces are copper.

Velocity Profile for Turbulent Flow

9.51M *For the flow of 12.9 L/min of water at 75°C in a ½-in Type K copper tube, compute the expected maximum velocity of flow from Eq. (9–14).*

9.52M *A large pipeline with a 1.200-m inside diameter carries oil similar to SAE 10 at 40°C (sg = 0.8). Compute the volume flow rate required to produce a Reynolds number of 3.60×10^4. Then, if the pipe is clean steel, compute several points on the velocity profile and plot the data in a manner similar to that shown in Fig. 9.7.*

FIGURE 9.20   Problem 9.48.

FIGURE 9.21   Problem 9.49.

FIGURE 9.22   Problem 9.50.

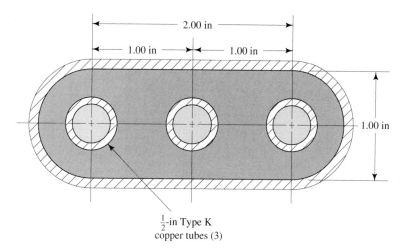

$\frac{1}{2}$-in Type K
copper tubes (3)

9.53M *Repeat Problem 9.52 if the oil is at 110°C but with the same flow rate. Discuss the differences in the velocity profile.*

9.54 Using Eq. (9–13), compute the distance y for which the local velocity U is equal to the average velocity v.

9.55E The result for Problem 9.54 predicts that the average velocity for turbulent flow will be found at a distance of $0.216(r_o)$ from the wall of the pipe. Compute this distance for a 24-in Schedule 40 steel pipe. Then, if the pipe carries water at 50°F at a flow rate of 16.75 ft³/s, compute the velocity at points 0.50 in on either side of the average velocity point.

9.56 Using Eq. (9–14), compute the ratio of the average velocity to the maximum velocity of flow in smooth pipes with Reynolds numbers of 4000, 10^4, 10^5, and 10^6.

9.57 Using Eq. (9–14), compute the ratio of the average velocity to the maximum velocity of flow for the flow of a liquid through a concrete pipe with an inside diameter of 8.00 in with Reynolds numbers of 4000, 10^4, 10^5, and 10^6.

9.58E Using Eq. (9–13), compute several points on the velocity profile for the flow of 400 gal/min of water at 50°F in a new, clean 4-in Schedule 40 steel pipe. Make a plot similar to Fig. 9.7 with a fairly large scale.

9.59E Repeat Problem 9.58 for the same conditions, except that the inside of the pipe is roughened by age

so that $\epsilon = 5.0 \times 10^{-3}$. Plot the results on the same graph as that used for the results of Problem 9.58.

9.60E For both situations described in Problems 9.58 and 9.59, compute the pressure drop that would occur over a distance of 250 ft of horizontal pipe.

Energy Loss Using the Hazen-Williams Formula

Use the design values for the coefficient C_h from Table 9.2 unless stated otherwise. Use either of the various forms of the formula or the nomograph in Fig. 9.8 as assigned.)

9.61E Water flows at a rate of 1.50 ft³/s through 550 ft of 6-in cement-lined ductile iron pipe. Compute the energy loss.

9.62M *Compute the energy loss as water flows in a 4-in type K copper tube at a rate of 1000 L/min over a length of 45 m.*

9.63E A water main is an 18-in diameter concrete pressure pipe. Calculate the energy loss over a 1-mile length if it carries 7.50 ft³/s of water.

9.64E A fire protection system includes 1500 ft of 10-in Schedule 40 steel pipe. Compute the energy loss in the pipe when it carries 1500 gal/min of water.

9.65M *A 4-in type K copper tube carries 900 L/min of water over a length of 80 m. Compute the energy loss.*

9.66E Compute the energy loss as 0.20 ft³/s of water flows through a length of 80 ft of 2½-in Schedule 40 steel pipe.

9.67E It is desired to flow 2.0 ft³/s of water through 2500 ft of 8-in pipe. Compute the head loss for both plain Schedule 40 steel pipe and ductile iron pipe coated with a centrifugally applied cement lining.

9.68E Specify a suitable size of new, clean Schedule 40 steel pipe that would carry 300 gal/min of water over a length of 1200 ft with no more than 10 ft of head loss. For the selected pipe, compute the actual expected head loss.

9.69E For the pipe selected in Problem 9.68, compute the head loss using the design value for C_h rather than that for new, clean pipe.

9.70E Compare the head loss that would result from the flow of 100 gal/min of water through 1000 ft of new, clean Schedule 40 steel pipe for 2-in and 3-in sizes.

COMPUTER PROGRAMMING ASSIGNMENTS

1. Write a program to compute the friction factor for the flow of any fluid through pipes and tubes using Eqs. (9–3) and (9–9). The program must compute the Reynolds number and the relative roughness. Then, decisions must be made as follows:

   a. If $N_R < 2000$, use $f = 64/N_R$ [Eq. (9–3)].

   b. If $2000 < N_R < 4000$, the flow is in the critical range and no reliable value can be computed for f. Print a message to the user of the program.

   c. If $N_R > 4000$, the flow is turbulent. Use Eq. (9–9) to compute f.

   d. Print out N_R, D/ϵ, and f.

2. Incorporate Program 1 into an enhanced program to compute the pressure drop for the flow of any fluid through a pipe of any size. The two points of interest can be any distance apart, and one end can be at any elevation relative to the other. The program should be able to complete such analyses as required for Problems 9.1, 9.2, and 9.5. The program also can be set up to determine the energy loss only in order to solve problems such as 9.3.

3. Write a program to compute the velocity profile for turbulent flow using Eqs. (9–12), (9–13), and (9–14). Specified increments of position within the pipe can be input by the operator. Program 1 could be incorporated.

4. Write a program to solve the Hazen-Williams formula in any of its forms listed in Table 9.3. Allow the program operator to specify the unit system to be used, which values are known, and which values are to be solved.

■■■■ 10 Minor Losses

10.1 OBJECTIVES

Here we continue to develop the techniques for analyzing real pipe line problems in which several types of flow system components exist. These techniques are part of the overall goal of computing the energy losses in a flow system. In Chapter 9, we developed the method of computing the energy loss due to friction as fluid flows in straight sections of pipe. This chapter deals with the *minor losses* due to valves, fittings, changes in the size of the flow path, and changes in direction of the flow.

After completing this chapter, you should be able to:

1. Recognize the sources of minor losses.
2. Define *resistance coefficient*.
3. Determine the energy loss for flow through the following types of minor losses:
    a. Sudden enlargement of the flow path.
    b. Exit loss when fluid leaves a pipe and enters a static reservoir.
    c. Gradual enlargement of the flow path.
    d. Sudden contraction of the flow path.
    e. Gradual contraction of the flow path.
    f. Entrance loss when fluid enters a pipe from a static reservoir.
4. Define the term *vena contracta*.
5. Define and use the *equivalent length technique* for computing energy losses in valves, fittings, and pipe bends.

10.2 SOURCES OF MINOR LOSSES

In most pipe flow systems the primary energy loss is due to pipe friction, as described in Chapter 9. Other types of losses are usually small by comparison, and they are therefore referred to as *minor losses*. Minor losses occur when there is a change in the cross section of the flow path or in the direction of flow or where the flow path is obstructed, as with a valve. Energy is lost under these conditions due to rather complex physical phenomena. Theoretical prediction of the magnitude of these losses is also complex, and therefore, experimental data are normally used.

The procedures developed in this chapter for analyzing minor losses were taken from many sources, some of which are listed as references at the end of the chapter. The various sets of data are presented in a form that you can easily use in the analysis of pipe flow problems.

10.3 RESISTANCE COEFFICIENT

Energy losses are proportional to the velocity head of the fluid as it flows around an elbow, through an enlargement or contraction of the flow section, or through a valve. Experimental values for energy losses are usually reported in terms of a resistance coefficient, K, as follows:

$$h_L = K(v^2/2g) \tag{10-1}$$

In Eq. (10–1), h_L is the minor loss, K is the resistance coefficient, and v is the average velocity of flow in the pipe in the vicinity where the minor loss occurs. In some cases, there may be more than one velocity of flow, as with enlargements or contractions. It is most important for you to know which velocity is to be used with each resistance coefficient.

If the velocity head $v^2/2g$ in Eq. (10–1) is expressed in the units of meters, then the energy loss h_L will also be in meters or N·m/N of fluid flowing. The resistance coefficient is unitless, as it represents a constant of proportionality between the energy loss and the velocity head. The magnitude of the resistance coefficient depends on the geometry of the device that causes the loss and sometimes on the velocity of flow. In the following sections, we will describe the process for determining the value of K and for calculating the energy loss for many types of minor loss conditions.

10.4 SUDDEN ENLARGEMENT

As a fluid flows from a smaller pipe into a larger pipe through a sudden enlargement, its velocity abruptly decreases, causing turbulence that generates an energy loss (see Fig. 10.1). The amount of turbulence, and therefore the amount of energy loss, is dependent on the ratio of the sizes of the two pipes.

FIGURE 10.1 Sudden enlargement.

Region of turbulence

v_1

D_1

D_2

The minor loss is calculated from the equation

$$h_L = K(v_1^2/2g) \tag{10-2}$$

where v_1 is the average velocity of flow in the smaller pipe ahead of the enlargement. Tests have shown that the value of the loss coefficient K is

FIGURE 10.2   Resistance coefficient—Sudden enlargement.

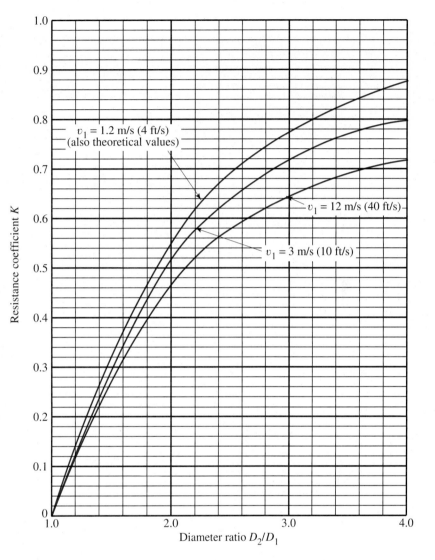

dependent on both the ratio of the sizes of the two pipes and the magnitude of the flow velocity. This is illustrated graphically in Fig. 10.2 and in tabular form in Table 10.1.

By making some simplifying assumptions about the character of the flow stream as it expands through the sudden enlargement, it is possible to analytically predict the value of K from the following equation:

$$K = [1 - (A_1/A_2)]^2 = [1 - (D_1/D_2)^2]^2 \qquad \textbf{(10–3)}$$

The subscripts 1 and 2 refer to the smaller and larger sections, respectively, as shown in Fig. 10.1. Values for K from this equation agree well with experimental data when the velocity v_1 is approximately 1.2 m/s. At higher

TABLE 10.1   Resistance coefficient—Sudden enlargement

D_2/D_1	Velocity, v_1						
	0.6 m/s 2 ft/s	1.2 m/s 4 ft/s	3 m/s 10 ft/s	4.5 m/s 15 ft/s	6 m/s 20 ft/s	9 m/s 30 ft/s	12 m/s 40 ft/s
1.0	0.0	0.0	0.0	0.0	0.0	0.0	0.0
1.2	0.11	0.10	0.09	0.09	0.09	0.09	0.08
1.4	0.26	0.25	0.23	0.22	0.22	0.21	0.20
1.6	0.40	0.38	0.35	0.34	0.33	0.32	0.32
1.8	0.51	0.48	0.45	0.43	0.42	0.41	0.40
2.0	0.60	0.56	0.52	0.51	0.50	0.48	0.47
2.5	0.74	0.70	0.65	0.63	0.62	0.60	0.58
3.0	0.83	0.78	0.73	0.70	0.69	0.67	0.65
4.0	0.92	0.87	0.80	0.78	0.76	0.74	0.72
5.0	0.96	0.91	0.84	0.82	0.80	0.77	0.75
10.0	1.00	0.96	0.89	0.86	0.84	0.82	0.80
∞	1.00	0.98	0.91	0.88	0.86	0.83	0.81

Source: H. W. King and E. F. Brater. 1963. *Handbook of Hydraulics*, 5th ed. New York: McGraw-Hill. (Table 6–7. Velocities converted to SI units.)

velocities, the actual values of K are lower than the theoretical values. We recommend that experimental values be used if the velocity of flow is known.

☐ **EXAMPLE PROBLEM 10.1**   Determine the energy loss that will occur as 100 L/min of water flows through a sudden enlargement from a 1-in copper tube (Type K) to a 3-in tube (Type K). See Appendix H for tube dimensions.

Solution   Using the subscript 1 for the section just ahead of the enlargement and 2 for the section downstream from the enlargement, we get

$$D_1 = 25.3 \text{ mm} = 0.0253 \text{ m}$$

$$A_1 = 5.017 \times 10^{-4} \text{ m}^2$$

$$D_2 = 73.8 \text{ mm} = 0.0738 \text{ m}$$

$$A_2 = 4.282 \times 10^{-3} \text{ m}^2$$

$$v_1 = \frac{Q}{A_1} = \frac{100 \text{ L/min}}{5.017 \times 10^{-4} \text{ m}^2} \times \frac{1 \text{ m}^3/\text{s}}{60\ 000 \text{ L/min}} = 3.32 \text{ m/s}$$

$$\frac{v_1^2}{2g} = \frac{(3.32)^2}{(2)(9.81)} \text{ m} = 0.56 \text{ m}$$

To find a value for K, the diameter ratio is needed. We find that

$$D_2/D_1 = 73.8/25.3 = 2.92$$

From Fig. 10.2, $K = 0.72$. Then we have

$$h_L = K(v_1^2/2g) = (0.72)(0.56 \text{ m}) = 0.40 \text{ m}$$

This result indicates that 0.40 N·m of energy is dissipated from each newton of water that flows through the sudden enlargement. The following problem illustrates the calculation of the pressure difference between points 1 and 2.

☐ **EXAMPLE PROBLEM 10.2**   Determine the difference between the pressure ahead of a sudden enlargement and the pressure downstream from the enlargement. Use the data from Example Problem 10.1.

Solution   First, the energy equation must be written:

$$\frac{p_1}{\gamma} + z_1 + \frac{v_1^2}{2g} - h_L = \frac{p_2}{\gamma} + z_2 + \frac{v_2^2}{2g}$$

Solving for $p_1 - p_2$ gives

$$p_1 - p_2 = \gamma[(z_2 - z_1) + (v_2^2 - v_1^2)/2g + h_L]$$

If the enlargement is horizontal, $z_2 - z_1 = 0$. Even if it were vertical, the distance between points 1 and 2 is so small that it is considered negligible. Now, calculating the velocity in the larger pipe, we get

$$v_2 = \frac{Q}{A_2} = \frac{100 \text{ L/min}}{4.282 \times 10^{-3} \text{ m}^2} \times \frac{1 \text{ m}^3/\text{s}}{60\,000 \text{ L/min}} = 0.39 \text{ m/s}$$

Using $\gamma = 9.81$ kN/m³ for water and $h_L = 0.40$ m from Example Problem 10.1, we have

$$p_1 - p_2 = \frac{9.81 \text{ kN}}{\text{m}^3} \left[0 + \frac{(0.39)^2 - (3.32)^2}{(2)(9.81)} \text{ m} + 0.40 \text{ m} \right]$$

$$= -1.51 \text{ kN/m}^2 = -1.51 \text{ kPa}$$

Therefore, p_2 is 1.51 kPa greater than p_1.

**10.5
EXIT LOSS**

As a fluid flows from a pipe into a large reservoir or tank, as shown in Fig. 10.3, its velocity is decreased to very nearly zero. In the process, the kinetic energy that the fluid possessed in the pipe, indicated by the velocity head

FIGURE 10.3 Exit loss as fluid flows from a pipe into a static reservoir.

$v_1^2/2g$, is dissipated. Therefore, the energy loss for this condition is

$$h_L = 1.0(v_1^2/2g) \qquad \text{(10–4)}$$

This is called the *exit loss*. The value of $K = 1.0$ is used regardless of the form of the exit where the pipe connects to the tank wall.

☐ **EXAMPLE PROBLEM 10.3**   Determine the energy loss that will occur as 100 L/min of water flows from a 1-in copper tube (Type K) into a large tank.

Solution   Using Eq. (10–4), we have

$$h_L = 1.0(v_1^2/2g)$$

From the calculations in Example Problem 10.1, we know that

$$v_1 = 3.32 \text{ m/s}$$
$$v_1^2/2g = 0.56 \text{ m}$$

Then the energy loss is

$$h_L = (1.0)(0.56 \text{ m}) = 0.56 \text{ m}$$

**10.6
GRADUAL
ENLARGEMENT**

If the transition from a smaller to a larger pipe can be made less abrupt than the square-edged sudden enlargement, the energy loss is reduced. This is normally done by placing a conical section between the two pipes as shown in Fig. 10.4. The sloping walls of the cone tend to guide the fluid during the deceleration and expansion of the flow stream.

FIGURE 10.4   Gradual enlargement.

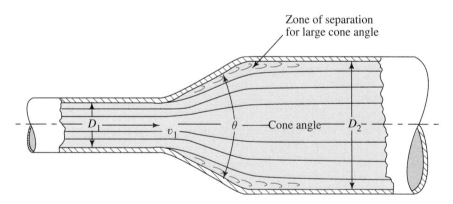

The energy loss for a gradual enlargement is calculated from

$$h_L = K(v_1^2/2g) \qquad \text{(10–5)}$$

where v_1 is the velocity in the smaller pipe ahead of the enlargement. The magnitude of K is dependent on both the diameter ratio D_2/D_1 and the cone angle θ. Data for various values of θ and D_2/D_1 are given in Fig. 10.5 and Table 10.2.

FIGURE 10.5 Resistance coefficient—Gradual enlargement.

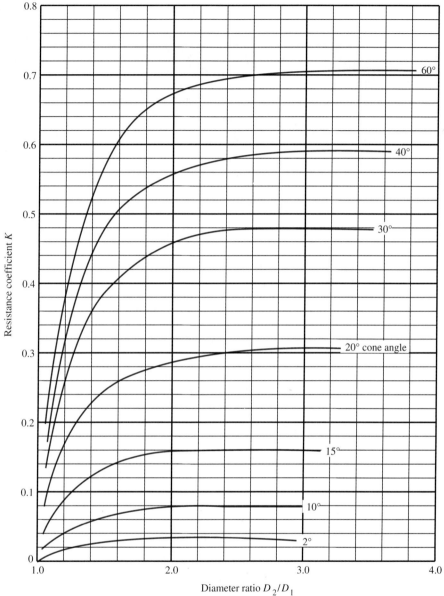

The energy loss calculated from Eq. (10–5) does not include the loss due to friction at the walls of the transition. For relatively steep cone angles, the length of the transition is short and therefore the wall friction loss is negligible. However, as the cone angle decreases, the length of the transition increases and wall friction becomes significant. Taking both wall friction loss and the loss due to the enlargement into account, we can obtain the minimum energy loss with a cone angle of about 7°.

TABLE 10.2   Resistance coefficient—Gradual enlargement

D_2/D_1	Angle of Cone, θ											
	2°	6°	10°	15°	20°	25°	30°	35°	40°	45°	50°	60°
1.1	0.01	0.01	0.03	0.05	0.10	0.13	0.16	0.18	0.19	0.20	0.21	0.23
1.2	0.02	0.02	0.04	0.09	0.16	0.21	0.25	0.29	0.31	0.33	0.35	0.37
1.4	0.02	0.03	0.06	0.12	0.23	0.30	0.36	0.41	0.44	0.47	0.50	0.53
1.6	0.03	0.04	0.07	0.14	0.26	0.35	0.42	0.47	0.51	0.54	0.57	0.61
1.8	0.03	0.04	0.07	0.15	0.28	0.37	0.44	0.50	0.54	0.58	0.61	0.65
2.0	0.03	0.04	0.07	0.16	0.29	0.38	0.46	0.52	0.56	0.60	0.63	0.68
2.5	0.03	0.04	0.08	0.16	0.30	0.39	0.48	0.54	0.58	0.62	0.65	0.70
3.0	0.03	0.04	0.08	0.16	0.31	0.40	0.48	0.55	0.59	0.63	0.66	0.71
∞	0.03	0.05	0.08	0.16	0.31	0.40	0.49	0.56	0.60	0.64	0.67	0.72

Source: King, H. W., and E. F. Brater. 1963. *Handbook of Hydraulics*, 5th ed. New York: McGraw-Hill. (Table 6–8.)

☐ **EXAMPLE PROBLEM 10.4**   Determine the energy loss that will occur as 100 L/min of water flows from a 1-in copper tube (Type K) into a 3-in copper tube (Type K) through a gradual enlargement having an included cone angle of 30°.

Solution   Using data from Appendix H and the results of some calculations in preceding example problems, we know that

$$v_1 = 3.32 \text{ m/s}$$
$$v_1^2/2g = 0.56 \text{ m}$$
$$D_2/D_1 = 73.8/25.3 = 2.92$$

From Fig. 10.5, we find that $K = 0.48$. Then we have

$$h_L = K(v_1^2/2g) = (0.48)(0.56 \text{ m}) = 0.27 \text{ m}$$

Compared with the sudden enlargement described in Example Problem 10.1, the energy loss decreases by 33 percent when the 30° gradual enlargement is used.

**10.7
SUDDEN
CONTRACTION**

The energy loss due to a sudden contraction, such as that sketched in Fig. 10.6, is calculated from

$$h_L = K(v_2^2/2g) \tag{10–6}$$

where v_2 is the velocity in the smaller pipe downstream from the contraction. The resistance coefficient K is dependent on the ratio of the sizes of the two pipes and on the velocity of flow, as Fig. 10.7 and Table 10.3 show.

The mechanism by which energy is lost due to a sudden contraction is quite complex. Figure 10.8 illustrates what happens as the flow stream converges. The lines in the figure represent the paths of various parts of the flow stream called *streamlines*. As the streamlines approach the contraction, they assume a curved path and the total stream continues to neck down for

FIGURE 10.6 Sudden contraction.

FIGURE 10.7 Resistance coefficient—Sudden contraction.

TABLE 10.3 Resistance coefficient—Sudden contraction

D_1/D_2	Velocity, v_2								
	0.6 m/s 2 ft/s	1.2 m/s 4 ft/s	1.8 m/s 6 ft/s	2.4 m/s 8 ft/s	3 m/s 10 ft/s	4.5 m/s 15 ft/s	6 m/s 20 ft/s	9 m/s 30 ft/s	12 m/s 40 ft/s
1.0	0.0	0.0	0.0	0.0	0.0	0.0	0.0	0.0	0.0
1.1	0.03	0.04	0.04	0.04	0.04	0.04	0.05	0.05	0.06
1.2	0.07	0.07	0.07	0.07	0.08	0.08	0.09	0.10	0.11
1.4	0.17	0.17	0.17	0.17	0.18	0.18	0.18	0.19	0.20
1.6	0.26	0.26	0.26	0.26	0.26	0.25	0.25	0.25	0.24
1.8	0.34	0.34	0.34	0.33	0.33	0.32	0.31	0.29	0.27
2.0	0.38	0.37	0.37	0.36	0.36	0.34	0.33	0.31	0.29
2.2	0.40	0.40	0.39	0.39	0.38	0.37	0.35	0.33	0.30
2.5	0.42	0.42	0.41	0.40	0.40	0.38	0.37	0.34	0.31
3.0	0.44	0.44	0.43	0.42	0.42	0.40	0.39	0.36	0.33
4.0	0.47	0.46	0.45	0.45	0.44	0.42	0.41	0.37	0.34
5.0	0.48	0.47	0.47	0.46	0.45	0.44	0.42	0.38	0.35
10.0	0.49	0.48	0.48	0.47	0.46	0.45	0.43	0.40	0.36
∞	0.49	0.48	0.48	0.47	0.47	0.45	0.44	0.41	0.38

Source: King, H. W., and E. F. Brater. 1963. *Handbook of Hydraulics*. 5th ed. New York: McGraw-Hill. (Table 6–9. Velocities converted to SI units.)

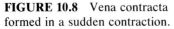

FIGURE 10.8 Vena contracta formed in a sudden contraction.

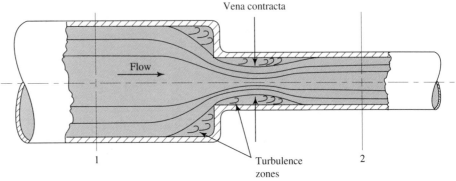

some distance beyond the contraction. Thus, the minimum cross section of the flow is smaller than that of the smaller pipe. The section were this minimum flow area occurs is called the *vena conracta*. Beyond the vena contracta, the flow stream must decelerate and expand again to fill the pipe. The turbulence caused by the contraction and the subsequent expansion generates the energy loss.

☐ **EXAMPLE PROBLEM 10.5**   Determine the energy loss that will occur as 100 L/min of water flows from a 3-in copper tube (Type K) into a 1-in copper tube (Type K) through a sudden contraction.

Solution   From Eq. (10–6), we have

$$h_L = K(v_2^2/2g)$$

For the copper tube we know that $D_1 = 73.8$ mm, $D_2 = 25.3$ mm, and $A_2 = 5.017 \times 10^{-4}$ m^2. Then we can find the following values:

$$D_1/D_2 = 73.8/25.3 = 2.92$$

$$v_2 = \frac{Q}{A_2} = \frac{100 \text{ L/min}}{5.017 \times 10^{-4} \text{ m}^2} \times \frac{1 \text{ m}^3/\text{s}}{60\,000 \text{ L/min}} = 3.32 \text{ m/s}$$

$$v_2^2/2g = 0.56 \text{ m}$$

From Fig. 10.7 we can find $K = 0.42$. Then we have

$$h_L = K(v_2^2/2g) = (0.42)(0.56 \text{ m}) = 0.24 \text{ m}$$

10.8 GRADUAL CONTRACTION

The energy loss in a contraction can be decreased substantially by making the contraction more gradual. Figure 10.9 shows such a gradual contraction, formed by a conical section between the two diameters with sharp breaks at the junctions. The angle θ is called the *cone angle*.

Figure 10.10 shows the data (from Reference 5) for the resistance coefficient versus the diameter ratio for several values of the cone angle. The energy loss is computed from Eq. (10–6), where the resistance coefficient is based on the velocity head in the smaller pipe after the contraction. These data are for Reynolds numbers greater than 1.0×10^5. Note that for angles

FIGURE 10.9 Gradual contraction.

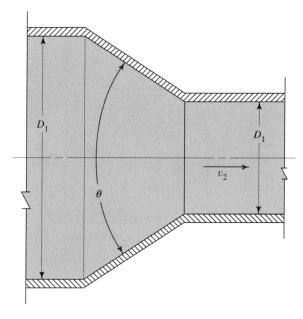

FIGURE 10.10 Resistance coefficient—Gradual contraction.

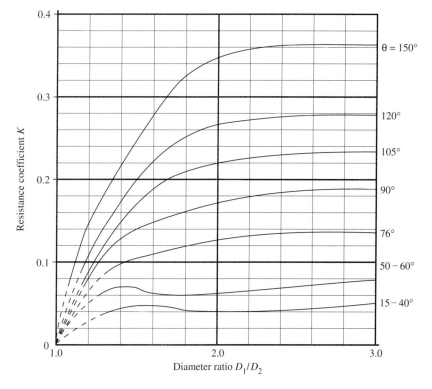

FIGURE 10.11 Resistance coefficient—Gradual contraction.

FIGURE 10.12 Gradual contraction with rounded end at small diameter.

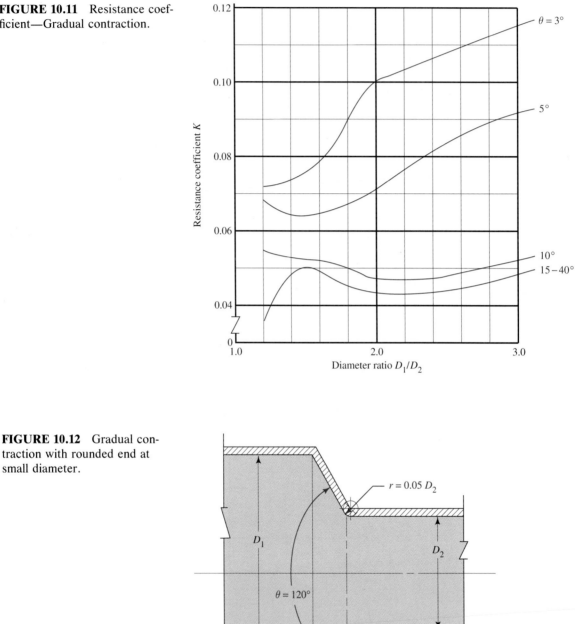

over the wide range of 15° to 40°, $K = 0.05$ or less, a very low value. For angles as high as 60°, K is less than 0.08.

As the cone angle of the contraction decreases below 15 degrees, the resistance coefficient actually increases, as shown in Fig. 10.11. The reason is that the data include the effects of both the local turbulence caused by flow separation and pipe friction. For the smaller cone angles, the transition between the two diameters is very long, which increases the friction losses.

Rounding the end of the conical transition to blend it with the smaller pipe can decrease the resistance coefficient to below the values shown in Fig. 10.10. For example, in Fig. 10.12, which shows a contraction with a 120° included angle, the value of K decreases from approximately 0.27 to 0.10 with a radius of only $0.05(D_2)$, where D_2 is the inside diameter of the smaller pipe.

**10.9
ENTRANCE LOSS**

A special case of a contraction occurs when a fluid flows from a relatively large reservoir or tank into a pipe. The fluid must accelerate from a negligible velocity to the flow velocity in the pipe. The ease with which the acceleration is accomplished determines the amount of energy loss, and therefore, the value of the entrance resistance coefficient is dependent on the geometry of the entrance. Figure 10.13 shows four different configurations and the suggested value of K for each. The streamlines illustrate the flow of fluid into the pipe and show that the turbulence associated with the formation of a vena contracta in the tube is a major cause of the energy loss. This condition is most severe for the inward-projecting entrance, for which a conservative value of $K = 1.0$ is recommended for problems in this book. Reference 2 recommends $K = 0.78$. A more precise estimate of the resistance coefficient for an inward-projecting entrance is given in Reference 5. For a *well-rounded entrance* with $r/D_2 > 0.15$, no vena contracta is formed, the energy loss is quite small, and we use $K = 0.04$. In summary, after selecting a value for the resistance coefficient from Fig. 10.13, we can calculate the energy loss at an entrance from

$$h_L = K(v_2^2/2g) \tag{10-7}$$

where v_2 is the velocity of flow in the pipe.

☐ **EXAMPLE PROBLEM 10.6**

Determine the energy loss that will occur as 100 L/min of water flows from a reservoir into a 1-in copper tube (Type K) (a) through an inward projecting tube and (b) through a well-rounded inlet.

Solution

Part (a): For the tube, $D_2 = 25.3$ mm and $A_2 = 5.017 \times 10^{-4}$ m². Then we get

$$v_2 = Q/A_2 = 3.32 \text{ m/s} \quad \text{(from Example Problem 10.1)}$$
$$v_2^2/2g = 0.56 \text{ m}$$

For an inward projecting entrance, $K = 1.0$. Then we have

$$h_L = (1.0)(0.56 \text{ m}) = 0.56 \text{ m}$$

FIGURE 10.13   Entrance resistance coefficients.

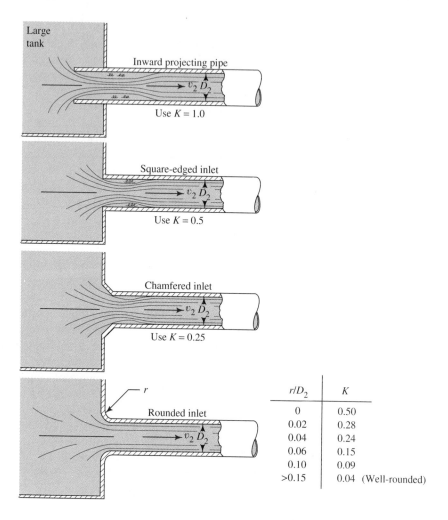

r/D_2	K	
0	0.50	
0.02	0.28	
0.04	0.24	
0.06	0.15	
0.10	0.09	
>0.15	0.04	(Well-rounded)

Part (b): For a well-rounded inlet, $K = 0.04$. Then we have

$$h_L = (0.04)(0.56 \text{ m}) = 0.02 \text{ m}$$

**10.10
RESISTANCE
COEFFICIENTS FOR
VALVES AND FITTINGS**

Many different kinds of valves and fittings are available from several manufacturers for specification and installation into fluid flow systems. Valves are used to control the amount of flow and may be globe valves, angle valves, gate valves, butterfly valves, any of several types of check valves, and many more. See Figs. 10.14 to 10.19 for some examples. Fittings direct the path of flow or cause a change in the size of the flow path. Included are elbows of several designs, tees, reducers, nozzles, and orifices. See Figs. 10.20 and 10.21.

It is important to determine the resistance data for the particular type and size chosen because the resistance is dependent on the geometry of the

FIGURE 10.14 Globe valve. (Source: Crane Valves, Joliet, IL)

(a)

(b)

FIGURE 10.15 Angle valve. (Source: Crane Valves, Joliet, IL)

(a)

(b)

FIGURE 10.16 Gate valve. (Source: Crane Valves, Joliet, IL)

FIGURE 10.17 Check valve—Swing type. (Source: Crane Valves, Joliet, IL)

(a)

(b)

FIGURE 10.18 Check valve—
Ball type. (Source: Crane Valves,
Joliet, IL)

FIGURE 10.19 Butterfly valve.
(Source: Crane Valves, Joliet,
IL)

FIGURE 10.20 Pipe elbows.
(Source: Crane Valves, Joliet,
IL)

90° elbow

(a)

90° long radius elbow

(b)

45° elbow

(c)

90° street elbow

(d)

45° street elbow

(e)

Return bend

(f)

FIGURE 10.21 Standard tees.
(Source: Crane Valves, Joliet,
IL)

(a) Flow through run

(b) Flow through branch

valve for fitting. Also, different manufacturers may report data in different forms.

Data reported here are taken from Reference 2, which includes a much more extensive list.

Energy loss incurred as fluid flows through a valve or fitting is computed from Eq. 10–1 as used for the minor losses already discussed. However, the method of determining the resistance coefficient K is different. The value of K is reported in the form

$$K = (L_e/D) f_T \qquad \text{(10–8)}$$

The value of L_e/D, called the equivalent length ratio, is reported in Table 10.4 and it is considered to be constant for a given type of valve or fitting. The value of L_e itself is called the equivalent length and is the length of straight pipe of the same nominal diameter as the valve that would have the same resistance as the valve. The term D is the actual inside diameter of the pipe.

The term f_T is the friction factor in the pipe to which the valve or fitting is connected, *taken to be in the zone of complete turbulence*. Note in Fig.

TABLE 10.4   Resistance in valves and fittings expressed as equivalent length in pipe diameters, L_e/D

Type	Equivalent Length in Pipe Diameters, L_e/D
Globe valve—fully open	340
Angle valve—fully open	150
Gate valve—fully open	8
—¾ open	35
—½ open	160
—¼ open	900
Check valve—swing type	100
Check valve—ball type	150
Butterfly valve—fully open	45
90° standard elbow	30
90° long radius elbow	20
90° street elbow	50
45° standard elbow	16
45° street elbow	26
Close return bend	50
Standard tee—with flow through run	20
—with flow through branch	60

Source: Crane Valves, Joliet, IL

9.2, the Moody diagram, that the zone of complete turbulence lies in the far right area where the friction factor is independent of Reynolds number. The dashed line running generally diagonally across the diagram divides the zone of complete turbulence from the transition zone to the left.

Values for f_T vary with the size of the pipe and valve, causing the value of the resistance coefficient K to also vary. Table 10.5 lists the values of f_T for standard sizes of new, clean, commercial steel pipe.

TABLE 10.5 Friction factor in zone of complete turbulence for new, clean commercial steel pipe

Nominal Pipe Size (in)	Friction Factor, f_T	Nominal Pipe Size (in)	Friction Factor, f_T
½	0.027	4	0.017
¾	0.025	5	0.016
1	0.023	6	0.015
1¼	0.022	8–10	0.014
1½	0.021	12–16	0.013
2	0.019	18–24	0.012
2½, 3	0.018		

Some system designers prefer to compute the equivalent length of pipe for a valve and combine that value with the actual length of pipe. Equation (10–8) can be solved for L_e.

$$L_e = K \, D/f_T \qquad (10\text{–}9)$$

Note, however, that this would be valid only if the flow in the pipe is in the zone of complete turbulence.

☐ **EXAMPLE PROBLEM 10.7**   Determine the resistance coefficient K for a fully open globe valve placed in a 6-in Schedule 40 steel pipe.

Solution   From Table 10.4 we find that the equivalent length ratio L_e/D for a fully open globe valve is 340. From Table 10.5 we find $f_T = 0.015$ for a 6-in pipe. Then,

$$K = (L_e/D) \, f_T = (340)(0.015) = 5.10$$

Using $D = 0.5054$ ft for the pipe, the equivalent length is

$$L_e = K \, D/f_T = (5.10)(0.5054 \text{ ft})/(0.015) = 172 \text{ ft}$$

The physical appearances of typical valves and fittings listed in Table 10.4 are shown in Figs. 10.14 through 10.21. The magnitude of the equivalent length ratio L_e/D, and therefore of the energy loss, is dependent on the complexity of the fluid path through the device.

The internal construction of the angle valve is similar to the globe valve except that the fluid flows directly through the seat and then turns 90° as it leaves the valve. In the gate valve, the gate is lifted out of the flow stream as it is opened. When fully open, there is only a very minor obstruction. The function of the check valve is to allow flow in only one direction.

In some cases, particularly with regard to control valves in fluid power systems, energy loss as such is not reported. Instead, the magnitude of the pressure drop as the fluid flows through the valve at a certain flow rate is reported.

☐ **EXAMPLE PROBLEM 10.8**   Calculate the pressure drop across a fully open globe valve placed in a 4-in Schedule 40 steel pipe carrying 400 gal/min of oil (sg = 0.87).

Solution   A sketch of the installation is shown in Fig. 10.22. In order to determine the pressure drop, the energy equation should be written for the flow between points 1 and 2:

$$\frac{p_1}{\gamma} + z_1 + \frac{v_1^2}{2g} - h_L = \frac{p_2}{\gamma} + z_2 + \frac{v_2^2}{2g}$$

The energy loss h_L is the minor loss due to the valve only. The pressure drop is the difference between p_1 and p_2. Solving the energy equation for this difference gives

$$p_1 - p_2 = \gamma \left[(z_2 - z_1) + \frac{v_2^2 - v_1^2}{2g} + h_L \right]$$

But $z_1 = z_2$ and $v_1 = v_2$. Then we have

$$p_1 - p_2 = \gamma h_L$$

Equation (10–1) is used to determine h_L:

$$h_L = K \times \frac{v^2}{2g} = f_T \times \frac{L_e}{D} \times \frac{v^2}{2g}$$

The velocity v is the average velocity of flow in the 4-in pipe. For the pipe, $D = 0.3355$ ft and $A = 0.0884$ ft^2. Then we have

$$v = \frac{Q}{A} = \frac{400 \text{ gal/min}}{0.0884 \text{ ft}^2} \times \frac{1 \text{ ft}^3/\text{s}}{449 \text{ gal/min}} = 10.08 \text{ ft/s}$$

From Table 10.5 we find $f_T = 0.017$ for a 4-in pipe. For the globe valve, $L_e/D = 340$. Then,

$$h_L = f_T \times \frac{L_e}{D} \times \frac{v^2}{2g} = (0.017)(340) \frac{(10.08)^2}{(2)(32.2)} \text{ ft} = 9.12 \text{ ft}$$

FIGURE 10.22 Globe valve for Example Problem 10.8.

4-in Schedule 40 pipe

• 1   • 2

Globe
valve

For the oil, $\gamma = (0.870)(62.4 \text{ lb/ft}^3)$. Then we have

$$p_1 - p_2 = \gamma h_L = \frac{(0.870)(62.4) \text{ lb}}{\text{ft}^3} \times 9.12 \text{ ft} \times \frac{1 \text{ ft}^2}{144 \text{ in}^2}$$

$$p_1 - p_2 = 3.4 \text{ psi}$$

Therefore, the pressure in the oil drops by 3.4 psi as it flows through the valve. Also, an energy loss of 9.12 lb-ft is dissipated as heat from each pound of oil that flows through the valve.

**10.11
PIPE BENDS**

It is frequently more convenient to bend a pipe or tube than to install a commercially made elbow. The resistance to flow of a bend is dependent on the ratio of the bend radius r to the pipe inside diameter D. Figure 10.23 shows that the minimum resistance occurs when the ratio r/D is approximately three. The resistance is given in terms of the equivalent length ratio L_e/D, and therefore, Eq. (10–8) must be used to calculate the resistance coefficient. The resistance shown in Fig. 10.23 includes both the bend resistance and the resistance due to the length of the pipe in the bend.

When we compute the r/D ratio, r is defined as the radius to the *centerline* of the pipe or tube, called the *mean radius* (see Fig. 10.24). That is, if R_o is the radius to the outside of the bend, R_i is the radius to the inside of the bend, and D_o is the *outside diameter* of the pipe or tube

$$r = R_i + D_o/2$$
$$r = R_o - D_o/2$$
$$r = (R_o + R_i)/2$$

FIGURE 10.23 Resistance due to 90° pipe bends. (Source: Beij, K. H. Pressure Losses for Fluid Flow in 90 Degree Pipe Bends. *Journal of Research of the National Bureau of Standards* 21 [July 1938]: 1–18.)

Equivalent length ratio L_e/D

Relative radius r/D

FIGURE 10.24 90° pipe bend.

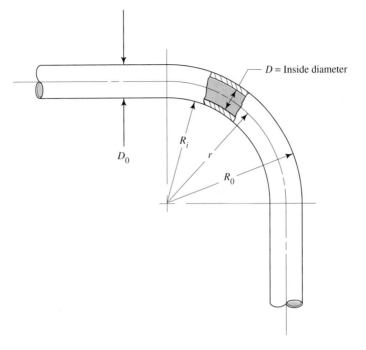

☐ **EXAMPLE PROBLEM 10.9**   A distribution system for liquid propane is made from 1¼-in steel tubing with a wall thickness of 0.083 in. Several 90° bends are required to fit the tubes to the other equipment in the system. The specifications call for the radius to the inside of each bend to be 200 mm. When the system carries 160 L/min of propane at 25°C, compute the energy loss in each bend.

Solution   Darcy's equation should be used to compute the energy loss with the L_e/D ratio for the bends found from Fig. 10.23. First, let's determine r/D, recalling that D is the inside diameter of the tube and r is the radius to the centerline of the tube. From Appendix G we find $D = 27.5$ mm $= 0.0275$ m. The radius r must be computed from

$$r = R_i + D_o/2$$

where $D_o = 31.75$ mm, the outside diameter of the tube as found from Appendix G. Completion of the calculation gives

$$r = 200 \text{ mm} + (31.75 \text{ mm})/2 = 215.9 \text{ mm}$$

and

$$r/D = 215.9 \text{ mm}/27.5 \text{ mm} = 7.85$$

From Fig. 10.23 we find the equivalent length ratio to be 23.

We now must compute the velocity to complete the evaluation of the energy loss from Darcy's equation:

$$v = \frac{Q}{A} = \frac{160 \text{ L/min}}{5.954 \times 10^{-4} \text{ m}^2} \frac{1.0 \text{ m}^3/\text{s}}{60\,000 \text{ L/min}} = 4.48 \text{ m/s}$$

The relative roughness is

$$D/\epsilon = (0.0275 \text{ m})/(4.6 \times 10^{-5} \text{ m}) = 598$$

Then, we can find $f_T = 0.022$ from Moody's diagram (Fig. 9.2) in the zone of complete turbulence. Now the energy loss can be computed:

$$h_L = f_T \frac{L_e}{D} \frac{v^2}{2g} = (0.022)(23) \frac{(4.48)^2}{(2)(9.81)} = 0.517 \text{ m} = 0.517 \text{ N} \cdot \text{m/N}$$

REFERENCES

1. Beij, K. H. 1938. Pressure Losses for Fluid Flow in 90 Degree Pipe Bends. *Journal of Research of the National Bureau of Standards* 21:1–18.

2. Crane Co. 1988. *Flow of Fluids Through Valves, Fittings, and Pipe.* Technical paper No. 410. Joliet, Illinois: Author.

3. Dodge, L. 1968. How to Compute and Combine Fluid Flow Resistances in Components. *Hydraulics and Pneumatics* 21:118–21.

4. The Hydraulic Institute. 1961. *Pipe Friction Manual.* 3rd ed. New York: Author.

5. Idelchik, I. E. 1986. *Handbook of Hydraulic Resistance.* 2nd ed. New York: Harper & Row.

6. King, H. W., and E. F. Brater. 1976. *Handbook of Hydraulics.* 6th ed. New York: McGraw-Hill.

7. King, R. C., and Sabin Crocker. 1967. *Piping Handbook.* 5th ed. New York: McGraw-Hill.

8. Simpson, L. L. 1968. Sizing Piping for Process Plants. *Chemical Engineering* 75:193–214.

PRACTICE PROBLEMS

10.1M *Determine the energy loss due to a sudden enlargement from a 50-mm pipe to a 100-mm pipe when the velocity of flow is 3 m/s in the smaller pipe.*

10.2M *Determine the energy loss due to a sudden enlargement from a standard 1-in Schedule 80 pipe to a 3½-in Schedule 80 pipe when the rate of flow is 3×10^{-3} m³/s.*

10.3E Determine the energy loss due to a sudden enlargement from a standard 1-in Schedule 80 pipe to a 3½-in Schedule 80 pipe when the rate of flow is 0.10 ft³/s.

10.4E Determine the pressure difference between two points on either side of a sudden enlargement from a tube with a 2-in inside diameter to one with a 6-in inside diameter when the velocity of flow of water is 4 ft/s in the smaller tube.

10.5E Determine the pressure difference for the conditions in Problem 10.4 if the enlargement was gradual with a cone angle of 15°.

10.6M *Determine the energy loss due to a gradual enlargement from a 25-mm pipe to a 75-mm pipe*

when the velocity of flow is 3 m/s in the smaller pipe and the cone angle of the enlargement is 20°.

10.7M *Determine the energy loss for the conditions in Problem 10.6 if the cone angle is increased to 60°.*

10.8E Compute the energy loss for gradual enlargements with cone angles from 2° to 60° in the increments shown in Fig. 10.5. For each case, water at 60°F is flowing at 85 gal/min in a 2-in Schedule 40 steel pipe that enlarges to a 6-in Schedule 40 pipe.

10.9E Plot a graph of energy loss versus cone angle for the results of Problem 10.8.

10.10E For the data of Problem 10.8, compute the length required to achieve the enlargement for each cone angle. Then compute the energy loss due to friction in that length using the velocity, diameter, and Reynolds number for the midpoint between the ends of the enlargement. Use water at 60°F.

10.11E Add the energy loss due to friction from Problem 10.10 to the energy loss for the enlargement from Problem 10.8 and plot the total versus the cone angle on the same graph used in Problem 10.9.

10.12M *Another term for an enlargement is a diffuser. A* *diffuser is used to convert kinetic energy ($v^2/2g$) to pressure energy (p/γ). An ideal diffuser is one in which no energy losses occur and Bernoulli's equation can be used to compute the pressure after the expansion. Compute the pressure after the expansion for an ideal diffuser for the flow of water at 20°C from a 1-in Type K copper tube to a 3-in Type K tube. The volume flow rate is 150 L/min, and the pressure before the expansion is 500 kPa.*

10.13M *Compute the resulting pressure after a "real" diffuser in which the energy loss due to the enlargement is considered for the data presented in Problem 10.12. The enlargement is sudden.*

10.14M *Compute the resulting pressure after a "real" diffuser in which the energy loss due to the enlargement is considered for the data presented in Problem 10.12. The enlargement is gradual with cone angles of (a) 60°, (b) 30°, and (c) 10°. Compare the results with those of Problems 10.12 and 10.13.*

10.15M *Determine the energy loss when 0.04 m^3/s of water flows from a 6-in standard Schedule 40 pipe into a large reservoir.*

10.16E Determine the energy loss when 1.50 ft^3/s of water flow from a 6-in standard Schedule 40 pipe into a large reservoir.

10.17E Determine the energy loss when oil with a specific gravity of 0.87 flows from a 4-in pipe to a 2-in pipe through a sudden contraction if the velocity of flow in the larger pipe is 4.0 ft/s.

10.18E For the conditions in Problem 10.17, if the pressure before the contraction were 80 psig, calculate the pressure in the smaller pipe.

10.19 True or False: For a sudden contraction with a diameter ratio of 3.0, the energy loss decreases as the velocity of flow increases.

10.20M *Determine the energy loss for a sudden contraction from a 5-in Schedule 80 steel pipe to a 2-in Schedule 80 pipe for a flow rate of 500 L/min.*

10.21M *Determine the energy loss for a gradual contraction from a 5-in Schedule 80 steel pipe to a 2-in Schedule 80 pipe for a flow rate of 500 L/min. The cone angle for the contraction is 105°.*

10.22E Determine the energy loss for a sudden contraction from a 4-in Schedule 80 steel pipe to a 1½-in Schedule 80 pipe for a flow rate of 250 gal/min.

10.23E Determine the energy loss for a gradual contrac-

tion from a 4-in Schedule 80 steel pipe to a 1½-in Schedule 80 pipe for a flow rate of 250 gal/min. The cone angle for the contraction is 76°.

10.24E For the data in Problem 10.22, compute the energy loss for gradual contractions with each of the cone angles listed in Figs. 10.10 and 10.11. Plot energy loss versus the cone angle.

10.25E For each contraction described in Problems 10.22 and 10.24, make a scale drawing of the device to observe its physical appearance.

10.26E Note in Figs. 10.10 and 10.11 that the minimum energy loss for a gradual contraction ($K = 0.04$ approximately) occurs when the cone angle is in the range of 15° to 40°. Make scale drawings of contractions at both of these extremes for a reduction from a 6-in to a 3-in ductile iron pipe.

10.27E If the contraction from a 6-in to a 3-in ductile iron pipe described in Problem 10.26 was made with a cone angle of 120°, what would the resulting resistance coefficient be? Make a scale drawing of this reducer.

10.28E Compute the energy loss that would occur as 50 gal/min flows from a tank into a steel tube with an outside diameter of 2.0 in and a wall thickness of 0.065 in. The tube is installed flush with the inside of the tank wall with a square edge.

10.29M *Determine the energy loss that will occur if water flows from a reservoir into a pipe with a velocity of 3 m/s if the configuration of the entrance is (a) an inward projecting pipe (using $K = 1.0$), (b) a square-edged inlet, (c) a chamfered inlet, or (d) a well-rounded inlet.*

10.30M *Determine the equivalent length in meters of pipe of a fully open globe valve placed in a 10-in Schedule 40 pipe.*

10.31M *Repeat Problem 10.30 for a fully open gate valve.*

10.32E Calculate the resistance coefficient K for a ball-type check valve placed in a 2-in Schedule 40 steel pipe if water at 100°F is flowing with a velocity of 10 ft/s.

10.33E Calculate the pressure difference across a fully open angle valve placed in a 5-in Schedule 40 steel pipe carrying 650 gal/min of oil (sg = 0.90).

10.34M *Determine the pressure drop across a 90° standard elbow in a 2½-in Schedule 40 steel pipe if water at 15°C is flowing at the rate of 750 L/min.*

10.35M *Repeat Problem 10.34 for a street elbow.*

FIGURE 10.25   Problem 10.37.

10.36M *Repeat Problem 10.34 for a long radius elbow. Compare the results from Problems 10.34, 10.35, and 10.36.*

10.37E A simple heat exchanger is made by installing a close return bend on two ½-in Schedule 40 steel pipes as shown in Fig. 10.25. Compute the pressure difference between the inlet and outlet for a flow rate of 12.5 gal/min of ethylene glycol at 77°F.

10.38E A proposed alternate form for the heat exchanger described in Problem 10.37 is shown in Fig. 10.26.

The entire flow conduit is a ¾-in steel tube with a wall thickness of 0.065 in. Note that the inside diameter for this tube is 0.620 in, slightly smaller than that of the ½-in Schedule 40 pipe ($D = 0.622$ in). The return bend is formed by two 90° bends with a short length of straight tube between them. Compute the pressure difference between the inlet and the outlet of this design and compare it with the system from Problem 10.37.

10.39E A piping system for a pump contains a tee, as shown in Fig. 10.27, to permit the pressure at the outlet of the pump to be measured. However,

FIGURE 10.26   Problem 10.38.

FIGURE 10.27   Problem 10.39.

FIGURE 10.28 Problem 10.40.

FIGURE 10.29 Problem 10.43.

(a) Basic layout

(b) Proposal 1

(c) Proposal 2

there is no flow into the line leading to the gage. Compute the energy loss as 0.40 ft³/s of water as 50°F flows through the tee.

10.40M *A piping system for supplying heavy fuel oil at 25°C is arranged as shown in Fig. 10.28. The bottom leg of the tee is normally capped, but the cap can be removed to clean the pipe. Compute the energy loss as 0.08 m³/s flows through the tee.*

10.41M *A 1-in Type K copper tube supplies hot water (80°C) to a washing system in a factory at a flow rate of 250 L/min. At several points in the system, a 90° bend is required. Compute the energy loss in each bend if the radius to the outside of the bend is 300 mm.*

10.42M *Specify the radius in mm to the centerline of a 90° bend in a 1-in Type K copper tube to achieve the minimum energy loss. For such a bend carrying 250 L/min of water at 80°C, compute the energy loss. Compare the results with those of Problem 10.41.*

10.43M *The inlet and the outlet shown in Fig. 10.29(a) are to be connected with a 2-in Type K copper tube to carry 750 L/min of propyl alcohol at 25°C. Evaluate the two schemes shown in parts (b) and (c) of the figure with regard to the energy loss. Include the losses due to both the bend and the friction in the straight tube.*

10.44M *Compare the energy losses for the two proposals from Problem 10.43 with the energy loss for the proposal in Fig. 10.30.*

10.45M *Determine the energy loss that occurs as 40 L/min of water at 10°C flows around a 90° bend in a commercial steel tube having an outside diameter of ¾ in and a wall thickness of 0.065 in. The radius of the bend to the centerline of the tube is 150 mm.*

10.46M *Figure 10.31 shows a test setup to determine the energy loss due to a heat exchanger. Water at 50°C is flowing vertically upward at 6.0 × 10⁻³ m³/s. Calculate the energy loss between points 1 and 2. Determine the resistance coefficient for the heat exchanger based on the velocity in the inlet pipe.*

FIGURE 10.31   Problem 10.46.

FIGURE 10.30   Problem 10.44.

■ ■ ■ ■ 11 Series Pipe Line Systems

**11.1
OBJECTIVES**

This chapter is a capstone for the preceding chapters, which concerned the flow of fluids in pipes and tubes. We have developed the concepts of fluid flow rate, the continuity equation, Bernoulli's equation, and the general energy equation. Laminar and turbulent flow have been defined, and the Reynolds number has been used to determine the type of flow in a given system. The manner for computing energy losses due to friction has been presented. Also, we have discussed several types of minor losses for the flow of fluids through valves and fittings and for changes in the velocity or direction of the flow.

Of course, real fluid flow systems often contain several minor losses as well as the energy loss due to friction as the fluid is delivered from one point to another. More than one size of pipe may be used. This chapter presents the methods of analysis for real pipe line systems in which the fluid flows through a single continuous path. Such a system is called a *series pipe line system*.

After completing this chapter you should be able to:

1. Identify series pipe line systems.
2. Determine whether a given system is Class I, Class II, or Class III.
3. Compute the total energy loss, elevation differences, or pressure differences for Class I systems with any combination of pipes, minor losses, pumps, or reservoirs when the system carries a given flow rate.
4. Determine for Class II systems the velocity or volume flow rate through the system with known pressure differences and elevation heads.
5. Determine for Class III systems the size of pipe required to carry a given fluid flow rate with a specified limiting pressure drop or for a given elevation difference.

**11.2
SYSTEM
CLASSIFICATIONS**

Most pipe flow systems involve both major friction energy losses and minor losses. If the system is arranged so that the fluid flows through a continuous line without branching, it is referred to as a *series system*. Conversely, if the flow branches into two or more lines, it is referred to as a *parallel system*.

This chapter deals only with series systems such as the one illustrated in Fig. 11.1. If the energy equation is written for this system, using the surface of each reservoir as the reference points, it would appear as

$$\frac{p_1}{\gamma} + z_1 + \frac{v_1^2}{2g} + h_A - h_L = \frac{p_2}{\gamma} + z_2 + \frac{v_2^2}{2g} \tag{11-1}$$

FIGURE 11.1   Series pipe line system.

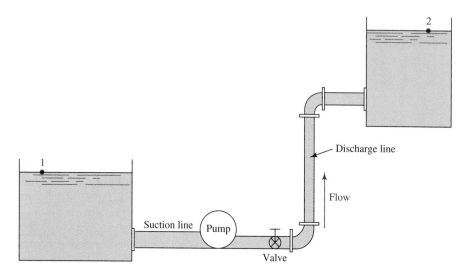

The first three terms on the left side of this equation represent the energy possessed by the fluid at point 1 in the form of pressure head, elevation head, and velocity head. Likewise, the terms on the right side of the equation represent the energy possessed by the fluid at point 2. The two terms h_A and h_L indicate the energy added to the fluid and the energy lost from the system anywhere between reference points 1 and 2. In this problem, h_A is the energy added by the pump. Energy is lost, however, due to several conditions. We can say that

$$h_L = h_1 + h_2 + h_3 + h_4 + h_5 + h_6 \qquad \text{(11–2)}$$

where h_L = total energy loss per unit weight of fluid flowing

           h_1 = entrance loss

           h_2 = friction loss in the suction line

           h_3 = energy loss in the valve

           h_4 = energy loss in the two 90° elbows

           h_5 = friction loss in the discharge line

           h_6 = exit loss

In a series pipe line the total energy loss is the sum of the individual major and minor losses. This statement is in agreement with the principle that the energy equation is a means of accounting for all of the energy in the system between the two reference points.

In designing or analyzing a pipe flow system there are six primary parameters involved, namely:

1. The energy losses from the system or energy additions to the system
2. The volume flow rate of the fluid or the velocity of flow
3. The size of the pipe

4. The length of pipe
5. The pipe wall roughness, ϵ
6. The fluid properties of specific weight, density, and viscosity

Usually, one of the first three parameters is to be determined, while the remaining items are known or can be specified by the designer. The method of performing the design or completing the analysis is different depending on what is unknown. The methods we describe in this book are classified as follows:

Class I : The energy losses or additions are to be determined

Class II : The volume flow rate is to be determined

Class III: The pipe diameter is to be determined

11.3
CLASS I SYSTEMS

Our approach to the analysis of Class I systems is identical to that used throughout the previous chapters except that generally many types of energy losses will exist and each energy loss must be evaluated and included in the energy equation. The total energy loss h_L is the sum of the individual major and minor losses. The following programmed example problem will illustrate the solution of a Class I problem.

PROGRAMMED EXAMPLE PROBLEM

☐ **EXAMPLE PROBLEM 11.1**

Calculate the power supplied to the pump shown in Fig. 11.2 if its efficiency is 76 percent. Methyl alcohol at 25°C is flowing at the rate of 0.015 m³/s. The suction line is a standard 4-in Schedule 40 steel pipe, 15 m long. The total length of 2-in Schedule 40 steel pipe in the discharge line is 200 m. Assume that the entrance from reservoir 1 is

FIGURE 11.2 System for Example Problem 11.1.

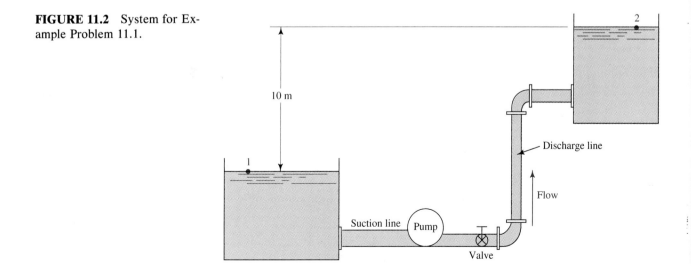

through a square-edged inlet and that the elbows are standard. The valve is a fully open globe valve.

To begin the solution, write the energy equation for the system.

Using the surfaces of the reservoirs as the reference points, you should have

$$\frac{p_1}{\gamma} + z_1 + \frac{v_1^2}{2g} + h_A - h_L = \frac{p_2}{\gamma} + z_2 + \frac{v_2^2}{2g}$$

Since $p_1 = p_2 = 0$ and v_1 and v_2 are approximately zero, the equation can be simplified to

$$z_1 + h_A - h_L = z_2$$

Because the objective of the problem is to calculate the power supplied to the pump, solve now for the total head on the pump.

The total head is

$$h_A = z_2 - z_1 + h_L$$

There are six components to the total energy loss. List them and write the formula for evaluating each one.

Your list should include the following items. The subscript s indicates the suction line and the subscript d indicates the discharge line.

$h_1 = K(v_s^2/2g)$ (entrance loss)

$h_2 = f_s(L/D)(v_s^2/2g)$ (friction loss in suction line)

$h_3 = f_{dT}(L_e/D)(v_d^2/2g)$ (valve)

$h_4 = f_{dT}(L_e/D)(v_d^2/2g)$ (two 90° elbows)

$h_5 = f_d(L/D)(v_d^2/2g)$ (friction loss in discharge line)

$h_6 = 1.0(v_d^2/2g)$ (exit loss)

Since the velocity head in the suction or discharge line is required for each energy loss, calculate these values now.

You should have $v_s^2/2g = 0.17$ m and $v_d^2/2g = 2.44$ m.

$$v_s = \frac{Q}{A_s} = \frac{0.015 \text{ m}^3}{\text{s}} \times \frac{1}{8.213 \times 10^{-3} \text{ m}^2} = 1.83 \text{ m/s}$$

$$\frac{v_s^2}{2g} = \frac{(1.83)^2}{2(9.81)} \text{ m} = 0.17 \text{ m}$$

$$v_d = \frac{Q}{A_d} = \frac{0.015 \text{ m}^3}{\text{s}} \times \frac{1}{2.168 \times 10^{-3} \text{ m}^2} = 6.92 \text{ m/s}$$

$$\frac{v_d^2}{2g} = \frac{(6.92)^2}{2(9.81)} \text{ m} = 2.44 \text{ m}$$

To determine the friction losses in the suction line and the discharge line, and the minor losses in the discharge line, we need the Reynolds number, relative roughness, and friction factor for each pipe, and the friction factor in the zone of complete turbulence for the discharge line that contains a valve and pipe fittings. Find these values now.

For methyl alcohol at 25°C, $\rho = 789$ kg/m³ and $\mu = 5.60 \times 10^{-4}$ Pa·s. Then, in the suction line, we have

$$N_R = \frac{vD\rho}{\mu} = \frac{(1.83)(0.1023)(789)}{5.60 \times 10^{-4}} = 2.64 \times 10^5$$

The value of f_s must be evaluated from the Moody diagram, Fig. 9.2. For steel pipe, $\epsilon = 4.6 \times 10^{-5}$ m.

$$D/\epsilon = 0.1023/4.6 \times 10^{-5} = 2224$$
$$N_R = 2.64 \times 10^5$$

Then $f_s = 0.018$.
In the discharge line, we have

$$N_R = \frac{vD\rho}{\mu} = \frac{(6.92)(0.0525)(789)}{5.60 \times 10^{-4}} = 5.12 \times 10^5$$

Therefore, the flow in both lines is turbulent.
Evaluating the friction factor f_d gives

$$D/\epsilon = 0.0525/4.6 \times 10^{-5} = 1141$$
$$N_R = 5.12 \times 10^5$$
$$f_d = 0.020$$

We can find from Table 10.5 that $f_{dT} = 0.019$ for the 2-in discharge pipe in the zone of complete turbulence.
Returning now to the energy loss calculations, evaluate h_1, the entrance loss, in N·m/N or m.

The result is $h_1 = 0.09$ m. For a square-edged inlet, $K = 0.5$ and

$$h_1 = 0.5(v_s^2/2g) = (0.5)(0.17 \text{ m}) = 0.09 \text{ m}$$

Now calculate h_2.

The result is $h_2 = 0.45$ m.

$$h_2 = f_s \times \frac{L}{D} \times \frac{v_s^2}{2g} = f_s\left(\frac{15}{0.1023}\right)(0.17) \text{ m}$$

$$h_2 = (0.018)\left(\frac{15}{0.1023}\right)(0.17) \text{ m} = 0.45 \text{ m}$$

Now calculate h_3.

From the data in Chapter 10, the equivalent length ratio L_e/D for a fully open globe valve is 340. The friction factor is $F_{dT} = 0.019$.
Then we have

$$h_3 = f_{dT} \times \frac{L_e}{D} \times \frac{v_d^2}{2g} = (0.019)(340)(2.44) \text{ m} = 15.76 \text{ m}$$

Now calculate h_4.

For standard 90° elbows, $L_e/D = 30$. The value of f_{dT} is the same as that used in the preceding panel. Then we have

$$h_4 = 2f_{dT} \times \frac{L_e}{D} \times \frac{v_d^2}{2g} = (2)(0.019)(30)(2.44) \text{ m} = 2.78 \text{ m}$$

Now calculate h_5.

The discharge line friction loss is

$$h_5 = f_d \times \frac{L}{D} \times \frac{v_d^2}{2g} = (0.020)\left(\frac{200}{0.0525}\right)(2.44) \text{ m} = 185.9 \text{ m}$$

Now calculate h_6.

The exit loss is

$$h_6 = 1.0(v_d^2/2g) = 2.44 \text{ m}$$

This concludes the calculation of the individual energy losses. The total loss h_L can now be determined.

$$\begin{aligned}
h_L &= h_1 + h_2 + h_3 + h_4 + h_5 + h_6 \\
&= (0.09 + 0.45 + 15.76 + 2.78 + 185.9 + 2.44) \text{ m} \\
&= 207.4 \text{ m}
\end{aligned}$$

From the energy equation the expression for the total head on the pump was found to be

$$h_A = z_2 - z_1 + h_L$$

Then we have

$$h_A = 10 \text{ m} + 207.4 \text{ m} = 217.4 \text{ m}$$

Now calculate the power supplied to the pump.

$$\text{Power} = \frac{h_A \gamma Q}{e_M} = \frac{(217.4 \text{ m})(7.74 \times 10^3 \text{ N/m}^3)(0.015 \text{ m}^3/\text{s})}{0.76}$$

$$= 33.2 \times 10^3 \text{ N} \cdot \text{m/s} = 33.2 \text{ kW}$$

This concludes the programmed example problem.

■

**Critique of the System Shown in Fig. 11.2
and Analyzed in Example Problem 11.1**
Problem solutions such as that just concluded can give you, the fluid flow system designer, much useful information on which you can evaluate the proposed design and make rational decisions about system improvement. The following are some observations:

1. The suction line between the first reservoir and the pump, given to be 15 m, appears to be excessively long. It is recommended that the pump be relocated closer to the reservoir so the suction line can be as short as practical. This is especially important when low pressure is observed at the inlet to the pump as will be discussed in Chapter 15 on pump application.

2. It may be desirable to place a valve in the suction line before the inlet to the pump to allow the pump to be removed or serviced without draining the reservoir. A gate valve should be used so the energy loss is small during normal operation with the valve completely open.

3. The size of the suction line is somewhat small. Note that the velocity is 1.83 m/s. Section 6.5 recommends that the velocity at the inlet to a pump be approximately 1.0 m/s. Increasing the suction line size from 4-in to 5-in Schedule 40 steel pipe would decrease the velocity to 1.16 m/s. The velocity head would decrease from 0.17 m to 0.069 m. Because the energy losses at the pipe entrance and the friction loss in the suction line are proportional to the velocity head, these losses would be correspondingly lower.

4. The 2-in discharge line is much too small as indicated by the velocity of 6.92 m/s. Section 6.5 recommends that the velocity in the discharge line of a fluid distribution system be approximately 3.0 m/s. Using a 3-in pipe would produce a velocity of 3.15 m/s and a 3½-in pipe results in $v_d = 2.35$ m/s. The final choice may be influenced by the design of the pump discharge port. Assuming the 3½-in pipe is used, the velocity head in the discharge line would decrease from 2.44 m to 0.282 m, a reduction of approximately *nine times*! Because the energy losses in the discharge line are proportional to the velocity head, a substantial reduction would result.

5. The globe valve in the discharge line should be replaced by a gate valve or some other more efficient valve if it is normally operated in the fully open position. However, if the valve is used to adjust the flow rate by throttling, the globe valve may be satisfactory provided the energy loss can be tolerated.

Summary of Design Changes
The following changes are proposed:

1. Decrease the suction line length from 15 m to 1.5 m. Assuming that the two reservoirs must stay in the same position, the extra 13.5 m of length will be added to the discharge line, making it a total of 213.5 m long.

2. Increase the size of the suction line from 4-in to 5-in Schedule 40 steel pipe. Then $v_s = 1.16$ m/s and the velocity head is 0.069 m.
3. Add a fully open gate valve in the suction line.
4. Increase the discharge line size from 2-in to 3½-in Schedule 40. Then, $v_d = 2.35$ m/s and the velocity head is 0.282 m.
5. Replace the globe valve in the discharge line with a fully open gate valve.

Making all of these changes would result in the reduction of the energy to be added by the pump from 217.4 m to 13.4 m. The power supplied to the pump would decrease from 33.2 kW to 2.05 kW, *a reduction of more than a factor of 16!*

11.4 CLASS II SYSTEMS

Whenever the volume flow rate in the system is unknown, we analyze the system performance by a procedure called *iteration*. This is required because there are too many unknown quantities to use the direct solution procedure described for Class I problems. Specifically, if the volume flow rate is unknown, then the velocity of flow is also unknown. It follows that the Reynolds number is unknown since it depends on the velocity. If the Reynolds number cannot be found, then the friction factor f cannot be determined directly. Since energy losses due to friction are dependent on both the velocity and the friction factor, the value of these losses cannot be calculated directly.

Iteration overcomes these difficulties. It is a type of trial-and-error solution method in which a trial value is assumed for the unknown friction factor f, allowing the calculation of a corresponding velocity of flow. The procedure provides a means of checking the accuracy of the trial value of f and also indicates the new trial value to be used if an additional calculation cycle is required. This procedure for solving Class II problems is presented in step-by-step form below. The programmed example problems which follow then illustrate how the procedure is applied.

SOLUTION PROCEDURE FOR CLASS II SYSTEMS WITH ONE PIPE

1. Write the energy equation for the system.
2. Evaluate known quantities such as pressure heads and elevation heads.
3. Express energy losses in terms of the unknown velocity v and friction factor f.
4. Solve for the velocity in terms of f.
5. Express the Reynolds number in terms of the velocity.
6. Calculate the relative roughness D/ϵ.
7. Select a trial value of f based on the known D/ϵ and a Reynolds number in the turbulent range.
8. Calculate the velocity, using the equation from Step 4.
9. Calculate the Reynolds number from the equation in Step 5.
10. Evaluate the friction factor f for the Reynolds number from Step 9 and the known value of D/ϵ, using the Moody diagram, Fig. 9.2.
11. If the new value of f is different from the value used in Step 8, repeat Steps 8 through 11 using the new value of f.

12. If there is no significant change in f from the assumed value, then the velocity found in Step 8 is correct.

PROGRAMMED EXAMPLE PROBLEM

☐ **EXAMPLE PROBLEM 11.2**   A lubricating oil must be pumped through a horizontal 6-in Schedule 40 steel pipe with a maximum pressure drop of 60 kPa per 100 m of pipe. The oil has a specific gravity of 0.88 and a dynamic viscosity of 9.5×10^{-3} Pa·s. Calculate the maximum allowable volume flow rate of oil.

Is this a Class II system?

Yes, it is. Because the volume flow rate is unknown, the Class II solution procedure described above must be used. To begin, draw a sketch of the system, write the energy equation, and simplify it as much as possible.

Figure 11.3 shows the two points of interest in the pipe. Then the energy equation is

$$\frac{p_1}{\gamma} + z_1 + \frac{v_1^2}{2g} - h_L = \frac{p_2}{\gamma} + z_2 + \frac{v_2^2}{2g}$$

Because $z_1 = z_2$ and $v_1 = v_2$, we use the equations

$$\frac{p_1}{\gamma} - h_L = \frac{p_2}{\gamma}$$

$$\frac{p_1 - p_2}{\gamma} = h_L$$

FIGURE 11.3   Reference points in pipe for Example Problem 11.2.

Evaluate the pressure head difference now.

You should have $(p_1 - p_2)/\gamma = 6.95$ m. Since the allowable pressure drop is 60 kPa, we get

$$\frac{p_1 - p_2}{\gamma} = \frac{60 \text{ kN}}{\text{m}^2} \times \frac{\text{m}^3}{(0.88)(9.81 \text{ kN})} = 6.95 \text{ m}$$

Then $h_L = 6.95$ m. The total energy loss in this case is due to friction. From Darcy's equation, we know that

$$h_L = f \times \frac{L}{D} \times \frac{v^2}{2g}$$

Step 4 of the solution procedure says to express the velocity in terms of the friction factor. Do this now.

The final form should be $v = \sqrt{0.21/f}$. To obtain this result we use the equations

$$h_L = f \times \frac{L}{D} \times \frac{v^2}{2g}$$

$$v = \sqrt{\frac{2gh_LD}{Lf}}$$

In the second equation we substitute the values $g = 9.81$ m/s^2, $h_L = 6.95$ m, $D = 0.1541$ m, and $L = 100$ m. Then we have

$$v = \sqrt{\frac{(2)(9.81)(6.95)(0.1541)}{(100)(f)}} = \sqrt{0.21/f}$$

Now do Step 5 of the procedure.

You should have $N_R = (1.43 \times 10^4)v$. We obtain the value of the Reynolds number by the equation

$$N_R = \frac{vD\rho}{\mu}$$

We know that $D = 0.1541$ m, $\mu = 9.5 \times 10^{-3}$ Pa·s, and $\rho = 0.88 \times 1000$ kg/m^3 = 880 kg/m^3. Then we have

$$N_R = \frac{v(0.1541)(880)}{9.5 \times 10^{-3}} = (1.43 \times 10^4)v$$

Now do Step 6.

Since $\epsilon = 4.6 \times 10^{-5}$ m for steel pipe, we get

$$D/\epsilon = 0.1541/4.6 \times 10^{-5} = 3350$$

These first six steps are all preliminary to the iteration portion of the procedure, Steps 7–12. The results we have just found will simplify the process of iteration.

Step 7 states that a trial value of the friction factor f should be selected. Moody's diagram, Fig. 9.2, can be used as an aid to the rational selection of a trial value. Since the relative roughness is known to be 3350, the range of possible values of f is from approximately 0.039 for $N_R = 4000$, to 0.015 for $N_R = 1.0 \times 10^7$ and larger. Any value in this range can be selected for the first trial. Use $f = 0.020$ and proceed with Step 8.

For $f = 0.020$, the velocity would be

$$v = \sqrt{0.21/f} = \sqrt{0.21/0.02} \text{ m/s}$$
$$= 3.24 \text{ m/s}$$

We can now calculate the corresponding Reynolds number.

$$N_R = (1.43 \times 10^4)v = (1.43 \times 10^4)(3.24)$$
$$= 4.63 \times 10^4$$

For this value of the Reynolds number and $D/\epsilon = 3350$, the new value of $f = 0.0225$. Since this is different from the initially assumed value, Steps 8 through 11 must be repeated. Do Steps 8, 9, and 10 before looking at the next panel.

The results are

$$v = \sqrt{\frac{0.21}{0.0225}} = 3.06 \text{ m/s}$$

$$N_R = (1.43 \times 10^4)(3.06) = 4.38 \times 10^4$$

The new value of f is 0.0225, unchanged from the previous value. Therefore, $v = 3.06$ m/s is the correct velocity. Now the volume flow rate can be calculated to complete the problem:

$$Q = Av = (1.864 \times 10^{-2} \text{ m}^2)(3.06 \text{ m/s}) = 0.057 \text{ m}^3/\text{s}$$

This programmed example problem is concluded.

∎

11.4.1
Class II Systems Having
Minor Losses

Example Problem 11.3 gives another example of a Class II system with one pipe. However, there are minor losses in addition to the pipe friction loss. The details of the solution are slightly different, but the general solution procedure is the same as that used for Example Problem 11.2.

PROGRAMMED EXAMPLE PROBLEM

☐ **EXAMPLE PROBLEM 11.3**

Water at 80°F is being supplied to an irrigation ditch from an elevated storage reservoir as shown in Fig. 11.4. Calculate the volume flow rate of water into the ditch.

Begin with Step 1 of the solution procedure by writing the energy equation. Use A and B as the reference points and simplify the equation as much as possible.

Compare this with your solution:

$$\frac{p_A}{\gamma} + z_A + \frac{v_A^2}{2g} - h_L = \frac{p_B}{\gamma} + z_B + \frac{v_B^2}{2g}$$

Since $p_A = p_B = 0$, and v_A is approximately zero, then

$$z_A - h_L = z_B + (v_B^2/2g)$$
$$z_A - z_B = (v_B^2/2g) + h_L \tag{11-3}$$

FIGURE 11.4 Pipe line system
for Example Problem 11.3.

Notice that the stream of water at point B has the same velocity as that inside the
pipe.

The elevation difference, $z_A - z_B$, is known to be 40 ft. However, the energy
losses which make up h_L all depend on the unknown velocity, v_B. Thus, iteration is
required. Do Step 3 of the solution procedure now.

There are four components of the total energy loss h_L:

$$h_L = h_1 + h_2 + h_3 + h_4$$

where

$$h_1 = 1.0(v_B^2/2g) \qquad \text{(entrance loss)}$$

$$\begin{aligned} h_2 &= f(L/D)(v_B^2/2g) \qquad \text{(pipe friction loss)} \\ &= f(330/0.3355)(v_B^2/2g) \\ &= 985f(v_B^2/2g) \end{aligned}$$

$$\begin{aligned} h_3 &= f_T(L_e/D)(v_B^2/2g) \qquad \text{(long radius elbow)} \\ &= 20f_T(v_B^2/2g) \end{aligned}$$

$$\begin{aligned} h_4 &= f_T(L_e/D)(v_B^2/2g) \qquad \text{(half-open gate valve)} \\ &= 160f_T(v_B^2/2g) \end{aligned}$$

From Table 10.5 we find $f_T = 0.017$ for a 4-in steel pipe.
Then we have

$$\begin{aligned} h_L &= (1.0 + 985f + 20f_T + 160f_T)(v_B^2/2g) \\ &= (4.06 + 985f)(v_B^2/2g) \end{aligned} \tag{11-4}$$

Now substitute this expression for h_L into Eq. (11–3) and solve for v_B in terms of f.

You should have $v_B = \sqrt{2580/(5.06 + 985f)}$.

$$z_A - z_B = (v_B^2/2g) + h_L$$

$$\begin{aligned} 40 \text{ ft} &= (v_B^2/2g) + (4.06 + 985f)(v_B^2/2g) \\ &= (5.06 + 985f)(v_B^2/2g) \end{aligned}$$

Solving for v_B, we get

$$v_B = \sqrt{\frac{2g(40)}{5.06 + 985f}} = \sqrt{\frac{2580}{5.06 + 985f}} \qquad \textbf{(11–5)}$$

Equation (11–5) represents the completion of Step 4 of the procedure. Now do Steps 5 and 6.

$$N_R = \frac{v_B D}{v} = \frac{v_B(0.3355)}{9.15 \times 10^{-6}} = (0.366 \times 10^5)v_B \qquad \textbf{(11–6)}$$

$$D/\epsilon = (0.3355/1.5 \times 10^{-4}) = 2235$$

Step 7 is the start of the iteration process. What is the possible range of values for the friction factor for this system?

Since $D/\epsilon = 2235$, the lowest possible value of f is 0.0155 for very high Reynolds numbers and the highest possible value is 0.039 for a Reynolds number of 4000. The initial trial value of f must be in this range. Use $f = 0.020$ and complete Steps 8 and 9.

We find the values for velocity and the Reynolds number by using Eqs. (11–5) and (11–6):

$$v_B = \sqrt{\frac{2580}{5.06 + (985)(0.02)}} = \sqrt{104} = 10.2 \text{ ft/s}$$

$$N_R = (0.366 \times 10^5)(10.2) = 3.73 \times 10^5$$

Now do Step 10.

You should have $f = 0.0175$. Since this is different from the initial trial value of f, Steps 8 through 11 must be repeated now.

Using $f = 0.0175$, we get

$$v_B = \sqrt{\frac{2580}{5.06 + (985)(0.0175)}} = \sqrt{116} = 10.8 \text{ ft/s}$$

$$N_R = (0.366 \times 10^5)(10.8) = 3.94 \times 10^5$$

The new value of f is 0.0175, which is unchanged. Therefore, we have

$$v_B = 10.8 \text{ ft/s}$$

$$Q = A_B v_B = (0.0884 \text{ ft}^2)(10.8 \text{ ft/s}) = 0.955 \text{ ft}^3/\text{s}$$

This programmed example problem is concluded.

■

11.4.2
Class II Systems
with Two Pipes

We will now present another Class II system which is more complex than the two systems just described. It includes minor losses in addition to friction losses and has two pipes of different sizes in series. These factors require that the solution procedure be modified. Because there are two pipes, there are two unknown friction factors and two unknown velocities. Although more calculations are required, the following solution procedure is a straightforward iteration process similar to the one we have just used. Under average conditions for pipe flow, the procedure will yield the final result in two iteration cycles.

SOLUTION PROCEDURE FOR CLASS II SYSTEMS WITH TWO PIPES

1. Write the energy equation for the system.
2. Evaluate known quantities such as pressure heads and elevation heads.
3. Express energy losses in terms of the two unknown velocities and the two friction factors.
4. Using the continuity equation, express the velocity in the smaller pipe in terms of that in the larger pipe:

$$A_1 v_1 = A_2 v_2$$
$$v_1 = v_2(A_2/A_1)$$

5. Substitute the expression in Step 4 into the energy equation, thereby eliminating one unknown velocity.
6. Solve for the remaining velocity in terms of the two friction factors.
7. Express the Reynolds number for each pipe in terms of the velocity in that pipe.
8. Calculate the relative roughness D/ϵ for each pipe.
9. Select trial values for f in each pipe, using the known values of D/ϵ as a guide. In general, the two friction factors will not be equal.
10. Calculate the velocity in the larger pipe, using the equation from Step 6.
11. Calculate the velocity in the smaller pipe, using the equation from Step 4.
12. Calculate the two Reynolds numbers.
13. Determine the new value of the friction factor for each pipe.
14. Compare the new values of f with those assumed in Step 9 and repeat Steps 9–14 until no significant changes can be detected. The velocities found in Steps 10 and 11 are then correct.

The following programmed example problem will illustrate the details of applying this procedure.

PROGRAMMED EXAMPLE PROBLEM

☐ **EXAMPLE PROBLEM 11.4**

The piping system shown in Fig. 11.5 is being used to transfer water at 15°C from one storage tank to another. Determine the volume flow rate of water through the system. The larger pipe is a standard 6-in Schedule 40 steel pipe having a total length of

FIGURE 11.5 Pipe line system for Example Problem 11.4.

30 m. The smaller pipe is a standard 2-in Schedule 40 steel pipe having a total length of 15 m. The elbows are standard.

Perform the first three steps of the solution procedure before looking at the next panel.

The correct results are summarized below. The subscripts A and B refer to the surfaces of the upper and lower tanks, respectively. The subscripts 6 and 2 refer to the 6-in and 2-in pipes, respectively. The energy equation is

$$\frac{p_A}{\gamma} + z_A + \frac{v_A^2}{2g} - h_L = \frac{p_B}{\gamma} + z_B + \frac{v_B^2}{2g}$$

Because $p_A = p_B = 0$ and v_A and v_B are approximately zero, we have

$$z_A - z_B = h_L = 10 \text{ m} \qquad (11\text{--}7)$$

The energy losses are

$$h_L = h_1 + h_2 + h_3 + h_4 + h_5 + h_6 + h_7$$
$$h_1 = 1.0(v_6^2/2g) \qquad \text{(entrance loss)}$$
$$h_2 = f_6(L/D)(v_6^2/2g) \qquad \text{(friction in 6-in pipe)}$$
$$\quad = f_6(30/0.1541)(v_6^2 2g) = 195 f_6(v_6^2/2g)$$
$$h_3 = 2 f_{6T}(L_e/D)(v_6^2/2g) \qquad \text{(two 90° elbows)}$$
$$\quad = 60 f_{6T}(v_6^2/2g)$$
$$h_4 = K(v_2^2/2g) \qquad \text{(sudden contraction)}$$

Using $D_6/D_2 = 0.1541/0.0525 = 2.93$ and assuming that v_2 is approximately 3 m/s, we find $K = 0.42$ from the data in Chapter 10. If v_2 is much different from 3 m/s, this value of K may have to be re-evaluated.

$$h_4 = 0.42(v_2^2/2g)$$

$$h_5 = f_2(L/D)(v_2^2/2g) \qquad \text{(friction in 2-in pipe)}$$
$$= f_2(15/0.0525)(v_2^2/2g) = 286f_2(v_2^2/2g)$$

$$h_6 = f_{2T}(L_e/D)(v_2^2/2g) \qquad \text{(gate valve—½ open)}$$
$$= 160f_{2T}(v_2^2/2g)$$

$$h_7 = 1.0(v_2^2/2g) \qquad \text{(exit loss)}$$

From Table 10.5 we find $f_{6T} = 0.015$ and $f_{2T} = 0.019$.
It is convenient to algebraically sum the energy losses as shown here:

$$h_L = (1.0 + 195f_6 + 60f_{6T})(v_6^2/2g)$$
$$+ (0.42 + 286f_2 + 160f_{2T} + 1.0)(v_2^2/2g)$$
$$= (1.9 + 195f_6)(v_6^2/2g) + (4.46 + 286f_2)(v_2^2/2g) \qquad \textbf{(11–8)}$$

This is the simplest form of the expression for the energy loss. Now do Step 4 of the solution procedure.

You should have the following:

$$v_2 = v_6(A_6/A_2) = v_6(1.864 \times 10^{-2}/2.168 \times 10^{-3}) = 8.60v_6 \qquad \textbf{(11–9)}$$

In order to substitute this into Eq. (11–8), the expression should be squared:

$$v_2^2 = 74.0v_6^2 \qquad \textbf{(11–10)}$$

Now do Steps 5 and 6.

The final expression for v_6 is

$$v_6 = \sqrt{\frac{196}{331.9 + 195f_6 + 21\,164f_2}}$$

Here is how it was found. Substituting Eq. (11–10) into Eq. (11–8) gives

$$h_L = (1.9 + 195f_6)(v_6^2/2g) + (4.46 + 286f_2)(74.0v_6^2/2g)$$
$$= (1.9 + 195f_6 + 330 + 21\,164f_2)(v_6^2/2g)$$
$$= (331.9 + 195f_6 + 21\,164f_2)(v_6^2/2g)$$

Solving for v_6, we get

$$v_6 = \sqrt{\frac{2gh_L}{331.9 + 195f_6 + 21\,164f_2}}$$

From Eq. (11–7), $h_L = 10$ m. Then we have

$$v_6 = \sqrt{\frac{196}{331.9 + 195f_6 + 21\,164f_2}} \qquad \textbf{(11–11)}$$

Now do Steps 7 and 8 of the solution procedure.

Using $v = 1.15 \times 10^{-6}$ m²/s for water at 15°C and $\epsilon = 4.6 \times 10^{-5}$ m for steel pipe, we get the following values:

$$(N_R)_6 = (1.34 \times 10^5)v_6$$

$$(N_R)_2 = (4.57 \times 10^4)v_2$$

$$(D/\epsilon)_6 = 3350$$

$$(D/\epsilon)_2 = 1141$$

Step 9 begins the iteration procedure. As initial trial values for the friction factors, use $f_6 = 0.02$ and $f_2 = 0.025$. These are in the range of possible values for f for the known relative roughness values. Now do Steps 10 through 13 as the first cycle of iteration.

Here are the correct solutions:

$$v_6 = \sqrt{\frac{196}{331.9 + (195)(0.02) + 21\ 164(0.025)}} = 0.476 \text{ m/s}$$

$$v_2 = 8.60 v_6 = 4.09 \text{ m/s}$$

$$(N_R)_6 = (1.34 \times 10^5)(0.476) = 6.38 \times 10^4$$

$$(N_R)_2 = (4.57 \times 10^4)(4.09) = 1.87 \times 10^5$$

Then the new $f_6 = 0.0210$ and the new $f_2 = 0.0208$. Since these are different from the initially assumed values, repeat Steps 10–13.

These are the results:

$$v_6 = 0.503 \text{ m/s}$$

$$v_2 = 4.32 \text{ m/s}$$

$$(N_R)_6 = 6.73 \times 10^4$$

$$(N_R)_2 = 1.97 \times 10^5$$

The values of f_6 and f_2 are unchanged. Therefore, the velocities listed above are correct. Using v_6 to calculate the volume flow rate gives

$$Q = A_6 v_6 = (1.864 \times 10^{-2} \text{ m}^2)(0.503 \text{ m}^3/\text{s}) = 9.38 \times 10^{-3} \text{ m}^3/\text{s}$$

∎

**11.4.3
Computational Approach to
Class II Systems**

The method presented here avoids the need to use iteration to solve Class II pipe line system problems when only energy loss due to friction is involved. The method was originally developed by Swamee and Jain (Reference 4). The maximum permissible volume flow rate that a given pipe can carry while limiting the energy loss to a specified maximum value is:

$$Q = -2.22\, D^2 \sqrt{\frac{gDh_L}{L}} \log\left[\frac{1}{3.7D/\epsilon} + \frac{1.784\,v}{D\sqrt{gDh_L/L}}\right] \qquad \textbf{(11–12)}$$

This procedure is illustrated in Example Problem 11.5, which uses the same data as Example Problem 11.2.

☐ **EXAMPLE PROBLEM 11.5**   Compute the maximum permissible velocity of flow of a lubricating oil through a horizontal 6-in Schedule 40 steel pipe with a maximum pressure drop of 60 kPa per 100 m of pipe. The oil has a specific gravity of 0.88 and a dynamic viscosity of 9.5×10^{-3} Pa·s.

Solution   Some key results from Example Problem 11.2 can be used.

$$L = 100 \text{ m} \qquad D/\epsilon = 3350$$
$$D = 0.1541 \text{ m} \qquad h_L = 6.95 \text{ m}$$

We need the kinematic viscosity, ν, which can be computed from

$$\nu = \mu/\rho = (9.5 \times 10^{-3} \text{ Pa·s})/(880 \text{ kg/m}^3) = 1.08 \times 10^{-5} \text{ m}^2/\text{s}$$

Placing these values into Eq. 11–12 gives $Q = 0.057$ m³/s, identical to that obtained from Example Problem 11.2.

11.5 CLASS III SYSTEMS

Systems that fall into Class III present true design problems. The requirements placed on the system are specified in terms of an allowable pressure drop or energy loss, a desired volume flow rate, the fluid properties, and the type of pipe to be used. Then the proper size pipe is determined which will meet these requirements.

Iteration is required to solve Class III system design problems because there are too many unknowns to allow a direct solution. The flow velocity, the Reynolds number, and the relative roughness D/ϵ are all dependent on the pipe diameter. Therefore, the friction factor cannot be determined directly.

The procedure for designing Class III systems is different depending on the complexity of the system. In the simplest case, only pipe friction loss must be considered. A step-by-step solution procedure for this kind of system follows, along with a programmed example problem. Systems that include minor losses are more complex. A solution procedure for this type of system is demonstrated through another example problem.

SOLUTION PROCEDURE FOR CLASS III SYSTEMS WITH PIPE FRICTION LOSS ONLY

The problem is to select the proper size pipe which will carry a given volume flow rate of a fluid with a certain maximum allowable pressure drop. The solution procedure is outlined below. The first seven steps represent an algebraic reduction of the problem to a simple form. Steps 8 through 13 comprise the iteration routine.

1. Write the energy equation for the system.
2. Solve for the total energy loss h_L and evaluate known pressure heads and elevations.

3. Express the energy loss in terms of the velocity, using Darcy's equation:

$$h_L = f \times \frac{L}{D} \times \frac{v^2}{2g}$$

4. Express the velocity in terms of the volume flow rate and the pipe diameter:

$$v = \frac{Q}{A} = \frac{4Q}{\pi D^2}$$

5. Substitute the expression for v into Darcy's equation:

$$h_L = f \times \frac{L}{D} \times \frac{16Q^2}{\pi^2 D^4 (2g)} = \frac{8LQ^2}{\pi^2 g} \times \frac{f}{D^5}$$

6. Solve for the diameter:

$$D = \left(\frac{8LQ^2}{\pi^2 g h_L} f \right)^{1/5} = (C_1 f)^{0.2}$$

Notice that the terms which make up C_1 are all known and are independent of the pipe diameter.

7. Express the Reynolds number in terms of the diameter:

$$N_R = \frac{vD\rho}{\mu} = \frac{vD}{\nu}$$

But $v = 4Q/\pi D^2$. Then we have

$$N_R = \frac{4Q}{\pi D^2} \times \frac{D}{\nu} = \frac{4Q}{\pi \nu} \times \frac{1}{D} = \frac{C_2}{D}$$

where $C_2 = 4Q/\pi \nu$.

8. Assume an initial trial value for f. Since both N_R and D/ϵ are unknown, there are no specific guidelines for selecting the initial value. Unless special conditions exist or experience dictates otherwise, assume $f = 0.02$.

9. Compute $D = (C_1 f)^{0.2}$.

10. Compute $N_R = (C_2/D)$.

11. Compute D/ϵ.

12. Determine a new value for the friction factor f from Moody's diagram, Fig. 9.2.

13. Compare the new value of f with that assumed in Step 8 and repeat Steps 8–12 until no significant change in f can be detected. The diameter calculated in Step 9 is then correct.

The following example problem illustrates the application of this procedure.

PROGRAMMED EXAMPLE PROBLEM

☐ **EXAMPLE PROBLEM 11.6**   A water line is to be run to the green of the seventh hole of a golf course as shown in Fig. 11.6. The supply is from a main line at point A where the pressure is 80 psig.

FIGURE 11.6 Water line for
Example Problem 11.6.

In order to ensure proper operation of the sprinklers at the green, the pressure at
point B must be at least 60 psig. Determine the smallest allowable size of standard
Schedule 40 steel pipe to supply 0.50 ft³/s of water at 60°F.

Do Steps 1 and 2 of the Class III system solution procedure now to determine
the allowable energy loss h_L.

You should have $h_L = 21.2$ ft. The energy equation is

$$\frac{p_A}{\gamma} + z_A + \frac{v_A^2}{2g} - h_L = \frac{p_B}{\gamma} + z_B + \frac{v_B^2}{2g}$$

But $v_A = v_B$. Then we have

$$h_L = \frac{p_A - p_B}{\gamma} + z_A - z_B$$

$$= \frac{(80 - 60)\ \text{lb}}{\text{in}^2} \times \frac{\text{ft}^3}{62.4\ \text{lb}} \times \frac{144\ \text{in}^2}{\text{ft}^2} - 25\ \text{ft}$$

$$= (46.2 - 25)\ \text{ft} = 21.2\ \text{ft}$$

It is now required to determine the proper size pipe which will limit the energy loss
due to friction to no more than 21.2 ft. Steps 3, 4, and 5 represent the reduction of
Darcy's equation. Since we are concerned only with friction losses in this problem,
these steps are identical. Therefore, go to Step 6 and evaluate the constant C_1.

You should get $C_1 = 0.178$ and $D = (0.178f)^{0.2}$.

$$C_1 = \frac{8LQ^2}{\pi^2 g h_L} = \frac{(8)(600)(0.50)^2}{\pi^2(32.2)(21.2)} = 0.178$$

Now evaluate C_2 in Step 7.

The correct value is $C_2 = 0.526 \times 10^5$.

$$C_2 = \frac{4Q}{\pi \nu} = \frac{(4)(0.5)}{\pi(1.21 \times 10^{-5})} = 0.526 \times 10^5$$

Then $N_R = (0.526 \times 10^5)/D$. Now, using $f = 0.02$ for the initial trial, complete Steps 8–12 before looking at the next panel.

Compare your results with these:

$$f = 0.02 \quad \text{(assumed)}$$
$$D = [(0.178)(0.02)]^{0.2} = (0.00356)^{0.2} = 0.324 \text{ ft}$$
$$N_R = (0.526 \times 10^5)/(0.324) = 1.625 \times 10^5$$
$$D/\epsilon = (0.324)/(1.5 \times 10^{-4}) = 2160 \quad \text{(steel pipe)}$$

New $f = 0.019$

Since the new value for f is different from the assumed value, repeat Steps 8–12 now.

Here are the revised values using $f = 0.019$:

$$D = 0.320 \text{ ft}$$
$$N_R = 1.65 \times 10^5$$
$$D/\epsilon = 2135$$

The new value of $f = 0.019$ is unchanged. Therefore, the minimum allowable size for the pipe is 0.320 ft. Now select a standard Schedule 40 pipe closest to this size.

A 4-in Schedule 40 steel pipe has an internal diameter of 0.3355 ft and is selected for this application.

This programmed example problem is concluded.

■

In the next example problem, a Class III system is to be designed. Minor losses of various kinds are included. The solution is not in a programmed format, but the details and the logic of the procedure should be carefully followed.

□ **EXAMPLE PROBLEM 11.7**   In a chemical processing system, propyl alcohol at 25°C is taken from the bottom of a large tank and transferred by gravity to another part of the system as shown in Fig. 11.7. The length of the line between the two tanks is 7.0 m. A filter is installed in the line and is known to have a resistance coefficient K of 8.5. Drawn stainless steel tubing is to be used for the transfer line. Use $\epsilon = 3.0 \times 10^{-5}$ m for the equivalent roughness of the tubing. Specify the standard size of tubing which would allow a volume flow rate of 150 L/min through this system.

Solution   Using the surfaces of the two tanks as the reference points, the energy equation is

$$\frac{p_A}{\gamma} + z_A + \frac{v_A^2}{2g} - h_L = \frac{p_B}{\gamma} + z_B + \frac{v_B^2}{2g}$$

FIGURE 11.7 System for Example Problem 11.7.

But $p_A = p_B = 0$ and $v_A = v_B = 0$. Then we have

$$h_L = z_A - z_B = 15 \text{ m}$$

There are four components to the total energy loss:

$$
\begin{aligned}
h_L = \; &1.0(v^2/2g) \; + \; 8.5(v^2/2g) \; + \; f(7/D)(v^2/2g) \; + \; 1.0(v^2/2g) \\
&\text{(entrance)} \qquad \text{(filter)} \qquad\quad \text{(friction)} \qquad\qquad \text{(exit)} \\
= \; &10.5(v^2/2g) \; + \; f(7/D)(v^2/2g)
\end{aligned}
$$

In this equation, v is the velocity of flow in the tube. Solving for the friction factor f gives

$$f\left(\frac{7}{D}\right)\left(\frac{v^2}{2g}\right) = h_L - 10.5\,\frac{v^2}{2g}$$

$$f = \frac{2gh_L D}{7v^2} - \left(\frac{10.5v^2}{2g}\right)\left(\frac{2gD}{7v^2}\right)$$

$$= \frac{2gh_L D}{7v^2} - 1.50D$$

Now let $v = 4Q/\pi D^2$:

$$f = \frac{2gh_L D \pi^2 D^4}{(7)(16)Q^2} - 1.50D$$

Expressing Q in m³/s, we get

$$Q = 150 \text{ L/min}\left(\frac{1 \text{ m}^3/\text{s}}{60\,000 \text{ L/min}}\right) = 2.50 \times 10^{-3} \text{ m}^3/\text{s}$$

Since $h_L = 15$ m, we have

$$f = \frac{(2)(9.81)(15)(\pi)^2}{(7)(16)(2.50 \times 10^{-3})^2} D^5 - 1.50D$$

$$= (4.15 \times 10^6)D^5 - 1.50D$$

This equation will be used to iterate for the diameter D. Since we cannot solve for D in terms of f, the iteration must proceed as follows:

1. Assume a value for D.
2. Calculate f.
3. Calculate D/ϵ.
4. Calculate N_R.
5. Evaluate f and compare with that calculated in Step 2.
6. Adjust D in a manner which will decrease the difference between the values of f, and repeat Steps 2–6 until agreement is found for successive values of f.

There is a greater probability of several steps of iteration with this procedure than before since judgment is required in the selection of trial values of D. Let's start with a trial value of 0.040 m for the inside diameter, corresponding approximately to a 1¾-in steel tube with a wall thickness of 0.083 in (see Appendix G). Then we have

$$f = (4.15 \times 10^6)(0.040)^5 - 1.50(0.040) = 0.365$$

$$D/\epsilon = (0.040)/(3.0 \times 10^{-5}) = 1333$$

$$N_R = \frac{vD\rho}{\mu} = \frac{4Q\rho D}{\pi D^2 \mu} = \frac{4Q\rho}{\pi\mu} \times \frac{1}{D} = \frac{(4)(2.50 \times 10^{-3})(802)}{(\pi)(1.92 \times 10^{-3})} \times \frac{1}{D}$$

$$= (1.33 \times 10^3)/D = (1.33 \times 10^3)/(0.040) = 3.33 \times 10^4$$

$$f = 0.0250 \quad \text{(from Fig. 9.2)}$$

A smaller value of D is required to cause the two values of f to be equal. Table 11.1 shows the results of successive trials for D including the one just completed.

TABLE 11.1

Trial	D (m)	f (Step 2)	D/ε	N_R	f (Step 5)	Required Change in D
1	0.040	0.365	1333	3.33×10^4	0.0250	decrease
2	0.030	0.56	1000	4.43×10^4	0.0247	decrease
3	0.027	0.19	900	4.93×10^4	0.0245	increase
4	0.028	0.29	933	4.75×10^4	0.0245	decrease
5	0.02753	0.24	918	4.83×10^4	0.0245	—

Trials 3 and 4 showed that the proper value for D is greater than 0.027 m but less than 0.028 m. From Appendix G we find that the 1¼-in tube with a 0.083-in wall thickness has an inside diameter of 0.02753 m. Using this value on the last trial yielded a reasonably good match between the values of f computed in Steps 2 and 5. Thus, the assumed diameter is the proper value.

**11.6
PIPE LINE
DESIGN AIDS**

The methods shown in this chapter have given the tools for analyzing and designing the three classes of series pipe line systems for cases involving friction losses in pipes and minor losses. Also included were the ability to consider changes in elevation and velocity along with the presence of minor losses.

In some design and analysis situations, it may be sufficient to include only the energy loss due to friction. If minor losses total less than about 5 percent of friction losses, they can be neglected because the inherent uncertainty about pipe surface roughness and friction factors is likely to be greater than that. (See Reference 3.) Then, it is relatively simple to compute the energy loss and the corresponding pressure drop that occur in a given length of pipe of a given size. References 1 and 2 include tables of data listing energy loss or pressure drop per 100 ft of length for the flow of water through wide ranges of steel pipe sizes at flow rates that result in an average velocity from approximately 0.50 to 30 ft/s. Reference 2 includes other types of pipe and fluids with viscosities different from water.

Computational Approach to Class III Problems

The solution procedure presented in Section 11.5 required iteration to determine the required size of pipe to carry a given amount of fluid while limiting the energy loss to a specified maximum amount. This may be inconvenient if such a computation were to be done within a computer program.

References 3 and 4 present the following equation for computing the required size of pipe directly.

$$D = 0.66 \left[\epsilon^{1.25} \left(\frac{LQ^2}{gh_L} \right)^{4.75} + \nu Q^{9.4} \left(\frac{L}{gh_L} \right)^{5.2} \right]^{0.04} \qquad \textbf{(11–13)}$$

☐ **EXAMPLE PROBLEM 11.8**

Compute the required size of new clean Schedule 40 steel pipe that will carry 0.50 ft³/s of water at 60°F and limit the energy loss due to friction to no more than 3.50 ft per 100 ft of pipe.

Solution

The following data are needed for Equation (11–13).

$Q = 0.50$ ft³/s $L = 100$ ft $g = 32.2$ ft/s
$h_L = 3.50$ ft $\epsilon = 1.5 \times 10^{-4}$ ft $\nu = 1.21 \times 10^{-5}$ ft²/s

This gives $D = 0.326$ ft (minimum). Then a 4-in Schedule 40 pipe should be specified.

REFERENCES

1. Crane Co. 1988. *Flow of Fluids Through Valves, Fittings, and Pipe*. Technical Paper No. 410. Joliet, Illinois: Author.

2. Heald, C. C., ed. 1988. *Cameron Hydraulic Data*. Woodcliff Lake, New Jersey: Ingersoll-Rand.

3. Streeter, Victor L., and E. Benjamin Wylie. 1985. *Fluid Mechanics,* 8th ed. New York: McGraw-Hill.

4. Swamee, P. K., and A. K. Jain. 1976. Explicit Equa-

tions for Pipe-flow Problems. *Journal of the Hydraulics Division,* 102 (HY5) 657–664. New York: American Society of Civil Engineers.

PRACTICE PROBLEMS

Class I Systems

11.1M *Water at 10°C flows from a large reservoir at the rate of 1.5×10^{-2} m³/s through the system shown in Fig. 11.8. Calculate the pressure at B.*

FIGURE 11.8   Problem 11.1.

11.2M *For the system shown in Fig. 11.9, kerosene (sg = 0.82) at 20°C is to be forced from tank A to reservoir B by increasing the pressure in the sealed tank A above the kerosene. The total length of 2-in Schedule 40 steel pipe is 38 m. The elbows are standard. Calculate the required pressure in tank A to cause a flow rate of 435 L/min.*

11.3E Figure 11.10 shows a portion of a hydraulic circuit. The pressure at point B must be 200 psig when the volume flow rate is 60 gal/min. The hydraulic fluid has a specific gravity of 0.90 and a dynamic viscosity of 6.0×10^{-5} lb-s/ft². The total length of pipe between A and B is 50 ft. The elbows are standard. Calculate the pressure at the outlet of the pump at A.

11.4E Figure 11.11 shows a hydraulic system in which the pressure at B must be 500 psig while the flow rate is 750 gal/min. The fluid is a medium machine tool hydraulic oil. The total length of 4-in pipe is 40 ft. The elbows are standard. Neglect the energy

FIGURE 11.9   Problem 11.2.

loss due to friction in the 6-in pipe. Calculate the required pressure at A if the oil is (a) at 104°F and (b) at 212°F.

11.5M *Oil is flowing at the rate of 0.015 m³/s in the system shown in Fig. 11.12. Data for the system are*

Oil specific weight = 8.80 kN/m³
Oil kinematic viscosity = 2.12×10^{-5} m²/s

FIGURE 11.10   Problem 11.3.

FIGURE 11.11   Problem 11.4.

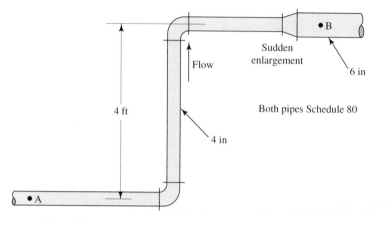

Length of 6-in pipe = 180 m
Length of 2-in pipe = 8 m
Elbows are long radius type
Pressure at B = 12.5 MPa

Considering all pipe friction and minor losses, calculate the pressure at A.

11.6M *For the system shown in Fig. 11.13, calculate the vertical distance between the surfaces of the two reservoirs when water at 10°C flows from A to B at the rate of 0.03 m³/s. The elbows are standard. The total length of the 3-in pipe is 100 m. For the 6-inch pipe it is 300 m. Use ε = 6.0 × 10⁻⁵ m for the roughness of the pipe.*

11.7M *A liquid refrigerant flows through the system shown in Fig. 11.14 at the rate of 1.70 L/min. The refrigerant has a specific gravity of 1.25 and a dynamic viscosity of 3 × 10⁻⁴ Pa·s. Calculate the pressure difference between points A and B. The tube is steel, with an outside diameter of ½ in, a wall thickness of 0.049 in, and a total length of 30 m.*

Class II Systems

11.8E Water at 100°F is flowing in a 4-in Schedule 80 steel pipe which is 25 ft long. Calculate the maximum allowable volume flow rate if the energy loss due to pipe friction is to be limited to 30 ft-lb/lb.

11.9M *A hydraulic oil is flowing in a steel tube with an outside diameter of 2 in and a wall thickness of 0.083 in. A pressure drop of 68 kPa is observed between two points in the tube, 30 m apart. The oil has a specific gravity of 0.90 and a dynamic viscosity of 3.0 × 10⁻³ Pa·s. Assume the tube wall roughness to be 3 × 10⁻⁵ m. Calculate the velocity of flow of oil.*

11.10E In a processing plant, ethylene glycol at 77°F is flowing in 5000 ft of 6-in ductile iron pipe. Over this distance, the pipe falls 55 ft and the pressure drops from 250 psig to 180 psig. Calculate the velocity of flow in the pipe. Use ε = 2.0 × 10⁻⁴ ft for the roughness of the pipe.

11.11M *Water at 15°C is flowing downward in a vertical tube 7.5 m long. The pressure is 550 kPa at the top*

FIGURE 11.12   Problem 11.5.

FIGURE 11.13 Problem 11.6.

FIGURE 11.14 Problem 11.7.

and 585 kPa at the bottom. A ball-type check valve is installed near the bottom. The tube is steel, with a 1¼-in outside diameter and a 0.083-in wall thickness. Compute the volume flow rate of the water.

11.12E Turpentine at 77°F is flowing from A to B in a 3-in asphalt-coated cast iron pipe (actual inside diameter = 3.58 in). Point B is 20 ft higher than A, and the total length of pipe is 60 ft. Two 90° long radius elbows are installed between A and B. Calculate the volume flow rate of turpentine if the pressure at A is 120 psig and the pressure at B is 105 psig.

11.13E A device designed to allow cleaning of walls and windows on the second floor of homes is similar to the system shown in Fig. 11.15. Determine the velocity of flow from the nozzle if the pressure at the bottom is (a) 20 psig and (b) 80 psig. The nozzle has a loss coefficient K of 0.15 based on the outlet velocity. The tube is smooth drawn alumi-

FIGURE 11.15 Problem 11.13.

FIGURE 11.16   Problem 11.14.

num and has an inside diameter of 0.5 in. The 90°
bend has a radius of 6 in. The total length of
straight tube is 20 ft. The fluid is water at 100°F.

11.14M *Kerosene at 25°C is flowing in the system shown in
Fig. 11.16. The total length of 2-in Type K copper
tubing is 30 m. The two 90° bends have a radius of*

*300 mm. Calculate the volume flow rate into tank
B if a pressure of 150 kPa is maintained above the
kerosene in tank A.*

11.15M *Water at 40°C is flowing from A to B through the
system shown in Fig. 11.17. Determine the volume
flow rate of water if the vertical distance between*

FIGURE 11.17   Problem 11.15.

FIGURE 11.18   Problem 11.16.

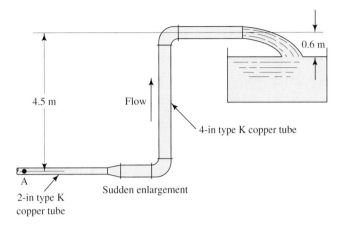

the surfaces of the two reservoirs is 10 m. Both
pipes are asphalt-coated cast iron. The elbows are
standard.

11.16M *Oil with a specific gravity of 0.93 and a dynamic
viscosity of 9.5 × 10⁻³ Pa·s is flowing into the
open tank shown in Fig. 11.18. The total length of
2-in tubing is 30 m. For the 4-in tubing the total
length is 100 m. The elbows are standard. Determine the volume flow rate into the tank if the pressure at point A is 175 kPa.*

Class III Systems

11.17E Determine the required size of new Schedule 80
steel pipe to carry water at 160°F with a maximum
pressure drop of 10 psi per 1000 ft when the flow
rate is 0.5 ft³/s.

11.18M *What size of standard Type K copper tube is required to transfer 0.06 m³/s of water at 80°C from a
heater where the pressure is 150 kPa to an open
tank? The tube is horizontal and 30 m long.*

11.19E Water at 60°F is to flow by gravity between two
points, 2 mi apart, at the rate of 13 500 gal/min.
The upper end is 130 ft higher than the lower
end. What size concrete pipe is required? Assume
that the pressure at both ends of the pipe is negligible.

11.20E The tank shown in Fig. 11.19 is to be drained to a
sewer. Determine the size of new Schedule 40
steel pipe which will carry at least 400 gal/min of
water at 80°F through the system shown. The total
length of pipe is 75 ft.

FIGURE 11.19   Problem 11.20.

h

$3\frac{1}{2}$-in Schedule 40
steel pipe

Gate valve
$\frac{1}{2}$ open

0.5 m

Truck

FIGURE 11.20 Problem 11.21.

Practice Problems for Any Class

11.21M *Figure 11.20 depicts gasoline flowing from a storage tank into a truck for transport. The gasoline has a specific gravity of 0.68 and the temperature is 25°C. Determine the required depth h in the tank to produce a flow rate of 1500 L/min into the truck. Since the pipes are short, neglect the energy losses due to pipe friction but do consider minor losses.*

Note: Figure 11.21 shows a system used to pump coolant from a collector tank to an elevated tank, where it is cooled. The pump delivers 30 gal/min. The coolant then flows back to the machines as needed, by gravity. The coolant has a specific gravity of 0.92 and a dynamic viscosity of 3.6×10^{-5} lb·s/ft². This system is used in Problems 11.22 through 11.24.

11.22E For the system in Fig. 11.21, compute the pressure at the inlet to the pump. The filter has a resistance coefficient of 1.85 based on the velocity head in the suction line.

20 GPM

1.0 ft

#1

10 GPM

#2

$L = 30$ ft

4.0 ft

Flow

4 ft

2.0 ft

Floor

Flow

18 ft

$1\frac{1}{4}$-in Schedule 40
steel pipe
$L = 20$ ft
Swing check valve

Flow

6.0 ft

3.0 ft

Filter

Pump

Fully open
gate valve

2-in Schedule 40
steel pipe
$L = 10.0$ ft

FIGURE 11.21 Problems 11.22, 11.23, and 11.24.

11.23E For the system in Fig. 11.21, compute the total head on the pump and the power delivered by the pump to the coolant.

11.24E For the system in Fig. 11.21, specify the size of Schedule 40 steel pipe required to return the fluid to the machines. Machine 1 requires 20 gal/min and Machine 2 requires 10 gal/min. The fluid leaves the pipes at the machines at 0 psig.

11.25E A manufacturer of spray nozzles specifies that the maximum pressure drop in the pipe feeding a nozzle be 10.0 psi per 100 ft of pipe. Compute the maximum allowable velocity of flow through a 1-in Schedule 80 steel pipe feeding the nozzle. The pipe is horizontal and the fluid is water at 60°F.

11.26E Specify the size of new Schedule 40 steel pipe required to carry gasoline at 77°F through 120 ft of horizontal pipe with no more than 8.0 psi of pressure drop at a volume flow rate of 100 gal/min.

11.27M *Refer to Fig. 11.22. Water at 80°C is being*

pumped from the tank at the rate of 475 L/min. Compute the pressure at the inlet of the pump.

11.28M *For the system shown in Fig. 11.22 and analyzed in Problem 11.27, it is desirable to change the system to increase the pressure at the inlet to the pump. The volume flow rate must stay at 475 L/min, but anything else can be changed. Redesign the system and recompute the pressure at the inlet to the pump for comparison with the result of Problem 11.27.*

11.29E In a water pollution control project, the polluted water is pumped vertically upward 80 ft and then sprayed into the air to increase the oxygen content in the water and to evaporate volatile materials. The system is sketched in Fig. 11.23. The polluted water has a specific weight of 64.0 lb/ft³ and a dynamic viscosity of 4.0×10^{-5} lb·s/ft². The flow rate is 0.50 ft³/s. The pressure at the inlet to the pump is 3.50 psi below atmospheric pressure. The total length of discharge pipe is 82 ft. The nozzle

FIGURE 11.22 Problems 11.27 and 11.28.

FIGURE 11.23   Problems 11.29 and 11.30.

FIGURE 11.24   Problems 11.31, 11.32, and 11.33.

FIGURE 11.25   Problems 11.34, 11.35, 11.36, and 11.37.

FIGURE 11.26 Problems 11.38, 11.39, 11.40, and 11.41.

has a resistance coefficient of 2.50 based on the velocity head at the outlet of the nozzle. Compute the power delivered by the pump to the fluid.

11.30E Repeat Problem 11.29, but use a 3-in Schedule 40 steel pipe for the discharge line instead of the 2½-in pipe. Compare the power delivered by the pump for the two designs.

11.31M *Water at 10°C is being delivered to a tank on the roof of a building, as shown in Fig. 11.24. The elbow is standard. What pressure must exist at Point A for 200 L/min to be delivered?*

11.32M *If the pressure at point A in Fig. 11.24 is 200 kPa, compute the volume flow rate of water of 10°C delivered to the tank.*

11.33M *Change the design of the system in Fig. 11.24 to replace the globe valve with a fully open gate valve. Then, if the pressure at point A is 200 kPa, compute the volume flow rate of water at 10°C delivered to the tank. Compare the result with that of Problem 11.32 to demonstrate the effect of the valve change.*

11.34E It is desired to deliver 250 gal/min of ethyl alcohol at 77°F from tank A to tank B in the system shown in Fig. 11.25. The total length of pipe is 110 ft. Compute the required pressure in tank A.

11.35E For the system shown in Fig. 11.25, compute the volume flow rate of ethyl alcohol at 77°F that would occur if the pressure in tank A is 100 psig. The total length of pipe is 110 ft.

11.36E Repeat Problem 11.35, but consider the valve to be fully open.

11.37E Repeat Problem 11.35, but consider the valve to be fully open and the elbows to be the long radius type instead of standard. Compare the results with those from Problems 11.35 and 11.36.

11.38M *Figure 11.26 depicts a pipe delivering water at 15°C from a main line to a factory. The pressure at the main is 415 kPa. Compute the maximum allowable flow rate if the pressure at the factory must be no less than 200 kPa.*

11.39M *Repeat Problem 11.38, but replace the globe valve with a fully open butterfly valve.*

11.40M *Repeat Problem 11.38, but use a 5-in Schedule 40 pipe.*

11.41M *Repeat Problem 11.38, but replace the globe valve with a butterfly valve and use a 5-in Schedule 40 steel pipe. Compare the results of Problems 11.38–11.41.*

11.42E It is desired to drive a small, positive displacement pump by chucking a household electric drill to the drive shaft of the pump. The pump delivers 1.0 in³ of water at 60°F per revolution, and the pump rotates at 2100 rpm. The outlet of the pump flows through a 100-ft smooth plastic hose with an inside diameter of 0.75 in. How far above the source can the outlet of the hose be if the maximum power available from the drill motor is 0.20 hp? The pump efficiency is 75 percent. Consider the friction loss in the hose, but neglect other losses.

COMPUTER PROGRAMMING ASSIGNMENTS

1. Write a program to analyze Class I pipe line systems, including energy losses due to friction and minor losses due to valves and fittings.

2. Write a program to determine the velocity of flow and the volume flow rate in a given pipe with a limited pressure drop, considering energy loss due to friction only.

Use the computational approach described in Section 11.4 and illustrated in Example Problem 11.5.

3. Write a program to determine the size of pipe required to carry a specified flow rate with a limited pressure drop, using the Class III solution procedure described in Example Problem 11.8.

4. Write a program to determine the size of pipe required to carry a specified flow rate with a limited pressure drop. Consider both energy loss due to friction and minor losses. Use a method similar to that described in Example Problem 11.7.

12 Parallel Pipe Line Systems

**12.1
OBJECTIVES**

If a pipe line system is arranged so that the fluid flows in a continuous line without branching, it is referred to as a *series system*. Conversely, if the system causes the flow to branch into two or more lines, it is referred to as a *parallel system*.

The nature of parallel systems requires that the approach to their analysis be different from that used for series systems. In general, a parallel system can have any number of branches. The system shown in Fig. 12.1, with three branches, will be used to illustrate the basic concepts. The flow in the main line at section 1 splits into three parts and then rejoins at section 2. For most problems of this type, the objective is to determine how much of the flow occurs in each of the branches and how much pressure drop occurs between sections 1 and 2.

After completing this chapter, you should be able to:

1. Discuss the differences between series pipe line systems and parallel pipe line systems.
2. State the general relationships for flow rates and head losses for parallel pipe line systems.
3. Compute the amount of flow that occurs in each branch of a two-branch parallel pipe line system and the head loss that occurs across the system when the total flow rate and the description of the system are known.
4. Determine the amount of flow that occurs in each branch of a two-branch parallel pipe line system and the total flow if the pressure drop across the system is known.
5. Use the Hardy Cross technique to compute the flow rates in all branches of a network having three or more branches.

**12.2
PRINCIPLES
GOVERNING
PARALLEL PIPE
LINE SYSTEMS**

The analysis of parallel pipe line systems requires the use of the general energy equation along with equations relating the volume flow rates in the various branches of the system and expressions for the head losses throughout the system. The following equations state the principles relating the volume flow rates and the head losses for a parallel system with three branches such as the one shown in Fig. 12.1.

$$Q_1 = Q_2 = Q_a + Q_b + Q_c \tag{12-1}$$

$$h_{L_{1-2}} = h_a = h_b = h_c \tag{12-2}$$

FIGURE 12.1 Example of a parallel pipe line system with three branches.

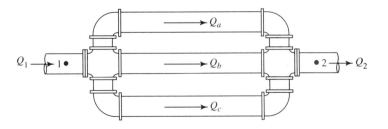

Equation (12–1) is the statement of the condition of continuity for steady flow in a parallel system. The total flow entering the system, Q_1, divides into the three branch flows, Q_a, Q_b, and Q_c. Then they merge into the outlet pipe where the flow rate is Q_2. By the principle of continuity, the outlet flow at section 2 is equal to the inlet flow at section 1. In Eq. (12–2), the term $h_{L_{1-2}}$ is the energy loss *per unit of fluid* between points 1 and 2 in the main lines. The terms h_a, h_b, and h_c are the energy losses *per unit of fluid* in each branch of the system. We can demonstrate that these quantities must all be equal by writing the energy equation, using points 1 and 2 as the reference points:

$$\frac{p_1}{\gamma} + z_1 + \frac{v_1^2}{2g} - h_L = \frac{p_2}{\gamma} + z_2 + \frac{v_2^2}{2g} \qquad \textbf{(12–3)}$$

In earlier chapters, the sum of the pressure head p/γ, the elevation head z, and the velocity head $v^2/2g$ was referred to as the total head E. This represents the energy possessed by each unit of the fluid at a particular point in a system. Substituting E into Eq. (12–3) gives

$$E_1 - h_L = E_2$$

and

$$h_L = E_1 - E_2 \qquad \textbf{(12–4)}$$

Then the term h_L represents the loss of head between points 1 and 2. In Fig. 12.1, each unit of fluid has the same total head at the point where the flow branches. As the fluid flows through the branches, some energy is lost. But at the point where the flow rejoins, the total head of each unit of fluid must again be the same. We can therefore draw the conclusion that the loss of head is the same, regardless of which path is taken between points 1 and 2. This conclusion is stated mathematically as Eq. (12–2).

The amount of fluid flowing in a particular branch of a parallel system is dependent on the resistance to flow in that branch relative to the resistance in other branches. The fluid will tend to follow the path of least resistance. As stated in Chapters 9 and 10, resistance to flow is due to pipe wall friction, to changes in the cross section of the flow path, to changes in the direction of flow, or to obstructions such as might exist in valves. All of these resistances are dependent on the velocity of flow. Therefore, in a parallel system, the flow divides in such a way that the velocities are different in the various branches and the loss of head in each branch is equal.

A common parallel piping system includes two branches arranged as shown in Fig. 12.2. The lower branch is added to allow some fluid to bypass the heat exchanger. The branch could also be used to isolate the heat exchanger, allowing continuous flow while the equipment is serviced. The analysis of this type of system is relatively simple and straightforward, as will be explained in Section 12.3.

FIGURE 12.2 Parallel system with two branches.

Parallel systems having more than two branches are more complex because there are more unknown quantities than there are equations relating the unknowns. A solution procedure requiring iteration is described in Section 12.4.

**12.3
SYSTEMS WITH
TWO BRANCHES**

We will use the system shown in Fig. 12.2 to illustrate the analysis of flow in two branches. The basic relationships which apply here are similar to Eqs. (12–1) and (12–2) except there are only two branches instead of three. These relationships are

$$Q_1 = Q_2 = Q_a + Q_b \tag{12–5}$$
$$h_{L_{1-2}} = h_a = h_b \tag{12–6}$$

The following example problems are presented in the programmed form. You should pay careful attention to the logic of the solution procedure as well as to the details performed.

SOLUTION METHOD FOR SYSTEMS WITH TWO BRANCHES WHEN THE TOTAL FLOW RATE AND THE DESCRIPTION OF THE BRANCHES ARE KNOWN

Example Problem 12.1 is of this type. The method of solution is

1. Equate the total flow rate to the sum of the flow rates in the two branches, as stated in Eq. (12–5). Then express the branch flows as the product of the flow area and the average velocity; that is,

$$Q_a = A_a v_a \quad \text{and} \quad Q_b = A_b v_b$$

2. Express the head loss in each branch in terms of the velocity of flow in that branch and the friction factor. Include all significant losses due to friction and minor losses.

3. Compute the relative roughness D/ϵ for each branch, estimate the value of the friction factor for each branch, and complete the calculation of head loss in each branch in terms of the unknown velocities.

4. Equate the expressions for the head losses in the two branches to each other as stated in Eq. (12–6).

5. Solve for one velocity in terms of the other from the equation in Step 4.

6. Substitute the result from Step 5 into the flow rate equation developed in Step 1, and solve for one of the unknown velocities.

7. Solve for the second unknown velocity from the relationship developed in Step 5.

8. If there is doubt about the accuracy of the value of the friction factor used in Step 2, compute the Reynolds number for each branch and re-evaluate the friction factor from the Moody diagram or compute the values for the friction factors from Eq. (9–9) in Chapter 9.

9. If the values for the friction factor have changed significantly, repeat Steps 3–8, using the new values for friction factor.

10. When satisfactory precision has been achieved, use the now-known velocity in each branch to compute the volume flow rate for that branch. Check the sum of the volume flow rates to ensure that it is equal to the total flow in the system.

11. Use the velocity in either branch to compute the head loss across that branch, employing the appropriate relationship from Step 3. This head loss is also equal to the head loss across the entire branched system. You can compute the pressure drop across the system, if desired, by using the relationship $\Delta p = \gamma h_L$.

PROGRAMMED EXAMPLE PROBLEM

☐ **EXAMPLE PROBLEM 12.1**   In Fig. 12.2, 100 gal/min of water at 60°F is flowing in a 2-in Schedule 40 steel pipe at section 1. The heat exchanger in branch a has a loss coefficient of $K = 7.5$ based on the velocity head in the pipe. All three valves are wide open. Branch b is a bypass line composed of 1¼-in Schedule 40 steel pipe. The elbows are standard. The length of pipe between points 1 and 2 in branch b is 20 ft. Because of the size of the heat exchanger, the length of pipe in branch a is very short and friction losses can be neglected. For this arrangement, determine (a) the volume flow rate of water in each branch and (b) the pressure drop between points 1 and 2.

If we apply Step 1 of the solution method, Eq. (12–5) relates the two volume flow rates. How many quantities are unknown in this equation?

The two velocities v_a and v_b are unknown. Since $Q = Av$, Eq. (12–5) can be expressed as

$$Q_1 = A_a v_a + A_b v_b \qquad (12\text{–}7)$$

From the given data, $A_a = 0.02333 \text{ ft}^2$, $A_b = 0.01039 \text{ ft}^2$, $Q_1 = 100 \text{ gal/min}$. Expressing Q_1 in the units of ft^3/s gives

$$Q_1 = 100 \text{ gal/min} \times \frac{1 \text{ ft}^3/\text{s}}{449 \text{ gal/min}} = 0.223 \text{ ft}^3/\text{s}$$

Generate another equation which also relates v_a and v_b using Step 2.

Equation (12–6) states that the head loss in each branch is equal. Since the head losses h_a and h_b are dependent on the velocities v_a and v_b, this equation can be used in conjunction with Eq. (12–7) to solve for the velocities. Now, express the head losses in terms of the velocities for each branch.

You should have something similar to this for branch a.

$$h_a = 2f_{aT}(L_e/D)(v_a^2/2g) + K(v_a^2/2g)$$
$$\text{(valves)} \qquad \text{(heat exchanger)}$$

The following data are known:

$$f_{aT} = 0.019 \text{ for a 2-in Schedule 40 pipe (Table 10.5)}$$

$$L_e/D = 8 \text{ for a fully open gate valve (Table 10.4)}$$

$$K = 7.5 \text{ from the problem statement}$$

Then,

$$h_a = (2)(0.019)(8)(v_a^2/2g) + 7.5(v_a^2/2g) = 7.80(v_a^2/2g) \qquad \textbf{(12–8)}$$

For branch b:
$$h_b = 2f_{bT}(L_e/D)(v_b^2/2g) + f_{bT}(L_e/D)(v_b^2/2g) + f_b(L/D)(v_b^2/2g)$$
$$\text{(elbows)} \qquad\qquad \text{(valve)} \qquad\qquad \text{(friction)}$$

The known data are

$$f_{bT} = 0.022 \text{ for a } 1\tfrac{1}{4}\text{-in Schedule 40 pipe (Table 10.5)}$$

$$L_e/D = 30 \text{ for each elbow (Table 10.4)}$$

$$L_e/D = 340 \text{ for a fully open globe valve (Table 10.4)}$$

Then,

$$h_b = (2)(0.022)(30)(v_b^2/2g) + (0.022)(340)(v_b^2/2g)$$
$$+ f_b(20/0.1150)(v_b^2/2g)$$
$$h_b = (8.80 + 173.9f_b)(v_b^2/2g)$$

This equation introduces the additional unknown, f_b. We can use an iteration procedure similar to that used for Class II series pipe line systems in Chapter 11. The relative roughness for branch b will aid in the estimation of the first trial value for f_b.
$$D/\epsilon = (0.1150/1.5 \times 10^{-4}) = 767$$

From the Moody diagram in Fig. 9.2, a logical estimate for the friction factor $f_b = 0.023$. Substituting this into the equation for h_b gives

$$h_b = [8.80 + 173.9(0.023)](v_b^2/2g) = 12.80(v_b^2/2g) \qquad \textbf{(12–9)}$$

We have now completed Step 3 of the solution procedure. Steps 4 and 5 can be done now to obtain an expression for v_a in terms of v_b.

You should have $v_a = 1.281 v_b$. Here is how it is done.

$$h_a = h_b$$
$$7.80(v_a^2/2g) = 12.80(v_b^2/2g)$$

Solving for v_a gives

$$v_a = 1.281 v_b \qquad \text{(12–10)}$$

At this time, you can combine Eqs. (12–7) and (12–10) to calculate the velocities (Steps 6 and 7).

The solutions are $v_a = 5.54$ ft/s and $v_b = 7.09$ ft/s. Here are the details:

$$Q_1 = A_a v_a + A_b v_b \qquad \text{(12–7)}$$
$$v_a = 1.281 v_b \qquad \text{(12–10)}$$

Then we have

$$Q_1 = A_a(1.281 v_b) + A_b v_b = v_b(1.281 A_a + A_b)$$

Solving for v_b, we get

$$v_b = \frac{Q_1}{1.281 A_a + A_b} = \frac{0.223 \text{ ft}^3/\text{s}}{[(1.281)(0.02333) + 0.01039] \text{ ft}^2}$$
$$= 5.54 \text{ ft/s}$$
$$v_a = (1.281)(5.54) \text{ ft/s} = 7.09 \text{ ft/s}$$

Since we made these calculations using an assumed value for f_b, we should check the accuracy of the assumption.

We can evaluate the Reynolds number for branch b.

$$N_{Rb} = v_b D_b/\nu$$

From Appendix Table A.2, we find $\nu = 1.21 \times 10^{-5}$ ft²/s. Then,

$$N_{Rb} = (5.54)(0.1150)/(1.21 \times 10^{-5}) = 5.26 \times 10^4$$

Using this value and the relative roughness of 767 from before in the Moody diagram yields a new value for $f_b = 0.025$. Because this is significantly different from the assumed value of 0.023, we can repeat the calculations for Steps 3 to 8. The results are summarized below.

$$h_b = [8.80 + 173.9(0.025)](v_b^2/2g) = 13.15(v_b^2/2g) \qquad \text{(12–9)}$$
$$h_a = 7.80(v_a^2/2g) \qquad \text{(Same as for first trial)}$$

Equating the head losses in the two branches,

$$h_a = h_b$$
$$7.80(v_a^2/2g) = 13.15(v_b^2/2g)$$

Solving for the velocities,

$$v_a = 1.298 v_b$$

Substituting this into the equation for v_b used before gives

$$v_b = \frac{0.223 \text{ ft}^3/\text{s}}{[(1.298)(0.02333) + 0.01039] \text{ ft}^2} = 5.48 \text{ ft/s}$$

$$v_a = 1.298 v_b = 1.298(5.48) = 7.12 \text{ ft/s}$$

Recomputing the Reynolds number for branch b gives

$$N_{Rb} = v_b D_b / \nu$$

$$N_{Rb} = (5.48)(0.1150)/(1.21 \times 10^{-5}) = 5.21 \times 10^4$$

There is no significant change in the value of f_b. Therefore the values of the two velocities computed above are correct. We can now complete Steps 10 and 11 of the procedure to find the volume flow rate in each branch and the head loss and pressure drop across the entire system.

Now calculate the volume flow rates Q_a and Q_b (Step 10.)

You should have

$$Q_a = A_a v_a = (0.02333 \text{ ft}^2)(7.12 \text{ ft/s}) = 0.166 \text{ ft}^3/\text{s}$$

$$Q_b = A_b v_b = (0.01039 \text{ ft}^2)(5.48 \text{ ft/s}) = 0.057 \text{ ft}^3/\text{s}$$

Converting these values to the units of gal/min gives $Q_a = 74.5$ gal/min and $Q_b = 25.5$ gal/min.

We are also asked to calculate the pressure drop. How can this be done?

We can write the energy equation using points 1 and 2 as reference points. Since the velocities and elevations are the same at these points, the energy equation is simply

$$\frac{p_1}{\gamma} - h_L = \frac{p_2}{\gamma}$$

Solving for the pressure drop, we get

$$p_1 - p_2 = \gamma h_L \qquad \textbf{(12–11)}$$

What can be used to calculate h_L?

Since $h_{L_{1-2}} = h_a = h_b$, we can use either Eq. (12–8) or (12–9). Using Eq. (12–8), we get

$$h_a = 7.80(v_a^2/2g) = (7.80)(7.12)^2/64.4 \text{ ft} = 6.14 \text{ ft}$$

Note that this neglects the minor losses in the two tees. Then we have

$$p_1 - p_2 = \gamma h_L = \frac{62.4 \text{ lb}}{\text{ft}^3} \times 6.14 \text{ ft} \times \frac{1 \text{ ft}^2}{144 \text{ in}^2} = 2.66 \text{ psi}$$

This example problem is concluded.

■

SOLUTION METHOD FOR SYSTEMS WITH TWO BRANCHES WHEN THE PRESSURE DROP ACROSS THE SYSTEM IS KNOWN AND THE VOLUME FLOW RATE IN EACH BRANCH AND THE TOTAL VOLUME FLOW RATE ARE TO BE COMPUTED

Example Problem 12.2 is of this type. The method of solution is

1. Compute the total head loss across the system using the known pressure drop, Δp, in the relation $h_L = \Delta p/\gamma$
2. Write expressions for the head loss in each branch in terms of the velocity in that branch and the friction factor.
3. Compute the relative roughness D/ϵ for each branch, assume a reasonable estimate for the friction factor, and complete the calculation for the head loss in terms of the velocity in each branch.
4. Letting the magnitude of the head loss in each branch equal the total head loss as found in Step 1, solve for the velocity in each branch by using the expression found in Step 3.
5. If there is doubt about the accuracy of the value of the friction factor used in Step 3, compute the Reynolds number for each branch and reevaluate the friction factor from the Moody diagram in Fig. 9.2 or compute the value of the friction factor from Eq. (9–9).
6. If the values for the friction factor have changed significantly, repeat Steps 3 and 4, using the new values for friction factor.
7. When satisfactory precision has been achieved, use the now-known velocity in each branch to compute the volume flow rate for that branch. Then, compute the sum of the volume flow rates, which is equal to the total flow in the system.

PROGRAMMED EXAMPLE PROBLEM

☐ **EXAMPLE PROBLEM 12.2**   The arrangement shown in Fig. 12.3 is used to supply lubricating oil to the bearings of a large machine. The bearings act as restrictions to the flow. The resistance coefficients are 11.0 and 4.0 for the two bearings. The lines in each branch are ½-in steel tubing with a wall thickness of 0.049 in. Each of the four bends in the tubing has a mean radius of 100 mm. Include the effect of these bends, but exclude the friction losses since the lines are short. Determine (a) the flow rate of oil in each bearing and (b) the total flow rate in L/min. The oil has a specific gravity of 0.881 and a kinematic viscosity of 2.50×10^{-6} m²/s. The system lies in one plane so all elevations are equal.

FIGURE 12.3 Parallel system for Example Problem 12.2.

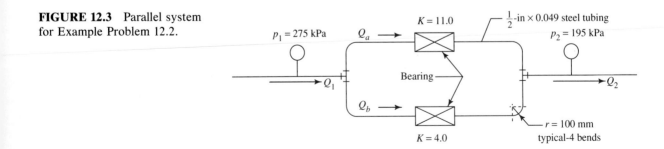

Write the equation that relates the head loss h_L across the parallel system to the head losses in each line h_a and h_b.

You should have

$$h_L = h_a = h_b \qquad\qquad \textbf{(12–12)}$$

They are all equal. Determine the magnitude of these head losses by using Step 1.

We can find h_L from the energy equation:

$$\frac{p_1}{\gamma} + z_1 + \frac{v_1^2}{2g} - h_L = \frac{p_2}{\gamma} + z_2 + \frac{v_2^2}{2g}$$

But $z_1 = z_2$ and $v_1 = v_2$. Then,

$$\frac{p_1}{\gamma} - h_L = \frac{p_2}{\gamma}$$

$$h_L = (p_1 - p_2)/\gamma \qquad\qquad \textbf{(12–13)}$$

Using the given data, we get

$$h_L = \frac{(275 - 195)\ \text{kN}}{\text{m}^2} \times \frac{\text{m}^3}{(0.881)(9.81)\ \text{kN}}$$

$$= 9.26\ \text{m}$$

Now write the expressions for h_a and h_b, Step 2.

Considering the losses in the bends and in the bearings, you should have

$$h_a = 2f_T(L_e/D)v_a^2/2g + 11.0v_a^2/2g \qquad\qquad \textbf{(12–14)}$$
$$h_b = 2f_T(L_e/D)v_b^2/2g + 4.0v_b^2/2g \qquad\qquad \textbf{(12–15)}$$

where f_T is the friction factor for the zone of complete turbulence in the tube. The equivalent length ratio for the bends can be found from Fig. 10.23 in Chapter 10. For each bend the relative radius is

$$r/D = 100\ \text{mm}/10.21\ \text{mm} = 9.79$$

Then $L_e/D = 29.5$. The friction factor in the zone of complete turbulence can be determined by using the relative roughness and the moody diagram, Fig. 9.2.

$$D/\epsilon = 0.010\ 21\ \text{m}/4.6 \times 10^{-5}\ \text{m} = 222$$

From the Moody diagram we can find the friction factor to be $f_T = 0.029$. Then we can complete Step 3 by computing h_a and h_b.

$$h_a = (2)(0.029)(29.5)(v_a^2/2g) + 11.0(v_a^2/2g)$$
$$= (1.71 + 11.0)v_a^2/2g = 12.71v_a^2/2g \qquad\qquad \textbf{(12–16)}$$

$$h_b = (2)(0.029)(29.5)(v_b^2/2g) + 4.0(v_b^2/2g)$$
$$= (1.71 + 4.0)v_b^2/2g = 5.71v_b^2/2g \qquad\qquad \textbf{(12–17)}$$

To complete Step 4, compute the velocities v_a and v_b.

We found earlier that $h_L = 9.26$ m. Since $h_L = h_a = h_b$, Eqs. (12–16) and (12–17) can be solved directly for v_a and v_b:

$$h_a = 12.71 v_a^2/2g$$

$$v_a = \sqrt{\frac{2gh_a}{12.71}} = \sqrt{\frac{(2)(9.81)(9.26)}{12.71}} \text{ m/s} = 3.78 \text{ m/s}$$

$$h_b = 5.71 v_b^2/2g$$

$$v_b = \sqrt{\frac{2gh_b}{5.71}} = \sqrt{\frac{(2)(9.81)(9.26)}{5.71}} \text{ m/s} = 5.64 \text{ m/s}$$

Now find the volume flow rates, as called for in Step 7.

You should have $Q_a = 18.6$ L/min, $Q_b = 27.7$ L/min, and the total volume flow rate = 46.3 L/min. The area of each tube is 8.189×10^{-5} m². Then we have

$$Q_a = A_a v_a = 8.189 \times 10^{-5} \text{ m}^2 \times 3.78 \text{ m/s} \times \frac{60\,000 \text{ L/min}}{\text{m}^3/\text{s}}$$

$$= 18.6 \text{ L/min}$$

Similarly,

$$Q_b = A_b v_b = 27.7 \text{ L/min}$$

Then the total flow rate is

$$Q_1 = Q_a + Q_b = (18.6 + 27.7)\text{L/min} = 46.3 \text{ L/min}$$

This example problem is concluded.

■

**12.4
SYSTEMS WITH
THREE OR MORE
BRANCHES—
NETWORKS**

When three or more branches occur in a pipe flow system, it is called a *network*. Networks are indeterminate because there are more unknown factors than there are independent equations relating the factors. For example, in Fig. 12.4 there are three unknown velocities, one in each pipe. The equations available to describe the system are

$$Q_1 = Q_2 = Q_a + Q_b + Q_c \qquad \text{(12–18)}$$

$$h_{L_{1-2}} = h_a = h_b = h_c \qquad \text{(12–19)}$$

A third independent equation is required to solve explicitly for the three velocities, and none is available.

A rational approach to complete the analysis of a system such as that shown in Fig. 12.4 employing an iteration procedure has been developed by Hardy Cross.* This procedure converges on the correct flow rates quite

---

* Hardy Cross, *Analysis of flow in networks of conduits or conductors,* University of Illinois Engineering Experiment Station Bulletin No. 286 (Urbana: University of Illinois, November 1936).

FIGURE 12.4 Network with three branches.

Note: Inlet and outlet pipes: 2-in Sch. 40
    Branch pipes a, b, and c: 1-in Sch. 40
    Elbows are standard

rapidly. Many calculations are still required, but they can be set up in an orderly fashion for use on a calculator or digital computer.

The Cross technique requires that the head loss terms for each pipe in the system be expressed in the form

$$h = kQ^n \tag{12-20}$$

where k is an equivalent resistance to flow for the entire pipe and Q is the flow rate in the pipe. We will illustrate the creation of such an expression in the example problem to follow this general discussion of the Hardy Cross technique. For now, you should recall that both friction losses and minor losses are proportional to the velocity head, $v^2/2g$. Then, using the continuity equation we can express the velocity in terms of the volume flow rate. That is,

$$v = Q/A$$

and

$$v^2 = Q^2/A^2$$

This will allow the development of an equation of the form shown in Eq. (12–20).

The Cross iteration technique requires that initial estimates for the volume flow rate in each branch of the system be made. Two factors that help in making these estimates are

1. At each junction in the network, the sum of the flow into the junction must equal the flow out.
2. The fluid tends to follow the path of least resistance through the network. Therefore, a pipe having a lower value of k will carry a higher flow rate than those having higher values.

The network should be divided into a set of closed loop circuits prior to beginning the iteration process. Figure 12.5 shows a schematic representation of a 3-pipe system such as that shown in Fig. 12.4. The dashed arrows drawn in a clockwise direction assist in defining the signs for the flow rates Q and the head losses h in the various pipes of each loop according to the following convention:

▪ If the flow in a given pipe of a circuit is clockwise, Q and h are positive.
▪ If the flow is counterclockwise, Q and h are negative.

Then for circuit 1 in Fig. 12.5, h_a and Q_a are positive, while h_b and Q_b are negative. The signs are critical to the correct calculation of adjustments to the volume flow rates, indicated by ΔQ, that are produced at the end of each iteration cycle. Notice that pipe b is common to both circuits. Therefore, the) adjustments ΔQ for each circuit must be applied to the flow rate in this pipe.

FIGURE 12.5   Closed loop circuits used for the Hardy Cross technique for analysis of pipe networks.

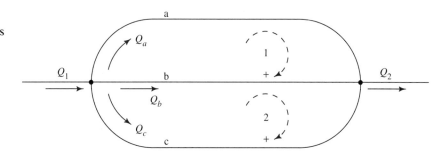

The Cross technique for analyzing the flow in pipe networks is presented in step-by-step form below. A programmed example problem follows to illustrate the application of the procedure.

CROSS TECHNIQUE FOR ANALYSIS OF PIPE NETWORKS

1. Express the energy loss in each pipe in the form $h = kQ^2$.
2. Assume a value for the flow rate in each pipe such that the flow into each junction equals the flow out of the junction.
3. Divide the network into a series of closed loop circuits.
4. For each pipe, calculate the head loss $h = kQ^2$, using the assumed value of Q.

5. Proceeding around each circuit, algebraically sum all values for h using the following sign convention:
   If the flow is clockwise, h and Q are positive.
   If the flow is counterclockwise, h and Q are negative.
   The resulting summation is referred to as Σh.

6. For each pipe, calculate $2kQ$.

7. Sum all values of $2kQ$ for each circuit, assuming all are positive. This summation is referred to as $\Sigma (2kQ)$.

8. For each circuit, calculate the value of ΔQ from

$$\Delta Q = \frac{\Sigma h}{\Sigma(2kQ)} \tag{12-21}$$

9. For each pipe, calculate a new estimate for Q from

$$Q' = Q - \Delta Q$$

10. Repeat Steps 4–8 until ΔQ from Step 8 becomes negligibly small. The Q' value is used for the next cycle of iteration.

PROGRAMMED EXAMPLE PROBLEM

☐ **EXAMPLE PROBLEM 12.3**   For the system shown in Fig. 12.4, determine the volume flow rate of water at 15°C through each branch if 600 L/min (0.01 m³/s) is flowing into and out of the system through the 2-in pipes.

The head loss in each pipe should now be expressed in the form $h = kQ^2$ as Step 1 of the procedure. Consider branch a first and write an expression for the head loss h_a.

The total head loss for the branch is due to the two elbows (each with $L_e/D = 30$), the restriction (with $K = 4.0$ based on the velocity head in the pipe), and friction in the pipe. Then,

$$h_a = 2(f_{aT})(30)(v_a^2/2g) + 4.0(v_a^2/2g) + f_a(L_a/D_a)(v_a^2/2g)$$
$$\text{(elbows)} \qquad \text{(restriction)} \qquad \text{(friction)}$$

The friction factor f_a for flow in the pipe is dependent on the Reynolds number and therefore on the volume flow rate. Because that is the objective of the network analysis, we cannot determine that value explicitly at this time. Furthermore, the flow rate will, in general, be different in each segment of the flow system resulting in different values for the friction factor. We will take that into account in the present analysis by computing the value of the friction factor after assuming the magnitude of the volume flow rate in each pipe, a step that is inherent in the Hardy Cross technique. We will use the Swamee-Jain method to compute the friction factor from Eq. (9–9). Then we will recompute the values of the friction factors for each trial as the value of the volume flow rate is refined.

First, let's simplify the equation for h_a by completing as many calculations as we can. What values can be determined?

The total length of pipe in branch a is 12 m and for the 1-in Schedule 40 pipe $D = 0.0266$ m and $A = 5.574 \times 10^{-4}$ m^2. From Table 10.5 we can find that the value of $f_{aT} = 0.023$ for a 1-in Schedule 40 steel pipe with flow in the completely turbulent zone. The water at 15°C has a kinematic viscosity $\nu = 1.15 \times 10^{-6}$ m^2/s. Also, we can introduce the volume flow rate Q into the equation by noting that, as shown before,

$$v_a^2 = Q_a^2/A_a^2$$

Substitute these values into the equation for h_a now and simplify it as much as possible.

You should have something like this.

$$h_a = [60(f_{aT}) + 4.0 + (f_a)(12/0.0266)](v_a^2/2g)$$
$$h_a = [60(f_{aT}) + 4.0 + 451(f_a)](Q_a^2/2gA^2)$$
$$h_a = [60(0.023) + 4.0 + 451(f_a)]\left[\frac{Q_a^2}{2(9.81)(5.574 \times 10^{-4})^2}\right]$$
$$h_a = [5.38 + 451(f_a)](1.64 \times 10^5)Q_a^2 \qquad (12\text{--}22)$$

It is also convenient to express the Reynolds number in terms of the volume flow rate, Q, and to compute the value for the relative roughness, D/ϵ. Do that now.

Because all three branches have the same size and type of pipe, these calculations apply to each branch. If different pipes are used throughout the network, these calculations must be redone for each pipe. For the 1-in steel pipe,

$$D/\epsilon = (0.0266 \text{ m})/(4.6 \times 10^{-5} \text{ m}) = 578$$

We should modify the Reynolds number formula as

$$N_{Ra} = \frac{v_a D_a}{\nu} = \frac{Q_a D_a}{A_a \nu} = \frac{Q_a(0.0266)}{(5.574 \times 10^{-4})(1.15 \times 10^{-6})}$$
$$N_{Ra} = (4.15 \times 10^7)Q_a \qquad (12\text{--}23)$$

Now create expressions for the head losses in the other two pipes, h_b and h_c, using similar procedures.

Compare your results with these. Note that the pipe size in branches b and c is the same as that in branch a. For branch b:

$$h_b = 8.0(v_b^2/2g) + f_b(L_b/D_b)(v_b^2/2g)$$
$$\qquad \text{(restriction)} \qquad \text{(friction)}$$
$$h_b = [8.0 + f_b(6/0.0266)](Q_b^2/2gA^2)$$
$$h_b = [8.0 + 225.6(f_b)](1.64 \times 10^5)Q_b^2 \qquad (12\text{--}24)$$

For branch c:

$$h_c = 2(f_{cT})(30)(v_c^2/2g) + 12.0(v_c^2/2g) + f_c(L_c/D_c)(v_c^2/2g)$$
$$\text{(elbows)} \qquad \text{(restriction)} \qquad \text{(friction)}$$

$$h_c = [60(f_{cT}) + 12.0 + f_c(12/0.0266)](v_c^2/2g)$$

$$h_c = [60(0.023) + 12.0 + 451f_c](Q_c^2/2gA^2)$$

$$h_c = [12.38 + 451(f_c)](1.64 \times 10^5)Q_c^2 \qquad \qquad \textbf{(12–25)}$$

Equations 12–22 to 12–25 will be used in the calculations of head losses as the Cross iteration process continues. When the values for the friction factors are known or assumed, the head loss equations can be reduced to the form of Eq. (12–20). Often it is satisfactory to assume reasonable values for the various friction factors, as minor changes have little effect on the flow distribution and the total head loss. However, we will demonstrate the more complete solution procedure in which new friction factors are calculated for each pipe for each trial.

Step 2 of the procedure calls for estimating the volume flow rate in each branch. Which pipe should have the greatest flow rate, and which should have the least?

Although the final values for the friction factors could affect the magnitudes of the resistances, it appears that pipe b has the least resistance and, therefore, it should carry the greatest flow. Pipe c has the most resistance and it should carry the least flow. Many different first estimates are possible for the flow rates, but we know that

$$Q_a + Q_b + Q_c = Q_1 = 0.01 \text{ m}^3/\text{s}$$

Let's use the initial assumptions,

$$Q_a = 0.0033 \text{ m}^3/\text{s} \qquad Q_b = 0.0036 \text{ m}^3/\text{s} \qquad Q_c = 0.0031 \text{ m}^3/\text{s}$$

Step 3 of the procedure is already shown in Fig. 12.5. To complete Step 4 we need values for the friction factor in each pipe. With the assumed values for the volume flow rates we can compute the Reynolds numbers and then the friction factors. Do that now.

You should have

$$N_{Ra} = (4.15 \times 10^7)Q_a = (4.15 \times 10^7)(0.0033 \text{ m}^3/\text{s}) = 1.37 \times 10^5$$

$$N_{Rb} = (4.15 \times 10^7)Q_b = (4.15 \times 10^7)(0.0036 \text{ m}^3/\text{s}) = 1.49 \times 10^5$$

$$N_{Rc} = (4.15 \times 10^7)Q_c = (4.15 \times 10^7)(0.0031 \text{ m}^3/\text{s}) = 1.29 \times 10^5$$

We now use Eq. (9–9) to compute the friction factor for each pipe.

$$f_a = \frac{0.25}{\left[\log_{10}\left(\dfrac{1}{3.7(D/\epsilon)} + \dfrac{5.74}{N_{Ra}^{0.9}}\right)\right]^2}$$

$$f_a = \frac{0.25}{\left[\log_{10}\left(\dfrac{1}{3.7(578)} + \dfrac{5.74}{(1.37 \times 10^5)^{0.9}}\right)\right]^2} = 0.0241$$

In a similar manner we compute $f_b = 0.0240$ and $f_c = 0.0242$. These values are quite close in magnitude and such precision may not be justified. However, with greater disparity among the pipes in the network, more sizable differences would occur and the accuracy of the iteration technique would depend on the accuracy of evaluating the friction factors.

Insert the friction factors and the assumed values for Q into Eqs. (12–22), (12–24), and (12–25) now to compute k_a, k_b, and k_c.

$$h_a = [5.38 + 451(f_a)](1.64 \times 10^5)Q_a^2 = k_aQ_a^2$$
$$h_a = [5.38 + 451(0.0241)](1.64 \times 10^5)Q_a^2 = 2.67 \times 10^6Q_a^2$$

Then $k_a = 2.67 \times 10^6$. Completing the calculation gives,

$$h_a = 2.67 \times 10^6(0.0033)^2 = 29.05$$

Similarly, for branch b:

$$h_b = [8.0 + 225.6(f_b)](1.64 \times 10^5)Q_b^2 = k_bQ_b^2$$
$$h_b = [8.0 + 225.6(0.0240)](1.64 \times 10^5)Q_b^2 = 2.20 \times 10^6Q_b^2$$
$$h_b = 2.20 \times 10^6(0.0036)^2 = 28.53$$

For branch c:

$$h_c = [12.38 + 451(f_c)](1.64 \times 10^5)Q_c^2 = k_cQ_c^2$$
$$h_c = [12.38 + 451(0.0242)](1.64 \times 10^5)Q_c^2 = 3.99 \times 10^6Q_c^2$$
$$h_c = 3.99 \times 10^6(0.0031)^2 = 38.31$$

This completes Step 4. Now do Step 5.

For circuit 1,

$$\Sigma h_1 = h_a - h_b = 29.05 - 28.53 = 0.52$$

For circuit 2,

$$\Sigma h_2 = h_b - h_c = 28.53 - 38.31 = -9.78$$

Now do Step 6.

Here are the correct values for the three pipes:

$$2k_aQ_a = (2)(2.67 \times 10^6)(0.0033) = 17\ 609$$
$$2k_bQ_b = (2)(2.20 \times 10^6)(0.0036) = 15\ 850$$
$$2k_cQ_c = (2)(3.99 \times 10^6)(0.0031) = 24\ 717$$

Roundoff differences may occur. Now do Step 7.

For circuit 1,

$$\Sigma\ (2kQ)_1 = 17\ 609 + 15\ 850 = 33\ 459$$

For circuit 2,

$$\Sigma\,(2kQ)_2 = 15\ 850 + 24\ 717 = 40\ 567$$

Now you can calculate the adjustment for the flow rates, ΔQ, for each circuit, using Step 8.

For circuit 1,

$$\Delta Q_1 = \frac{\Sigma\,h_1}{\Sigma\,(2kQ)_1} = \frac{0.52}{33\ 459} = 1.56 \times 10^{-5}$$

For circuit 2,

$$\Delta Q_2 = \frac{\Sigma\,h_2}{\Sigma\,(2kQ)_2} = \frac{-9.78}{40\ 567} = -2.41 \times 10^{-4}$$

The values for ΔQ are estimates of the error in the originally assumed values for Q. We recommend that the process be repeated until the magnitude of ΔQ is less than 1 percent of the assumed value of Q. Special circumstances may warrant using a different criterion for judging ΔQ.

Step 9 can now be completed. Calculate the new value for Q_a before looking at the next panel.

The calculation is as follows:

$$Q'_a = Q_a - \Delta Q_1 = 0.0033 - 1.56 \times 10^{-5}$$
$$= 0.003\ 28 \text{ m}^3/\text{s}$$

Calculate the new value for Q_c before Q_b. Pay careful attention to algebraic signs.

You should have

$$Q'_c = Q_c - \Delta Q_2 = -0.0031 - (-2.41 \times 10^{-4})$$
$$= -0.002\ 86 \text{ m}^3/\text{s}$$

Notice that Q_c is negative since it flows in a counterclockwise direction in circuit 2. We can interpret the calculation for Q'_c as indicating that the magnitude of Q_c must be decreased in absolute value.

Now calculate the new value for Q_b. Remember, pipe b is in each circuit.

Both ΔQ_1 and ΔQ_2 must be applied to Q_b. For circuit 1,

$$Q'_b = Q_b - \Delta Q_1 = -0.0036 - 1.56 \times 10^{-5}$$

This would result in an increase in the absolute value of Q_b. For circuit 2,

$$Q'_b = Q_b - \Delta Q_2 = +0.0036 - (-2.41 \times 10^{-4})$$

TABLE 12.1

Trial	Circuit	Pipe	Q	N_R	f	k	$h = kQ^2$	$2kQ$	ΔQ	% Change
1	1	a	3.300×10^{-3}	1.37×10^5	0.0241	2.67×10^6	29.054	17609		0.48
		b	-3.600×10^{-3}			2.20×10^6	-28.530	15850		-0.44
						Sum of h and $2kQ =$	0.524	33459	1.568×10^{-5}	
	2	b	3.600×10^{-3}	1.49×10^5	0.0240	2.20×10^6	28.530	15850		-6.70
		c	-3.100×10^{-3}	1.29×10^5	0.0242	3.99×10^6	-38.312	24717		7.78
						Sum of h and $2kQ =$	-9.782	40567	-2.411×10^{-4}	
2	1	a	3.284×10^{-3}	1.36×10^5	0.0241	2.67×10^6	28.784	17528		-3.46
		b	-3.857×10^{-3}			2.20×10^6	-32.700	16957		2.94
						Sum of h and $2kQ =$	-3.916	34485	-1.135×10^{-4}	
	2	b	3.857×10^{-3}	1.60×10^5	0.0239	2.20×10^6	32.700	16957		0.03
		c	-2.859×10^{-3}	1.19×10^5	0.0243	4.00×10^6	-32.654	22844		-0.04
						Sum of h and $2kQ =$	0.046	39801	1.151×10^{-6}	
3	1	a	3.398×10^{-3}	1.41×10^5	0.0241	2.67×10^6	30.770	18112		-0.03
		b	-3.742×10^{-3}			2.20×10^6	-30.802	16462		0.02
						Sum of h and $2kQ =$	-0.032	34574	-9.176×10^{-7}	
	2	b	3.742×10^{-3}	1.55×10^5	0.0240	2.20×10^6	30.802	16462		-1.28
		c	-2.860×10^{-3}	1.19×10^5	0.0243	4.00×10^6	-32.680	22853		1.67
						Sum of h and $2kQ =$	-1.878	39315	-4.776×10^{-5}	
4	1	a	3.399×10^{-3}	1.41×10^5	0.0241	2.67×10^6	30.787	18116		-0.66
		b	-3.789×10^{-3}			2.20×10^6	-31.570	16664		0.59
						Sum of h and $2kQ =$	-0.783	34780	-2.252×10^{-5}	
	2	b	3.789×10^{-3}	1.57×10^5	0.0240	2.20×10^6	31.570	16664		-0.03
		c	-2.812×10^{-3}	1.17×10^5	0.0244	4.00×10^6	-31.612	22482		0.04
						Sum of h and $2kQ =$	-0.042	39146	-1.073×10^{-6}	

This also results in increasing Q_b. Then Q_b is actually increased in absolute value by the sum of ΔQ_1 and ΔQ_2. That is,

$$Q'_b = 0.0036 + 1.56 \times 10^{-5} + 2.41 \times 10^{-4}$$
$$= 0.003\ 86\ \text{m}^3/\text{s}$$

Remember that the sum of the absolute values of the flow rates in the three pipes must equal $0.01\ \text{m}^3/\text{s}$, the total Q.

We can continue the iteration by using Q'_a, Q'_b, and Q'_c as the new estimates for the flow rates and repeating Steps 4–8. The results for four iteration cycles are summarized in Table 12.1, which is presented on p. 344. You should carry out the calculations yourself before looking at the table.

Notice that in Trial 4, the values of ΔQ are below 1 percent of the respective values of Q. This is an adequate degree of precision. The results show that $Q_a = 3.399 \times 10^{-3}\ \text{m}^3/\text{s}$, $Q_b = 3.789 \times 10^{-3}\ \text{m}^3/\text{s}$, and $Q_c = 2.812 \times 10^{-3}\ \text{m}^3/\text{s}$ in the directions shown in Fig. 12.5. The results, expressed more conveniently in L/min, are $Q_a = 204$ L/min, $Q_b = 227$ L/min, and $Q_c = 169$ L/min. The total $Q = 600$ L/min. Once again, observe that the pipes having the lower resistances carry the greater flow rates.

■

The results of the iteration process for the Hardy Cross technique for the data of Example Problem 12.3 as shown in Table 12.1 were found using a spreadsheet on a computer. This facilitated the sequential, repetitive calculations typically required in such problems. A computer program written in BASIC, FORTRAN, or other technical language can also be used to advantage, especially if a large number of pipes and circuits exist in the network to be analyzed.

Many network analysis computer programs are commercially available.

REFERENCE

1. Cross, Hardy. 1936 (November). *Analysis of Flow in Networks of Conduits or Conductors.* University of Illinois Engineering Experiment Station Bulletin No. 286. Urbana: University of Illinois.

PRACTICE PROBLEMS

12.1M *Figure 12.6 shows a branched system in which the pressure at A is 700 kPa and the pressure at B is 550 kPa. Each branch is 60 m long. Neglect losses at the junctions, but consider all elbows. If the system carries oil with a specific weight of 8.80 kN/m³, calculate the total volume flow rate. The oil has a kinematic viscosity of $4.8 \times 10^{-6}\ m^2/s$.*

12.2E Using the system shown in Fig. 12.2 and the data from Example Problem 12.1, determine (a) the

FIGURE 12.6   Problem 12.1.

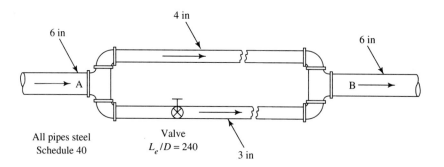

All pipes steel
Schedule 40

Valve
$L_e/D = 240$

volume flow rate of water in each branch and (b) the pressure drop between points 1 and 2 if the first gate valve is one-half closed and the other valves are wide open.

12.3M *In the branched pipe system shown in Fig. 12.7, 850 L/min of water at 10°C is flowing in a 4-in Schedule 40 pipe at A. The flow splits into two 2-in Schedule 40 pipes as shown and then rejoins at B. Calculate (a) the flow rate in each of the branches and (b) the pressure difference $p_A - p_B$. Include the effect of the minor losses in the lower branch of the system. The total length of pipe in the lower branch is 60 m. The elbows are standard.*

12.4E In the branched pipe system shown in Fig. 12.8, 1350 gal/min of benzene (sg = 0.87) at 140°F is flowing in the 8-in pipe. Calculate the volume flow rate in the 6-in and the 2-in pipes. All pipes are standard Schedule 40 steel pipes.

12.5M *A 150-mm pipe branches into a 100-mm and a 50-mm pipe as shown in Fig. 12.9. Both pipes are copper and 30 m long. Determine what the resistance coefficient K of the valve must be in order to obtain equal volume flow rates in each branch.*

12.6E For the system shown in Fig. 12.10, the pressure at A is maintained constant at 20 psig. The total volume flow rate exiting from the pipe at B depends on which valves are open or closed. Use $K = 0.9$ for each elbow, but neglect the energy losses in the tees. Also, since the length of each branch is short, neglect pipe friction losses. The pipe in branch 1 has a 2-in inside diameter, and branch 2 has a 4-in inside diameter. Calculate the volume flow rate of water for each of the following conditions:

a. Both valves open
b. Valve in branch 2 only open
c. Valve in branch 1 only open

12.7E Solve Problem 12.4, using the Hardy Cross technique.

12.8M *Solve Problem 12.3, using the Hardy Cross technique.*

12.9E Find the flow rate in each pipe of Fig. 12.11.

12.10M *Figure 12.12 represents a spray rinse system in which water at 15°C is flowing. All pipes are 3-in Type K copper tubing. Determine the flow rate in each pipe.*

FIGURE 12.7   Problems 12.3 and 12.8.

FIGURE 12.8 Problems 12.4 and 12.7.

FIGURE 12.9 Problem 12.5.

FIGURE 12.10 Problem 12.6.

FIGURE 12.11 Problem 12.9.

FIGURE 12.12 Problem 12.10.

6000 L/min

10 m   15 m   10 m

8 m   6 m   6 m   8 m

15 m   15 m   15 m

1500 L/min   1500 L/min   1500 L/min   1500 L/min

FIGURE 12.13 Problem 12.11.

15.5 ft³/s

1.5 ft³/s

1 ft³/s

4 ft³/s

3 ft³/s   3 ft³/s   3 ft³/s

Pipe Data
All pipes Schedule 40

Pipe no.	Length (ft)	Size (in)
1	1500	16
2	1500	16
3	2000	18
4	2000	12
5	2000	16
6	1500	16
7	1500	12
8	4000	14
9	4000	12
10	4000	8
11	1500	12
12	1500	8

FIGURE 12.14 Problem 12.12.

880 L/min

115 L/min

200 L/min   375 L/min

75 L/min

115 L/min   Transfer line

Pipe data
All pipes 7.5 m long
All pipes steel tubing
Wall thickness = 0.065 in

Pipe no.	Outside diameter (in)
1	2
2	$1\frac{1}{2}$
3	2
4	1
5	1
6	1
7	1

12.11E Figure 12.13 represents the water distribution network in a small industrial park. The supply of 15.5 ft³/s enters the system at A. Manufacturing plants draw off the indicated flows at points C, E, F, G, H, and I. Determine the flow in each pipe in the system.

12.12M *Figure 12.14 represents the network for delivering coolant to five different machine tools in an automated machining system. The grid is a rectangle 7.5 m by 15 m. All pipes are commercial steel tubing with a 0.065-in wall thickness. Pipes 1 and 3 are 2-in diameter; pipe 2 is 1½-in diameter; and all others are 1-in. Determine the flow in each pipe.*

COMPUTER PROGRAMMING ASSIGNMENTS

1. Write a program to analyze parallel pipe line systems with two branches of the type demonstrated in Example Problem 12.1. Part of the preliminary analysis, such as writing the expressions for head losses in the branches in terms of the velocities and the friction factors, may be done prior to entering data into the program. The program can be interactive, thereby permitting the operator to input successive values of the friction factor after the pipe velocities and the Reynolds numbers have been computed by the program.

2. Enhance the program from Assignment 1 so that it uses the equations from Chapter 9 to calculate the friction factor.

3. Write a program to analyze parallel pipe line systems with two branches of the type demonstrated in Example Problem 12.2. Use an approach similar to that described for Program Assignment 1.

4. Enhance the program from Assignment 3 so that it uses the equations from Chapter 9 to calculate the friction factor.

5. Write a program that uses the Hardy Cross technique, as described in Section 12.4 and illustrated in Example Problem 12.3, to perform the analysis of pipe flow networks. The following optional approaches may be taken:

   a. Consider single circuit networks with two branches as an alternative to the program from Assignment 1 or 2.

   b. Consider networks of two or more circuits similar to those described in Problems 12.9 through 12.12.

13 Open Channel Flow

13.1
OBJECTIVES

In contrast to the closed conduits that have been discussed in preceding chapters, an *open channel* is a flow system in which the top surface of the fluid is exposed to the atmosphere.

Many examples of open channels occur in nature and in systems designed to supply water to communities or to carry storm drainage and sewage safely away. See Fig. 13.1. Rivers and streams are obvious examples of natural channels. Rain gutters on buildings and at the sides of streets carry rainwater. Storm sewers, usually beneath the streets, collect the runoff from the streets and conduct it to a stream or to a larger man-made ditch or canal. In industry, open channels are often used to convey cooling water away from heat exchangers or coolants away from machining systems.

This chapter will present some of the methods of analyzing open channel flow. Complete coverage of the subject is an extensive undertaking and is treated in entire texts such as References 3, 4, 5, 6, and 7 (listed at the end of this chapter).

After completing this chapter, you should be able to:

1. Compute the hydraulic radius for open channels.
2. Describe *uniform flow* and *varied flow*.
3. Use Manning's equation to analyze uniform flow.
4. Define the slope of an open channel and compute its value.
5. Compute the normal discharge for an open channel.
6. Compute the normal depth of flow for an open channel.
7. Design an open channel to transmit a given discharge with uniform flow.
8. Define the *Froude number*.
9. Describe *critical flow*, *subcritical flow*, and *supercritical flow*.
10. Define the specific energy of the flow in open channels.
11. Define the terms *critical depth*, *alternate depth*, and *sequent depth*.
12. Describe the term *hydraulic jump*.

13.2
HYDRAULIC RADIUS

The characteristic dimension of open channels is the *hydraulic radius*, defined as the ratio of the net cross-sectional area of a flow stream to the wetted perimeter of the section. That is,

⇨ **HYDRAULIC RADIUS**

$$R = \frac{A}{WP} = \frac{\text{area}}{\text{wetted perimeter}} \tag{13-1}$$

351

FIGURE 13.1 Examples of cross sections of open channels.

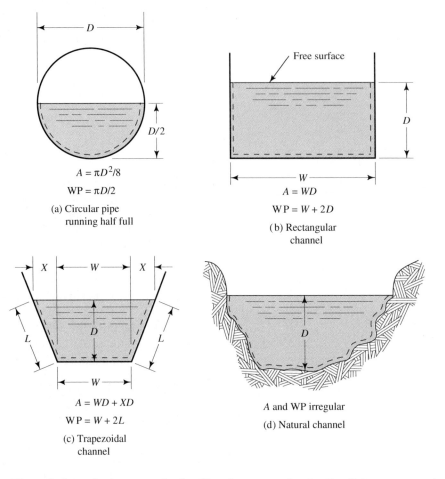

(a) Circular pipe running half full

$A = \pi D^2/8$

$WP = \pi D/2$

(b) Rectangular channel

$A = WD$

$WP = W + 2D$

(c) Trapezoidal channel

$A = WD + XD$

$WP = W + 2L$

(d) Natural channel

A and WP irregular

The unit for R is the meter in the SI unit system. In the English system, R is expressed in feet.

In the calculation of the hydraulic radius, the net cross-sectional area should be evident from the geometry of the section. The *wetted perimeter* is defined as the sum of the length of the boundaries of the section actually in contact with (that is, wetted by) the fluid. Expressions for the area A and the wetted perimeter WP are given in Fig. 13.1 for those sections illustrated. In each case, the fluid flows in the shaded portion of the section. A dashed line is shown adjacent to the boundaries that make up the wetted perimeter. Notice that the length of the free surface of an open channel is *not* included in WP.

☐ **EXAMPLE PROBLEM 13.1**   Determine the hydraulic radius of the trapezoidal section shown in Fig. 13.1(c) if $W = 4$ ft, $X = 1$ ft, and $D = 2$ ft.

Solution   The net flow area is

$$A = WD + 2(XD/2) = WD + XD$$
$$= (4)(2) + (1)(2) = 10 \text{ ft}^2$$

To find the wetted perimeter, we must determine the value of L:

$$WP = W + 2L$$
$$L = \sqrt{X^2 + D^2} = \sqrt{(1)^2 + (2)^2} = 2.24 \text{ ft}$$
$$WP = 4 + 2(2.24) = 8.48 \text{ ft}$$

Then we have

$$R = A/WP = 10 \text{ ft}^2/8.48 \text{ ft} = 1.18 \text{ ft}$$

13.3
CLASSIFICATION OF
OPEN CHANNEL FLOW

Open channel flow can be classified into several types.

Uniform steady flow occurs when the volume flow rate (typically called *discharge* in open channel flow analysis) remains constant in the section of interest and the depth of the fluid in the channel does not vary. To achieve uniform steady flow, the cross section of the channel may not change along its length. Such a channel is *prismatic*. Figure 13.2 shows uniform flow in a side view.

FIGURE 13.2 Uniform steady open channel flow—side view.

Varied steady flow occurs when the discharge remains constant but the depth of the fluid varies along the section of interest. This will occur if the channel is not prismatic.

Unsteady varied flow occurs when the discharge varies with time, resulting in changes in the depth of the fluid along the section of interest whether the channel is prismatic or not.

Varied flow can be further classified into *rapidly varying flow* or *gradually varying flow*. As the name implies, the difference refers to the rate of change in depth with position along the channel. Figure 13.3 illustrates a series of conditions in which varied flow occurs. The following discussion describes the flow in the various parts of this figure.

- *Section 1* The flow starts from a reservoir in which the fluid is virtually at rest. The *sluice gate* is a device that allows the fluid to flow from the reservoir at a point beneath the surface. Rapidly varying flow occurs near the gate as the fluid accelerates, and the velocity of flow is likely to be quite large in this area.
- *Section 2* If the channel downstream from the sluice gate is relatively short, and if its cross section does not vary much, then gradually varying

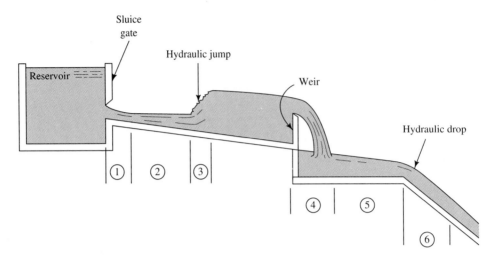

FIGURE 13.3　Conditions causing varied flow.

flow occurs. If the channel is prismatic and long enough, uniform flow could develop.

- *Section 3*　The formation of a *hydraulic jump* is a curious open channel flow phenomenon. The flow before the jump is quite rapid and relatively shallow. In the jump, the flow becomes very turbulent, and a large amount of energy is dissipated. Then, following the jump, the flow velocity is much lower, and the depth of the fluid is greater. More will be said later about the hydraulic jump.
- *Section 4*　A *weir* is an obstruction placed in the flow stream that causes an abrupt change in the cross section of the channel. Weirs can be used as control devices or to measure the volume flow rate. Typically the flow is rapidly varying as it travels over the weir, with a "waterfall" formed downstream.
- *Section 5*　As with Section 2, the flow downstream from a weir is usually gradually varying if the channel is prismatic.
- *Section 6*　The *hydraulic drop* occurs when the slope of the channel suddenly increases to a steep angle. The flow accelerates because of gravity, and rapidly varying flow occurs.

**13.4
REYNOLDS NUMBER
IN OPEN
CHANNEL FLOW**

Recall that the Reynolds number for closed circular cross sections running full is

$$N_R = \frac{vD}{\nu} \tag{13–2}$$

where v = average velocity of flow, D = pipe diameter, and ν = kinematic viscosity of the fluid. We have seen that laminar flow occurs when $N_R <$ 2000 and turbulent flow occurs when $N_R >$ 4000 for most practical pipe flow situations. The Reynolds number represents the effects of fluid viscosity relative to the inertia of the fluid.

In open channel flow, the characteristic dimension is the hydraulic radius, R. It was shown in Chapter 8 that, for a full circular cross section, $D = 4R$. For closed, noncircular cross sections, it was convenient to substitute $4R$ for D so that the Reynolds number would have the same order of magnitude as that for circular pipes and tubes. However, this is not usually done in open channel flow analysis. The Reynolds number for open channel flow is then

▷ **REYNOLDS NUMBER FOR OPEN CHANNELS**

$$N_R = \frac{vR}{\nu} \qquad (13\text{--}3)$$

Experimental evidence (Reference 3) shows that, in open channels, laminar flow occurs when $N_R < 500$. The range from 500 to 2000 is the transition region. Turbulent flow normally occurs when $N_R > 2000$.

13.5
KINDS OF OPEN CHANNEL FLOW

The Reynolds number and the terms *laminar* and *turbulent* are not sufficient to characterize all kinds of open channel flow. In addition to the viscosity versus inertial effects, the ratio of inertial forces to gravity forces is also important. The *Froude number*, N_F, is defined as

▷ **FROUDE NUMBER**

$$N_F = \frac{v}{\sqrt{g y_h}} \qquad (13\text{--}4)$$

where

▷ **HYDRAULIC DEPTH**

$$y_h = A/T \qquad (13\text{--}5)$$

and T is the width of the free surface of the fluid at the top of the channel.

When the Froude number is equal to 1.0, that is, when $v = \sqrt{g y_h}$, the flow is called *critical flow*. When $N_F < 1.0$, the flow is *subcritical*, and when $N_F > 1.0$, the flow is *supercritical*. See also Section 13.9.

Then, the following kinds of flow are possible:

1. Subcritical-laminar. $N_R < 500$, and $N_F < 1.0$.
2. Subcritical-turbulent. $N_R > 2000$, and $N_F < 1.0$.
3. Supercritical-turbulent. $N_R > 2000$, and $N_F > 1.0$.
4. Supercritical-laminar. $N_R < 500$, and $N_F > 1.0$.

In addition, flows can be in the transition region. However, such flows are unstable and very difficult to characterize.

In this discussion, the terms *laminar* and *turbulent* have the same significance as they did for pipe flow. In laminar flow, there is little or no mixing of the fluid, so that a stream of dye injected into the flow would remain virtually intact. In turbulent flow, however, chaotic intermixing occurs, and the dye stream rapidly dissipates throughout the fluid.

**13.6
UNIFORM STEADY
FLOW IN OPEN
CHANNELS**

Figure 13.2 is a schematic illustration of uniform steady flow in an open channel. The distinguishing feature of uniform flow is that the fluid surface is parallel to the slope of the channel bottom. We will use the symbol S to indicate the slope of the channel bottom and S_w for the slope of the water surface. Then for uniform flow, $S = S_w$. Theoretically, uniform flow can exist only if the channel is prismatic, that is, if its sides are parallel to an axis in the direction of flow. Examples of prismatic channels are rectangular, trapezoidal, triangular, and circular sections running partially full. Also, the channel slope S must be constant. If the cross section or slope of the channel is changing, then the flow stream would be either converging or diverging and varied flow would occur.

In uniform flow, the driving force for the flow is provided by the component of the weight of the fluid which acts along the channel as shown in Fig. 13.4. This force is $w \sin \theta$, where w is the weight of a certain element of fluid and θ is the angle of the slope of the channel bottom. If the flow is uniform, it cannot be accelerating. Therefore, there must be an equal opposing force acting along the channel surface. This is a friction force which depends on the roughness of the channel surfaces and on the cross-sectional size and shape.

FIGURE 13.4   Uniform open channel flow.

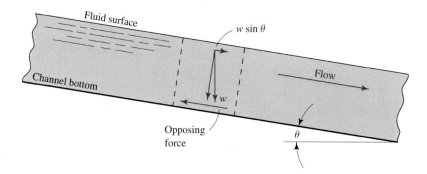

By equating the expressions for the driving force and the opposing force, an expression for the average velocity of uniform flow can be derived. A commonly used form of the resulting equation was developed by Robert Manning. In SI units, Manning's equation is written as

▷ **MANNING'S EQUATION—SI UNITS**

$$v = \frac{1.00}{n} R^{2/3} S^{1/2} \tag{13-6}$$

Units must be consistent in Eq. (13–6). The average velocity of flow v will be in m/s when the hydraulic radius R is in m. The channel slope S, which we will define later, is dimensionless. The final term n is a resistance factor sometimes called *Manning's n*. The value of n depends on the condition of the channel surface and is therefore somewhat analogous to the pipe wall roughness ϵ used previously. The form of Manning's equation in English units is given later in this section.

Typical design values of n are listed in Table 13.1 for materials commonly used for artificial channels and natural streams. A very extensive discussion of the determination of Manning's n and a more complete table of values are given by V. T. Chow (Reference 3). The values listed in Table 13.1 are average values which will give good estimates for use in design or for rough analysis of existing channels. Variations from the average should be expected.

TABLE 13.1 Values for Manning's n

Channel Description	n
Glass, copper, plastic, or other smooth surfaces	0.010
Smooth unpainted steel, planed wood	0.012
Painted steel or coated cast iron	0.013
Smooth asphalt, common clay drainage tile, trowel finished concrete, glazed brick	0.013
Uncoated cast iron, black wrought iron pipe, vitrified clay sewer tile	0.014
Brick in cement mortar, float finished concrete	0.015
Formed, unfinished concrete	0.017
Clean excavated earth	0.022
Corrugated metal storm drain	0.024
Earth with light brush	0.050
Earth with heavy brush	0.100

The slope S of a channel can be reported in several ways. It is ideally defined as the ratio of the vertical drop h to the horizontal distance over which the drop occurs. For small slopes, which are typical in open channel flow, it is more practical to use h/L, where L is the length of the channel as shown in Fig. 13.5. Normally, the magnitude of the slope for natural streams and drainage structures is very small, a typical value being 0.001. This number can also be expressed as a percentage, where 0.01 = 1 percent. Then, 0.001 = 0.1 percent. Since $\sin \theta = h/L$, the angle that the channel bottom makes with the horizontal could also be used. In summary, the slope of 0.001 could be reported as:

1. The channel falls 1 m per 1000 m of channel.
2. The slope is 0.1 percent.
3. $\sin \theta = 0.001$. Then $\theta = \sin^{-1}(0.001) = 0.057°$.

Because the angle is so small, it is rarely used as a measure of the slope.
We can calculate the volume flow rate in the channel from the continuity equation, which is the same as that used for pipe flow:

$$Q = Av \qquad \text{(13–7)}$$

FIGURE 13.5 Slope of a channel.

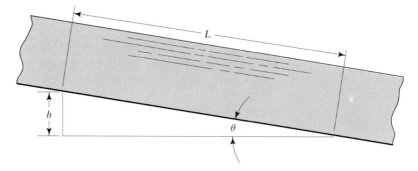

In open channel flow analysis, Q is typically called the *discharge*. Substituting Eq. (13–6) into (13–7) gives an equation which directly relates the discharge to the physical parameters of the channel:

NORMAL DISCHARGE—SI UNITS

$$Q = \left(\frac{1.00}{n}\right)AR^{2/3}S^{1/2} \qquad (13\text{–}8)$$

This is the only value of discharge for which uniform flow will occur for the given channel depth, and it is called the *normal discharge*. The units of Q are m³/s when the area A is expressed in square meters (m²) and the hydraulic radius in meters (m).

Another useful form of this equation is

$$AR^{2/3} = \frac{nQ}{S^{1/2}} \qquad (13\text{–}9)$$

The term on the left side of Eq. (13–9) is solely dependent on the geometry of the section. Therefore, for a given discharge, slope, and surface type, we can determine the geometrical features of a channel. Alternatively, for a given size and shape of channel, we can calculate the depth at which the normal discharge Q would occur. This depth is called the *normal depth*.

In analyzing uniform flow, typical problems encountered are the calculations of the normal discharge, the normal depth, the geometry of the channel section, the slope, or the value of Manning's n. We can make these calculations by using Eqs. (13–6) through (13–9).

**13.6.1
Manning's Equation in
U.S. Customary System**

Though not strictly true, it is conventional to take the values of Manning's n to be dimensionless so that the same data can be used in either the SI form of the equation [Eq. (13–6)] or the U.S. Customary System form. A careful conversion of units (see Reference 3) allows the use of the same values of n in the following equation:

MANNING'S EQUATION—U.S.
CUSTOMARY UNITS

$$v = \frac{1.49}{n} R^{2/3}S^{1/2} \qquad (13\text{–}10)$$

The velocity will then be expressed in feet per second (ft/s) when R is in ft. This is the form of Manning's equation for the U.S. Customary System.

We can also create forms of this equation parallel to Eqs. (13–8) and (13–9). That is,

NORMAL DISCHARGE—U.S.
CUSTOMARY UNITS

$$Q = AV = \left(\frac{1.49}{n}\right)AR^{2/3}S^{1/2} \qquad (13\text{–}11)$$

and

$$AR^{2/3} = \frac{nQ}{1.49S^{1/2}} \qquad (13\text{–}12)$$

In these equations, Q is the *normal discharge* in cubic feet per second (ft³/s) when A is the flow area in square feet (ft²) and R is expressed in feet (ft).

☐ **EXAMPLE PROBLEM 13.2**   Determine the normal discharge for a 200-mm inside diameter common clay drainage tile running half full if it is laid on a slope which drops 1 m over a run of 1000 m.

Solution   Equation (13–8) will be used:

$$Q = \left(\frac{1.00}{n}\right)AR^{2/3}S^{1/2}$$

The slope $S = 1/1000 = 0.001$. From Table 13.1 we find $n = 0.013$. Figure 13.6 shows the cross section of the tile half full.

$$A = \frac{1}{2}\left(\frac{\pi D^2}{4}\right) = \frac{\pi D^2}{8} = \frac{\pi (200)^2}{8}\ \text{mm}^2 = 5000\pi\ \text{mm}^2$$

$$= 15\ 708\ \text{mm}^2 = 0.0157\ \text{m}^2$$

$$WP = \pi D/2 = 100\pi\ \text{mm}$$

Then we have

$$R = A/WP = 5000\pi\ \text{mm}^2/100\pi\ \text{mm} = 50\ \text{mm}$$

or

$$R = 0.05\ \text{m}$$

Then in Eq. (13–8),

$$Q = \frac{(0.0157)(0.05)^{2/3}(0.001)^{1/2}}{0.013}$$

$$= 5.18 \times 10^{-3}\ \text{m}^3/\text{s}$$

FIGURE 13.6   Circular drain tile running half full for Example Problem 13.2.

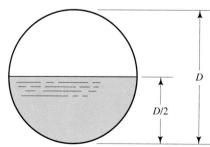

☐ **EXAMPLE PROBLEM 13.3**   Calculate the minimum slope on which the channel shown in Fig. 13.7 must be laid if it is to carry 50 ft³/s of water with a depth of 2 ft. The sides and bottom of the channel are made of formed, unfinished concrete.

FIGURE 13.7   Trapezoidal channel for Example Problem 13.3.

Solution   Equation (13–11) can be solved for the slope S:

$$Q = \left(\frac{1.49}{n}\right) AR^{2/3} S^{1/2}$$

$$S = \left(\frac{Qn}{1.49AR^{2/3}}\right)^2 \qquad \textbf{(13–13)}$$

From Table 13.1 we find $n = 0.017$. The values of A and R can be calculated from the geometry of the section:

$$A = (4)(2) + (2)(2)(2)/2 = 12 \text{ ft}^2$$
$$WP = 4 + 2\sqrt{4 + 4} = 9.66 \text{ ft}$$
$$R = A/WP = 12/9.66 = 1.24 \text{ ft}$$

Then from Eq. (13–13) we have

$$S = \left[\frac{(50)(0.017)}{(1.49)(12)(1.24)^{2/3}}\right]^2 = 0.00169$$

Therefore, the channel must drop at least 1.69 ft per 1000 ft of length.

☐ **EXAMPLE PROBLEM 13.4**   Design a rectangular channel to be made of formed, unfinished concrete to carry 5.75 m³/s of water when laid on a 1.2 percent slope. The normal depth should be one-half the width of the channel bottom.

Solution   Since the geometry of the channel is to be determined, Eq. (13–9) is most convenient:

$$AR^{2/3} = \frac{nQ}{S^{1/2}} = \frac{(0.017)(5.75)}{(0.012)^{1/2}} = 0.892$$

Figure 13.8 shows the cross section. Since $y = b/2$, only b must be determined. Both A and R can be expressed in terms of b:

$$A = by = \frac{b^2}{2}$$

$$WP = b + 2y = 2b$$

$$R = A/WP = \frac{b^2}{(2)(2b)} = \frac{b}{4}$$

Then we have

$$AR^{2/3} = 0.892$$

$$\frac{b^2}{2}\left(\frac{b}{4}\right)^{2/3} = 0.892$$

$$\frac{b^{8/3}}{5.04} = 0.892$$

$$b = (4.50)^{3/8} = 1.76 \text{ m}$$

The width of the channel must be 1.76 m.

FIGURE 13.8   Rectangular channel for Example Problem 13.4.

□ **EXAMPLE PROBLEM 13.5**   In the final design of the channel described in Example Problem 13.4, the width was made 2 m. The maximum expected discharge for the channel is 12 m³/s. Determine the normal depth for this discharge.

Solution   Equation (13–9) will be used again:

$$AR^{2/3} = \frac{nQ}{S^{1/2}} = \frac{(0.017)(12)}{(0.012)^{1/2}} = 1.86$$

Both A and R must be expressed in terms of the dimension y in Fig. 13.8, with $b = 2.0$ m:

$$A = 2y$$

$$WP = 2 + 2y$$

$$R = A/WP = 2y/(2 + 2y)$$

Then we have

$$1.86 = AR^{2/3} = 2y\left(\frac{2y}{2 + 2y}\right)^{2/3}$$

Algebraic solution for y is not simply done. A trial-and-error approach can be used. The results follow.

y	A	WP	R	$R^{2/3}$	$AR^{2/3}$	Required Change in y
2.0 m	4.0 m²	6.0 m	0.667 m	0.763	3.05	y too high
1.5	3.0	5.0	0.600	0.711	2.13	y too high
1.30	2.6	4.6	0.565	0.684	1.78	y too low

Therefore, the channel depth would be 1.35 m when the discharge is 12 m³/s.

13.7
THE GEOMETRY OF
TYPICAL OPEN
CHANNELS

Frequently used shapes for open channels include circular, rectangular, trapezoidal, and triangular. Table 13.2 gives the formulas for computing the geometric features pertinent to open channel flow calculations.

The trapezoid is popular for several reasons. It is an efficient shape because it gives a large flow area relative to the wetted perimeter. The sloped sides are convenient for channels made in the earth, because the slopes can be set at an angle at which the construction materials are stable.

The slope of the sides can be defined by the angle with respect to the horizontal or by means of the *pitch*, the ratio of the horizontal distance to the vertical distance. The pitch in Table 13.2 is indicated by the value of z, which is the horizontal distance corresponding to one unit of vertical distance. Practical earth channels made in the trapezoidal shape use values of z from 1.0 to 3.0.

The rectangle is a special case of the trapezoid with a side slope of 90° or $z = 0$. Formed concrete channels are often made in this shape. The triangular channel is also a special case of the trapezoid with a bottom width of zero. Simple ditches in earth are often made in this shape.

The computation of the data for circular sections at various depths can be facilitated by the graph in Fig. 13.9. At the left side of the figure is shown half of a circular section running partially full with the depth of the fluid called y. The vertical scale for the graph is the ratio, y/D. Curve A gives the ratio of A/A_f, in which A is the actual fluid flow area and A_f is the full cross-sectional area of the circle, easily calculated from $A_f = \pi D^2/4$. The use of Curve A is demonstrated by noting that the figure is drawn for the case $y/D = 0.65$. Follow the dashed horizontal line from this value on the vertical scale over to Curve A and then project down to the horizontal scale and read

TABLE 13.2 Geometry of open channel sections

Section	Area A	Wetted Perimeter WP	Hydraulic Radius R
Rectangle $b = T$, y	by	$b + 2y$	$\dfrac{by}{b + 2y}$
Triangle $T = 2zy$, y, z, 1	zy^2	$2y\sqrt{1 + z^2}$	$\dfrac{zy}{2\sqrt{1 + z^2}}$
Trapezoid $T = b + 2zy$, y, z, b, 1	$(b + zy)y$	$b + 2y\sqrt{1 + z^2}$	$\dfrac{(b + zy)y}{b + 2y\sqrt{1 + z^2}}$
Circle $T = 2\sqrt{y(D - y)}$, D, y, θ	$\dfrac{(\theta - \sin\theta)\, D^2}{8}$	$\theta D/2$	$\left[\dfrac{(\theta - \sin\theta)}{\theta}\right]\dfrac{D}{4}$

θ is in radians

Note: θ must be in radians.
For $y < D/2$, $\theta = \pi - 2\sin^{-1}[1 - (2y/D)]$
For $y > D/2$, $\theta = \pi + 2\sin^{-1}[(2y/D) - 1]$

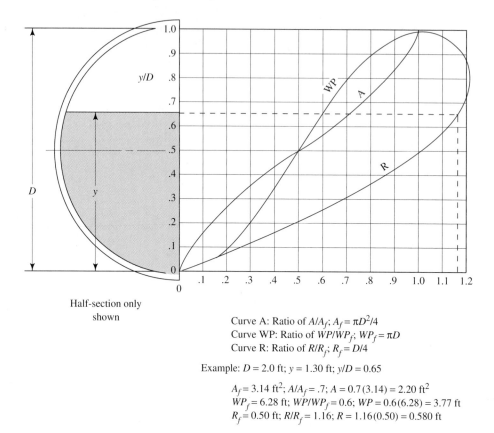

Half-section only
shown

Curve A: Ratio of A/A_f; $A_f = \pi D^2/4$
Curve WP: Ratio of WP/WP_f; $WP_f = \pi D$
Curve R: Ratio of R/R_f; $R_f = D/4$

Example: $D = 2.0$ ft; $y = 1.30$ ft; $y/D = 0.65$

$A_f = 3.14$ ft^2; $A/A_f = .7$; $A = 0.7(3.14) = 2.20$ ft^2
$WP_f = 6.28$ ft; $WP/WP_f = 0.6$; $WP = 0.6(6.28) = 3.77$ ft
$R_f = 0.50$ ft; $R/R_f = 1.16$; $R = 1.16(0.50) = 0.580$ ft

FIGURE 13.9   Geometry for partially full circular section.

the value of 0.70. This means $A/A_f = 0.70$ for $y/D = 0.65$. As an example, assume $D = 2.00$ ft. Then,

$$A_f = \pi D^2/4 = \pi (2.00 \text{ ft})^2/4 = 3.14 \text{ ft}^2$$
$$A = (0.70)A_f = (0.70)(3.14 \text{ ft}^2) = 2.20 \text{ ft}^2$$

In a similar manner, you should be able to read the wetted perimeter ratio to be $WP/WP_f = 0.60$ and the hydraulic radius ratio to be $R/R_f = 1.16$. Then,

$$WP_f = \pi D \text{ for a full circle} = \pi (2.00 \text{ ft}) = 6.28 \text{ ft}$$
$$WP = (0.60)WP_f = (0.60)(6.28 \text{ ft}) = 3.77 \text{ ft}$$

and,

$$R_f = D/4 \text{ for a full circle} = (2.00 \text{ ft})/4 = 0.50 \text{ ft}$$
$$R = (1.16)R_f = (1.16)(0.50 \text{ ft}) = 0.580 \text{ ft}$$

Thus, the curves in Fig. 13.9 enable you to compute the values of A, WP, and R for partially full circular sections with simple formulas using values of the three ratios read from the chart. Otherwise, the equations for direct calculation of A, WP, and R are quite complex.

Figure 13.10 shows three other shapes used for open channels. Natural streams frequently can be approximated as shallow parabolas. The triangle with a rounded bottom is more practical to make in the earth than the sharp-V triangle. The round-cornered rectangle performs somewhat better than the square-cornered rectangle and is easier to maintain. However, it is more difficult to form. Reference 3 gives formulas for the geometric features of these types of cross sections.

FIGURE 13.10   Other shapes for open channels.

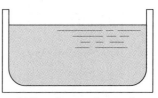

(a) Round-cornered rectangle

(b) Round-bottom triangle

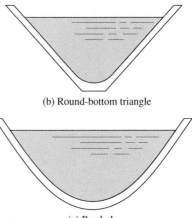

(c) Parabola

13.8
THE MOST EFFICIENT
SHAPES FOR
OPEN CHANNELS

The term *conveyance* is used to indicate the carrying capacity of open channels. Its value can be deduced from Manning's equation. In SI metric units,

$$Q = \left(\frac{1.00}{n}\right) A R^{2/3} S^{1/2} \tag{13–8}$$

Everything on the right side of this equation is dependent on the design of the channel except the slope. We can then define the conveyance K to be

$$K = \left(\frac{1.00}{n}\right) A R^{2/3} \tag{13–14}$$

In U.S. Customary units,

$$K = \left(\frac{1.49}{n}\right) A R^{2/3} \tag{13–15}$$

TABLE 13.3 Most efficient sections for open channels

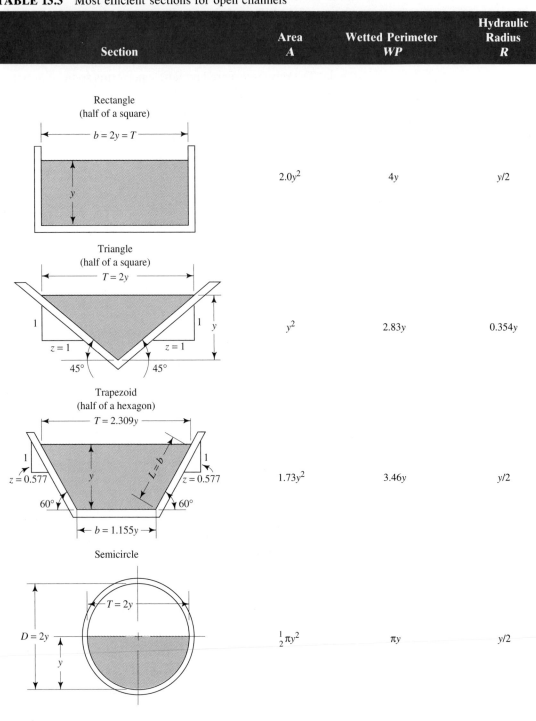

Section	Area A	Wetted Perimeter WP	Hydraulic Radius R
Rectangle (half of a square)	$2.0y^2$	$4y$	$y/2$
Triangle (half of a square)	y^2	$2.83y$	$0.354y$
Trapezoid (half of a hexagon)	$1.73y^2$	$3.46y$	$y/2$
Semicircle	$\frac{1}{2}\pi y^2$	πy	$y/2$

Manning's equation is then

$$Q = KS^{1/2} \tag{13-16}$$

The conveyance of a channel would be maximum when the wetted perimeter is the least for a given area. Using this criterion, we find that the most efficient shape is the semicircle, that is, the circular section running half full. Table 13.3 shows the most efficient designs of other shapes.

**13.9
CRITICAL FLOW AND
SPECIFIC ENERGY**

Consideration of energy in open channel flow usually involves a determination of the energy possessed by the fluid at a particular section of interest. The total energy is measured relative to the channel bottom and is composed of potential energy due to the depth of the fluid plus kinetic energy due to its velocity.

Letting E denote the total energy, we get

$$E = y + v^2/2g \tag{13-17}$$

where y is the depth and v is the average velocity of flow. As with the energy equation used previously, the terms in Eq. (13–17) have the units of energy per unit weight of fluid flowing. In open channel flow analysis, E is usually referred to as the *specific energy*. For a given discharge Q, the velocity is Q/A. Then

$$E = y + Q^2/2gA^2 \tag{13-18}$$

Since the area can be expressed in terms of the depth of flow, Eq. (13–18) relates the specific energy to the depth of flow. A graph of the depth y versus the specific energy E is useful in visualizing the possible regimes of flow in a channel. For a particular channel section and discharge, the specific energy curve appears as shown in Fig. 13.11.

Several features of this curve are important. The 45° line on the graph represents the plot of $E = y$. Then for any point on the curve, the horizontal distance to this line from the y axis represents the potential energy y. The remaining distance to the specific energy curve is the kinetic energy $v^2/2g$. A definite minimum value of E appears, and we can show that this occurs when the flow is at the critical state, that is, when $N_F = 1$. See Section 13.5, Eq. (13–4) for the definition of the Froude number, N_F.

The depth corresponding to the minimum specific energy is therefore called the *critical depth* y_c. For any depth greater than y_c, the flow is subcritical. Conversely, for any depth lower than y_c, the flow is supercritical. Notice that for any energy level greater than the minimum, there can exist two different depths. In Fig. 13.12, both y_1 below the critical depth y_c, and y_2 above y_c, have the same energy. In the case of y_1, the flow is supercritical and much of the energy is kinetic energy due to the high velocity of flow. At the greater depth y_2, the flow is slower and only a small portion of the energy is kinetic energy. The two depths y_1 and y_2 are called the *alternate depths* for the specific energy E.

FIGURE 13.11 Variation of specific energy with depth.

FIGURE 13.12 Critical depth and alternate depths.

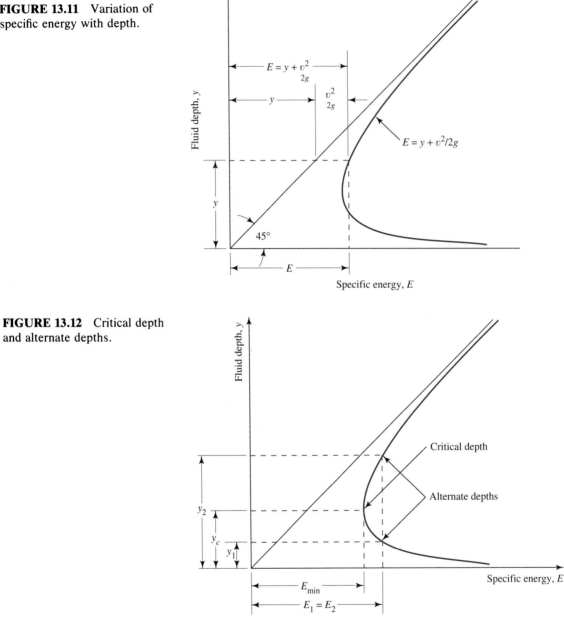

**13.10
HYDRAULIC JUMP**   To understand the significance of the phenomenon known as *hydraulic jump*, consider one of its most practical uses illustrated in Fig. 13.13. The water flowing over the spillway normally has a high velocity in the supercritical range when it reaches the bottom of the relatively steep slope at section 1. If this velocity were to be maintained into the natural stream bed beyond the paved spillway structure, the sides and bottom of the stream would be severely eroded. Instead, good design would cause a hydraulic jump to

FIGURE 13.13 Hydraulic jump at the bottom of a spillway.

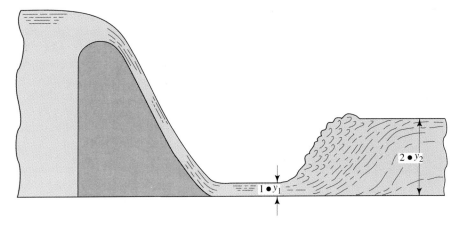

occur as shown, where the depth of flow abruptly changes from y_1 to y_2. Two beneficial effects result from the hydraulic jump. First, the velocity of flow is decreased substantially, decreasing the tendency for the flow to erode the stream bed. Second, much of the excess energy contained in the high velocity flow is dissipated in the jump. Energy dissipation occurs because the flow in the jump is extremely turbulent.

In order for a hydraulic jump to occur, the flow before the jump must be the supercritical range. That is, at section 1 in Fig. 13.13, y_1 is less than the critical depth for the channel and the Froude number N_{F_1} is greater than 1.0. The depth at section 2 after the jump can be calculated from the equation

$$y_2 = (y_1/2)(\sqrt{1 + 8N_{F_1}^2} - 1) \tag{13–19}$$

Also, the energy loss in the jump is dependent on the two depths y_2 and y_1:

$$E_1 - E_2 = \Delta E = (y_2 - y_1)^3/4y_1y_2 \tag{13–20}$$

Figure 13.14 illustrates what happens in a hydraulic jump by using a specific energy curve. The flow enters the jump with an energy E_1 corre-

FIGURE 13.14 Energy and depths in a hydraulic jump.

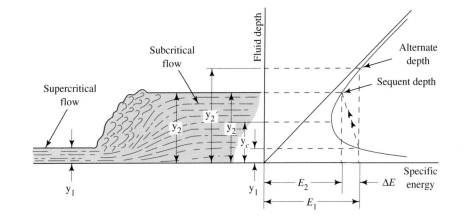

sponding to a supercritical depth y_1. In the jump, the depth abruptly increases. If no energy were lost, the new depth would be y'_2, which is the alternate depth for y_1. However, since there was some energy dissipated, ΔE, the actual new depth y_2 corresponds to the energy level E_2. Still, y_2 is in the subcritical range and tranquil flow would be maintained downstream from the jump. The name given to the actual depth y_2 after the jump is the *sequent depth*.

The following example problem illustrates another practical case in which hydraulic jump might occur.

☐ **EXAMPLE PROBLEM 13.6**   As shown in Fig. 13.15, water is being discharged from a reservoir under a sluice gate at the rate of 18 m³/s into a horizontal rectangular channel, 3 m wide, made of unfinished formed concrete. At a point where the depth is 1 m, a hydraulic jump is observed to occur. Determine the following:

a. The velocity before the jump
b. The depth after the jump
c. The velocity after the jump
d. The energy dissipated in the jump

FIGURE 13.15 Hydraulic jump for Example Problem 13.6.

Solution   **a.** The velocity before the jump is

$$v_1 = Q/A_1$$
$$A_1 = (3)(1) = 3 \text{ m}^2$$
$$v_1 = (18 \text{ m}^3/\text{s})/3 \text{ m}^2 = 6.0 \text{ m/s}$$

b. Equation (13–19) can be used to determine the depth after the jump, y_2:

$$y_2 = (y_1/2)(\sqrt{1 + 8N_{F_1}^2} - 1)$$
$$N_{F_1} = v_1/\sqrt{g y_h}$$

The hydraulic depth is equal to A/T, where T is the width of the free surface. Then for a rectangular channel $y_h = y$. Then we have

$$N_{F_1} = 6.0/\sqrt{(9.81)(1)} = 1.92$$

The flow is in the supercritical range.

$$y_2 = (1/2)(\sqrt{1 + (8)(1.92)^2} - 1) = 2.26 \text{ m}$$

c. Because of continuity,

$$v_2 = Q/A_2 = (18 \text{ m}^3/\text{s})/(3)(2.26) \text{ m}^2 = 2.65 \text{ m/s}$$

d. From Eq. (13–20), we get

$$\Delta E = (y_2 - y_1)^3/4y_1y_2$$

$$= \frac{(2.26 - 1.0)^3}{(4)(1.0)(2.26)} \text{ m} = 0.221 \text{ m}$$

This means that 0.221 N·m of energy is dissipated from each newton of water as it flows through the jump.

REFERENCES

1. Albertson, M. L., J. R. Barton, and D. B. Simons. 1960. *Fluid Mechanics for Engineers.* Englewood Cliffs, NJ: Prentice-Hall.
2. Binder, R. C. 1973. *Fluid Mechanics.* 5th ed. Englewood Cliffs, NJ: Prentice-Hall.
3. Chow, V. T. 1959. *Open Channel Hydraulics.* New York: McGraw-Hill.
4. Henderson, F. M. 1966. *Open Channel Flow.* New York: Macmillan.
5. Murdock, J. W. 1976. *Fluid Mechanics.* Boston: Houghton Mifflin.
6. Roberson, John A., John J. Cassidy, and M. Hanif Chaudhry. 1988. *Hydraulic Engineering.* Boston: Houghton Mifflin.
7. Simon, A. L. 1986. *Hydraulics.* 3rd ed. New York: John Wiley & Sons.

PRACTICE PROBLEMS

13.1M *Compute the hydraulic radius for a circular drain pipe running half full if its inside diameter is 300 mm.*

13.2M *A rectangular channel has a bottom width of 2.75 m. Compute the hydraulic radius when the fluid depth is 0.50 m.*

13.3E A drainage structure for an industrial park has a trapezoidal cross section similar to that shown in Fig. 13.1(c). The bottom width is 3.50 ft, and the sides are inclined at an angle of 60° from the horizontal. Compute the hydraulic radius for this channel when the fluid depth is 1.50 ft.

13.4E Repeat Problem 13.3 if the side slope is 45°.

13.5E Compute the hydraulic radius for a trapezoidal channel with a bottom width of 150 mm and with sides that pitch 15 mm horizontally for a vertical change of 10 mm. That is, the ratio of *X/D* in Fig. 13.1(c) is 1.50. The depth of the fluid in the channel is 62 mm.

13.6E Compute the hydraulic radius for the section shown in Fig. 13.16 if water flows at a depth of 2.0 in. The section is that of a rain gutter for a house.

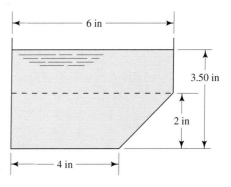

FIGURE 13.16 Problems 13.6, 13.7, and 13.11.

FIGURE 13.17   Problems 13.8, 13.9, and 13.14.

13.7E Repeat Problem 13.6 for a depth of 3.50 in.

13.8M *Compute the hydraulic radius for the channel shown in Fig. 13.17 if the water depth is 0.50 m.*

13.9M *Compute the hydraulic radius for the channel shown in Fig. 13.17 if the water depth is 2.50 m.*

13.10M *Water is flowing in a formed unfinished concrete rectangular channel 3.5 m wide. For a depth of 2.0 m, calculate the normal discharge and the Froude number of the flow. The channel slope is 0.1 percent.*

13.11E Determine the normal discharge for an aluminum rain spout with the shape shown in Fig. 13.16 that runs at the depth of 3.50 in. Use $n = 0.013$. The spout falls 4 in over a length of 60 ft.

13.12E A circular culvert under a highway is 6 ft in diameter and is made of corrugated metal. It drops 1 ft over a length of 500 ft. Calculate the normal discharge when the culvert runs half full.

13.13M *A wooden flume is being built to temporarily carry 5000 L/min of water until a permanent drain can be installed. The flume is rectangular, with a 205-mm bottom width and a maximum depth of 250 mm. Calculate the slope required to handle the expected discharge.*

13.14M *A storm drainage channel in a city where heavy sudden rains occur has the shape shown in Fig. 13.17. It is made of unfinished concrete and has a slope of 0.5 percent. During normal times, the wa-*

ter remains in the small rectangular section. The upper section allows large volumes to be carried by the channel. Determine the normal discharge for depths of 0.5 m and 2.5 m.

13.15E Figure 13.18 represents the approximate shape of a natural stream channel with levees built on either side. The channel is earth with grass cover. Use $n = 0.04$. If the average slope is 0.000 15, determine the normal discharge for depths of 3 ft and 6 ft.

13.16E Calculate the depth of flow of water in a rectangular channel 10 ft wide, made of brick in cement mortar for a discharge of 150 ft³/s. The slope is 0.1 percent.

13.17M *Calculate the depth of flow in a trapezoidal channel with a bottom width of 3 m and whose walls slope 45° with the horizontal. The channel is made of unfinished concrete and is laid on a 0.1 percent slope. The discharge is 15 m³/s.*

13.18M *A rectangular channel must carry 2.0 m³/s of water from a water-cooled refrigeration condenser to a cooling pond. The available slope is 75 mm over a distance of 50 m. The maximum depth of flow is 0.40 m. Determine the width of the channel if its surface is trowel-finished concrete.*

13.19M *The channel shown in Fig. 13.19 has a surface of float-finished concrete and is laid on a slope which falls 0.1 m per 100 m of length. Calculate the nor-*

FIGURE 13.18   Problem 13.15.

FIGURE 13.19 Problem 13.19.

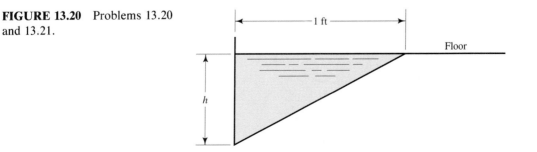

mal discharge and the Froude number for a depth of 1.5 m. For that discharge, calculate the critical depth.

13.20E A square storage room is equipped with automatic sprinklers for fire protection which spray 1000 gal/min of water. The floor is designed to drain this flow evenly to troughs near each outside wall. The troughs are shaped as shown in Fig. 13.20. Each trough carries 250 gal/min, is laid on a 1 percent slope, and is formed of unfinished concrete. Determine the minimum depth h.

13.21E The flow from two of the troughs described in Problem 13.20 passes into a sump, from which a round common clay drainage tile carries it to a storm sewer. Determine the size of tile required to carry the flow (500 gal/min) when running half full. The slope is 0.1 percent.

13.22M *For a rectangular channel with a bottom width of 1.00 m, compute the flow area and hydraulic radius for depths ranging from 0.10 m to 2.0 m. Plot a graph of area and hydraulic radius versus depth.*

13.23M *It is desired to carry 2.00 m³/s of water at a velocity of 3.0 m/s in a rectangular open channel. The bottom width is 0.80 m. Compute the depth of the flow and the hydraulic radius.*

13.24M *For the channel designed in Problem 13.23, compute the required slope if the channel is float-finished concrete.*

13.25M *It is desired to carry 2.00 m³/s of water at a velocity of 3.0 m/s in a rectangular open channel. Compute the depth and hydraulic radius for a range of designs for the channel, with bottom widths of 0.50 m to 2.00 m. Plot depth and hydraulic radius versus bottom width.*

13.26M *For each of the channels designed in Problem 13.25, compute the required slope if the channel is float-finished concrete. Plot slope versus bottom width.*

13.27E A trapezoidal channel has a bottom width of 2.00 ft and a pitch of its sides of $z = 1.50$. Compute the flow area and hydraulic radius for a depth of 20 in.

13.28E For the channel described in Problem 13.27, compute the normal discharge that would be expected for a slope of 0.005 if the channel is made from formed unfinished concrete.

13.29E Repeat Problem 13.28, except that the channel is lined with smooth plastic sheets.

13.30E A trapezoidal channel has a bottom width of 2.00 ft and a pitch of its sides of $z = 1.50$. Compute the

FIGURE 13.20 Problems 13.20 and 13.21.

flow area and hydraulic radius for depths ranging from 6.00 in to 24.00 in. Plot flow area and hydraulic radius versus depth.

13.31E For each channel designed in Problem 13.30, compute the normal discharge that would be expected for a slope of 0.005 if the channel is made from formed unfinished concrete.

13.32M *Compute the flow area and hydraulic radius for a circular drain pipe 375 mm in diameter for a depth of 225 mm.*

13.33M *Repeat Problem 13.32 for a depth of 135 mm.*

13.34M *For the channel designed in Problem 13.32, compute the normal discharge that is expected for a slope of 0.12 percent if the channel is made from painted steel.*

13.35M *For the channel designed in Problem 13.33, compute the normal discharge that is expected for a slope of 0.12 percent if the channel is made from painted steel. Compare the result with that from Problem 13.34.*

13.36E It is desired to carry 1.25 ft³/s of water at a velocity of 2.75 ft/s. Design the channel cross section for each of the shapes shown in Table 13.3, which lists the most efficient sections for open channels.

13.37E For each section designed in Problem 13.36, compute the required slope if the channel is made of float-finished concrete. Compare the results.

13.38E For each section designed in Problem 13.36, compute the Froude number and tell whether the flow is subcritical or supercritical.

13.39M *A rectangular channel 2 m wide carries 5.5 m³/s of water and is made of unfinished concrete. Perform the following operations:*

    a. *Calculate the critical depth.*
    b. *Calculate the minimum specific energy.*
    c. *Plot the specific energy curve.*
    d. *Determine the specific energy when y = 0.5 m and the alternate depth for this energy.*
    e. *Determine the velocity of flow and the Froude numbers for each depth in (d).*
    f. *Calculate the required slopes of the channel if the depths from (d) are to be normal depths for 5.5 m³/s flow.*

COMPUTER PROGRAMMING ASSIGNMENTS

1. Write a program to compute the geometry features for each section shown in Table 13.2. Include the area, wetted perimeter, and hydraulic radius. These programs could be used as subroutines in more comprehensive open channel flow analysis problems.

2. Write a program to compute the geometry features for each section shown in Table 13.3. Include the area, wetted perimeter, and hydraulic radius.

3. Write a program to compute the normal discharge for a given open channel and slope. Include the ability to compute the geometry features of the channel. The program can be written for a given channel shape (e.g., rectangular, circular, etc.), or it can include a set of subroutines for a variety of shapes. The program can include the table of values for Manning's n from which the user can select a design value. This program will then perform the computations of the type shown in Example Problem 13.2.

4. Write a program that computes the slope required for a channel of a given shape to carry a given discharge by using a method similar to that in Example Problem 13.3.

5. Write a program that computes the normal depth for a rectangular channel of a given width carrying a given normal discharge at a given slope. This program requires a trial-and-error solution method similar to that shown in Example Problem 13.5.

■ ■ ■ ■ 14 Flow Measurement

**14.1
OBJECTIVES**

Four primary reasons for using flow measurement systems are accounting, performance evaluation, research, and process control. Whenever the custody of a fluid is transferred, there is a need to account for the quantity involved. In commerce, examples of custody transfer are numerous. Gasoline flow is measured as it is pumped into the fuel tank of a car. Natural gas for cooking and heating is metered, as is water use. In industry, both the raw material input and finished product output must be continuously monitored for internal accounting purposes. Examples of performance evaluation are the measurement of fuel flow into an engine, air flow in a heating system, blood flow during surgery, or water flow through a heat exchanger. In process control, the success of a continuous operation is very dependent on flow measurement and control. For example, in paper making the flow of the pulp slurry into the machine and the flow of steam to the drying rolls must be monitored and controlled to ensure a uniform product. In the manufacture of urethane foam insulation, the density and thus the insulating quality of the foam is dependent on the precise metering of two primary constituents.

After completing this chapter, you should be able to:

1. Describe six factors that should be considered when specifying a flow measurement system.
2. Describe four types of variable head meters: the *venturi tube*, the *flow nozzle*, the *orifice*, and the *flow tube*.
3. Compute the velocity of flow and the volume flow rate for variable head meters, including the determination of the discharge coefficient.
4. Describe the *rotameter variable area meter*.
5. Describe the *turbine flowmeter*.
6. Describe the *magnetic flowmeter*.
7. Describe the *vortex flowmeter*.
8. Describe two methods of measuring mass flow rate.
9. Describe the *pitot-static tube*, and compute the velocity of flow using data acquired from such a device.
10. Define the term *anemometer* and describe two kinds.
11. Describe *weirs* and *flumes* as they are used for measuring flow in open channels, and perform the necessary computations.   .

14.2
FLOWMETER
SELECTION FACTORS

Many devices are available for measuring flow. Some measure volume flow rate directly, while others measure an average velocity of flow which can then be converted to volume flow rate by using $Q = Av$. Also, some provide direct primary measurements, while others require calibration or the application of a discharge coefficient to the observed output of the device. The form of the flow meter output also varies considerably from one type to another. The indication can be a pressure, a liquid level, a mechanical counter, the position of an indicator in the fluid stream, a continuous electrical signal, or a series of electrical pulses. The choice of the basic type of fluid meter and its indication system depends on several factors, some of which we will discuss here.

14.2.1
Range

Commercially available meters can measure flows from a few milliliters per second (mL/s) for precise laboratory experiments to several thousand cubic meters per second (m³/s) for irrigation water or municipal water and sewage systems. Then, for a particular meter installation, the general order of magnitude of the flow rate must be known as well as the range of the expected variations.

14.2.2
Accuracy Required

Virtually any flow measuring device properly installed and operated can produce an accuracy within 5 percent of the actual flow. Most commercially made meters are capable of 2 percent accuracy, and several claim better than 0.5 percent. Cost usually becomes an important factor when great accuracy is desired.

14.2.3
Pressure Loss

Because the construction details of the various meters are quite different, they produce differing amounts of energy loss or pressure loss as the fluid flows through them. Except for a few types, fluid meters accomplish the measurement by placing a restriction or a mechanical device in the flow stream, thus causing the energy loss.

14.2.4
Type of Indication

Factors to consider when choosing the type of flow indication include whether remote sensing or recording is required, whether automatic control is to be actuated by the output, whether an operator needs to monitor the output, and whether severe environmental conditions exist.

14.2.5
Type of Fluid

The performance of some fluid meters is affected by the properties and condition of the fluid. A basic consideration is whether the fluid is a liquid or a gas. Other factors which may be important are viscosity, temperature, corrosiveness, electrical conductivity, optical clarity, lubricating properties, and homogeneity. Slurries and multiphase fluids require special meters.

14.2.6
Calibration

Calibration is required for some types of flowmeters. Some manufacturers provide a calibration in the form of a graph or chart of actual flow versus indicator reading. Some are equipped for direct reading, with scales calibrated in the desired units of flow. In the case of the more fundamental types of meters, such as the variable head types, standard geometrical forms and dimensions have been determined for which empirical data are available. These data relate flow to an easily measured variable such as a pressure difference or fluid level. References at the end of this chapter give many of these calibration factors.

If calibration is required by the user of the device, he or she may use another precision meter as a standard against which the reading of the test device can be compared. Alternatively, primary calibration can be performed by adjusting the flow to a constant rate through the meter and then collecting the output during a fixed time interval. The fluid thus collected can either be weighed for a weight per unit time calibration, or its volume can be measured for a volume flow rate calibration. Figure 14.1 shows a commercially available flow calibrator in which a precision piston moves at a controlled rate to move the test fluid through the flowmeter being calibrated. The meter output is compared with the known flow rate by a computer data acquisition and analysis system to prepare calibration charts and graphs.

FIGURE 14.1 Flow calibration system. (Source: EG&G Flow Technology, Inc., Phoenix, AZ)

14.2.7
Other Factors

In most cases the physical size of the meter, cost, the system pressure, and the operator's skill should also be considered.

14.3
VARIABLE HEAD
METERS

The basic principle on which variable head meters are based is that when a fluid stream is restricted, its pressure decreases by an amount which is dependent on the rate of flow through the restriction. Therefore, the pressure difference between points before and after the restriction can be used to indicate flow rate. The most common types of variable head meters are the venturi tube, the flow nozzle, the orifice, and the flow tube. The derivation of the relationship between the pressure difference and the volume flow rate is the same regardless of which type of device is used. The venturi tube will be used as an example.

14.3.1 Venturi Tube

Figure 14.2 shows the basic appearance of a venturi tube. The flow from the main pipe at section 1 is caused to accelerate through a narrow section called the *throat*, where the fluid pressure is decreased. The flow then expands through the diverging portion to the same diameter as the main pipe. Pressure taps are located in the pipe wall at section 1 and in the wall of the throat, which we will call section 2. These pressure taps are attached to the two sides of a differential manometer so that the deflection h is an indication of the pressure difference $p_1 - p_2$. Of course, other types of differential pressure gages could be used.

The energy equation and the continuity equation can be used to derive the relationship from which we can calculate the flow rate. Using sections 1 and 2 in Fig. 14.2 as the reference points, we can write the following equations:

$$\frac{p_1}{\gamma} + z_1 + \frac{v_1^2}{2g} - h_L = \frac{p_2}{\gamma} + z_2 + \frac{v_2^2}{2g} \qquad (14\text{--}1)$$

$$Q = A_1 v_1 = A_2 v_2 \qquad (14\text{--}2)$$

These equations are valid only for incompressible fluids, that is, liquids. For the flow of gases, we must give special consideration to the variation of the specific weight γ with pressure. The algebraic reduction of Eqs. (14–1) and (14–2) proceeds as follows:

$$\frac{v_2^2 - v_1^2}{2g} = \frac{p_1 - p_2}{\gamma} + (z_1 - z_2) - h_L$$

$$v_2^2 - v_1^2 = 2g[(p_1 - p_2)/\gamma + (z_1 - z_2) - h_L]$$

But $v_1^2 = v_2^2(A_2/A_1)^2$. Then we have

$$v_2^2[1 - (A_2/A_1)^2] = 2g[(p_1 - p_2)/\gamma + (z_1 - z_2) - h_L]$$

$$v_2 = \sqrt{\frac{2g[(p_1 - p_2)/\gamma + (z_1 - z_2) - h_L]}{1 - (A_2/A_1)^2}} \qquad (14\text{--}3)$$

We can make two simplifications at this time. First, the elevation difference $(z_1 - z_2)$ is very small, even if the meter is installed vertically. Therefore, this term is neglected. Second, the term h_L is the energy loss from the fluid as it flows from section 1 to section 2. The value of h_L must be determined experimentally. But it is more convenient to modify Eq. (14–3) by dropping h_L and introducing a discharge coefficient C:

$$v_2 = C\sqrt{\frac{2g(p_1 - p_2)/\gamma}{1 - (A_2/A_1)^2}} \qquad (14\text{--}4)$$

Equation (14–4) can be used to calculate the velocity of flow in the throat of the meter. However, normally we want to calculate the volume flow rate.

(a)

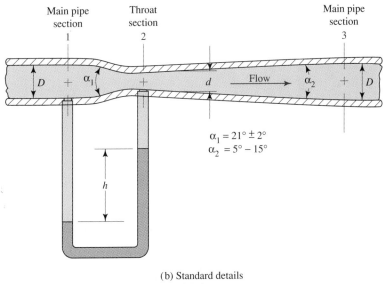

$\alpha_1 = 21° \pm 2°$
$\alpha_2 = 5° - 15°$

(b) Standard details
of design

FIGURE 14.2 Venturi tube. (Source of photo: BARCO, Barrington, IL)

Since $Q = A_2 v_2$, we have

$$Q = CA_2 \sqrt{\frac{2g(p_1 - p_2)/\gamma}{1 - (A_2/A_1)^2}} \qquad (14-5)$$

The value of the coefficient C is dependent on the Reynolds number of the flow and on the actual geometry of the meter. Figure 14.3 shows a typical curve of C versus Reynolds number in the main pipe.

FIGURE 14.3 Venturi tube discharge coefficient. (Source: ASME Research Committee on Fluid Meters. 1959. *Fluid Meters: Their Theory and Application.* 5th ed. New York: The American Society of Mechanical Engineers, p. 125.)

Reference 3 recommends $C = 0.984$ for a rough cast or fabricated venturi tube for the following conditions:

$$100 \text{ mm} \leq D \leq 1200 \text{ mm}$$
$$4 \text{ in} \leq D \leq 48 \text{ in}$$
$$0.30 \leq \beta \leq 0.75$$
$$2 \times 10^5 \leq N_R \leq 6 \times 10^6 \qquad \text{(in the main pipe)}$$

where β is defined as the ratio of the throat diameter to the diameter of the main pipe section. That is, $\beta = d/D$.

For a machined venturi tube, $C = 0.995$ is recommended for the following conditions:

$$50 \text{ mm} \leq D \leq 250 \text{ mm}$$
$$2 \text{ in} \leq D \leq 10 \text{ in}$$
$$0.30 \leq \beta \leq 0.75$$
$$2 \times 10^5 \leq N_R \leq 2 \times 10^6 \qquad \text{(in the main pipe)}$$

References 3, 5, and 9 give extensive information on the proper selection and application of venturi tubes.

Equation (14–5) is used for the flow nozzle and the orifice as well as for the venturi tube.

**14.3.2
Flow Nozzle**

The *flow nozzle* is a gradual contraction of the flow stream followed by a short, straight cylindrical section as illustrated in Fig. 14.4. Several standard geometries for flow nozzles have been presented and adopted by organizations such as the American Society of Mechanical Engineers (ASME) and the International Organization for Standardization. Because of the smooth, gradual contraction, there is very little energy loss between points 1 and 2 for a flow nozzle. A typical curve of C versus Reynolds number is shown in Fig. 14.5. At high Reynolds numbers C is above 0.99. At lower Reynolds numbers the sudden expansion outside the nozzle throat causes greater energy loss and a lower value for C.

FIGURE 14.4 Flow nozzle.

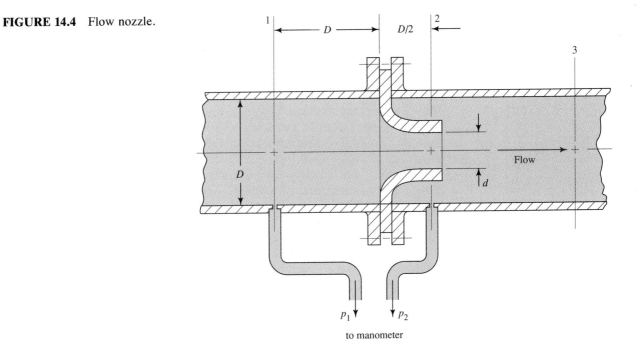

Reference 3 recommends Eq. (14–6) for C:

$$C = 0.9975 - 0.00653(10^6/N_R)^a \qquad \textbf{(14–6)}$$

where $a = 0.5$ for Reynolds numbers less than 10^6 and $a = 0.2$ for larger values. The Reynolds number is based on the *throat diameter* of the nozzle in Eq. (14–6).

References 3, 5, and 9 give extensive information on the proper selection and application of flow nozzles.

FIGURE 14.5 Flow nozzle discharge coefficient. (Source: ASME Research Committee on Fluid Meters. 1959. *Fluid Meters: Their Theory and Application.* 5th ed. New York: The American Society of Mechanical Engineers, p. 134.)

14.3.3 Orifice

A flat plate with an accurately machined, sharp-edged hole is referred to as an *orifice*. When placed concentrically inside a pipe as shown in Fig. 14.6(b) it causes the flow to suddenly contract as it approaches the orifice and then suddenly expand back to the full pipe diameter. The stream flowing through the orifice forms a vena contracta and the rapid flow velocity results in a decreased pressure downstream from the orifice. Pressure taps before and after the orifice (sections 1 and 2) allow the measurement of the differential pressure across the meter that is related to the volume flow rate by Eq. (14–5).

Figure 14.6(a) shows a commercially available unit that incorporates all of the major systems needed to measure flow. The orifice plate is part of an *integral flow orifice assembly* that also includes:

- pressure taps accurately located on both sides of the plate
- a manifold that facilitates the mounting of the differential producing (d/p cell)
- A d/p cell and transmitter to transmit the signal to a remote location
- a set of valves that allow fluid to bypass the d/p cell for service
- straight lengths of pipe into and out of the orifice to ensure predictable flow conditions at the orifice
- flanges to connect the unit to process piping
- a built-in microprocessor in the d/p cell that linearizes the output signal across the entire range of the meter giving a signal that is directly proportional to the flow. It performs the square root operation called for in Eq. (14–5).

The actual value of the discharge coefficient C depends on the location of the pressure taps. Three possible locations are listed in Table. 14.1.

FIGURE 14.6 Square-edged orifice with pressure taps at D and $D/2$. (Source of photo: The Foxboro Company, Foxboro, MA)

(a)

(b) Sketch of orifice installation

TABLE 14.1 Location of pressure taps for orifice meters

	Inlet Pressure Tap, p_1	Outlet Pressure Tap, p_2
1	One pipe diameter upstream from plate	One-half pipe diameter downstream from inlet face of plate
2	One pipe diameter upstream from plate	At vena contracta (see Reference 3)
3	In flange, 1 inch upstream from plate	In flange, 1 inch downstream from outlet face of plate

The value of C also is affected by small variations in the geometry of the edge of the orifice. Typical curves for sharp-edged orifices are shown in Fig. 14.7, where D is the pipe diameter and d is the orifice diameter. The value of C is much lower than that for the venturi tube or the flow nozzle since the fluid is forced to make a sudden contraction followed by a sudden expansion. Also, because measurements are based on the orifice diameter, the decrease in the diameter of the flow stream at the vena contracta tends to reduce the value of C.

FIGURE 14.7 Orifice discharge coefficient. (Source: ASME Research Committee on Fluid Meters. 1959. *Fluid Meters: Their Theory and Application.* 5th ed. New York: The American Society of Mechanical Engineers, p. 148.)

References 3, 5, and 9 give extensive information on the proper selection and application of orifices.

14.3.4 Flow Tubes

Several proprietary designs for modified variable head flow meters called flow tubes are available. These can be used for applications similar to those for which the venturi, nozzle, or orifice meters are used, but flow tubes have somewhat lower pressure loss (higher pressure recovery). Figure 14.8 is a photograph of one manufacturer's flow tube.

FIGURE 14.8 Flow tube. (Source: Fischer & Porter Company, Warminster, PA)

14.3.5
Pressure Recovery

In each of the four types of variable head meters just described, the flow stream expands back to the main pipe diameter after passing the restriction. This is indicated as section 3 in Figs. 14.2, 14.4, and 14.6(b). Then the difference between the pressure p_1 and p_3 is due to the meter. The difference can be evaluated by considering the energy equation:

$$\frac{p_1}{\gamma} + z_1 + \frac{v_1^2}{2g} - h_L = \frac{p_3}{\gamma} + z_3 + \frac{v_3^2}{2g}$$

Because the pipe sizes are the same at both sections, $v_1 = v_3$. We may also assume $z_1 = z_3$. Then,

$$p_1 - p_3 = \gamma h_L$$

The pressure drop is proportional to the energy loss. The careful streamlining of the venturi tube and the long gradual expansion after the throat cause very little excess turbulence in the flow stream. Therefore, energy loss is low and pressure recovery is high. The lack of a gradual expansion causes the nozzle to have a lower pressure recovery, while that for an orifice is still lower. The best pressure recovery is obtained from the flow tube. Figure

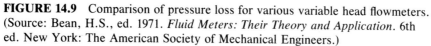

FIGURE 14.9   Comparison of pressure loss for various variable head flowmeters. (Source: Bean, H.S., ed. 1971. *Fluid Meters: Their Theory and Application.* 6th ed. New York: The American Society of Mechanical Engineers.)

14.9 shows the comparison among the several types of variable head meters with regard to pressure recovery.

**14.4
VARIABLE AREA
METERS**

The *rotameter* is a common type of variable area meter. Figure 14.10 shows a typical geometry. The fluid flows upward through a clear tube which has an accurate taper on the inside. A float is suspended in the flowing fluid at a position proportional to the flow rate. The upward forces due to the fluid dynamic drag on the float and buoyancy just balance the weight of the float.

FIGURE 14.10 Rotameter. (Source: Fischer & Porter Company, Warminster, PA)

A different flow rate causes the float to move to a new position, changing the clearance area between the float and the tube until equilibrium is achieved again. The position of the float is measured against a calibrated scale graduated in convenient units of volume flow rate or weight flow rate.

14.5 TURBINE FLOWMETER

Figure 14.11 shows a turbine flowmeter in which the fluid causes the turbine rotor to rotate at a speed dependent on the flow rate. As each blade of the rotor passes the magnetic coil, a voltage pulse is generated which can be input to a frequency meter, electronic counter, or other similar device whose readings can be converted to flow rate. Flow rates from as low as 0.02 L/min to several thousand L/min can be measured with turbine flowmeters of various sizes.

FIGURE 14.11 Turbine flowmeter. (Source: EG&G Flow Technology, Inc., Phoenix, AZ)

14.6
VORTEX FLOWMETER

Figure 14.12 shows a *vortex flowmeter*, in which a blunt obstruction placed in the flow stream causes vortexes to be created and shed from the body at a frequency that is proportional to the flow velocity. A sensor in the flowmeter detects the vortexes and creates an indication for the meter readout device (see Reference 6).

(a) Photograph of a vortex flowmeter

(b) Sketch of vortices shedding from a blunt body

FIGURE 14.12   Vortex flowmeter. (Source of photo: Fischer & Porter Company, Warminster, PA)

Part (b) of Fig. 14.12 shows a sketch of the vortex shedding phenomenon. The shape of the blunt body, also called the *vortex shedding element*, may vary from manufacturer to manufacturer. As the flow approaches the front face of the shedding element, it divides into two streams. Fluid close to the body has a low velocity relative to that in the main streamlines. The difference in velocity causes shear layers to form which eventually break down into vortices alternately on the two sides of the shedding element. The frequency of the vortices created is directly proportional to the flow velocity and, therefore, to the volume flow rate. Sensors in the meter detect the pressure variations around the vortices and generate a voltage signal that alternates at the same frequency as the vortex shedding frequency. The output signal is either a stream of voltage pulses or a dc (direct current) analog signal. Standard instrumentation systems often use an analog signal that varies from 4 to 20 mA dc (milliamps dc). For the pulse output, the manufacturer supplies a flowmeter K-factor that indicates pulses per unit volume through the meter.

Vortex meters can be used for a wide range of fluids including clean and dirty liquids and gases and steam.

14.7 MAGNETIC FLOWMETER

Totally unobstructed flow is one of the advantages of a magnetic flowmeter like that shown in Fig. 14.13. The fluid must be slightly conducting because the meter operates on the principle that when a moving conductor cuts across a magnetic field, a voltage is induced. The primary components of the magnetic flowmeter include a tube lined with a nonconducting material, two electromagnetic coils, and two electrodes mounted 180° apart in the tube wall. The electrodes detect the voltage generated in the fluid. Since the generated voltage is directly proportional to the fluid velocity, a greater flow rate generates a greater voltage. An important feature of this type of meter is that its output is completely independent of temperature, viscosity, specific gravity, or turbulence. Tube sizes from 5 mm to several meters in diameter are available.

FIGURE 14.13 Magnetic flowmeter. (Source: The Foxboro Company, Foxboro, MA)

One electrode

Electromagnetic coils

14.8 MASS FLOW MEASUREMENT

The flowmeters discussed thus far in this chapter are designed to produce an output signal that is proportional to the average velocity of flow or the volume flow rate. This is satisfactory when only the *volume* delivered through the meter is needed. However, some processes require a measurement of the *mass* of fluid delivered. For example, in food processing plants the production is often indicated as the amount delivered in kilograms, pounds-mass, or slugs. Some chemical processes are sensitive to the mass of the various constituents that are blended or that are introduced into a reaction. Two-phase fluids, such as steam, may be difficult to measure accurately if the temperature and pressure vary enough to cause significant changes in the amount of liquid and vapor in the steam.

One way to obtain mass flow rate measurements is to use a flowmeter of the type just discussed that indicates volume flow rate and then simultaneously measure the density of the fluid. Then the mass flow rate would be

$$M = \rho Q$$

That is, mass flow rate equals density times volume flow rate as discussed in Chapter 6 of this book. If the density of the fluid is known or can be conveniently measured, this is a simple calculation. For some fluids, the density can be calculated if the temperature of the fluid is known. Sometimes, particularly with gases, the pressure is also needed. Temperature probes and pressure transducers are readily available to provide the necessary data. Specific gravity can be measured with a device called a *gravitometer*. Density can be measured directly for some fluids with a *densitometer*. The signals related to volume flow rate, temperature, pressure, specific gravity, or density can all be input to special electronic devices that effectively perform the calculation of $M = \rho Q$. This is shown schematically in Fig. 14.14. This process, while straightforward, requires several separate mea-

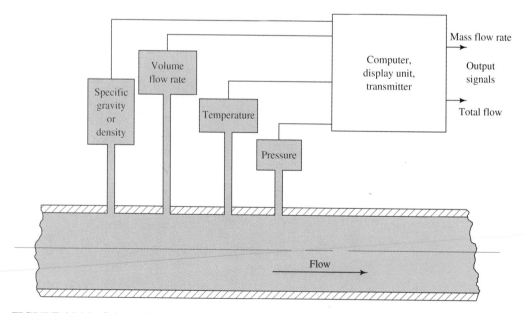

FIGURE 14.14   Schematic representation of mass flow measurement using multiple sensors.

FIGURE 14.15 Coriolis mass flowtube. (Source: The Foxboro Company, Foxboro, MA)

(a) External view with programmer and indicator

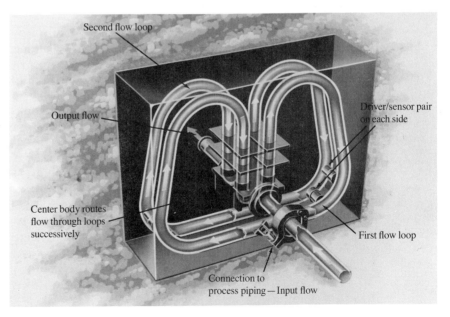

(b) Internal view

surements to be made, each of which is subject to small errors. Then the errors are compounded in the final calculation.

True mass flowmeters avoid the problems discussed above by generating a signal proportional to the mass flow rate directly. One such mass flowmeter is called the *Coriolis mass flowtube*, shown in Fig. 14.15. The fluid enters the flowmeter from the process piping and is directed through a continuous path of the same size that routes the fluid first through one loop, through a center body, then through the second loop, and then out to the exit pipe. Two electromagnetic drivers bridge both loops at opposite extremities, equidistant from the center. The vibratory motion created moves the two parallel loops alternately toward each other and then apart. Fluid in the tubes is simultaneously following the path of the loops and moving perpendicular to that path because of the action of the drivers. A Coriolis acceleration (and a corresponding Coriolis force) is produced that is proportional to the mass of fluid flowing through the tubes. Sensors mounted near the drivers detect the Coriolis force and transmit a signal that can be related to the true mass flow rate through the meter. Accuracy is reported to be 0.2 percent of the indicated rate or 0.02 percent of full-scale capacity, whichever is greater.

Density of the fluid can also be measured with the Coriolis mass flowtube because the driving frequency of the tubes is dependent on the density of the fluid flowing through the tubes. A temperature probe is also included in the system, completing a comprehensive set of fluid properties and mass flow rate data.

**14.9
VELOCITY PROBES**

Several devices are available which measure velocity of flow at a specific location rather than an average velocity. These are referred to as *velocity probes*. Some of the more common types will be described in this section.

**14.9.1
Pitot Tube**

When a moving fluid is caused to stop because it encounters a stationary object, a pressure is created which is greater than the pressure of the fluid stream. The magnitude of this increased pressure is related to the velocity of the moving fluid. The *pitot tube* uses this principle to indicate velocity, as illustrated in Fig. 14.16. The pitot tube is a hollow tube positioned so that the open end points directly into the fluid stream. The pressure at the tip causes a column of fluid to be supported. The fluid at or just inside the tip is then stationary or stagnant, and this point is referred to as the *stagnation point*. We can use the energy equation to relate the pressure at the stagnation point with the fluid velocity. If point 1 is in the undisturbed stream ahead of the tube and point s is at the stagnation point, then

$$\frac{p_1}{\gamma} + z_1 + \frac{v_1^2}{2g} - h_L = \frac{p_s}{\gamma} + z_s + \frac{v_s^2}{2g} \qquad (14\text{--}7)$$

Observe that $v_s = 0$, $z_1 = z_2$ or very nearly so, and $h_L = 0$ or very nearly so. Then we have

$$\frac{p_1}{\gamma} + \frac{v_1^2}{2g} = \frac{p_s}{\gamma} \qquad (14\text{--}8)$$

FIGURE 14.16 Pitot tube.

The names given to the terms in Eq. (14–8) are as follows:

$$p_1 = \text{static pressure in the main fluid stream}$$
$$p_1/\gamma = \text{static pressure head}$$
$$p_s = \text{stagnation pressure or total pressure}$$
$$p_s/\gamma = \text{total pressure head}$$
$$v_1^2/2g = \text{velocity pressure head}$$

The total pressure head is equal to the sum of the static pressure head and the velocity pressure head. Solving Eq. (14–8) for the velocity gives

$$v_1 = \sqrt{2g(p_s - p_1)/\gamma} \qquad \textbf{(14–9)}$$

Notice that only the difference between p_s and p_1 is required to calculate the velocity. For this reason, most pitot tubes are made as shown in Fig. 14.17, providing for the measurement of both pressures with the same device.

The device shown in Fig. 14.17 facilitates the measurement of both the static pressure and the stagnation pressure simultaneously and so it is sometimes called a *pitot-static* tube. Its construction shown in part (b) is actually a tube within a tube. The small central tube is open at the end and functions in the same manner as the single pitot tube shown in Fig. 14.16. Thus the stagnation pressure, also called the *total pressure*, is sensed through this tube. The *total pressure tap* at the end of this tube allows connection to a pressure-measuring device.

The larger outer tube is sealed around the central tube at its end, thus creating a closed annular cavity between the central and the outer tube. Section A-A shows that a series of small radial holes are drilled through the outer tube but not through the central tube. When the tube is aligned with the direction of flow, these radial holes are perpendicular to the flow and thus they sense the local static pressure which we have called p_1. Notice that a static pressure tap is affixed at the end of the tube to allow connection to a measuring instrument.

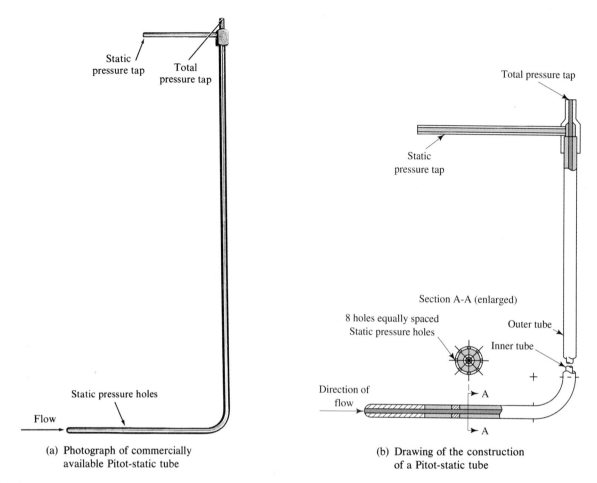

(a) Photograph of commercially
available Pitot-static tube

(b) Drawing of the construction
of a Pitot-static tube

FIGURE 14.17   Pitot-static tube. (Source: Dwyer Instruments, Inc., Michigan
City, IN)

The measuring instrument need not measure either p_s or p_1 because it is
the *difference* $(p_s - p_1)$ that is needed in Eq. (14–9). Several manufacturers
make differential pressure-measuring devices for such applications.

If a differential manometer is used as shown in Fig. 14.18, the manometer deflection h can be related directly to the velocity. We can write the
equation describing the difference between p_s and p_1 by starting at the static
pressure holes in the side of the tube, proceeding through the manometer,
and ending at the open tip of the tube at point s:

$$p_1 - \gamma x + \gamma y + \gamma_g h - \gamma h - \gamma y + \gamma x = p_s$$

The terms involving the unknown distances x and y drop out. Then, solving
for the pressure difference, we get

$$p_s - p_1 = \gamma_g h - \gamma h = h(\gamma_g - \gamma) \qquad \textbf{(14–10)}$$

FIGURE 14.18 Differential manometer used with a pitot-static tube.

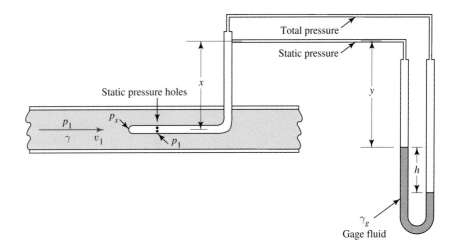

Substituting this into Eq. (14–9) gives

$$v_1 = \sqrt{2gh(\gamma_g - \gamma)/\gamma} \qquad \textbf{(14–11)}$$

The velocity calculated by either Eq. (14–9) or (14–11) is the local velocity at the particular location of the tip of the tube. In Chapter 8 we found that the velocity of flow varies from point to point across a pipe. Therefore, if the average velocity of flow is desired, a traverse of the pipe

FIGURE 14.19 Velocity measurement points within a pipe for computing average velocity.

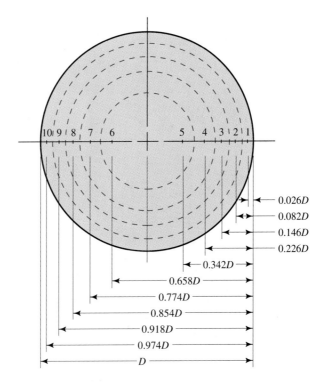

should be made with the tip of the tube placed at the ten points indicated in Fig. 14.19. The dashed circles define concentric annular rings which have equal areas. The velocity at each point can be calculated, using Eq. (14–11). Then the average velocity of flow is the average of these ten values. We can find the volume flow rate from $Q = Av$, using the average velocity.

□ **EXAMPLE PROBLEM 14.1**   For the apparatus shown in Fig. 14.18, the fluid in the pipe is water at 60°C and the manometer fluid is mercury with a specific gravity of 13.54. If the manometer deflection h is 264 mm, calculate the velocity of the water.

Solution   Equation (14–11) will be used:

$$v_1 = \sqrt{2gh(\gamma_g - \gamma)/\gamma}$$
$$\gamma = 9.65 \text{ kN/m}^3 \qquad \text{(water at 60°C)}$$
$$\gamma_g = (13.54)(9.81 \text{ kN/m}^3) = 132.8 \text{ kN/m}^3 \qquad \text{(mercury)}$$
$$h = 264 \text{ mm} = 0.264 \text{ m}$$

Since all terms are in SI units, the velocity is in m/s:

$$v_1 = \sqrt{\frac{(2)(9.81)(0.264)(132.8 - 9.65)}{9.65}}$$
$$= 8.13 \text{ m/s}$$

The pressure differential created by a pitot tube can also be read by an electronic device such as that shown in Fig. 14.20. The individual readings

FIGURE 14.20   Electronic readout device for pitot tubes. (Source: Alnor Instruments Co., Skokie IL)

taken during a traverse of a pipe or duct can be recorded on the portable printer. Then the average is automatically computed and printed in either SI units or U.S. Customary System units.

14.9.2
Cup Anemometer

Air velocity is often measured with a *cup anemometer* such as that shown in Fig. 14.21. The moving air strikes the open cups, causing rotation of the shaft on which they are mounted. The rotational speed of the shaft is proportional to the air velocity, which is indicated on a meter or transmitted electrically.

FIGURE 14.21 Rotating cup anemometer.

14.9.3
Hot Wire Anemometer

This type of velocity probe employs a very thin wire, about 12 μm in diameter, through which an electric current is passed. The wire is suspended on two supports as shown in Fig. 14.22 and inserted into the fluid stream. The wire tends to heat because of the current flowing in it, but it is cooled by convection heat transfer to the moving fluid stream. The amount of cooling depends on the velocity of the fluid. In one type of hot wire anemometer, a constant current is applied to the wire. A variation in the flow velocity causes a change in the wire temperature, and therefore its resistance changes. The electronic measurement of the resistance change can be related to flow velocity. Another type senses a change in the resistance of the wire, but then varies the current flow to maintain a set wire temperature regardless of fluid velocity. The magnitude of the current flow is then related to fluid velocity.

FIGURE 14.22 Hot wire anemometer tip.

Hot wire

14.10
COMPUTER-BASED
DATA ACQUISITION
AND PROCESSING

Microcomputers, programmable controllers, and other microprocessor-based electronic instrumentation greatly simplify the acquisition, processing, and recording of flow measurement data. As shown in this chapter, many of the flowmeters produce an electrical signal that is proportional to the flow velocity. The signal is either an analog voltage that varies with the velocity or a pulse frequency that can be counted electronically. Analog signals can be converted to digital signals by analog-to-digital converters, often called *A-D converters*, for input to digital computers.

The computers can total the fluid flow rate over time to determine the total quantity of fluid transferred to a given location. Figure 14.23 shows an indicating totalizer that is battery powered for remote operation.

FIGURE 14.23   Flow totalizer. (Source: Fischer & Porter Company, Warminster, PA)

A comprehensive measurement and control system can consist of pressure, temperature, and flow measurement devices; automatic process controllers; interface units; operator control stations; and large host computers. Figure 14.24 shows an interface device that can communicate with up to 32 controllers, combining and reformatting data for transmission to an operator's station like the one shown in Fig. 14.25. The video terminal can display the status of several measurements simultaneously for the operator while monitoring the data for values that fall outside prescribed levels. The host

FIGURE 14.24 Electronic interface device for flowmeter data acquisition system. (Source: Fischer & Porter Company, Warminster, PA)

FIGURE 14.25 Operator's station for computer-based process control system. (Source: Fischer & Porter Company, Warminster, PA)

computer can acquire data from several places within the plant and maintain the central data base for quality control, production data, and inventory control.

14.11 OPEN CHANNEL FLOW MEASUREMENT

Two widely used devices for open channel flow measurement are *weirs* and *flumes*. Each causes the area of the stream to change, which in turn changes the level of the fluid surface. The resulting level of the surface relative to some feature of the device is related to the quantity of flow. Large volume flow rates of liquids can be measured with weirs and flumes.

14.11.1 Weirs

A *weir* is a barrier or dam placed in the channel so that the fluid backs up behind it and then falls through a notch cut into the face of the weir. Two common notch geometries are the rectangular and the triangular. Figure 14.26 shows a side view of a weir in operation. Front views showing different notch geometries are shown in Fig. 14.27. The discharge over the weir is dependent on the dimensions of the notch and on the head H of the fluid. Figure 14.26 shows that the fluid surface is somewhat curved as it passes over the crest of the weir. In order to ensure consistent measurements, the value of H should be the difference between the height to the crest H_c and the total height to the liquid surface H_t, with H_t measured upstream from the weir plate where the surface profile is undisturbed. Normally, this upstream distance is approximately six times the maximum expected head H.

FIGURE 14.26 Flow over a weir.

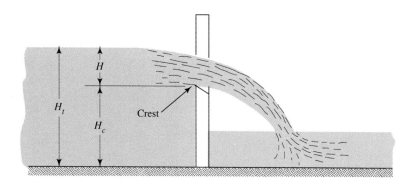

The theoretical equation for discharge over a rectangular notched weir is

$$Q = \tfrac{2}{3}L\sqrt{2gH^3} \qquad (14\text{–}12)$$

where L is the length of the crest between the sides of the notch and H is the head above the crest. If L and H are measured in ft, then Q is in cubic feet per second (ft^3/s).

The actual discharge is different from the theoretical for a variety of reasons; therefore, relationships that are more accurate and easier to use have been developed. Most formulas take the form

$$Q = CLH^{3/2} \tag{14-13}$$

where C is a discharge coefficient, L is the effective length of the crest, and H is the head above the weir crest. For the full width rectangular weir in Fig. 14.27(a), the following equation can be used:

$$Q = (3.27 + 0.40H/H_c)LH^{3/2} \tag{14-14}$$

The two end contractions in the contracted weir in Fig. 14.27(b) cause the stream to be curved in from the sides, decreasing the effective length of the crest. The discharge for this type of weir can be calculated from

$$Q = (3.27 + 0.40H/H_c)(L - 0.2H)H^{3/2} \tag{14-15}$$

FIGURE 14.27 Notch geometry for weirs.

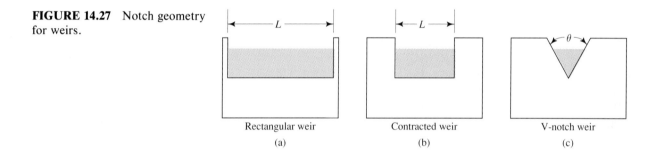

Rectangular weir	Contracted weir	V-notch weir
(a)	(b)	(c)

The triangular weir is used primarily for low flow rates since the V-notch produces a larger head H than can be obtained with a rectangular notch. The angle of the V-notch is a factor in the discharge equation. Angles from 35° to 120° are satisfactory, but angles of 60° and 90° are quite commonly used. The theoretical equation for a triangular weir is

$$Q = \tfrac{8}{15}C\sqrt{2g}\,\tan(\theta/2)H^{5/2} \tag{14-16}$$

where θ is the total included angle between the sides of the notch. An additional reduction of this equation gives

$$Q = 4.28C\,\tan(\theta/2)H^{5/2} \tag{14-17}$$

The value of C is somewhat dependent on the head H, but a nominal value is 0.58. Using this and the common values of 60° and 90° for θ, we get

$$Q = 1.43H^{5/2} \quad (60° \text{ notch}) \tag{14-18}$$

$$Q = 2.48H^{5/2} \quad (90° \text{ notch}) \tag{14-19}$$

14.11.2 Flumes

Critical flow flumes are contractions in the stream which cause the flow to achieve its critical depth within the structure. There is a definite relationship between depth and discharge when critical flow exists. A widely used type of critical flow flume is the *Parshall flume*, the geometry of which is shown in Fig. 14.28. The discharge is dependent on the width of the throat section L and the head H, where H is measured at a specific location along the converging section of the flume.

FIGURE 14.28   Parshall flume.

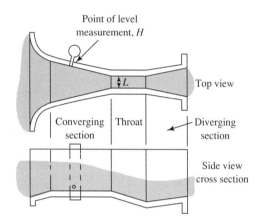

The discharge equations for the Parshall flume were developed empirically for flumes designed and constructed in dimensions in the U.S. customary system. Table 14.2 lists the discharge equations for several sizes of

TABLE 14.2   Discharge equations for Parshall flumes

Throat Width L	Flow Range (ft³/s) Min.	Flow Range (ft³/s) Max.	Equation (H and L in ft, Q in ft³/s)
3 in	0.03	1.9	$Q = 0.992 H^{1.547}$
6 in	0.05	3.9	$Q = 2.06 H^{1.58}$
9 in	0.09	8.9	$Q = 3.07 H^{1.53}$
1 ft	0.11	16.1	
2 ft	0.42	33.1	
4 ft	1.3	67.9	$Q = 4LH^{1.522L^{0.26}}$
6 ft	2.6	103.5	
8 ft	3.5	139.5	
10 ft	6	200	
20 ft	10	1000	
30 ft	15	1500	$Q = (3.6875L + 2.5)H^{1.6}$
40 ft	20	2000	
50 ft	25	3000	

flumes. The resulting value of Q can be converted to SI units by use of the factor

$$1.0 \text{ ft}^3/\text{s} = 0.028\ 32 \text{ m}^3/\text{s}$$

Chow (Reference 8) gives additional data about flumes and weirs.

REFERENCES

1. ASME Research Committee on Fluid Meters. 1961. *Flowmeter Computation Handbook*. New York: The American Society of Mechanical Engineers.
2. ————. 1959. *Fluid Meters: Their Theory and Application*. 5th ed. New York: The American Society of Mechanical Engineers.
3. ————. 1971. *Fluid Meters: Their Theory and Application*. Ed. Howard S. Bean. 6th ed. New York: The American Society of Mechanical Engineers.
4. ASME Standards Committee for the Measurement of Fluid Flow in Closed Conduits. 1979 (Reaffirmed 1986). *Glossary of Terms Used in the Measurement of Fluid Flow in Pipes*. ANSI/ASME Standard MFC-1M-1979. New York: The American Society of Mechanical Engineers.
5. ————. 1985. *Measurement of Fluid Flow in Pipes Using Orifice, Nozzle, and Venturi*. ASME standard MFC-3M-1985. New York: The American Society of Mechanical Engineers.
6. ————. 1987. *Measurement of Fluid Flow in Pipes Using Vortex Flow Meters*. ASME/ANSI MFC-6M-1987. New York: The American Society of Mechanical Engineers.
7. Baker, Roger C. 1989. *An Introductory Guide to Flow Measurement*. London: Mechanical Engineering Publications Limited.
8. Chow, V. T. 1959. *Open-Channel Hydraulics*. New York: McGraw-Hill.
9. DeCarlo, J. P. 1984. *Fundamentals of Flow Measurement*. Research Triangle Park, North Carolina: The Instrument Society of America.
10. Miller, R. W. 1983. *Flow Measurement Engineering Handbook*. New York: McGraw-Hill.

REVIEW QUESTIONS

1. List six factors that affect the selection and use of flowmeters.
2. Define *range* as it relates to flowmeters.
3. Describe three methods for calibrating flowmeters.
4. Name four types of variable head meters.
5. Describe the venturi tube.
6. What is meant by the *throat* of a venturi tube?
7. What is the nominal included angle of the convergent section for a venturi tube?
8. What is the nominal included angle of the divergent section for a venturi tube?
9. Why is there such a difference between the angles of the convergent and the divergent sections for a venturi tube?
10. Describe the term *discharge coefficient* as it relates to variable head meters.
11. Describe a flow nozzle and how it is used.
12. Describe an orifice meter and how it is used.
13. Describe a flow tube and how it is used.
14. Of the venturi, the flow nozzle, the flow tube, and the orifice, which has the lowest discharge coefficient? Why?
15. Describe *pressure recovery* as it relates to flowmeters.
16. Rank the venturi, the flow nozzle, the orifice, and the flow tube on the basis of pressure recovery.
17. Describe a rotameter variable area meter.
18. Describe a turbine flowmeter and how it is used.
19. Describe a vortex flowmeter and how it is used.
20. Describe a magnetic flowmeter and how it is used.
21. Describe how mass flow rate can be measured.
22. Describe a pitot tube and how it is used.
23. Define *stagnation pressure* and show how it can be derived from Bernoulli's equation.
24. Define *static pressure head*.
25. Define *velocity pressure head*.
26. Why is a differential manometer a convenient device for use with a pitot tube?
27. Describe the method used to measure the average velocity of flow in a pipe using a pitot tube.
28. Describe a cup anemometer.

29. Describe a hot wire anemometer and how it is used.
30. Describe the use of an analog-digital converter in the practice of flow measurement.
31. Name two devices used to measure the flow rate in open channels.

32. Describe a weir and three different geometries used for weirs.
33. What is the parameter that must be measured to determine flow using a weir?
34. Describe a Parshall flume and how it is used.

PRACTICE PROBLEMS

14.1M *A venturi meter similar to the one in Fig. 14.2 has a pipe diameter of 100 mm and a throat diameter of 50 mm. While it is carrying water at 80°C, a pressure difference of 55 kPa is observed between sections 1 and 2. Calculate the volume flow rate of water.*

14.2M *Air with a specific weight of 12.7 N/m³ and a kinematic viscosity of 1.3 × 10⁻⁵ m²/s is flowing through a flow nozzle similar to that shown in Fig. 14.4. A manometer using water as the gage fluid reads 81 mm of deflection. Calculate the volume flow rate if the nozzle diameter is 50 mm. Pipe inside diameter = 100 mm.*

14.3E The flow of kerosene is being measured with an orifice meter similar to that shown in Fig. 14.6. The pipe is a 2-in Schedule 40 pipe and the orifice diameter is 1.00 in. The kerosene is at 77°F. For a pressure difference of 0.53 psi across the orifice, calculate the volume flow rate of kerosene.

14.4E A sharp-edged orifice is placed in a 10-in diameter pipe carrying ammonia. If the volume flow rate is 25 gal/min, calculate the deflection of a water manometer (a) if the orifice diameter is 1.0 in and (b) if the orifice diameter is 7.0 in. The ammonia has a specific gravity of 0.83 and a dynamic viscosity of $2.5 × 10^{-6}$ lb-s/ft².

14.5M *A pitot-static tube is inserted into a pipe carrying methyl alcohol at 25°C. A differential manometer using mercury as the gage fluid is connected to the tube and shows a deflection of 225 mm. Calculate the velocity of flow of the alcohol.*

14.6M *A pitot-static tube is connected to a differential manometer using water at 40°C as the gage fluid. The velocity of air at 40°C and atmospheric pressure is to be measured, and it is expected that the maximum velocity will be 25 m/s. Calculate the expected manometer deflection.*

14.7M *A pitot-static tube is inserted in a pipe carrying water at 10°C. A differential manometer using*

mercury as the gage fluid shows a deflection of 106 mm. Calculate the velocity of flow.

14.8E Determine the maximum possible flow rate over a 60° V-notch weir if the width of the notch at the top is 12 in.

14.9E Determine the required length of a contracted weir similar to that shown in Fig. 14.27(b) to pass 15 ft³/s of water. The height of the crest is to be 3 ft from the channel bottom, and the maximum head above the crest is to be 18 in.

14.10E Plot a graph of Q versus H for a full width weir with a crest length of 6 ft and whose crest is 2 ft from the channel bottom. Consider values of the head H from zero to 12 inches in 2-in steps.

14.11E Repeat the calculations of Q versus H for a weir with the same dimensions as used in Problem 14.10 except that it is placed in a channel wider than 6 ft. It thus becomes a contracted weir.

14.12E Compare the discharges over the following weirs when the head H is 18 in:
   a. Full width rectangular: $L = 3$ ft, $H_c = 4$ ft
   b. Contracted rectangular: $L = 3$ ft, $H_c = 4$ ft
   c. 90° V-notch (top width also 3 ft)

14.13E Plot a graph of Q versus H for a 90° V-notch weir for values of the head from zero to 12 inches in 2-in steps.

14.14E For a Parshall flume with a throat width of 9 in, calculate the head H corresponding to the minimum and maximum flows.

14.15E For a Parshall flume with a throat width of 8 ft, calculate the head H corresponding to the minimum and maximum flows. Plot a graph of Q versus H, using five values of H spaced approximately equally between the minimum and maximum.

14.16E A flow rate of 50 ft³/s falls within the range of both the 4-ft and the 10-ft wide Parshall flume. Compare the head H for this flow rate in each size.

COMPUTER PROGRAMMING ASSIGNMENTS

1. Write a program using Eq. (14–5) to compute the volume flow rate for any variable head meter. Include the computation of the area in the main pipe, the area in the throat, the diameter ratio β, and the Reynolds number. Include the discharge coefficient C for the venturi tube if the dimensions of the venturi tube are in the standard ranges given in Section 14.3.1. Otherwise ask the user to input a value for C. Use Eq. (14–6) to compute the discharge coefficient for a nozzle. For the orifice, prompt the user to find the value of C from Fig. 14.7 when given the Reynolds number and the diameter ratio. Permit the user to input the differential pressure in SI units (Pascals), U.S. Customary System units (psi), or in terms of the deflection of a differential manometer with a known gage fluid.

2. Write a program to accept data for the ten measurements required to complete a traverse of a circular pipe using a pitot tube as shown in Fig. 14.19. Compute the velocity of flow for each point by using Eq. (14–11). Then compute the average of ten values to determine the average velocity. Finally, compute the volume flow rate from $Q = Av$.

3. Write a program to compute the discharge through a rectangular weir by using Eq. (14–14); through a contracted weir by using Eq. (14–15); and through a V-notch weir by using Eqs. (14–18) and (14–19).

4. Write a program to compute the discharge through any of the standard Parshall flumes listed in Table 14.2.

5. Use the program from Assignment 3 to solve Practice Problems 14.10, 14.11, 14.12, and 14.13.

6. Use the program from Assignment 4 to solve Practice Problems 14.14, 14.15, and 14.16.

■■■■ 15 Pump Selection and Application

A wide variety of pumps is available to transport liquids in fluid flow systems. The proper selection and application of pumps require an understanding of their performance characteristics and typical uses.

After completing this chapter, you should be able to:

1. List the parameters involved in pump selection.
2. List the types of information that must be specified for a given pump.
3. Describe the basic pump classifications.
4. List four types of rotary positive displacement pumps.
5. List three types of reciprocating positive displacement pumps.
6. List three types of kinetic pumps.
7. Describe the main features of centrifugal pumps.
8. Describe *deep well jet pumps* and *shallow well jet pumps*.
9. Describe the typical performance curve for rotary positive displacement pumps.
10. Describe the typical performance curve for centrifugal pumps.
11. State the *affinity laws* for centrifugal pumps as they relate to the relationships among speed, impeller diameter, capacity, total head capability, and power required to drive the pump.
12. Describe how the operating point of a pump is related to the system resistance curve.
13. Define the *net positive suction head* (*NPSH*) for a pump, and discuss its significance in pump performance.
14. Describe the importance of the vapor pressure of the fluid in relation to the NPSH.
15. Compute the NPSH available for a given suction line design and a given fluid.
16. Define the *specific speed* for a centrifugal pump, and discuss its relationship to pump selection.
17. Describe the effect of increased viscosity on the performance of centrifugal pumps.
18. Describe the performance of parallel pumps and pumps connected in series.
19. Describe the features of a desirable suction line design.
20. Describe the features of a desirable discharge line design.

15.2
PARAMETERS
INVOLVED IN PUMP
SELECTION

When selecting a pump for a particular application, the following factors must be considered:

1. The nature of the liquid to be pumped
2. The required capacity (volume flow rate)
3. The conditions on the suction (inlet) side of the pump
4. The conditions on the discharge (outlet) side of the pump
5. The total head on the pump (the term h_a from the energy equation)
6. The type of system to which the pump is delivering the fluid
7. The type of power source (electric motor, diesel engine, steam turbine, etc.)
8. Space, weight, and position limitations
9. Environmental conditions
10. Cost of pump purchase and installation
11. Cost of pump operation
12. Governing codes and standards

The nature of the fluid is characterized by its temperature at the pumped condition, its specific gravity, its viscosity, its tendency to corrode or erode the pump parts, and its vapor pressure at the pumping temperature. The term *vapor pressure* is used to define the pressure at the free surface of a fluid due to the formation of a vapor. The vapor pressure gets higher as the temperature of the liquid increases, and it is essential that the pressure at the pump inlet stay above the vapor pressure of the fluid. We will learn more about vapor pressure in Section 15.10.

After pump selection, the following items must be specified:

1. Type of pump and manufacturer
2. Size of pump
3. Size of suction connection and type (flanged, screwed, etc.)
4. Size and type of discharge connection
5. Speed of operation
6. Specifications for driver (for example: for an electric motor—power required, speed, voltage, phase, frequency, frame size, enclosure type)
7. Coupling type, manufacturer, and model number
8. Mounting details
9. Special materials and accessories required, if any
10. Shaft seal design and seal materials

Pump catalogs and manufacturer's representatives will supply the necessary information to assist in the selection and specification of pumps and accessory equipment.

15.3
TYPES OF PUMPS

The types of pumps commonly used for fluid delivery can be classified as shown in Table 15.1.

15.3.1
Positive Displacement
Pumps

Positive displacement pumps deliver a fixed quantity of fluid with each revolution of the pump rotor. Therefore, except for minor slippage because of clearances between the rotor and the casing, the delivery or capacity of

TABLE 15.1 Classification of types of pumps

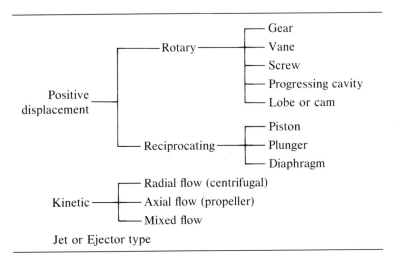

the pump is unaffected by changes in the pressure it must develop. Most positive displacement pumps can also handle liquids with high viscosities.

In Chapter 7, Figs. 7.2 and 7.3 show two types of positive displacement pumps commonly used for fluid power applications. The gear pump (Fig. 7.2) is comprised of two counter-rotating, tightly meshing gears rotating within a housing. The outer periphery of the gear teeth fit closely with the inside surface of the housing. Fluid is drawn in from the supply reservoir at the suction port and carried by the spaces between teeth to the discharge port where it is delivered at high pressure to the system. The delivery pressure is dependent on the resistance of the system. A cutaway of a commercially available gear pump is shown in part (a) of the figure. Gear pumps develop system pressures in the range of 1500 psi to 4000 psi (10.3 MPa to 27.6 MPa). Delivery varies with the size of the gears and the rotational speed which can be up to 4000 rpm. Deliveries from 1 to 50 gallons per minute (4 to 190 L/min) are possible with different size units.

Figure 7.3 shows an axial piston pump that uses a rotating swashplate that acts like a cam to reciprocate the pistons. The pistons alternately draw fluid into their cylinders through suction valves and then force it out the discharge valves against system pressure. Delivery can be varied from zero to maximum by changing the angle of the swashplate and thus changing the stroke of the pistons. Pressure capacity ranges up to 5000 psi (34.5 MPa).

Also used for fluid power, the vane pump (Fig. 15.1) consists of an eccentric rotor containing a set of sliding vanes that ride inside a housing. A cam ring in the housing controls the radial position of the vanes. Fluid enters the suction port at the left, is then captured in a space between two successive vanes, and is thus carried to the discharge port at the system pressure. The vanes then are retracted into their slots in the rotor as they travel back to the inlet, or suction, side of the pump. Variable displacement vane pumps can deliver from zero to the maximum flow rate by varying the position of the rotor with respect to the cam ring and the housing. The setting of the

FIGURE 15.1 Vane pump. (Source of drawing: *Machine Design Magazine*)

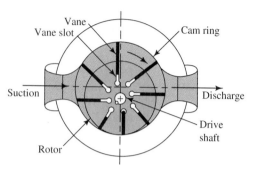

variable delivery can be manual, electric, hydraulic, or pneumatically actuated to tailor the performance of the fluid power unit to the needs of the system being driven. Typical pressure capacities are from 2000 to 4000 psi (13.8 to 27.6 MPa).

One disadvantage of the gear, piston, and vane pumps is that they deliver a pulsating flow to the output because each functional element moves a set, captured volume of fluid from suction to discharge. Screw pumps and progressing cavity pumps do not have this problem. Figure 15.2 shows a screw pump in which the central, thread-like power rotor meshes closely with the two idler rotors creating an enclosure inside the housing that moves axially from suction to discharge providing a continuous uniform flow. Screw pumps operate at nominally 3000 psi (20.7 MPa), can be run at high speeds, and run more quietly than most other types of hydraulic pumps.

The Moyno® progressing cavity pump shown in Fig. 15.3 also produces a smooth, nonpulsating flow and is used mostly for the delivery of process fluids rather than hydraulic applications. As the long central rotor turns within the stator, cavities are formed which progress toward the discharge end of the pump carrying the material being handled. The rotor is typically made from steel plated with heavy layers of hard chrome to increase resistance to abrasion. For most applications, stators are made from natural rubber or any of several types and formulations of synthetic rubbers. A compression fit exists between the metal rotor and the rubber stator to reduce slippage and improve efficiency. Delivery for a given pump is dependent on the dimensions of the rotor/stator combination and is proportional to the speed of rotation. Flow capacities range up to 1860 gal/min (7040 L/min) and pressure capability is up to 900 psi (6.2 MPa). This type of pump can handle a wide variety of fluids including clear water, slurries with heavy solids content, highly viscous liquids like adhesives and cement grout, abrasive fluids such as slurries of silicon carbide or ground limestone, pharmaceuticals such as shampoo and skin cream, corrosive chemicals such as cleaning solutions and fertilizers, and foods such as applesauce and even bread dough.

The lobe pump (Fig. 15.4), sometimes called a cam pump, operates in a similar fashion to the gear pump. The two counter-rotating rotors may have

FIGURE 15.2 Screw pump. (Source: IMO Industries Inc., IMO Pump Division, Monroe, NC)

(a) Cutaway of pump assembly

(b) Power rotor, idler rotors, and housing

two, three, or more lobes that mesh with each other and fit closely with the housing. Fluid is conducted around by the cavity formed between successive lobes.

Piston pumps used for fluid transfer are classified as either single-acting simplex or double-acting duplex types as shown in Fig. 15.5. In principle these are similar to the fluid power piston pumps but they typically have a larger flow capacity and operate at lower pressures. Also, they are usually driven through a crank-type drive rather than the swashplate described before.

FIGURE 15.3   Moyno® progressing cavity pump. (Source: Robbins & Myers, Inc., Fluids Handling Group, Springfield, OH)

FIGURE 15.4   Lobe pump.

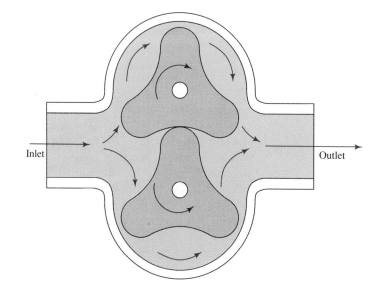

FIGURE 15.5   Piston pumps for fluid transfer.

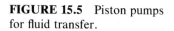

(a) Single acting—simplex

(b) Double acting—duplex

In the diaphragm pump shown in Fig. 15.6, a reciprocating rod moves a flexible diaphragm within a cavity, alternately discharging fluid as the rod moves to the left and drawing fluid in as it moves to the right. One advantage of this type of pump is that only the diaphragm contacts the fluid, eliminating contamination from the drive elements.

FIGURE 15.6 Diaphragm-type pump.

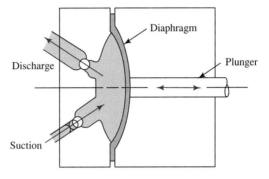

15.3.2 Kinetic Pumps

Kinetic pumps add energy to the fluid by accelerating it through the action of a rotating *impeller*. Figure 15.7 shows the basic configuration of a radial flow centrifugal pump, the most common type of kinetic pump. The fluid is drawn into the center of the impeller and then thrown outward by the vanes. Leaving the impeller, the fluid passes through a spiral-shaped volute where it is gradually slowed, causing part of the kinetic energy to be converted to fluid pressure.

The propeller type of pump (axial flow) depends on the hydrodynamic action of the propeller blades to lift and accelerate the fluid axially, along a path parallel to the axis of the propeller. The mixed flow pump incorporates some actions from both the centrifugal and propeller types.

Figure 15.8 shows the basic design of radial, axial, and mixed flow impellers.

Jet pumps, frequently used for household water systems, are composed of a centrifugal pump along with a jet or ejector assembly. Figure 15.9 on p. 416 shows a typical deep-well jet pump configuration where the main pump and motor are located above ground at the top of the well and the jet assembly is down near the water level. The pump delivers water under pressure down into the well through the pressure pipe to a nozzle. The jet issuing from the nozzle creates a vacuum behind it, which causes well water to be drawn up along with the jet. The combined stream passes through a diffuser where the flow is slowed, thus converting some of the kinetic energy of the water to pressure. Since the diffuser is inside the suction pipe, the

(a) Pump and
motor

(b) Cutaway of pump

(c) Impeller

FIGURE 15.7   Centrifugal pump. (Source: Gould Pumps, Inc., Seneca Falls, NY)

FIGURE 15.8 Impellers for kinetic pumps.

(a) Radial flow impeller

(b) Mixed flow impeller

(c) Axial flow impeller (propeller)

water is carried to the inlet of the pump where it is acted on by the impeller. Part of the output is discharged to the system being supplied, while the remainder is recirculated to the jet to continue the operation.

If the well is shallow, with less than about 6.0 m (20 ft) from the pump to the water level, the jet assembly can be built into the pump body. Then the water is lifted through a single suction pipe, as shown in Fig. 15.10.

Submersible pumps are designed so the entire assembly of the centrifugal pump, the drive motor, and the suction and discharge apparatus can be submerged in the fluid to be pumped. Figure 15.11 shows one design that is portable and that can be installed in a pipe casing because of its small-diameter cylindrical housing. These pumps are useful for removing unwanted water from construction sites, mines, utility manholes, industrial tanks, and shipboard cargo holds. The suction for the pump is at the bottom where the water flows through a strainer and into the eye of the abrasion resistant impeller. The discharge flows upward through an annular passage between the shell and the motor housing. At the top of the unit the flow rejoins and flows into a centrally located discharge pipe or hose. The dry motor is sealed in the center of the pump.

FIGURE 15.9 Deep-well jet pump. (Source: Goulds Pumps, Inc. Seneca Falls, NY)

FIGURE 15.10 Shallow-well jet pump. (Source: Goulds Pumps, Inc. Seneca Falls, NY)

FIGURE 15.11   Portable submersible pump. (Source: Burks Pumps Inc., Piqua, OH)

Discharge port

Electrical connection

Discharge flow path

Motor

Impeller

Suction strainer

(a) Cutaway view   (b) Exterior view

Although most of the centrifugal pump styles discussed thus far have been fairly large and designed for commercial and industrial applications, small units are available for use in small appliances such as clothes washers and dishwashers and for small-scale products. Figure 15.12 shows one such design that is only about 4 inches in diameter with a capacity up to 30 gal/min (114 L/min). Even smaller pumps are available.

FIGURE 15.12   Small centrifugal pump with integral motor. (Source: Gorman-Rupp Industries, Bellville, OH)

15.4
PERFORMANCE DATA
FOR POSITIVE
DISPLACEMENT PUMPS

In this section we will discuss the general characteristics of direct-acting reciprocating pumps, and rotary pumps.

The operating characteristics of positive displacement pumps make them useful for handling such fluids as water, hydraulic oils in fluid power systems, chemicals, paint, gasoline, greases, adhesives, and some food products. Since delivery is proportional to the rotational speed of the rotor, these pumps can be used for metering. Some disadvantages of some designs include pulsating output, susceptibility to damage by solids and abrasives, and need for a relief valve. In general, they are used for high pressure applications requiring a relatively constant delivery.

15.4.1
Reciprocating Pumps

In its simplest form, the reciprocating pump (Figure 15.5) employs a piston that draws fluid into a cylinder through an intake valve as the piston draws away from the valve. Then, as the piston moves forward, the intake valve closes and the fluid is pushed out through the discharge valve. Such a pump is called *simplex,* and its curve of discharge versus time would look like that shown in Fig. 15.13(a). The resulting intermittent delivery is often undesirable. If the piston is *double acting* or *duplex,* one side of the piston delivers fluid while the other takes fluid in, resulting in the performance curve shown in Fig. 15.13(b). The delivery can be smoothed even more by having three or more pistons. Piston pumps for hydraulic systems often have five or six pistons.

FIGURE 15.13   Simplex and duplex pump delivery.

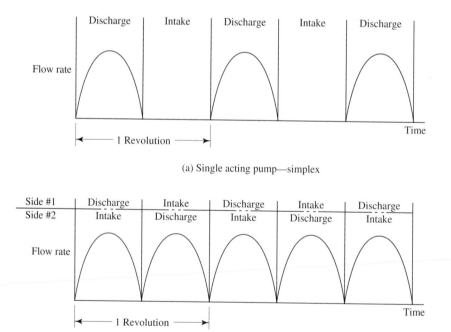

(a) Single acting pump—simplex

(b) Double acting pump—duplex

15.4.2
Rotary Pumps

Figure 15.14 shows a typical performance curve for rotary pumps such as gear, vane, screw, and lobe pumps. It is a plot of capacity and power versus discharge pressure. As pressure is increased, a slight decrease in capacity occurs, due to internal leakage from the high pressure side to the low pressure side. This is often insignificant. The power required to drive the pump varies almost linearly with pressure. Also, because of the positive displacement designs for rotary pumps, capacity varies almost linearly with the rotational speed, provided the suction conditions allow free flow into the pump.

FIGURE 15.14 Performance curves for a positive displacement rotary pump.

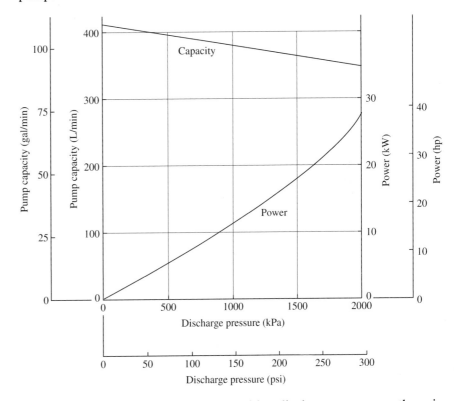

15.5
PERFORMANCE DATA FOR CENTRIFUGAL PUMPS

Because centrifugal pumps are not positive displacement types, there is a strong dependency between capacity and the pressure that must be developed by the pump. This makes their performance ratings somewhat more complex. The typical rating curve plots the total head on the pump h_a versus the capacity or discharge Q, as shown in Fig. 15.15. The total head h_a is calculated from the general energy equation, as described in Chapter 7. It represents the amount of energy added to a unit weight of the fluid as it passes through the pump.

As shown in Fig. 15.7, there are large clearances between the rotating impeller and the casing of the pump. This accounts for the decrease in capacity as the total head increases. Indeed, at a cut-off head, the flow is stopped completely when all of the energy input from the pump goes to

FIGURE 15.15 Performance curve for a centrifugal pump—Total head versus capacity.

maintain the head. Of course, the typical operating head is well below the cut-off head so that high capacity can be achieved.

Efficiency and power required are also important to the successful operation of a pump. Figure 15.16 shows a more complete performance rating of a pump, superimposing head, efficiency, and power curves and plotting all three versus capacity. Normal operation should be in the vicinity

FIGURE 15.16 Centrifugal pump performance curves.

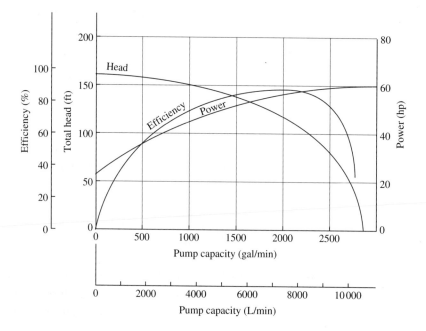

of the peak of the efficiency curve, with efficiencies in the range of 60 percent to 80 percent being typical for centrifugal pumps.

15.6 AFFINITY LAWS FOR CENTRIFUGAL PUMPS

Most centrifugal pumps can be operated at different speeds to obtain varying capacities. Also, a given size pump casing can accommodate impellers of differing diameters. It is important to understand the manner in which capacity, head, and power vary when either speed or impeller diameter is varied. These relationships, called *affinity laws,* are listed here. The symbol N refers to the rotational speed of the impeller, usually in revolutions per minute (r/min, or rpm).

When *speed varies,*

a. Capacity varies directly with speed:

$$\frac{Q_1}{Q_2} = \frac{N_1}{N_2} \quad\quad \textbf{(15–1)}$$

b. The total head capability varies with the square of the speed:

$$\frac{h_{a_1}}{h_{a_2}} = \left(\frac{N_1}{N_2}\right)^2 \quad\quad \textbf{(15–2)}$$

c. The power required by the pump varies with the cube of the speed:

$$\frac{P_1}{P_2} = \left(\frac{N_1}{N_2}\right)^3 \quad\quad \textbf{(15–3)}$$

When *impeller diameter varies,*

a. Capacity varies directly with impeller diameter:

$$\frac{Q_1}{Q_2} = \frac{D_1}{D_2} \quad\quad \textbf{(15–4)}$$

b. The total head varies with the square of the impeller diameter:

$$\frac{h_{a_1}}{h_{a_2}} = \left(\frac{D_1}{D_2}\right)^2 \quad\quad \textbf{(15–5)}$$

c. The power required by the pump varies with the cube of the impeller diameter:

$$\frac{P_1}{P_2} = \left(\frac{D_1}{D_2}\right)^3 \quad\quad \textbf{(15–6)}$$

Efficiency remains nearly constant for speed changes and for small changes in impeller diameter. (See Reference 2.)

□ **EXAMPLE PROBLEM 15.1** Assume that the pump for which the performance data are plotted in Fig. 15.16 was operating at a rotational speed of 1750 rpm and that the impeller diameter was 13 inches. First determine the head that would result in a capacity of 1500 gal/min and the power required to drive the pump. Then, compute the performance at a speed of 1250 rpm.

Solution From Fig. 15.16, projecting upward from $Q_1 = 1500$ gal/min gives:

Total head = 130 ft = h_{a_1}

Power required = 50 hp = P_1

When the speed is changed to 1500 rpm, the new performance can be computed by using the affinity laws.

Capacity: $Q_2 = Q_1(N_2/N_1) = 1500(1250/1750) = 1071$ gal/min

Head: $h_{a_2} = h_{a_1}(N_2/N_1)^2 = 130(1250/1750)^2 = 66.3$ ft

Power: $P_2 = P_1(N_2/N_1)^3 = 50(1250/1750)^3 = 18.2$ hp

Note the significant decrease in the power required to run the pump. If the capacity and the available head are adequate, large savings in energy costs can be obtained by varying the speed of operation of a pump.

**15.7
MANUFACTURER'S
DATA FOR
CENTRIFUGAL PUMPS**

Because they are able to use different impeller diameters and speeds, pump manufacturers can cover a wide range of requirements for capacity and head with a few basic pump sizes. Figure 15.17 shows a composite rating chart for one line of pumps which allows the quick determination of the pump size. Then, for each pump size, more complete performance charts are prepared as shown next.

**15.7.1
Effect of Impeller Size**

Figure 15.18 shows how the performance of a given pump varies as the size of the impeller varies. The $2 \times 3 - 10$ centrifugal pump is one with a 2-in discharge connection, a 3-in suction connection, and a casing that can accommodate an impeller with a diameter of 10 in or smaller. Shown are the

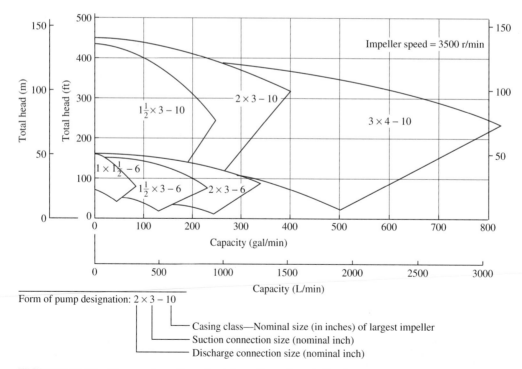

Form of pump designation: $2 \times 3 - 10$

— Casing class—Nominal size (in inches) of largest impeller
— Suction connection size (nominal inch)
— Discharge connection size (nominal inch)

FIGURE 15.17   Composite rating chart for a line of centrifugal pumps.

FIGURE 15.18 Illustration of pump performance for different impeller diameters.

capacity versus head curves for five different sizes of impellers in this same casing. The operating speed is 3500 rpm, which corresponds to the full load speed of a 2-pole electric motor.

15.7.2 Effect of Speed

Figure 15.19 is the performance of the same 2 × 3 − 10 pump operating at 1750 rpm (a standard 4-pole motor speed) instead of 3500 rpm. If we compare the maximum total heads for each impeller size, we illustrate the affinity law; that is, doubling the speed increases the total head capability by a factor of 4 (the square of the speed ratio). If the curves are extrapolated down to the zero total head point where the maximum capacity occurs, we see demonstrated that capacity doubles as the speed doubles.

15.7.3 Power Required

Figure 15.20 is the same as Fig. 15.18, except that the curves showing the power required to drive the pump have been added. For example, the pump with an 8-in impeller would deliver 220 gal/min against a total head of 250 ft of fluid. Under these conditions, the pump would draw 25 hp. The same pump would deliver 290 gal/min at 200 ft of head and would draw 28 hp.

15.7.4 Efficiency

Figure 15.21 is the same as Fig. 15.18, except that curves of constant efficiency have been added. The maximum efficiency for this pump is approximately 57 percent. Of course, it is desirable to operate a given pump near its peak efficiency point.

15.7.5 Net Positive Suction Head Required

Net positive suction head required ($NPSH_R$) is an important factor to consider in applying a pump, as will be discussed in Section 15.10. $NPSH_R$ is related to the pressure at the inlet to the pump. For this discussion, it is

FIGURE 15.19 Pump performance for 2 × 3 − 10 centrifugal pump operating at 1750 rpm.

FIGURE 15.20 Illustration of pump performance for different impeller diameters with power required. Performance chart for 2 × 3 − 10 centrifugal pump at 3500 rpm.

FIGURE 15.21 Illustration of pump performance for different impeller diameters with efficiency. Performance chart for $2 \times 3 - 10$ centrifugal pump at 3500 rpm.

sufficient to say that a low $NPSH_R$ is desirable. For the pump in Fig. 15.22, the range is from about 4.5 ft of fluid at low capacities to over 12 ft of fluid at higher capacities.

15.7.6 Composite Performance Chart

Figure 15.23 puts all these data together on one chart so the user can see all important parameters at the same time. The chart seems complicated at first, but, considering each individual part separately helps to interpret it correctly.

☐ **EXAMPLE PROBLEM 15.2**

A centrifugal pump must deliver at least 250 gal/min of water at a total head of 300 ft of water. Specify a suitable pump. List its performance characteristics.

Solution

One possible solution can be found from Fig. 15.23. The $2 \times 3 - 10$ pump with a 9-in impeller will deliver approximately 275 gal/min at 300 ft of head. At this operating point, the efficiency would be 57 percent, near the maximum for this type of pump. Approximately 37 hp would be required. The $NPSH_R$ at the suction inlet to the pump is approximately 9.2 ft of water.

15.7.7 Additional Performance Charts

Figures 15.24 to 15.29 show the composite performance charts for six other centrifugal pumps. They range in size from $1\frac{1}{2} \times 3 - 6$ to $6 \times 8 - 17$. Maximum capacities range from approximately 130 gal/min (492 L/min) up to nearly 4000 gal/min (15 140 L/min). A total head up to 700 ft (213 m) of fluid can be developed within the pumps in these figures. Note that Figs. 15.24 to 15.27 are for pumps operating at approximately 1750 rpm while Figs. 15.28 and 15.29 are for 3560 rpm.

FIGURE 15.22 Illustration of pump performance for different impeller diameters with net positive suction head required. Performance chart for 2 × 3 − 10 centrifugal pump at 3500 rpm.

FIGURE 15.23 Composite pump performance chart for 2 × 3 − 10 centrifugal pump at 3500 rpm.

FIGURE 15.24 Performance for $1\frac{1}{2} \times 3 - 6$ centrifugal pump at 1750 rpm. (Source: Goulds Pumps, Inc., Seneca Falls, NY)

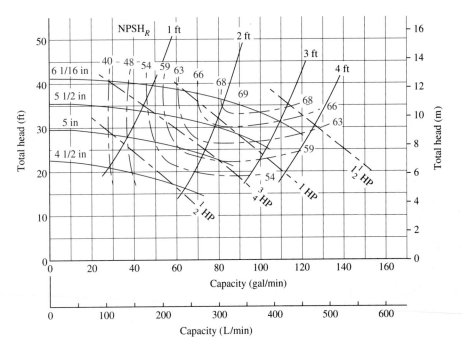

FIGURE 15.25 Performance for $3 \times 4 - 10$ centrifugal pump at 1750 rpm. (Source: Goulds Pumps, Inc., Seneca Falls, NY)

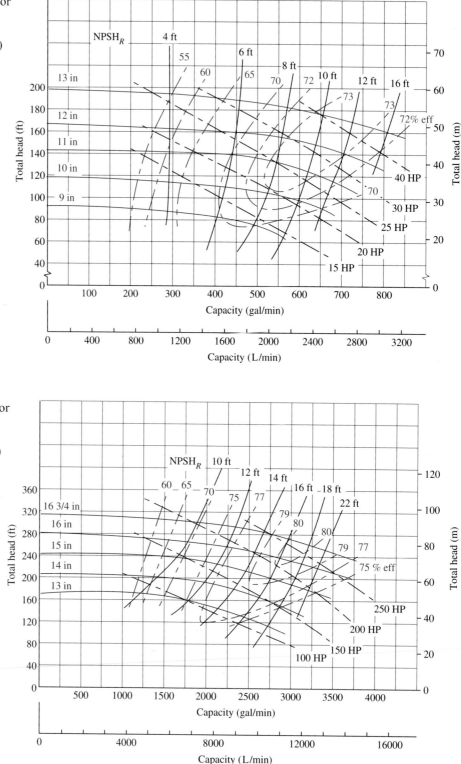

FIGURE 15.26 Performance for 3 × 4 − 13 centrifugal pump at 1780 rpm. (Source: Goulds Pumps, Inc., Seneca Falls, NY)

FIGURE 15.27 Performance for 6 × 8 − 17 centrifugal pump at 1780 rpm. (Source: Goulds Pumps, Inc., Seneca Falls, NY)

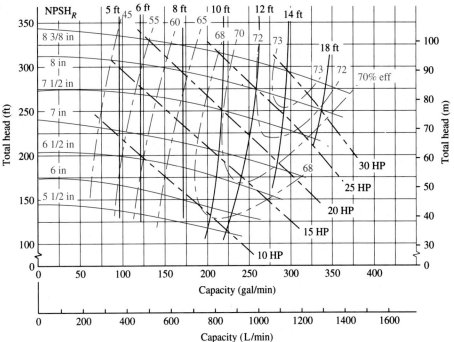

FIGURE 15.28 Performance for 2 × 3 − 8 centrifugal pump at 3560 rpm. (Source: Goulds Pumps, Inc., Seneca Falls, NY)

FIGURE 15.29 Performance for 1½ × 3 − 13 centrifugal pump at 3560 rpm. (Source: Goulds Pumps, Inc., Seneca Falls, NY)

15.8
THE OPERATING
POINT OF A PUMP

The operating point of a pump gives its capacity at a given total head. Curve A in Fig. 15.30 illustrates how a given piping system would behave in terms of the friction in the system as a function of the flow rate Q. Recall that the energy loss in a piping system, and therefore the head which must be developed by a pump, increases approximately as the square of the flow rate. Assume that curve A represents a system with all regulating valves open. As one or more valves is partially closed, a greater resistance is developed and the system curve would look like curve B.

FIGURE 15.30 Operating point of a pump.

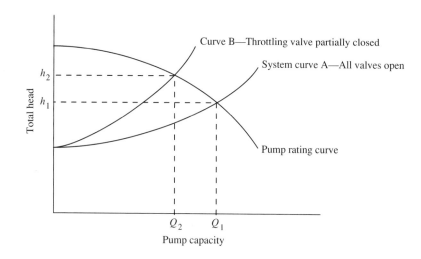

Curve B—Throttling valve partially closed

System curve A—All valves open

h_2

h_1

Total head

Pump rating curve

Q_2 Q_1

Pump capacity

Superimposing the system curves onto a typical pump rating curve shows how flow control is accomplished. With all valves open, the intersection between system curve A and the pump curve would be the operating point, a capacity of Q_1 at a head h_1 in this case. Partially closing a valve would move the operating point to the intersection of curve B and the pump curve, Q_2 at h_2. In the process, energy would be dissipated at the valve and wasted. Therefore, you should select a pump with care so that little, if any, throttling with valves is required.

15.8.1
Variable Speed Drives

When a fluid transfer system must operate at a variety of capacities, a throttling valve is often used as depicted in Fig. 15.30. The pump is sized for the largest capacity anticipated. If the delivery is lowered from Q_1 to Q_2, the energy represented by $h_2 - h_1$ is dissipated as the fluid passes through the valve. This energy is first delivered to the pump by the drive motor, then transferred to the fluid in the pump, and thus is wasted. High energy costs make it desirable to modify this manner of operation.

Variable speed drives offer an attractive alternative to throttling. Several types of mechanical variable speed drives and a variable frequency electronic control for a standard AC electric motor are available. The standard frequency for AC power in the United States and many other countries is 60 hertz (Hz), or 60 cycles per second. In Europe and some other countries, 50 Hz is standard. Because the speed of an AC motor is directly proportional to the frequency of the AC current, varying the frequency causes the motor speed to vary. Also, because of the affinity laws, as the motor speed decreases, its capacity decreases, which allows the pump to operate at the desired delivery without throttling. Further benefit is obtained because the power required by the pump decreases in proportion to the speed reduction ratio cubed. Of course, the variable speed drive is more expensive than a standard motor alone, and the overall economics of the system with time should be evaluated.

**15.8.2
Effect of Fluid Viscosity**

The performance rating curves for centrifugal pumps, such as those shown in Figs. 15.23 to 15.29, are generated from test data using water as the fluid. These curves are reasonably accurate for any fluid with a viscosity similar to that of water. However, pumping fluids with higher viscosities causes the capacity of the pump at a given head to decrease. Also the power required to drive the pump increases, and the efficiency decreases. Figure 15.31 illustrates the effect of increasing viscosity. Reference 4 gives data for correction factors that can be used to compute expected performance with fluids of different viscosities.

FIGURE 15.31   Effect of increased viscosity on pump performance.

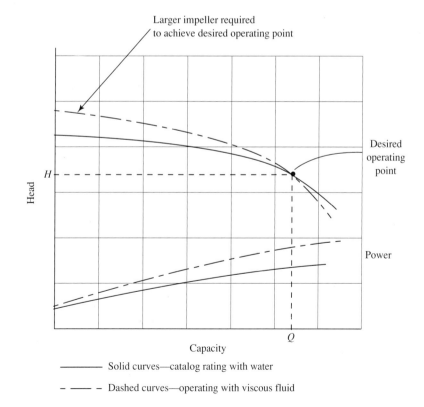

Larger impeller required
to achieve desired operating point

Desired
operating
point

Power

H

Head

Capacity

Q

———— Solid curves—catalog rating with water

– – – – Dashed curves—operating with viscous fluid

15.8.3
Operating Pumps in Parallel

Using two identical pumps to draw fluid from the same source and deliver it against the same system head doubles the flow rate delivered (see Fig. 15.32). This method is used when the desired capacity is beyond the range of any single suitable pump. This method also provides flexibility in the operation of the system, because one of the pumps can be shut down during low demand times or for service.

FIGURE 15.32   Performance of two pumps in parallel.

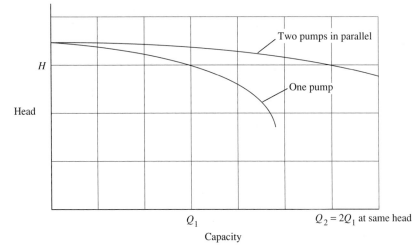

15.8.4
Operating Pumps in Series

Directing the output of one pump to the inlet of a second pump allows the same capacity to be obtained at a total head equal to the sum of the ratings of the two pumps. This method permits operation against unusually high heads.

15.8.5
Multistage Pumps

A performance similar to that achieved by using pumps in series can be obtained by using multistage pumps. Two or more impellers are arranged in the same housing in such a way that the fluid flows successively from one to the next. Each stage increases the fluid pressure, so that a high total head can be developed.

15.9
PUMP SELECTION

Figure 15.33 shows one method for deciding what type of pump is suitable for a given service. Some general conclusions can be drawn from such a chart, but it should be emphasized that boundaries between zones are approximate. Two or more types of pumps may give satisfactory service under the same conditions. Such factors as cost, physical size, suction conditions, and the type of fluid may dictate a particular choice. In general:

1. Reciprocating pumps are used for flow rates up to about 500 gal/min and from very low heads to as high as 50 000 ft of head.
2. Centrifugal pumps are used over a wide range of conditions, mostly in high capacity, moderate head applications.

FIGURE 15.33 Pump selection chart.

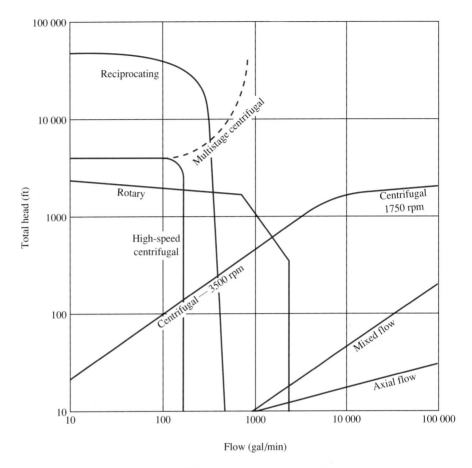

(Adapted from data in reference 5)

3. Single-stage centrifugal pumps operating at 3500 rpm are economical at lower flow rates and moderate heads.
4. Multistage pumps are desirable at high head conditions.
5. Rotary pumps (e.g., gear, vane, etc.) are used in applications requiring moderate capacities and high heads or for fluids with high viscosities.
6. Special high speed centrifugal pumps operating well above the 3500-rpm speed of standard electrical motors are desirable for high heads and moderate capacities. Such pumps are sometimes driven by steam turbines or gas turbines.
7. Mixed flow and axial flow pumps are used for very high flow rates and low heads.

Another parameter that is useful in selecting the type of pump for a given application is the *specific speed,* defined as

$$N_s = \frac{N\sqrt{Q}}{H^{3/4}} \qquad \textbf{(15–7)}$$

where N = rotational speed of the impeller (rpm)
             Q = flow rate through the pump (gal/min)
             H = total head on the pump (ft)

Other units can be used.

The specific speed is often combined with the *specific diameter* to produce a chart like that shown in Fig. 15.34. The specific diameter is

$$D_s = \frac{DH^{1/4}}{\sqrt{Q}} \qquad (15\text{--}8)$$

where D is the impeller diameter in inches. The other terms are as defined before.

From Fig. 15.34 we can see that radial flow centrifugal pumps are recommended for specific speeds from about 400 to 4000. Mixed flow pumps are used from 4000 to about 7000. Axial flow pumps are used from 7000 to over 60 000.

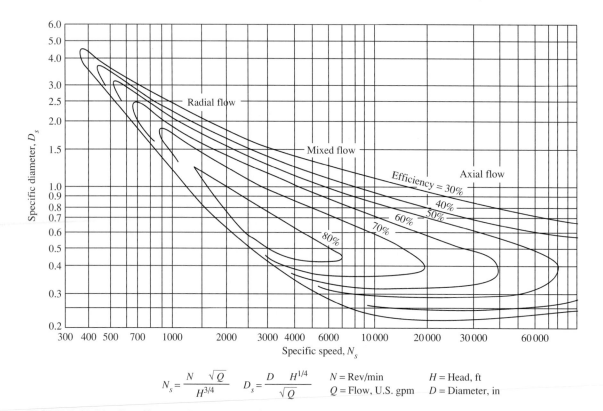

$$N_s = \frac{N\,\sqrt{Q}}{H^{3/4}} \qquad D_s = \frac{D\,H^{1/4}}{\sqrt{Q}} \qquad \begin{array}{ll} N = \text{Rev/min} & H = \text{Head, ft} \\ Q = \text{Flow, U.S. gpm} & D = \text{Diameter, in} \end{array}$$

FIGURE 15.34 Specific speed versus specific diameter for centrifugal pumps— An aid to pump selection. (Excerpted by special permission from *Chemical Engineering,* April 3, 1978. Copyright © 1978 by McGraw-Hill, Inc., New York, N.Y. 10020.)

**15.10
NET POSITIVE
SUCTION HEAD**

In addition to the total head, capacity, power, and efficiency requirements we have discussed, we must also consider that the condition at the inlet of a pump is critical. The inlet, or suction, system must allow a smooth flow of liquid to enter the pump at a sufficiently high pressure to avoid creating vapor bubbles in the fluid. As the pressure on a fluid decreases, the temperature at which vapor bubbles form (like boiling) also decreases. Therefore, it is essential that the suction pressure at the pump inlet be well above the pressure at which vaporization would occur for the operating temperature of the liquid. This is called providing a *net positive suction head,* sometimes abbreviated *NPSH.*

If the suction pressure is allowed to decrease to the point where vaporization occurs, *cavitation* is created inside the pump. Instead of a steady flow of liquid, the pump will draw a mixture of liquid and vapor, causing the delivery to decrease. Furthermore, as the vapor bubbles proceed through the pump, they encounter higher pressures which cause the bubbles to collapse rapidly. Excessive noise, vibration, and greatly increased wear of pump parts would result.

Pump manufacturers supply data about the *required* net positive suction head for satisfactory operation. The person selecting the pump must then ensure that there is a sufficiently high *NPSH available.* That is,

$$NPSH \text{ available} > NPSH \text{ required} \qquad \textbf{(15–9)}$$

The value of *NPSH* available is dependent on the nature of the fluid being pumped, the suction piping, the location of the fluid reservoir, and the pressure applied to the fluid in the reservoir. This can be expressed as

$$NPSH_a = h_{sp} \pm h_s - h_f - h_{vp} \qquad \textbf{(15–10)}$$

Refer to Fig. 15.35 for an illustration of these terms:

h_{sp} = static pressure head (absolute) applied to the fluid, expressed in *meters (or feet) of the liquid*

h_s = elevation difference from the level of fluid in the reservoir to the pump inlet, expressed in *meters or feet*
If pump is below reservoir, h_s is positive. (Preferred.)
If pump is above reservoir, h_s is negative.

h_f = friction loss in suction piping, expressed in *meters or feet*

h_{vp} = vapor pressure of the liquid at the pumping temperature in *meters (or feet) of the liquid*

Figure 15.36 shows a graph of the vapor pressure of water versus temperature.

Equation (15–10) can be developed by applying the energy equation to the suction piping system. Be careful to include pressure head in terms of *absolute* pressure. Strictly, Eq. (15–10) should have the velocity head for

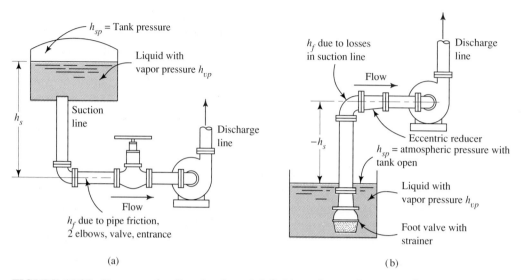

FIGURE 15.35   Pump suction line details and definition of terms for computing *NPSH*.

the fluid in the suction line added to the right side of the equation. However, most U.S. pump manufacturers already account for the velocity head in reporting the *NPSH required* by the pump. Therefore, it is not necessary to include it in the calculation of *NPSH available* (Reference 5, page 322). In critical installations, this matter should be checked with the pump supplier.

FIGURE 15.36   Vapor pressure of water.

15.10.1
Effect of Pump Speed
on *NPSH*

The data given in pump catalogs for *NPSH* are for water and apply only to the listed operating speed. If the pump is operated at a different speed, the *NPSH* required at the new speed can be calculated from

$$(NPSH_R)_2 = \left(\frac{N_2}{N_1}\right)^2 (NPSH_R)_1$$

where the subscript 1 refers to catalog data and the subscript 2 refers to conditions at the new operating speed. The pump speed in rpm is *N*.

☐ **EXAMPLE PROBLEM 15.3**

Determine the available *NPSH* for the system shown in Fig. 15.35(a). The fluid reservoir is a closed tank with a pressure of −20 kPa above water at 70°C. The atmospheric pressure is 100.5 kPa. The water level in the tank is 2.5 m above the pump inlet. The pipe is a 1½-in Schedule 40 pipe with a total length of 12.0 m. The elbow is standard and the valve is a fully open globe valve. The flow rate is 95 L/min.

Solution

Using Eq. (15–10), first find h_{sp}:

$$\text{absolute pressure} = \text{atmospheric pressure} - \text{tank pressure}$$
$$p_{abs} = 100.5 \text{ kPa} - 20 \text{ kPa} = 80.5 \text{ kPa}$$

But we know that

$$h_{sp} = p_{abs}/\gamma$$
$$= \frac{80.5 \times 10^3 \text{ N/m}^2}{9.59 \times 10^3 \text{ N/m}^3} = 8.39 \text{ m}$$

Now, based on the elevation of the tank, we have

$$h_s = +2.5 \text{ m}$$

To find the friction loss h_f, we must find the velocity, Reynolds number, and friction factor:

$$v = \frac{Q}{A} = \frac{95 \text{ L/min}}{1.314 \times 10^{-3} \text{ m}^2} \times \frac{1.0 \text{ m}^3/\text{s}}{60\,000 \text{ L/min}} = 1.21 \text{ m/s}$$

$$N_R = \frac{vD}{\nu} = \frac{(1.21)(0.0409)}{4.11 \times 10^{-7}} = 1.20 \times 10^5 \quad \text{(turbulent)}$$

$$\frac{D}{\epsilon} = \frac{0.0409 \text{ m}}{4.6 \times 10^{-5} \text{ m}} = 889$$

Then, from Fig. 9.2, $f = 0.0225$. From Table 10.5, $f_T = 0.021$. Now we have

$$h_f = f(L/D)(v^2/2g) + 2f_T(30)(v^2/2g) + f_T(340)(v^2/2g) + 1.0(v^2/2g)$$
$$\qquad \text{(pipe)} \qquad\qquad \text{(elbows)} \qquad\qquad \text{(valve)} \qquad\quad \text{(entrance)}$$

The velocity head is

$$\frac{v^2}{2g} = \frac{(1.21 \text{ m/s})^2}{2(9.81 \text{ m/s}^2)} = 0.0746 \text{ m}$$

Then, factoring $v^2/2g$ out from each term, we have

$$h_f = (0.0225)(12/0.0409)(0.0746) + (0.021)(60)(0.0746)$$
$$+ (0.021)(340)(0.0746) + 0.0746$$
$$= (0.0746 \text{ m})[(0.0225)(12/0.0409) + (0.021)(60) + (0.021)(340) + 1.0]$$
$$= 1.19 \text{ m}$$

Finally, from Fig. 15.36, we get

$$h_{vp} = 3.2 \text{ m}$$

Combining these terms gives

$$NPSH_a = 8.39 \text{ m} + 2.5 \text{ m} - 1.19 \text{ m} - 3.2 \text{ m} = 6.50 \text{ m}$$

Thus, a pump operating in this system must have a *required NPSH* less than 6.50 m.

15.11
SUCTION LINE
DETAILS

The *suction line* refers to all parts of the flow system from the source of the fluid to the inlet of the pump. Great care should be exercised in designing the suction line to ensure an adequate net positive suction head, as we discussed in Section 15.10. Also, special conditions may require auxiliary devices.

Figure 15.35 shows two methods of providing fluid to a pump. In part (a), a positive head is created by placing the pump below the supply reservoir. This is an aid in ensuring a satisfactory *NPSH*. Also, the pump will always be primed with a column of liquid at start-up.

In part (b) of Fig. 15.35, a *suction lift* condition occurs because the pump must draw liquid from below. Most positive displacement pumps can lift fluids about 8 m and are called *self-priming*. For centrifugal pumps, however, the pump must be artificially primed by filling the suction line with fluid. This can be done by providing an auxiliary supply of liquid during start-up or by drawing a vacuum on the pump casing, causing the fluid to be sucked up from the source. Then, with the pump running, it will maintain the flow.

Unless the fluid is known to be very clean, a strainer should be installed either at the inlet or elsewhere in the suction piping to keep debris out of the pump and out of the process to which the fluid is to be delivered. A foot valve at the inlet allows free flow to the pump but shuts when the pump stops. This maintains a column of liquid up to the pump and precludes the need to prime the pump each time it is started. If a valve is used near the pump, a gate valve, offering very little flow resistance when fully open, is preferred. The valve stem should be horizontal to avoid air pockets.

Although the pipe size for the suction line should never be smaller than the inlet connection on the pump, it can be somewhat larger to reduce flow velocity and friction losses. Pipe alignment should eliminate the possibility of forming air bubbles or air pockets in the suction line, as this will cause the pump to lose capacity and possibly to lose prime. Noise and vibration will also result. Elbows in a horizontal plane should be avoided. If a reducer is required, it should be of the eccentric type, as shown in Fig. 15.35(b). Concentric reducers place part of the supply pipe above the pump inlet where an air pocket could form.

Reference 5 recommends that suction velocities be in the range of 0.50 meters per second (1.6 feet per second) for smaller pipe sizes, and lower flow rates to about 1.50 meters per second (5.0 feet per second) for larger sizes and flows. Table 15.2 shows examples of the velocities that result from given flow rates for a variety of standard Schedule 40 pipe sizes. Such a table can be used to select the appropriate size pipe for a certain application.

TABLE 15.2 Suction pipe size versus flow rate and velocity

Flow Rate (gal/min)	Velocity (ft/s)	Pipe Size (Schedule 40)	Flow Rate (L/min)	Velocity (m/s)
12	1.89	$1\frac{1}{2}$	45	0.57
25	2.39	2	95	0.73
35	2.34	$2\frac{1}{2}$	130	0.71
50	2.17	3	190	0.66
75	2.43	$3\frac{1}{2}$	285	0.74
125	3.15	4	475	0.96
175	2.80	5	660	0.85
250	2.78	6	950	0.85
500	3.21	8	1 890	0.98
1 000	4.06	10	3 800	1.24
1 250	3.58	12	4 730	1.09
1 750	4.15	14	6 625	1.26
2 500	4.54	16	9 450	1.38
3 500	5.02	18	13 250	1.53

15.12 DISCHARGE LINE DETAILS

In general, the discharge line should be as short and direct as possible to minimize the head on the pump. Elbows should be standard or long radius type if possible. Pipe size should be chosen according to velocity or allowable friction losses.

The specification of the discharge pipe size is largely a matter of economics. Particularly for long pipe runs, the cost of the pipe increases significantly as the pipe size is increased. However, smaller pipe sizes for a given flow rate cause high velocities and correspondingly higher energy losses. Recall that energy losses are proportional to the velocity head and therefore are proportional to the *square* of the velocity. As energy losses increase, the total head on the pump increases, which requires more power to drive the pump. Also, a higher head usually requires a more expensive pump.

Recommended velocities for discharge lines can then be only approximate, because the overall economics of a given installation must be considered. Table 15.3 gives some examples of reasonable pipe sizes for the given

TABLE 15.3 Discharge pipe
size versus flow rate and velocity

Flow Rate (gal/min)	Velocity (ft/s)	Pipe Size (Schedule 40)	Flow Rate (L/min)	Velocity (m/s)
12	4.45	1	45	1.35
25	5.36	1¼	95	1.63
35	5.51	1½	130	1.68
50	7.88	1½	190	2.40
75	7.16	2	285	2.18
125	8.37	2½	475	2.55
175	7.59	3	660	2.31
250	6.30	4	950	1.92
500	8.01	5	1 890	2.44
1 000	11.10	6	3 800	3.38
1 000	6.41	8	3 800	1.95
1 250	8.02	8	4 730	2.44
1 750	11.22	8	6 625	3.42
1 750	7.11	10	6 625	2.17
2 500	10.16	10	9 450	3.10
2 500	7.16	12	9 450	2.18
3 500	10.03	12	13 250	3.06

flow rates with the resulting velocity of flow (Reference 5, pages 322 and 336). The velocities are in the range of from 1.35 meters per second (4.45 feet per second) for the smaller pipes and lower flow rates to 3.42 meters per second (11.22 feet per second) for the larger pipes and higher flow rates. Two pipe sizes are listed for some flow rates, because the increments in size are larger and either would produce a reasonable velocity.

The discharge line should contain a valve close to the pump to allow service or pump replacement. This valve acts with the valve in the suction line to isolate the pump. For low resistance, a gate or butterfly valve is preferred. If flow must be regulated during service, a globe valve is better because it allows a smooth throttling of the discharge. This, in effect, increases the system head and causes the pump delivery to decrease. As shown in Fig. 15.37, other elements may be added to the discharge line as required. A pressure relief valve will protect the pump and other equipment in case of a blockage of the flow or accidental shut-off of a valve. A check valve prevents flow back through the pump when it is not running. A tap into the discharge line for a gage with its shut-off valve may be necessary. Also, a sample cock will allow a small flow of the fluid to be drawn off for testing without disrupting operation. Figure 7.1 in Chapter 7 shows a photograph of an actual installation.

FIGURE 15.37 Discharge line details.

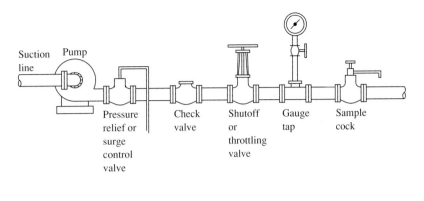

Suction line · Pump · Pressure relief or surge control valve · Check valve · Shutoff or throttling valve · Gauge tap · Sample cock

---

REFERENCES

1. Cheremisinoff, N. P. 1984. *Fluid Flow Pocket Handbook*. Houston: Gulf Publishing Co.

2. Goulds Pumps, Inc. 1993. *Goulds Pump Manual*. Seneca Falls, New York: Goulds Pumps, Inc.

3. Hicks, T.G., and Edwards, T. W. 1971. *Pump Application Engineering*. New York: McGraw-Hill.

4. Hydraulic Institute. 1983. *Hydraulic Institute Standards for Centrifugal, Rotary & Reciprocating Pumps*. 14th ed. Cleveland, OH: Hydraulic Institute.

5. Mately, J., ed. 1979. *Fluid Movers: Pumps, Compressors, Fans and Blowers*. New York: Chemical Engineering McGraw-Hill Publications Co.

6. Stewart, H. L. (Revised by T. Philbin). *Pumps*. 1986. New York: Macmillan Publishing Co.

7. Walker, R. 1972. *Pump Selection*. Ann Arbor, MI: Ann Arbor Science Publishers.

PRACTICE PROBLEMS

15.1 List twelve factors that should be considered when selecting a pump.

15.2 List ten items that must be specified for pumps.

15.3 Describe a positive displacement pump.

15.4 Name four examples of rotary positive displacement pumps.

15.5 Name three types of reciprocating positive displacement pumps.

15.6 Describe a kinetic pump.

15.7 Name three classifications of kinetic pumps.

15.8 Describe the action of the impellers and the general path of flow in the three types of kinetic pumps.

15.9 Describe a jet pump.

15.10 Distinguish between a shallow well jet pump and a deep well jet pump.

15.11 Describe the difference between a simplex reciprocating pump and a duplex type.

15.12 Describe the general shape of the plot of pump capacity versus discharge pressure for a positive displacement rotary pump.

15.13 Describe the general shape of the plot of total head versus pump capacity for centrifugal pumps.

15.14 To the head versus capacity plot of Practice Problem 15.13, add plots for efficiency and power required.

15.15 To what do the *affinity laws* refer in regard to pumps?

15.16 For a given centrifugal pump, if the speed of rotation of the impeller is cut in half, how does the capacity change?

15.17 For a given centrifugal pump, if the speed of rotation of the impeller is cut in half, how does the total head capability change?

15.18 For a given centrifugal pump, if the speed of rotation of the impeller is cut in half, how does the power required to drive the pump change?

15.19 For a given size of centrifugal pump casing, if the diameter of the impeller is reduced by 25 percent, how much does the capacity change?

15.20 For a given size of centrifugal pump casing, if the diameter of the impeller is reduced by 25 percent, how much does the total head capability change?

15.21 For a given size of centrifugal pump casing, if the diameter of the impeller is reduced by 25 percent, how much does the power required to drive the pump change?

15.22 Describe each part of this centrifugal pump designation: $1\frac{1}{2} \times 3 - 6$.

15.23 For the line of pumps shown in Fig. 15.17, specify a suitable size for delivering 100 gal/min of water at a total head of 300 ft.

15.24 For the line of pumps shown in Fig. 15.17, specify a suitable size for delivering 600 L/min of water at a total head of 25 m.

15.25 For the $2 \times 3 - 10$ centrifugal pump performance curve shown in Fig. 15.23, describe the performance that can be expected from a pump with an 8-in impeller operating against a system head of 200 ft. Give the expected capacity, the power required, the efficiency, and the required *NPSH*.

15.26 For the $2 \times 3 - 10$ centrifugal pump performance curve shown in Fig. 15.23, at what head will the pump having an 8-in impeller operate at its highest efficiency? List the pump's capacity, power required, efficiency, and the required *NPSH* at that head.

15.27 Using the result from Problem 15.26, describe how the performance of the pump changes if the system head increases by 15 percent.

15.28 For the $2 \times 3 - 10$ centrifugal pump performance curve shown in Fig. 15.23, list the total head and capacity at which the pump will operate at maximum efficiency for each of the impeller sizes shown.

15.29 For a given centrifugal pump and impeller size, describe how the *NPSH* required varies as the capacity increases.

15.30 State some advantages of using a variable speed drive for a centrifugal pump that supplies fluid to a process requiring varying flow rates of a fluid, as compared with adjusting throttling valves.

15.31 Describe how the capacity, efficiency, and power required for a centrifugal pump vary as the viscosity of the fluid pumped increases.

15.32 If two identical centrifugal pumps are connected in parallel and operated against a certain head, how would the total capacity compare with that of a single pump operating against the same head?

15.33 Describe the effect of operating two pumps in series.

15.34 For each of the following sets of operating conditions, list at least one appropriate type of pump. See Figure 15.33.
  a. 500 gal/min of water at 80 ft of total head
  b. 500 gal/min of water at 800 ft of head
  c. 500 gal/min of a viscous adhesive at 80 ft of head
  d. 80 gal/min of water at 8000 ft of head
  e. 80 gal/min of water at 800 ft of head
  f. 8000 gal/min of water at 200 ft of head
  g. 8000 gal/min of water at 60 ft of head
  h. 8000 gal/min of water at 12 ft of head

15.35 For the $1\frac{1}{2} \times 3 - 13$ centrifugal pump performance curve shown in Fig. 15.29, determine the capacity that can be expected from a pump with a 12-in impeller operating against a system head of 550 ft. Then, compute the specific speed and specific diameter and locate the corresponding point on Fig. 15.34.

15.36 For the $6 \times 8 - 17$ centrifugal pump performance curve shown in Fig. 15.27, determine the capacity that can be expected from a pump with a 15-in impeller operating against a system head of 200 ft. Then, compute the specific speed and specific diameter and locate the corresponding point on Fig. 15.34.

15.37 Figure 15.33 shows that a mixed flow pump is recommended for delivering 10 000 gal/min of water at a head of 40 ft. If such a pump operates with a specific speed of 5000, compute the appropriate operating speed of the pump.

15.38 Compute the specific speed for a pump operating at 1750 rpm delivering 5000 gal/min of water at a total head of 100 ft.

15.39 Compute the specific speed for a pump operating at 1750 rpm delivering 12 000 gal/min of water at a total head of 300 ft.

15.40 Compute the specific speed for a pump operating at 1750 rpm delivering 500 gal/min of water at a total head of 100 ft.

15.41 Compute the specific speed for a pump operating at 3500 rpm delivering 500 gal/min of water at a total head of 100 ft. Compare the result with that of Problem 15.40 and with Fig. 15.34.

15.42 It is desired to operate a pump at 1750 rpm by driving it with a 4-pole electric motor. For each of the following conditions, compute the specific speed using Eq. (15–7). Then, recommend whether you should use an axial pump, a mixed flow pump, a radial flow pump, or none of these, based on the discussion related to Fig. 15.34.

   a. 500 gal/min of water at 80 ft of total head
   b. 500 gal/min of water at 800 ft of head
   c. 3500 gal/min of water at 80 ft of head
   d. 80 gal/min of water at 8000 ft of head
   e. 80 gal/min of water at 800 ft of head
   f. 8000 gal/min of water at 200 ft of head
   g. 8000 gal/min of water at 60 ft of head
   h. 8000 gal/min of water at 12 ft of head

15.43 Define *net positive suction head (NPSH)*.

15.44 Distinguish between *NPSH available* and *NPSH required*.

15.45 Describe what happens to the vapor pressure of water as the temperature increases.

15.46 Describe why it is important to consider *NPSH* when designing and operating a pumping system.

15.47 For what point in a pumping system is the *NPSH* computed? Why?

15.48 Discuss why it is desirable to elevate the reservoir from which a pump draws liquid.

15.49 Discuss why it is desirable to use relatively large pipe sizes for the suction lines in pumping systems.

15.50 Discuss why an *eccentric reducer* should be used when it is necessary to decrease the size of a suction line as it approaches a pump.

15.51 If we assume that a given pump requires 7.50 ft of *NPSH* when operating at 3500 rpm, what would be the *NPSH* required at 2850 rpm?

15.52E Determine the available *NPSH* for the pump in Practice Problem 7.14 if the water is at 80°F and the atmospheric pressure is 14.4 psia.

15.53E Find the available *NPSH* when a pump draws water at 140°F from a tank whose level is 4.8 ft below the pump inlet. The suction line losses are 2.2 lb-ft/lb and the atmospheric pressure is 14.7 psia.

15.54M *A pump draws benzene at 25°C from a tank whose level is 2.6 m above the pump inlet. The suction line has a head loss of 0.8 N·m/N. The atmospheric pressure is measured to be 98.5 kPa(abs). Find the available NPSH. The vapor pressure of benzene is 13.3 kPa.*

15.55M *Determine the available NPSH for the system shown in Fig. 15.35(b). The fluid is water at 80°C and the atmospheric pressure is 101.8 kPa. The water level in the tank is 2.0 m below the pump inlet. The vertical leg of the suction line is a 3-in Schedule 40 steel pipe, while the horizontal leg is a 2-in Schedule 40 pipe, 1.5 m long. The elbow is of the long radius type. Neglect the loss in the reducer. The foot valve and strainer have a resistance coefficient of 3.0, based on the velocity in the 3-in line. The flow rate is 300 L/min.*

DESIGN PROBLEMS

Several situations are presented here in which a system is being designed to pump a fluid from some source to a given destination. In each case, the objective is to completely define the configuration of the system, including:

- Pipe sizes and types
- Location of the pump
- Length of pipe for all parts of the system
- Valves and fittings

Although the scope of the design problems can be adjusted, the following are suggested features of the exercises.

1. Specify pipe types suitable to the application.

2. Use Tables 15.2 and 15.3 to estimate the desirable sizes for pipes.

3. Select valves and fittings from the data given in Chapter 10.

4. If the head and flow characteristics of the system are appropriate, use the pump performance data from Figs. 15.23 through 15.29 to specify a suitable pump and evaluate its performance: efficiency, power required, and *NPSH* required.

5. Consider using other manufacturers' catalogs to select pumps for problems for which Figs. 15.23 to 15.29 are not suitable.

6. Analyze each system to determine the total head on the pump. This will require the calculation of energy losses throughout the system.

7. Compute the pressure at the inlet to the pump.

8. Compute the *NPSH* available at the pump inlet.

9. Where data are available, ensure that the *NPSH* available is greater than the *NPSH* required by the pump.

10. Make a list of materials required for the system.

11. Analyze the pressure at other points pertinent to the problem.

Problem Statements

1E Design a system to pump water at 140°F from a sump below a heat exchanger to the top of a cooling tower, as sketched in Fig. 15.38. The desired minimum flow rate is 200 gal/min.

2M *Design a system to pump water at 80°C from a water heater to a washing system, as sketched in Fig. 15.39. The desired minimum flow rate is 750 L/min (198 gal/min).*

3E Design a system to pump water at 60°F from a river to a tank elevated 55 ft above the surface of the river. The desired minimum flow rate is 1500 gal/min. The tank is to be set back 125 ft from the river bank.

4E Design the water system for Professor Crocket's cabin, as described in Fig. 7.38. The desired minimum flow rate is 40 gal/min, and the distribution tank is to be maintained at a pressure of 30 psig above the water. The cabin sits 150 ft from the side of the stream from which the water is to be drawn. The slope of the hillside is approximately 30° from the vertical.

5M *Design a system similar to that shown in Fig. 7.35, in which air pressure at 400 kPa above the kerosene at 25°C is used to cause the flow. The horizontal distance between the two tanks is 32 m. The desired minimum flow rate is 500 L/min.*

6E Design a system similar to that shown in Fig. 9.10, which must supply at least 1500 gal/min of water at 60°F for a fire protection system. The pressure at point B must be at least 85 psig. The depth of water in the tank is 5.0 ft. Ignore the specified sizes for the pipes and make your own decisions.

7E Design a system similar to that shown in Fig. 9.16 and described in Problem 9.18. Ignore the given pipe sizes and the given pressure at the pump inlet. The pump

FIGURE 15.38 Design Problem 1.

FIGURE 15.39   Design Problem 2.

draws the polluted water from a still pond whose surface is 30 in below the centerline of the pump inlet.

8M *Design a system similar to that shown in Fig. 7.22 to deliver 60 L/min of a water-based cutting fluid (sg = 0.95) to the cutter of a milling machine. Assume that the viscosity is 10 percent greater than that for water at 15°C. Assume that the pump is submerged and that*

the depth above the suction inlet is 75 mm. The total length of the path required for the discharge line is 1.75 m.

9M *Design a system similar to that shown in Fig. 7.21 to deliver 840 L/min of crude oil (sg = 0.85) from an underground storage tank to a pressurized storage tank.*

COMPREHENSIVE DESIGN PROBLEM

Consider yourself to be a plant engineer for a company that is planning a new manufacturing facility. As a part of the new plant, there will be an automated machining line in which five machines will be supplied with coolant from the same reservoir. You are responsible for the design of the system to handle the coolant from the time it reaches the plant in railroad tank cars until the dirty coolant is removed from the premises by a contract firm for reclaim.

The layout of the planned facility is shown in Fig. 15.40. The following data, design requirements, and limitations apply.

1. New coolant is delivered to the plant by tank cars carrying 15 000 gal each. A holding tank for new coolant must be specified.

2. The reservoir for the automated machining system must have a capacity of 1000 gal.

3. The 1000-gal tank is normally emptied once per week. Emergency dumps are possible if the coolant becomes overly contaminated prior to the scheduled emptying.

4. The dirty fluid is picked up by truck only once per month.

FIGURE 15.40 Plot plan for factory building for comprehensive design problem.

General manufacturing area

Machining area

600 ft

Driveway

100 ft

200 ft

Railroad siding

Highway

Note: Layout of machining area similar to Fig. 12.14, Problem 12.12.

5. A holding tank for the dirty fluid must be specified.

6. The plant is being designed to operate two shifts per day, 7 days per week.

7. Maintenance is normally performed during the third shift.

8. The building is one-story with a concrete floor.

9. The floor level is at the same elevation as the railroad track.

10. No storage tank can be inside the plant or under the floor except the 1000-gal reservoir that supplies the machining system.

11. The roof top is 32 ft from the floor level and the roof can be designed to support a storage tank.

12. The building is to be located in Dayton, Ohio, where the outside temperature may range from −20°F to +105°F.

13. The frost line is 30 in below the surface.

14. The coolant is a solution of water and a soluble oil with a specific gravity of 0.94 and a freezing point of 0°F. Its corrosiveness is approximately the same as that of water.

15. Assume that the viscosity of the coolant is 1.50 times that of water at any temperature.

16. The coolant pumped to the machines is to be at 80°F, and a flow rate of 230 gal/min is required.

17. The coolant drains from the machines by gravity and is at approximately 100°F.

18. The coolant draining from the machines must be conducted to a catch basin and pumped back to the reservoir. Prior to entering the reservoir, it should be filtered to remove metal chips and other debris.

19. The basic coolant storage and delivery system is to have the functional design sketched in the block diagram in Fig. 15.41.

The following tasks are to be completed by you, the system designer:

a. Specify the location and size of all storage tanks.

b. Specify the layout of the piping system, the types and sizes of all pipes, and the lengths required.

c. Specify the number, type, and size of all valves, elbows, and fittings.

d. Specify the number of pumps, their types, capacities, head requirements, and power required.

e. Specify the installation requirements for the pumps, including the complete suction line system. Evaluate the net positive suction head (*NPSH*) available for your

FIGURE 15.41 Block diagram
of coolant system.

design, and demonstrate that your pump has an acceptable *NPSH* required.

f. Determine the time required to fill and empty all tanks.

g. Sketch the layout of your design in both a plan view
(top) and an elevation view (side). An isometric sketch
also may be used.

h. Include the analysis of all parts of the system, including
energy losses due to friction and minor losses.

i. Submit the results of your design in a neat and complete
report, including a narrative description of the system,
the sketches, a list of materials, and the analysis to
show that your design meets the specifications.

■ ■ ■ ■ 16 Forces Due to Fluids in Motion

16.1 OBJECTIVES

There are numerous examples in fluid mechanics in which an external force is required to cause a change in the velocity of the flow of fluids. The nozzle of a fire hose shooting a high velocity stream of water must be firmly held or it will thrash wildly due to the force of the water on the nozzle. Wind striking a flat sign exerts a force on the sign because the direction of the air flow is being changed. High velocity water or steam striking the buckets of a turbine exert a force, causing the turbine to rotate and generate power. In piping installations, forces are exerted on elbows where the direction of fluid flow is changed and on expansions and contractions where the velocity is changed. These elements must be firmly anchored to resist these forces, especially if high velocity flows or rapid surges of flow are experienced.

After completing this chapter, you should be able to:

1. Use Newton's second law of motion, $F = ma$, to develop the *force equation,* which is used to compute the force exerted by a fluid as its direction of motion or its velocity is changed.
2. Relate the force equation to *impulse-momentum*.
3. Use the force equation to compute the force exerted on a stationary object that causes the change in direction of a fluid flow stream.
4. Use the force equation to compute the force exerted on bends in pipe lines.
5. Use the force equation to compute the force on moving objects, such as the vanes of a pump impeller.

16.2 FORCE EQUATION

Whenever the magnitude or direction of the velocity of a body is changed, a force is required to accomplish the change. Newton's second law of motion is often used to express this concept in mathematical form; the most common form is

$$F = ma \tag{16-1}$$

Force equals mass times acceleration. Acceleration is the time rate of change of velocity. However, since velocity is a vector quantity having both magnitude and direction, changing either the magnitude or the direction will result in an acceleration. According to Eq. (16–1), an external force is required to cause the change.

Equation (16–1) is convenient for use with solid bodies since the mass remains constant and the acceleration of the entire body can be determined. In fluid flow problems, a continuous flow is caused to undergo the acceleration, and a different form of Newton's equation is desirable. Because acceleration is the time rate of change of velocity, Eq. (16–1) can be written as

$$F = ma = m\,\frac{\Delta v}{\Delta t} \qquad\qquad (16\text{–}2)$$

The term $m/\Delta t$ can be interpreted as the mass flow rate, that is, the amount of mass flowing in a given amount of time. In the discussion of fluid flow in Chapter 6, mass flow rate was indicated by the symbol M. Also, M is related to the volume flow rate Q by the relationship

$$M = \rho Q \qquad\qquad (16\text{–}3)$$

where ρ is the density of the fluid. Then Eq. (16–2) becomes

GENERAL FORM OF FORCE EQUATION

$$F = (m/\Delta t)\Delta v = M\,\Delta v = \rho Q\,\Delta v \qquad\qquad (16\text{–}4)$$

This is the general form of the force equation for use in fluid flow problems because it involves the velocity and volume flow rate, items generally known in a fluid flow system.

16.3 IMPULSE-MOMENTUM EQUATION

The force equation, Eq. (16–4), is related to another principle of fluid dynamics, the *impulse-momentum equation*. Impulse is defined as a force acting on a body for a period of time and it is indicated by

$$\text{Impulse} = F(\Delta t)$$

This form, relying on the total change in time Δt, is suitable for dealing with steady flow conditions. When conditions vary, the instantaneous form of the equation is used.

$$\text{Impulse} = F(dt)$$

where dt is the differential amount of change in time.

Momentum is defined as the product of the mass of a body and its velocity. The *change* in momentum is

$$\text{Change in momentum} = m(\Delta v)$$

In an instantaneous sense,

$$\text{Change in momentum} = m(dv)$$

Now Eq. (16–2) can be rearranged to the form

$$F(\Delta t) = m(\Delta v)$$

Here we have shown the impulse-momentum equation for steady flow conditions. In an instantaneous sense,

$$F(dt) = m(dv)$$

**16.4
PROBLEM-SOLVING
METHOD USING THE
FORCE EQUATION**

We emphasize that problems involving forces must account for the directions in which the forces act. In Eq. (16–4), force and velocity are both vector quantities. The equation is valid only when all terms have the same direction. For this reason, different equations are written for each direction of concern in a particular case. In general, if three perpendicular directions are called x, y, and z, a separate equation can be written for each direction:

➩ **FORCE EQUATIONS IN x, y, AND z
DIRECTIONS**

$$F_x = \rho Q\, \Delta v_x = \rho Q(v_{2_x} - v_{1_x}) \tag{16–5}$$

$$F_y = \rho Q\, \Delta v_y = \rho Q(v_{2_y} - v_{1_y}) \tag{16–6}$$

$$F_z = \rho Q\, \Delta v_z = \rho Q(v_{2_z} - v_{1_z}) \tag{16–7}$$

This is the form of the force equation which will be used in this book, with the directions chosen according to the physical situation. In a particular direction, say x, the term F_x refers to the net external force which acts on the fluid in that direction. Therefore, it is the algebraic sum of *all* external forces, including that exerted by a solid surface and forces due to fluid pressure. The term Δv_x refers to the change in velocity in the x direction. Also, v_1 is the velocity as the fluid enters the device and v_2 is the velocity as it leaves. Then v_{1_x} is the component of v_1 in the x direction and v_{2_x} is the component of v_2 in the x direction.

The specific approach to problems using the force equation depends somewhat on the nature of the data given. A general procedure follows:

PROCEDURE FOR USING THE FORCE EQUATION

1. Identify a portion of the fluid stream to be considered a free body. This will be the part where the fluid is changing direction or where the geometry of the flow stream is changing.
2. Establish reference axes for directions of forces. Usually one axis is chosen to be parallel to one part of the flow stream. In the example problems to follow, the positive x and y directions are chosen to be in the same direction as the reaction forces.
3. Identify and show on the free body diagram all external forces acting on the fluid. All solid surfaces which affect the direction of the flow stream exert forces. Also, the fluid pressure acting on the cross-sectional area of the stream exerts a force in a direction parallel to the stream at the boundary of the free body.
4. Show the direction of the velocity of flow as it enters the free body and as it leaves the free body.
5. Using the data thus shown for the free body, write the force equation in the pertinent directions. Use Eqs. (16–5), (16–6), or (16–7).
6. Substitute data and solve for the desired quantity.

The example problems presented in the following sections illustrate this procedure.

**16.5
FORCES ON
STATIONARY OBJECTS**

When free streams of fluid are deflected by stationary objects, external forces must be exerted to maintain the object in equilibrium. Some examples follow.

☐ **EXAMPLE PROBLEM 16.1**   A 1-in diameter jet of water having a velocity of 20 ft/s is deflected by a curved vane 90°, as shown in Figure 16.1. The jet flows freely in the atmosphere in a horizontal plane. Calculate the x and y forces exerted on the water by the vane.

FIGURE 16.1   Water jet deflected by a curved vane.

Solution   Using the force diagram of Fig. 16.2, the force equation for the x direction is

$$F_x = \rho Q(v_{2_x} - v_{1_x})$$
$$R_x = \rho Q[0 - (-v_1)] = \rho Q v_1$$

But, we know that

$$Q = Av = (0.00545 \text{ ft}^2)(20 \text{ ft/s}) = 0.109 \text{ ft}^3/\text{s}$$

Then, assuming $\rho = 1.94$ slugs/ft³ = 1.94 lb·s²/ft⁴,

$$R_x = \rho Q v_1 = \frac{1.94 \text{ lb} \cdot \text{s}^2}{\text{ft}^4} \times \frac{0.109 \text{ ft}^3}{\text{s}} \times \frac{20 \text{ ft}}{\text{s}} = 4.23 \text{ lb}$$

For the y direction, assuming $v_2 = v_1$, the force is

$$F_y = \rho Q(v_{2_y} - v_{1_y})$$
$$R_y = \rho Q(v_2 - 0) = (1.94)(0.109)(20) \text{ lb} = 4.23 \text{ lb}$$

FIGURE 16.2   Force diagram for the fluid deflected by the vane.

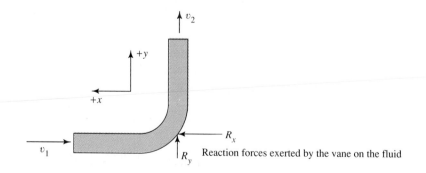

Reaction forces exerted by the vane on the fluid

□ **EXAMPLE PROBLEM 16.2**   In a decorative fountain, 0.05 m³/s of water having a velocity of 8 m/s is being deflected by the angled chute shown in Fig. 16.3. Determine the reactions on the chute in the x and y directions shown. Also, calculate the total resultant force and the direction in which it acts. Neglect elevation changes.

FIGURE 16.3 Decorative fountain deflecting a water jet.

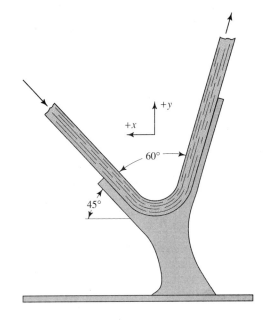

Solution   Figure 16.4 shows the x and y components of the velocity vectors and the assumed directions for R_x and R_y. The force equation in the x direction is:

$$F_x = \rho Q(v_{2_x} - v_{1_x})$$

But, we know that

$$v_{2_x} = -v_2 \sin 15° \qquad \text{(toward the right)}$$
$$v_{1_x} = -v_1 \cos 45° \qquad \text{(toward the right)}$$

Neglecting friction in the chute, we can assume that $v_2 = v_1$. Also, the only external force is R_x. Then we have

$$R_x = \rho Q[-v_2 \sin 15° - (-v_1 \cos 45°)]$$
$$= \rho Q v(-\sin 15° + \cos 45°) = 0.448 \, \rho Q v$$

Using $\rho = 1000$ kg/m³ for water, we get

$$R_x = \frac{(0.448)(1000 \text{ kg})}{\text{m}^3} \times \frac{0.05 \text{ m}^3}{\text{s}} \times \frac{8 \text{ m}}{\text{s}} = \frac{179 \text{ kg} \cdot \text{m}}{\text{s}^2} = 179 \text{ N}$$

In the y direction, the force equation is

$$F_y = \rho Q(v_{2_y} - v_{1_y})$$

FIGURE 16.4 Force diagram for the fluid deflected by the vane.

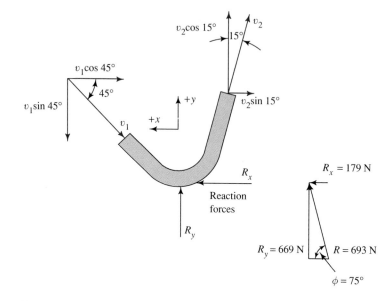

But, we know that

$$v_{2_y} = v_2 \cos 15° \qquad \text{(upward)}$$
$$v_{1_y} = -v_1 \sin 45° \qquad \text{(downward)}$$

Then we have

$$R_y = \rho Q[v_2 \cos 15° - (-v_1 \sin 45°)]$$
$$= \rho Q v(\cos 15° + \sin 45°)$$
$$= (1000)(0.05)(8)(0.966 + 0.707) \text{ N}$$
$$R_y = 669 \text{ N}$$

The resultant force R is

$$R = \sqrt{R_x^2 + R_y^2} = \sqrt{179^2 + 669^2} = 693 \text{ N}$$

For the direction of R, we get

$$\tan \phi = R_y/R_x = 669/179 = 3.74$$
$$\phi = 75.0°$$

Therefore, the resultant force that the chute must exert on the water is 693 N acting 75° from the horizontal, as shown in Fig. 16.4.

16.6
FORCES ON BENDS
IN PIPE LINES

Figure 16.5 shows a typical 90° elbow in a pipe carrying a steady volume flow rate Q. In order to ensure proper installation, it is important to know how much force is required to hold it in equilibrium. The following problem demonstrates an approach to this type of situation.

☐ **EXAMPLE PROBLEM 16.3**
Calculate the force that must be exerted on the pipe elbow shown in Fig. 16.5 to hold it in equilibrium. The elbow is in a horizontal plane and is connected to two 4-in

FIGURE 16.5 Pipe elbow.

Schedule 40 pipes carrying 3000 L/min of water at 15°C. The inlet pressure is 550 kPa.

Solution
The problem may be visualized by considering the fluid within the elbow to be a free body, as shown in Fig. 16.6. Forces are shown as black vectors, while the direction of the velocity of flow is shown by blue vectors. A convention must be set for the directions of all vectors. Here we assume that the positive x direction is to the left and the positive y direction is up. The forces R_x and R_y are the external reactions required to maintain equilibrium. The forces p_1A_1 and p_2A_2 are the forces due to the fluid pressure. The two directions will be analyzed separately.

FIGURE 16.6 Force diagram on the fluid in the elbow.

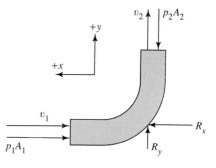

We find the net external force in the x direction by using the equation

$$F_x = \rho Q(v_{2_x} - v_{1_x})$$

But, we know that

$$F_x = R_x - p_1A_1$$
$$v_{2_x} = 0$$
$$v_{1_x} = -v_1$$

Then we have

$$R_x - p_1 A_1 = \rho Q[0 - (-v_1)]$$
$$R_x = \rho Q v_1 + p_1 A_1 \qquad (16\text{–}8)$$

From the given data, $p_1 = 550$ kPa, $\rho = 1000$ kg/m³, and $A_1 = 8.213 \times 10^{-3}$ m². Then,

$$Q = 3000 \text{ L/min} \times \frac{1 \text{ m}^3/\text{s}}{60\ 000 \text{ L/min}} = 0.05 \text{ m}^3/\text{s}$$

$$v_1 = \frac{Q}{A_1} = \frac{0.05 \text{ m}^3/\text{s}}{8.213 \times 10^{-3} \text{ m}^2} = 6.09 \text{ m/s}$$

$$\rho Q v_1 = \frac{1000 \text{ kg}}{\text{m}^3} \times \frac{0.05 \text{ m}^3}{\text{s}} \times \frac{6.09 \text{ m}}{\text{s}} = 305 \text{ kg} \cdot \text{m/s}^2 = 305 \text{ N}$$

$$p_1 A_1 = \frac{550 \times 10^3 \text{ N}}{\text{m}^2} \times (8.213 \times 10^{-3} \text{ m}^2) = 4517 \text{ N}$$

Substituting these values into Eq. (16–8) gives

$$R_x = (305 + 4517) \text{ N} = 4822 \text{ N}$$

In the y direction, the equation for the net external force is

$$F_y = \rho Q(v_{2_y} - v_{1_y})$$

But, we know that

$$F_y = R_y - p_2 A_2$$
$$v_{2_y} = + v_2$$
$$v_{1_y} = 0$$

Then we have

$$R_z - p_2 A_2 = \rho Q v_2$$
$$R_z = \rho Q v_2 + p_2 A_2$$

If energy losses in the elbow are neglected, $v_2 = v_1$ and $p_2 = p_1$ since the sizes of the inlet and outlet are equal. Then,

$$\rho Q v_2 = 305 \text{ N}$$
$$p_2 A_2 = 4517 \text{ N}$$
$$R_y = (305 + 4517) \text{ N} = 4822 \text{ N}$$

The forces R_x and R_y are the reactions caused at the elbow as the fluid turns 90°. They may be supplied by anchors on the elbow or taken up through the flanges into the main pipes.

☐ **EXAMPLE PROBLEM 16.4**   Linseed oil with a specific gravity of 0.93 enters the reducing bend shown in Fig. 16.7 with a velocity of 3 m/s and a pressure of 275 kPa. The bend is in a horizontal plane. Calculate the x and y forces required to hold the bend in place. Neglect energy losses in the bend.

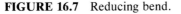

FIGURE 16.7 Reducing bend.

Solution The fluid in the bend is shown as a free body in Fig. 16.8. We must first develop the force equations for the x and y directions shown.

FIGURE 16.8 Force diagram for fluid in the reducing bend.

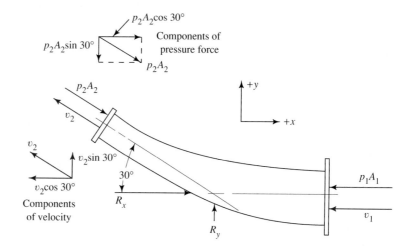

The force equation for the x direction is

$$F_x = \rho Q(v_{2_x} - v_{1_x})$$

$$R_x - p_1 A_1 + p_2 A_2 \cos 30° = \rho Q[-v_2 \cos 30° - (-v_1)] \qquad \textbf{(16–9)}$$

$$R_x = p_1 A_1 - p_2 A_2 \cos 30° - \rho Q v_2 \cos 30° + \rho Q v_1$$

Algebraic signs must be carefully included according to the sign convention shown in Fig. 16.8. Notice that all forces and velocity terms are the components *in the x direction*.

In the y direction, the force equation is

$$F_y = \rho Q(v_{2_y} - v_{1_y})$$

$$R_y - p_2 A_2 \sin 30° = \rho Q(v_2 \sin 30° - 0)$$

$$R_y = p_2 A_2 \sin 30° + \rho Q v_2 \sin 30° \qquad \textbf{(16–10)}$$

The numerical values of several items must now be calculated. For the entering and leaving pipes, $A_1 = 1.767 \times 10^{-2}$ m² and $A_2 = 4.418 \times 10^{-3}$ m².

$$\rho = (sg)(\rho_w) = (0.93)(1000 \text{ kg/m}^3) = 930 \text{ kg/m}^3$$

$$\gamma = (sg)(\gamma_w) = (0.93)(9.81 \text{ kN/m}^3) = 9.12 \text{ kN/m}^3$$

$$Q = A_1 v_1 = (1.767 \times 10^{-2} \text{ m}^2)(3 \text{ m/s}) = 0.053 \text{ m}^3/\text{s}$$

Because of continuity, $A_1 v_1 = A_2 v_2$. Then we have

$$v_2 = v_1(A_1/A_2) = (3 \text{ m/s})(1.767 \times 10^{-2}/4.418 \times 10^{-3}) = 12 \text{ m/s}$$

Bernoulli's equation can be used to find p_2:

$$\frac{p_1}{\gamma} + z_1 + \frac{v_1^2}{2g} = \frac{p_2}{\gamma} + z_2 + \frac{v_2^2}{2g}$$

But $z_1 = z_2$. Then we have

$$p_2 = p_1 + \gamma(v_1^2 - v_2^2)/2g$$

$$= 275 \text{ kPa} + \left[\frac{(9.12)(3^2 - 12^2)}{(2)(9.81)} \times \frac{\text{kN}}{\text{m}^3} \times \frac{\text{m}^2}{\text{s}^2} \times \frac{\text{s}^2}{\text{m}} \right]$$

$$= 275 \text{ kPa} - 62.8 \text{ kPa}$$

$$p_2 = 212.2 \text{ kPa}$$

The quantities needed for Eqs. (16–9) and (16–10) are

$$p_1 A_1 = (275 \text{ kN/m}^2)(1.767 \times 10^{-2} \text{ m}^2) = 4859 \text{ N}$$

$$p_2 A_2 = (212.2 \text{ kN/m}^2)(4.418 \times 10^{-3} \text{ m}^2) = 938 \text{ N}$$

$$\rho Q v_1 = (930 \text{ kg/m}^3)(0.053 \text{ m}^3/\text{s})(3 \text{ m/s}) = 148 \text{ N}$$

$$\rho Q v_2 = (930 \text{ kg/m}^3)(0.053 \text{ m}^3/\text{s})(12 \text{ m/s}) = 591 \text{ N}$$

From Eq. (16–9), we get

$$R_x = (4859 - 938 \cos 30° - 591 \cos 30° + 148) \text{ N} = 3683 \text{ N}$$

From Eq. (16–10), we get

$$R_y = (938 \sin 30° + 591 \sin 30°) \text{ N} = 765 \text{ N}$$

16.7
FORCES ON MOVING OBJECTS

The vanes of turbines and other rotating machinery are familiar examples of moving objects that are acted upon by high velocity fluids. A jet of fluid with a velocity greater than that of the blades of the turbine exerts a force on the blades, causing them to accelerate or to generate useful mechanical energy. When dealing with forces on moving bodies, the *relative motion* of the fluid with respect to the body must be considered.

□ **EXAMPLE PROBLEM 16.5**   Figure 16.9(a) shows a jet of water with a velocity v_1 striking a vane that is moving with a velocity v_0. Determine the forces exerted by the vane on the water if $v_1 = 20$ m/s and $v_0 = 8$ m/s. The jet is 50 mm in diameter.

FIGURE 16.9 Flow deflected by a moving vane.

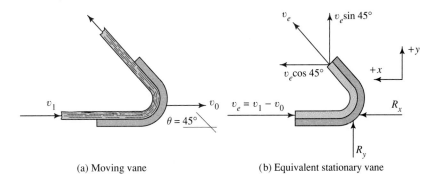

(a) Moving vane   (b) Equivalent stationary vane

Solution   The system with a moving vane can be converted into an equivalent stationary system as shown in Fig. 16.9(b) by defining an effective velocity v_e and an effective volume flow rate Q_e. We then have

▷ EFFECTIVE VELOCITY AND VOLUME FLOW RATE

$$v_e = v_1 - v_0 \qquad \text{(16–11)}$$
$$Q_e = A_1 v_e \qquad \text{(16–12)}$$

where A_1 is the area of the jet as it enters the vane. It is only the difference between the jet velocity and the vane velocity that is effective in creating a force on the vane. The force equations can be written in terms of v_e and Q_e. In the x direction,

$$R_x = \rho Q_e v_e \cos\theta - (-\rho Q_e v_e)$$
$$= \rho Q_e v_e (1 + \cos\theta) \qquad \text{(16–13)}$$

In the y direction,

$$R_y = \rho Q_e v_e \sin\theta - 0 \qquad \text{(16–14)}$$

But, we know that

$$v_e = v_1 - v_0 = (20 - 8)\ \text{m/s} = 12\ \text{m/s}$$
$$Q_e = A_1 v_e = (1.964 \times 10^{-3}\ \text{m}^2)(12\ \text{m/s}) = 0.0236\ \text{m}^3/\text{s}$$

Then the reactions are calculated from Eqs. (16–13) and (16–14):

$$R_x = (1000)(0.0236)(12)(1 + \cos 45°) = 483\ \text{N}$$
$$R_y = (1000)(0.0236)(12)(\sin 45°) = 200\ \text{N}$$

PRACTICE PROBLEMS

16.1M *Calculate the force required to hold a flat plate in equilibrium perpendicular to the flow of water at 25 m/s issuing from a 75-mm diameter nozzle.*

16.2E What must be the velocity of flow of water from a 2-in diameter nozzle to exert a force of 300 lb on a flat wall?

16.3E Calculate the force exerted on a stationary curved vane that deflects a 1-in diameter stream of water through a 90° angle. The volume flow rate is 150 gal/min.

16.4M *A highway sign is being designed to withstand winds of 125 km/h. Calculate the total force on a* *sign 4 m by 3 m if the wind is flowing perpendicular to the face of the sign. Calculate the equivalent pressure on the sign in Pa. The air is at −10°C. (See Chapter 17 and Problem 17.9 for a more complete analysis of this problem.)*

16.5E Compute the forces in the vertical and horizontal directions on the block shown in Fig. 16.10. The fluid stream is a 1.75-in diameter jet of water at 60°F with a velocity of 25 ft/s. The velocity leaving the block is also 25 ft/s.

FIGURE 16.11   Problem 16.6.

FIGURE 16.10   Problem 16.5.

16.6E Figure 16.11 shows a free stream of water at 180°F being deflected by a stationary vane through a 130° angle. The entering stream has a velocity of 22.0 ft/s. The cross sectional area of the stream is con-

stant at 2.95 in² throughout the system. Compute the forces in the horizontal and vertical directions exerted on the water by the vane.

16.7M *Compute the horizontal and vertical forces exerted on the vane shown in Fig. 16.12 due to a flow of water at 50°C. The velocity is constant at 15 m/s.*

16.8E In a plant where hemispherical cup-shaped parts are made, an automatic washer is being designed to clean the parts prior to shipment. One scheme being evaluated uses a stream of water at 180°F shooting vertically upward into the cup. The stream has a velocity of 30 ft/s and a diameter of 1.00 in. As shown in Fig. 16.13, the water leaves the cup vertically downward in the form of an annular ring having an outside diameter of 4.00 in and an inside diameter of 3.80 in. Compute the external force required to hold the cup down.

FIGURE 16.12   Problem 16.7.

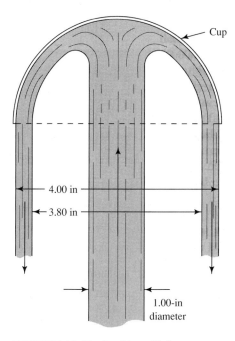

Cup

4.00 in

3.80 in

1.00-in
diameter

FIGURE 16.13   Problem 16.8.

16.9M *A stream of oil (sg = 0.90) is directed onto the center of the underside of a flat metal plate to keep it cool during a welding operation. The plate weighs 550 N. If the stream is 35 mm in diameter, calculate the velocity of the stream that will lift the plate. The stream strikes the plate perpendicularly.*

16.10E A 2-in diameter stream of water having a velocity of 40 ft/s strikes the edge of a flat plate such that half the stream is deflected downward as shown in Fig. 16.14. Calculate the force on the plate and the moment due to the force at point A.

16.11E Figure 16.15 represents a type of flowmeter in which the flat vane is rotated on a pivot as it deflects the fluid stream. The fluid force is counterbalanced by a spring. Calculate the spring force required to hold the vane in a vertical position when water at 100 gal/min flows from the 1-in Schedule 40 pipe to which the meter is attached.

16.12E Water is piped vertically from below a boat and discharged horizontally in a 4-in diameter jet with a velocity of 60 ft/s. Calculate the force on the boat.

16.13E A 2-in nozzle is attached to a hose with an inside diameter of 4 in. The resistance coefficient K of the nozzle is 0.12 based on the outlet velocity head. If the jet issuing from the nozzle has a velocity of 80 ft/s, calculate the force exerted by the water on the nozzle.

16.14M *Seawater (sg = 1.03) enters a heat exchanger through a reducing bend connecting a 4-in Type K copper tube with a 2-in Type K tube. The pressure upstream from the bend is 825 kPa. Calculate the force required to hold the bend in equilibrium. Consider the energy loss in the bend, assuming it has a resistance coefficient K of 3.5 based on the inlet velocity. The flow rate is 0.025 m³/s.*

16.15E A reducer connects a standard 6-in Schedule 40 pipe to a 3-in Schedule 40 pipe. The walls of the conical reducer are tapered at an included angle of 40°. The flow rate of water is 500 gal/min and the pressure ahead of the reducer is 125 psig. Considering the energy loss in the reducer, calculate the force exerted on the reducer by the water.

FIGURE 16.14   Problem 16.10.

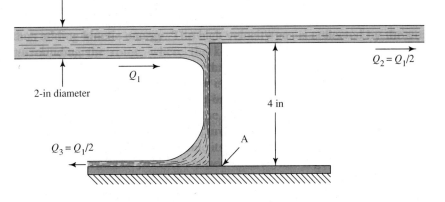

Q_1

2-in diameter

$Q_2 = Q_1/2$

4 in

A

$Q_3 = Q_1/2$

FIGURE 16.15   Problem 16.11.

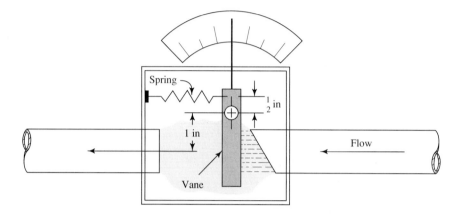

16.16E Calculate the force on a 45° elbow attached to an 8-in Schedule 80 steel pipe carrying water at 80°F at 6.5 ft³/s. The outlet of the elbow discharges into the atmosphere. Consider the energy loss in the elbow.

16.17M *Calculate the force required to hold a 90° elbow in place when attached to 6-in Schedule 40 pipes carrying water at 0.125 m³/s and 1050 kPa. Neglect energy lost in the elbow.*

16.18M *Calculate the force required to hold a 180° close return bend in equilibrium. The bend is in a horizontal plane and is attached to a 4-in Schedule 80 steel pipe carrying 2000 L/min of a hydraulic fluid at 2.0 MPa. The fluid has a specific gravity of 0.89. Neglect energy losses.*

16.19M *A bend in a tube causes the flow to turn through an angle of 135°. The pressure ahead of the bend is 275 kPa. If the 6-in Type K copper tube carries 0.12 m³/s of carbon tetrachloride at 25°C, determine the force on the bend. Neglect energy losses.*

16.20M *A vehicle is to be propelled by a jet of water impinging on a vane as shown in Fig. 16.16. The jet*

has a velocity of 30 m/s and issues from a nozzle with a diameter of 200 mm. Calculate the force on the vehicle (a) if it is stationary and (b) if it is moving at 12 m/s.

16.21M *A part of an inspection system in a packaging operation uses a jet of air to remove imperfect cartons from a conveyor line, as shown in Fig. 16.17. The jet is initiated by a sensor and timed so that the product to be rejected is in front of the jet at the right moment. The product is to be tipped over a ledge on the side of the conveyor as shown in the figure. Compute the required velocity of the air jet to tip the carton off the conveyor. The density of the air is 1.20 kg/m³. The carton has a mass of 0.10 kg. The jet has a diameter of 10.0 mm.*

16.22M *Shown in Fig. 16.18 is a small decorative wheel fitted with flat paddles so the wheel turns about its axis when acted on by a blown stream of air. Assuming that all the air in a 15-mm diameter stream moving at 0.35 m/s strikes one paddle and is deflected by it at right angles, compute the force exerted on the wheel initially when it is stationary. the air has a density of 1.20 kg/m³.*

FIGURE 16.16   Problem 16.20.

FIGURE 16.17 Problem 16.21.

FIGURE 16.18 Problems 16.22 and 16.23.

FIGURE 16.19 Problems 16.24, 16.25, and 16.26.

Pivot

5.0 in typical

Louver

Incoming air stream

θ

Louvers are 20.0 in long

FIGURE 16.20 Problems 16.27 and 16.28.

Air jet

16.23M *For the wheel described in Problem 16.22, compute the force exerted on the paddle when the wheel rotates at 40 rpm.*

16.24E A set of louvers deflects a stream of warm air onto painted parts, as illustrated in Fig. 16.19. The louvers are rotated slowly to distribute the air evenly over the parts. Compute the torque required to rotate the louvers toward the stream of air when it is flowing at a velocity of 10 ft/s. Assume that all the air that approaches a given louver is deflected to the angle of the louver. The air has a density of 2.06×10^{-3} slugs/ft³. Use $\theta = 45°$.

16.25E For the louvers shown in Fig. 16.19 and described in Problem 16.24, compute the torque required to rotate the louvers when the angle $\theta = 20°$.

16.26E For the louvers shown in Fig. 16.19 and described in Problem 16.24, compute the torque required to rotate the louvers for several settings of the angle θ from 10° to 90°. Plot a graph of torque versus angle.

16.27E Figure 16.20 shows a device for clearing debris using a 1½-in diameter jet of air issuing from a blower nozzle. As shown, the jet is striking a rectangular box-shaped object sitting on a floor. If the air velocity is 25 ft/s and the entire jet is deflected

FIGURE 16.21 Problems 16.29, 16.30, and 16.31.

by the box, what is the heaviest object that could be moved? Assume that the box slides rather than tumbling over and that the coefficient of friction is 0.60. The air has a density of 2.40×10^{-3} slugs/ft^3.

16.28E Repeat Problem 16.27, except change the jet to water at 50°F and the diameter to 0.75 in.

16.29M *Figure 16.21 is a sketch of a turbine in which the incoming stream of water at 15°C has a diameter of 7.50 mm and is moving with a velocity of 25 m/s. Compute the force on one blade of the turbine if the stream is deflected through the angle shown and the blade is stationary.*

16.30M *Repeat Problem 16.29 with the blade rotating as a part of the wheel at a radius of 200 mm and with a linear tangential velocity of 10 m/s. Also compute the rotational speed of the wheel in rpm.*

16.31M *Repeat Problem 16.29, except with the blade rotating as a part of the wheel at a radius of 200 mm and with a linear tangential velocity ranging from 0 to 25 m/s in 5 m/s steps.*

■■■■ ■ 17 Drag and Lift

17.1
OBJECTIVES

A moving body immersed in a fluid experiences forces caused by the action of the fluid. The total effect of these forces is quite complex. However, for the purposes of design or for the analysis of the behavior of a body in a fluid, two resultant forces—drag and lift—are the most important. Lift and drag forces are the same regardless of whether the body is moving in the fluid or the fluid is moving over the body.

Drag is the force on a body caused by the fluid which resists motion in the direction of travel of the body. The most familiar applications requiring the study of drag are in the transportation fields. *Wind resistance* is the term often used to describe the effects of drag on aircraft, automobiles, trucks, and trains. The drag force must be opposed by a propulsive force in the opposite direction to maintain or increase the velocity of the vehicle. Since the production of the propulsive force requires added power, it is desirable to minimize drag.

Lift is a force caused by the fluid in a direction perpendicular to the direction of travel of the body. Its most important application is in the design and analysis of aircraft wings called *airfoils*. The geometry of an airfoil is such that a lift force is produced as air passes over and under it. Of course, the magnitude of the lift must at least equal the weight of the aircraft in order for it to fly.

The study of the performance of bodies in moving air streams is called *aerodynamics*. Gases other than air could be considered in this field, but due to the obvious importance of the applications in aircraft design, the majority of work has been done with air as the fluid.

Hydrodynamics is the name given to the study of moving bodies immersed in liquids, particularly water. Many concepts concerning lift and drag are similar regardless of whether the fluid is a liquid or a gas. This is not true, however, at high velocities where the effects of the compressibility of the fluid must be taken into account. Liquids can be considered incompressible in the study of lift and drag. Conversely, a gas such as air is readily compressible.

Much of the practical data concerning lift and drag has been generated experimentally. We will report some of these data here to illustrate the concepts. The references listed at the end of this chapter include more comprehensive treatments of the subject.

After completing this chapter, you should be able to:

1. Define *drag*.
2. Define *lift*.
3. Write the expression for computing the drag force on a body moving relative to a fluid.
4. Define the *drag coefficient*.
5. Define the term *dynamic pressure*.
6. Describe the stagnation point for a body moving relative to a fluid.
7. Distinguish between *pressure drag* and *friction drag*.
8. Discuss the importance of flow separation on pressure drag.
9. Determine the value of the pressure drag coefficient for cylinders, spheres, and other shapes.
10. Discuss the effect of Reynolds number and surface geometry on the drag coefficient.
11. Compute the magnitude of the pressure drag force on bodies moving relative to a fluid.
12. Compute the magnitude of the friction drag force on smooth spheres.
13. Discuss the importance of drag on the performance of ground vehicles.
14. Discuss the effects of compressibility and cavitation on drag and the performance of bodies immersed in fluids.
15. Define the lift coefficient for a body immersed in a fluid.
16. Compute the lift force on a body moving relative to a fluid.
17. Describe the effects of friction drag, pressure drag, and induced drag on airfoils.

**17.2
DRAG FORCE
EQUATION**

Drag forces are usually expressed in the form

▷ **DRAG FORCE**

$$F_D = \text{drag} = C_D(\rho v^2/2)A \qquad (17\text{--}1)$$

The term C_D is the drag coefficient, a dimensionless factor. Its magnitude depends primarily on the physical shape of the object and its orientation relative to the fluid stream. The quantity $\rho v^2/2$ is the dynamic pressure as defined below. The term v is the velocity of the free stream of the fluid relative to the body. The term A refers to a characteristic area of the body, either the surface area or the maximum cross-sectional area perpendicular to the direction of flow as discussed in Sections 17.3, 17.5, and 17.8.

You can visualize the influence of the dynamic pressure on drag by referring to Fig. 17.1, which shows a sphere in a fluid stream. The streamlines depict the path of the fluid as it approaches and flows around the sphere. At point *s* on the surface of the sphere, the fluid stream is at rest or "stagnant." The term *stagnation point* is used to describe this point. The relationship between the pressure p_s and that in the undisturbed stream at point 1 can be found using Bernoulli's equation:

$$\frac{p_1}{\gamma} + \frac{v_1^2}{2g} = \frac{p_s}{\gamma} \qquad (17\text{--}2)$$

FIGURE 17.1 Sphere in a fluid stream showing the stagnation point on the front surface and the turbulent wake behind.

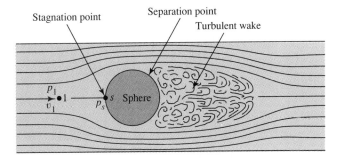

Solving for p_s, we get

$$p_s = p_1 + \gamma v_1^2/2g$$

But, since $\rho = \gamma/g$, we have

$$p_s = p_1 + \rho v_1^2/2 \qquad \textbf{(17–3)}$$

The stagnation pressure is greater than the static pressure in the free stream by the magnitude of the dynamic pressure $\rho v_1^2/2$. The kinetic energy of the moving stream is transformed into a kind of potential energy in the form of pressure.

The increase in pressure at the stagnation point can be expected to produce a force on the body opposing its motion, that is, a drag force. However, the magnitude of the force is dependent not only on the stagnation pressure but also on the pressure at the back side of the body. Since it is difficult to predict the actual variation in pressure on the back side, the drag coefficient is typically used.

The total drag on a body is due to two components. (For a lifting body such as an airfoil, a third component exists as described in Section 17.8.) *Pressure drag* (also called *form drag*) is due to the disturbance of the flow stream as it passes the body, creating a turbulent wake. The characteristics of the disturbance are dependent on the form of the body and sometimes on the Reynolds number of flow and the roughness of the surface. *Friction drag* is due to shearing stresses in the thin layer of fluid near the surface of the body called the *boundary layer*. These two types of drag are described in the following sections.

**17.3
PRESSURE DRAG**

As a fluid stream flows around a body, it tends to adhere to the surface for a portion of the length of the body. Then at a certain point, the thin boundary layer separates from the surface, causing a turbulent wake to be formed (see Fig. 17.1). The pressure in the wake is significantly lower than that at the stagnation point at the front of the body. A net force is thus created which acts in a direction opposite to that of the motion. This force is the pressure drag.

If the point of separation can be caused to occur farther back on the body, the size of the wake can be decreased and the pressure drag will be lower. This is the reason for streamlining. Figure 17.2 illustrates the change

FIGURE 17.2 Effect of streamlining on the wake.

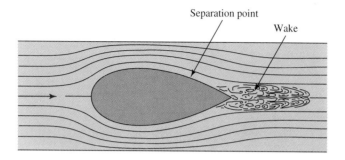

in the wake caused by the elongation and tapering of the tail of the body. Thus, the amount of pressure drag is dependent on the form of the body, and the term *form drag* is often used.

The pressure drag force is calculated from Eq. (17–1) in which *A is taken to be the maximum cross-sectional area of the body perpendicular to the flow.* The coefficient C_D is the pressure drag coefficient.

As an illustration of the importance of streamlining, the value of C_D for the drag on a smooth sphere moving through air with a Reynolds number of approximately 10^5 is 0.5. A highly streamlined shape like that used in most airships (blimps) has a C_D of approximately 0.04, a reduction by more than a factor of 10! (See Reference 2.)

17.3.1
Properties of Air

Drag on bodies moving in air is often the goal for drag analysis. In order to use Eq. (17–1) to calculate the drag forces, we need to know the density of the air. As with all gases, the properties of air change drastically with temperature. Also, as altitude above sea level increases, the density decreases. Appendix E presents the properties of air at various temperatures and altitudes.

17.4
DRAG COEFFICIENT

The magnitude of the drag coefficient for pressure drag depends on many factors, most notably the shape of the body, the Reynolds number of the flow, the surface roughness, and the influence of other bodies or surfaces in the vicinity. Two of the simpler shapes, the sphere and the cylinder, are discussed first.

17.4.1
Drag Coefficient for Spheres and Cylinders

Data plotted in Fig. 17.3 give the value of the drag coefficient versus Reynolds number for *smooth* spheres and cylinders. For spheres and cylinders, the Reynolds number is computed from the familiar *looking* relation

$$N_R = \frac{\rho v D}{\mu} = \frac{v D}{\nu} \qquad \textbf{(17–4)}$$

However, the diameter, D, is the diameter of the body itself, rather than the diameter of a flow conduit, which D represented earlier.

Note the very high values of C_D for low Reynolds numbers, over 100 for a smooth sphere at $N_R = 0.10$. This corresponds to motion through very

FIGURE 17.3 Drag coefficients for spheres and cylinders.

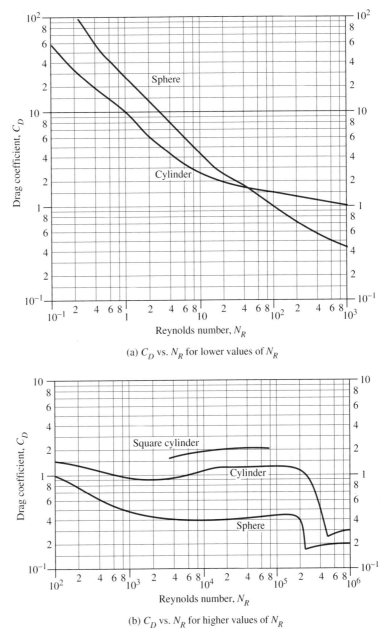

(a) C_D vs. N_R for lower values of N_R

(b) C_D vs. N_R for higher values of N_R

viscous fluids. It drops rapidly to a value of about 4 for $N_R = 10$ and then to
1.0 for $N_R = 100$. The value of C_D ranges from about 0.38 to 0.46 for the
higher Reynolds numbers from 1000 to 10^5.

For cylinders, $C_D \approx 60$ for the very low Reynolds number of 0.10. It
drops to a value of 10 for $N_R = 1.0$ and to a value of 1.0 for $N_R = 1000$. In the
higher ranges of Reynolds number, C_D ranges from about 0.90 to 1.30 for N_R
from 1000 to 10^5.

For very small Reynolds numbers ($N_R < 1.0$ approximately), the drag is almost entirely due to friction and will be discussed later. At higher Reynolds numbers, the importance of flow separation and the turbulent wake behind the body make pressure drag predominant. The following discussion relates only to pressure drag.

At a value of the Reynolds number of about 2×10^5, the drag coefficient for spheres drops sharply from approximately 0.42 to 0.17. This is caused by the abrupt change in the nature of the boundary layer from laminar to turbulent. Concurrently, the point on the sphere where separation occurs moves farther back, decreasing the size of the wake. For cylinders, a similar phenomenon occurs at approximately $N_R = 4 \times 10^5$ where C_D changes from about 1.2 to 0.30.

Either roughening the surface or increasing the turbulence in the flow stream can decrease the value of the Reynolds number at which the transition from a laminar to a turbulent boundary layer occurs, as illustrated in Fig. 17.4. This graph is meant to show typical curve shapes only and should not be used for numerical values.

FIGURE 17.4   Effect of turbulence and roughness on C_D for spheres.

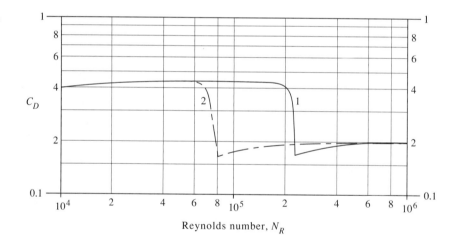

Golf balls are dimpled to optimize the turbulence of the air as it flows around the ball and to cause the abrupt decrease in the drag coefficient to occur at a low velocity (low Reynolds number), resulting in longer flights. A perfectly smooth golf ball could be driven only about 100 yd by even the best golfers, whereas the familiar dimpled design allows the average golfer to far exceed this distance. Highly skilled professional golfers can make 300-yd drives (Reference 3).

**17.4.2
Drag Coefficients
for Other Shapes**

Also shown in Fig. 17.3 is the drag coefficient for a square cylinder with a flat side perpendicular to the flow for Reynolds numbers from 3.5×10^3 to 8×10^4. The values range from approximately 1.60 to 2.05, somewhat higher than for the circular cylinder. Significant reductions can be obtained by small to moderate corner radii bringing values of C_D down to as low as 0.55 at high Reynolds numbers. However, the values tend to be highly affected by changes in Reynolds numbers for such designs. Testing is advised.

Figure 17.5 gives data for C_D for three versions of elliptical cylinders for Reynolds numbers from 3.0×10^4 to 2×10^5. These shapes have an ellipse for a cross section with different ratios of cross-sectional length to maximum thickness, sometimes called *fineness ratio*. Also shown for comparison is the circular cylinder that can be considered as a special case of the elliptical cylinder with a fineness ratio of $1:1$. Note the dramatic reduction of drag coefficient to about 0.21 for the elliptical cylinders of high fineness ratio.

FIGURE 17.5 Drag coefficients for elliptical cylinders and struts.

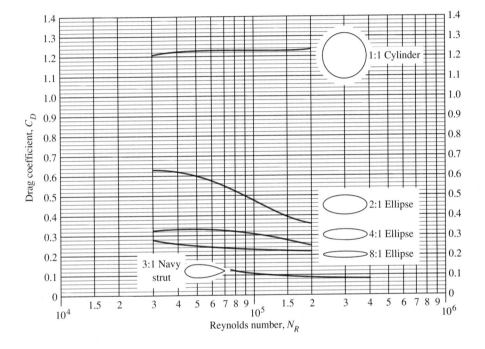

But even more reduction in drag coefficient can be made with the familiar "teardrop" shape, also shown in Fig. 17.5. This is a standard shape called a *Navy strut* that has values for C_D in the range of 0.07 to 0.11. Figure 17.6 on p. 476 shows the strut geometry. (See Reference 1.)

Table 17.1 lists values of the drag coefficients for several simple shapes. Note the orientation of the shape relative to the direction of the oncoming flow. The C_D values for such shapes are nearly independent of Reynolds numbers because they have sharp edges that cause the boundary layer to separate at the same place. Most of the testing for these shapes was done in the range of Reynolds numbers from 10^4 to 10^5.

For the square cylinder, semitubular cylinders, and triangular cylinders, the data are for models that are long relative to the major thickness

TABLE 17.1 Typical drag coefficients

Shape of Body	Orientation		C_D
Square cylinder	Flow →	◇	1.60
Semitubular cylinders	→	)	1.12
	→	(	2.30
Triangular cylinders	→	◁ 30°	1.05
	→	30° ◁	1.85
	→	◁	1.39
	→	▷ 60°	2.20
	→	◁	1.60
	→	▷ 90°	2.15
	→	◁	1.75
	→	▷ 120°	2.05
Rectangular plate	Flow →	b ▭ a	a/b 1 1.16 4 1.17 8 1.23 12.5 1.34 25 1.57 50 1.76 ∞ 2.00
Tandem disks L = spacing d = diameter	→	d L	L/d 1 0.93 1.5 0.78 2 1.04 3 1.52
One circular disk	→	d ○	1.11

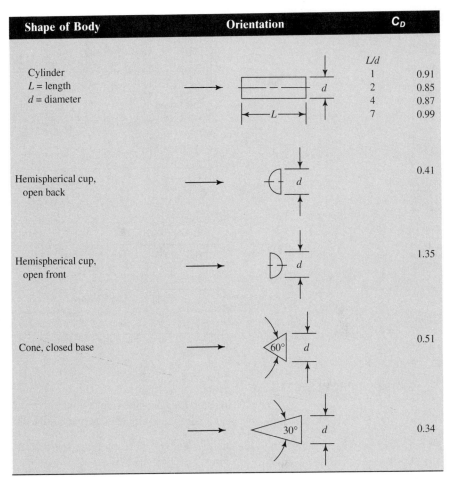

Shape of Body	Orientation	C_D
Cylinder L = length d = diameter	L/d 1 2 4 7	0.91 0.85 0.87 0.99
Hemispherical cup, open back		0.41
Hemispherical cup, open front		1.35
Cone, closed base	60°	0.51
	30°	0.34

Note: Reynolds numbers are based on the length of the body parallel to the flow direction except for the semitubular cylinders, for which the characteristic length is the diameter. Data adapted from Eugene A. Avallone and Theodore Baumeister III, eds., *Marks' standard handbook for mechanical engineers,* 9th ed. (New York: McGraw-Hill, 1987), Table 4, and W. F. Lindsey, *Drag of cylinders of simple shapes,* Report No. 619 (National Advisory Committee for Aeronautics, 1938).

dimension. For short cylinders of all shapes, the modified flow around the ends will tend to decrease the values for C_D below those listed in Table 17.1.

The computation of the Reynolds number for the shapes shown in Table 17.1 uses the *length of the body parallel to the flow* as the characteristic dimension for the body. The formula then becomes

$$N_R = \frac{\rho v L}{\mu} = \frac{v L}{\nu} \qquad \textbf{(17–5)}$$

FIGURE 17.6 Geometry of the Navy strut.

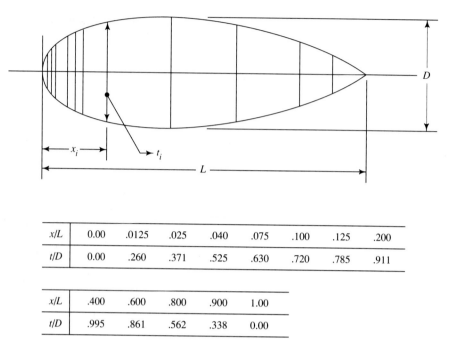

x/L	0.00	.0125	.025	.040	.075	.100	.125	.200
t/D	0.00	.260	.371	.525	.630	.720	.785	.911

x/L	.400	.600	.800	.900	1.00
t/D	.995	.861	.562	.338	0.00

☐ **EXAMPLE PROBLEM 17.1**   Compute the drag force on a 6.00-ft long square bar with a cross section of 4.00 in × 4.00 in when the bar is moving at 4.00 ft/s through water at 40°F. The long axis of the bar and a flat face are placed perpendicular to the flow.

Solution   We can use Eq. (17–1) to compute the drag force:

$$F_D = C_D(\rho v^2/2)A$$

Figure 17.3 shows that the drag coefficient depends on the Reynolds number found from Eq. (17–5).

$$N_R = \frac{vL}{\nu}$$

where L is the length of the bar parallel to the flow: 4.0 in or 0.333 ft. The kinematic viscosity of the water at 40°F is 1.67×10^{-5} ft²/s. Then

$$N_R = \frac{(4.00)(0.333)}{1.67 \times 10^{-5}} = 8.0 \times 10^4$$

Then, the drag coefficient $C_D = 2.05$. The maximum area perpendicular to the flow, A, can now be computed. A can also be described as the projected area seen if you look directly at the bar. In this case, then, the bar is a rectangle 0.333 ft high and 6.00 ft long. That is,

$$A = (0.333 \text{ ft})(6.00 \text{ ft}) = 2.00 \text{ ft}^2$$

The density of the air is 1.94 slugs/ft^3. Equivalent units are 1.94 lb · s^2/ft^4. We can now compute the drag force:

$$F_D = (2.05)(\tfrac{1}{2})(1.94 \text{ lb} \cdot \text{s}^2/\text{ft}^4)(4.00 \text{ ft/s})^2(2.00 \text{ ft}^2) = 63.6 \text{ lb}$$

**17.5
FRICTION DRAG
ON SPHERES IN
LAMINAR FLOW**

An exception to the method of analysis just presented for computing friction drag is used for spheres moving at low velocities in a viscous fluid, which results in very low Reynolds numbers. An important application of this phenomenon is the *falling ball viscometer,* discussed in Section 2.6.4. As a sphere falls through a viscous fluid, no separation occurs, and the boundary layer remains attached to the entire surface. Therefore, virtually all the drag is due to friction rather than to pressure drag.

In Reference 8, George G. Stokes presents important research on spheres moving through viscous fluids. He found that for Reynolds numbers less than about 1.0, the relationship between the drag coefficient and Reynolds number is $C_D = 24/N_R$. Special forms of the drag force equation can then be developed. The general form of the drag force equation is

$$F_D = C_D \left(\frac{\rho v^2}{2}\right) A$$

Letting $C_D = 24/N_R$ and letting $N_R = vD\rho/\mu$, we get

$$C_D = \frac{24}{N_R} = \frac{24\mu}{vD\rho}$$

Then, the drag force becomes

$$F_D = \frac{24\mu}{vD\rho}\left(\frac{\rho v^2}{2}\right) A = \frac{12\mu vA}{D} \qquad (17\text{--}6)$$

When computing friction drag, we use the surface area of the object. For a sphere, the surface area is $A = \pi D^2$. Then

$$F_D = \frac{12\mu vA}{D} = \frac{12\mu v(\pi D^2)}{D} = 12\pi\mu vD \qquad (17\text{--}7)$$

To correlate drag in the low Reynolds number range with that already presented in Section 17.4 dealing with pressure drag, we must redefine the area to be the maximum cross-sectional area of the sphere, $A = \pi D^2/4$. Equation (17–6) then becomes

⇨ **STOKES'S LAW**

$$F_D = \frac{12\mu vA}{D} = \left(\frac{12\mu v}{D}\right)\left(\frac{\pi D^2}{4}\right) = 3\pi\mu vD \qquad (17\text{--}8)$$

This form for the drag on a sphere in a viscous fluid is commonly called *Stokes's law.* As shown in Fig. 17.3, the relation $C_D = 24/N_R$ plots as a straight line for the low Reynolds numbers.

17.6
VEHICLE DRAG

Decreasing drag is a major goal in designing most kinds of vehicles, because a significant amount of energy is required to overcome drag as vehicles move through fluids. You are familiar with the streamlined shapes of aircraft bodies and the hulls of ships. Race cars and sports cars have long had the sleek styling characteristic of low aerodynamic drag. More recently, passenger cars and highway trucks have been redesigned to decrease drag.

Many factors affect the overall drag coefficient for vehicles, including the following:

1. The shape of the forward end, or *nose*, of the vehicle
2. The smoothness of the surfaces of the body
3. Such appendages as mirrors, door handles, antennas, and so forth
4. The shape of the tail section of the vehicle
5. The effect of nearby surfaces, such as the ground beneath an automobile
6. Discontinuities, such as wheels and wheel wells
7. The effect of other vehicles nearby
8. The direction of the vehicle with respect to prevailing winds
9. Air intakes to provide engine cooling or ventilation
10. The ultimate purpose of the vehicle (critical for commercial trucks)
11. The accommodation of passengers
12. Visibility afforded to operators and passengers
13. Stability and control of the vehicle
14. Aesthetics (the attractiveness of the design)

17.6.1
Automobiles

The overall drag coefficient, as defined in Eq. (17–1) based on the maximum projected frontal area, varies widely for passenger cars. Reference 5 lists a nominal mean value of 0.45, with a range of 0.30 to 0.60. Experimental shapes for cars have shown values as low as 0.175. An approximate value of 0.25 is practical for a "low drag" design.

The basic principles of drag reduction for automobiles include providing rounded, smooth contours for the forward part; elimination or streamlining of appendages; blending of changes in contour (such as at the hood/windshield interface); and rounding of rear corners.

□ **EXAMPLE PROBLEM 17.2**

A prototype automobile body has an overall drag coefficient of 0.35. Compute the total drag as it moves at 25 m/s through still air at 20°C. The maximum projected frontal area is 2.50 m².

Solution

We will use the drag force equation:

$$F_D = C_D \left(\frac{\rho v^2}{2}\right) A$$

From Appendix E, $\rho = 1.204$ kg/m³. Then

$$F_D = 0.35 \left[\frac{(1.204)(25)^2}{2}\right] (2.50) = 329 \text{ kg} \cdot \text{m/s}^2 = 329 \text{ N}$$

**17.6.2
Power Required to
Overcome Drag**

Power is defined as the rate of doing work. When a force is continuously exerted on an object while the object is moving at a constant velocity, power equals force times velocity. Then, the power required to overcome drag is

$$P_D = F_D v$$

Using the data from Example Problem 17.2, we get

$$P_D = (329 \text{ N})(25 \text{ m/s}) = 8230 \text{ N} \cdot \text{m/s} = 8230 \text{ W} = 8.23 \text{ kW}$$

In U.S. Customary System units, this would convert to 11.0 hp, a sizable power loss.

**17.6.3
Trucks**

The shapes commonly used for trucks fall into the category called *bluff bodies*. Reference 5 indicates that the approximate contribution of various parts of a truck to its total drag are

70 percent—the design of the front
20 percent—the design of the rear
10 percent—friction drag on body surfaces

As with automobiles, rounded smooth contours offer large improvements. For trucks with box-shaped cargo containers, designing corners with a large radius can assist in keeping the boundary layer from separating at the corners, consequently reducing the size of the turbulent wake behind the vehicle and reducing drag. In theory, providing a long, streamlined tail similar to the shape of an aircraft fuselage will reduce drag. However, such a vehicle would be too long to be practical or useful. Newer large highway trucks have drag coefficients in the range from 0.55 to 0.75.

**17.6.4
Trains**

Early locomotives had drag coefficients in the range of 0.80 to 1.05 (Reference 1). High-speed, streamlined trains can have values of approximately 0.40. For long passenger and freight trains, skin friction can be significant.

**17.6.5
Aircraft**

As with automobiles, wide variations in the overall drag coefficients of aircraft are to be expected with changes in the size and shape to accommodate different uses. For subsonic aircraft, the typical rounded, fairly blunt-nosed design with smooth blends at wings and tail structures and a long-tapered tail section results in drag coefficients of approximately 0.12 to 0.22. At supersonic speeds, the nose is usually sharp to diminish the effect of the shock wave. Operating at much lower speeds, the airship (dirigible or blimp) has a drag coefficient in the range of 0.04.

**17.6.6
Ships**

The total resistance to the motion of floating ships through water is due to skin friction, pressure or form drag, and wave-making resistance. Wave-making resistance, a large contributor to the total resistance, makes analyzing drag on ships quite different from analyzing drag on ground vehicles or aircraft. Reference 1 defines the *total ship resistance*, R_{ts}, as the force required to overcome all forms of drag. To normalize the values for different sizes of ships within a given class, values are reported as the ratio R_{ts}/Δ, where Δ is the displacement of the ship. Representative values of R_{ts}/Δ are given in Table 17.2. The resistance values can be combined with the speed of the ship (v) to compute the effective power required to propel it through the water:

$$P_E = R_{ts}v \qquad\qquad (17–9)$$

TABLE 17.2　Resistance of ships

Ship Type	R_{ts}/Δ
Ocean freighter	0.001
Passenger liner	0.004
Tugboat	0.006
Fast navy ship	0.01 to 0.12

□ **EXAMPLE PROBLEM 17.3**　Assume that a tugboat has a displacement of 625 long tons (1 long ton = 2240 lb) and is moving through water at 35 ft/s. Compute the total ship resistance and the total effective power required to drive the boat.

Solution　From Table 17.2, we find the specific resistance ratio to be $R_{ts}/\Delta = 0.006$. Then, the total ship resistance is

$$\Delta = (625 \text{ tons})(2240 \text{ lb/ton}) = 1.4 \times 10^6 \text{ lb}$$

$$R_{ts} = (0.006)(\Delta) = (0.006)(1.4 \times 10^6 \text{ lb}) = 8400 \text{ lb}$$

The power required is

$$P_E = R_{ts}v = (8400 \text{ lb})(35 \text{ ft/s}) = 0.294 \times 10^6 \text{ lb-ft/s}$$

Using 550 lb-ft/s = 1.0 hp, we get

$$P_E = (0.294 \times 10^6)/550 = 535 \text{ hp}$$

17.6.7 Submarines　A floating submarine's resistance can be computed in the same way as can a ship's. However, when completely submerged, none of the submarine's motion causes surface waves, and the computation of resistance is similar to that for an aircraft. The hull shape is similar to the shape of an aircraft fuselage, and skin friction plays a major role in the total resistance. Of course, the total *magnitude* of the drag for a submarine is significantly greater than that for an aircraft, because the density of water is far greater than that of air.

17.7 COMPRESSIBILITY EFFECTS AND CAVITATION

The results reported in Section 17.4 are for conditions in which the compressibility of the fluid (usually air) has little effect on the drag coefficient. These data are valid if the velocity of flow is less than about one-half the speed of sound in the fluid. Above that speed for air, the character of the flow changes and the drag coefficient increases rapidly.

When the fluid is a liquid such as water, we need not consider compressibility since liquids are very slightly compressible. However, we must consider another phenomenon called *cavitation*. As the liquid flows past a body, the static pressure decreases. If the pressure becomes sufficiently low, the liquid vaporizes, forming bubbles. Since the region of low pressure is generally small, the bubbles burst when they leave that region. When the collapsing of the vapor bubbles occurs near a surface of the body, rapid erosion or pitting results. Cavitation has other adverse effects when it occurs near control surfaces of boats or on propellers. The bubbles in the water decrease the forces exerted on rudders and control vanes and decrease thrust and performance of propellers.

17.8 LIFT AND DRAG ON AIRFOILS

We define lift as a force acting on a body in a direction perpendicular to that of the flow of the fluid. We will discuss the concepts concerning lift with reference to airfoils. The shape of the airfoil comprising the wings of an airplane determines its performance characteristics.

The manner in which an airfoil produces lift when placed in a moving air stream (or when moving in still air) is illustrated in Fig. 17.7. As the air flows over the airfoil, it achieves a high velocity on the top surface with a corresponding decrease in pressure. At the same time the pressure on the lower surface is increased. The net result is an upward force called *lift*. We express the *lift force* F_L as a function of a lift coefficient C_L in a manner similar to that presented for drag:

$$F_L = C_L(\rho v^2/2)A \qquad (17\text{--}10)$$

 LIFT FORCE

FIGURE 17.7 Pressure distribution on an airfoil.

$$\frac{\text{Pressure on airfoil}}{\text{Dynamic pressure}} = \frac{p}{\frac{1}{2}\rho v^2}$$

Upper surface (negative pressure)

Lower surface (positive pressure)

F_L Net lift

Flow

The velocity v is the velocity of the free stream of fluid relative to the airfoil. In order to achieve uniformity in the comparison of one shape with another, we usually define the area A as the product of the span of the wing and the length of the airfoil section called the *chord*. In Fig. 17.8, the span is b and the chord length is c.

FIGURE 17.8 Span and chord lengths for an airfoil.

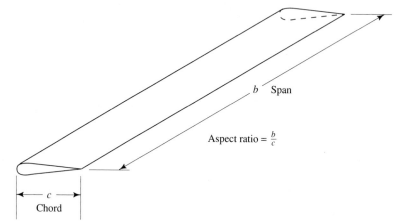

The value of the lift coefficient C_L is dependent on the shape of the airfoil and also on the angle of attack. Figure 17.9 shows that the angle of attack is the angle between the chord line of the airfoil and the direction of the fluid velocity. Other factors affecting lift are the Reynolds number, the surface roughness, the turbulence of the air stream, the ratio of the velocity of the fluid stream to the speed of sound, and the aspect ratio. *Aspect ratio* is the name given to the ratio of the span b of the wing to the chord length c. It is important because the characteristics of the flow at the wing tips are different from those toward the center of the span.

FIGURE 17.9 Induced drag.

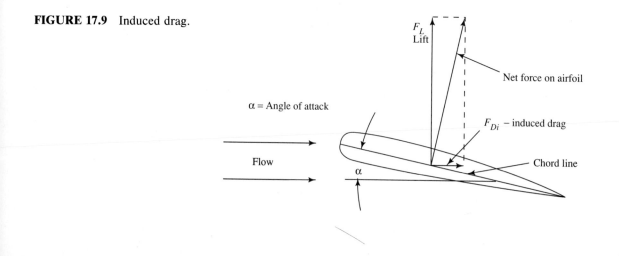

The total drag on an airfoil has three components. Friction drag and pressure drag occur as described before. The third component is called *induced drag*, which is a function of the lift produced by the airfoil. At a particular angle of attack, the net resultant force on the airfoil acts essentially perpendicular to the chord line of the section, as shown in Fig. 17.9. Resolving this force into vertical and horizontal components produces the true lift force F_L and the induced drag F_{Di}. Expressing the induced drag as a function of a drag coefficient gives

$$F_{Di} = C_{Di}(\rho v^2/2)A \qquad (17\text{--}11)$$

It can be shown that C_{Di} is related to C_L by the relation

$$C_{Di} = \frac{C_L^2}{\pi(b/c)} \qquad (17\text{--}12)$$

The total drag is then

$$F_D = F_{Df} + F_{Dp} + F_{Di} \qquad (17\text{--}13)$$

Normally, it is the total drag which is of interest in design. We determine a single drag coefficient C_D for the airfoil, from which the total drag can be calculated using the relation

$$F_D = C_D(\rho v^2/2)A \qquad (17\text{--}14)$$

As before, the area A is the product of the span b and the chord length c.

We use two methods to present the performance characteristics of airfoil profiles. In Fig. 17.10, the values of C_L, C_D, and the ratio of lift to drag

FIGURE 17.10 Airfoil performance curves.

F_L/F_D are all plotted versus the angle of attack as the abscissa. Note that the scale factors are different for each variable. The airfoil to which the data apply has the designation NACA 2409 according to a system established by the National Advisory Committee for Aeronautics. NACA Technical Report 610 explains the code used to describe airfoil profiles. NACA Reports 586, 647, 669, 708, and 824 present the performance characteristics of several airfoil sections.

The second method of presenting data for airfoils is shown in Fig. 17.11. This is called the *polar diagram* and is constructed by plotting C_L versus C_D with the angle of attack indicated as points on the curve.

FIGURE 17.11   Airfoil polar diagram.

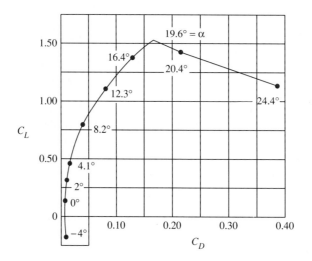

In both Fig. 17.10 and Fig. 17.11 it can be seen that the lift coefficient increases with increasing angle of attack up to a point where it abruptly begins to decrease. This point of maximum lift is called the *stall point*; at this angle of attack, the boundary layer of the air stream separates from the upper side of the airfoil. A large turbulent wake is created, greatly increasing drag and decreasing lift.

REFERENCES

1. Avallone, Eugene A., and Theodore Baumeister III, eds. 1987. *Marks' Standard Handbook for Mechanical Engineers*. 9th ed. New York: McGraw-Hill.

2. Daugherty, R. L., and J. B. Franzini. 1985. *Fluid Mechanics With Engineering Applications*. 8th ed. New York: McGraw-Hill.

3. Dunlop Sports Company. 1982. *MAXFLI DDH: The Technical Story*. Greenville, SC: Dunlop Sports Company.

4. Lindsey, W. F. 1938. *Drag of Cylinders of Simple Shapes*. Report No. 619. National Advisory Committee for Aeronautics.

5. Morel, T., and C. Dalton, eds. 1979. *Aerodynamics of Transportation*. New York: The American Society of Mechanical Engineers.

6. Prandtl, L. 1952. *Essentials of Fluid Dynamics*. New York: Hafner Publishing Co.

7. Siuru, Bill. 1988. Technology Keeps 18-wheelers

Truckin'. *Mechanical Engineering Magazine.* 110 (8):30–33.

8. Stokes, George G. 1901. *Mathematical and Physical Papers.* Vol. 3. London: Cambridge University Press.

9. Streeter, V. L. 1985. *Fluid Mechanics.* 8th ed. New York: McGraw-Hill.

10. von Mises, R. 1959. *Theory of Flight.* New York: Dover Publications. (First published in 1945 by the McGraw-Hill Book Co., New York.)

PRACTICE PROBLEMS

17.1M *A cylinder 25 mm in diameter is placed perpendicular to a fluid stream with a velocity of 0.15 m/s. If the cylinder is 1 m long, calculate the total drag force if the fluid is (a) water at 15°C and (b) air at 10°C and atmospheric pressure.*

17.2M *As part of an advertising sign on the top of a tall building, a 2-m diameter sphere called a "weather ball" glows different colors if the temperature is predicted to drop, rise, or remain about the same. Calculate the force on the weather ball due to winds of 15, 30, 60, and 120 km/h if the air is at 0°C.*

17.3M *Determine the terminal velocity (see Section 2.6.4) of a 75-mm diameter sphere made of solid aluminum (specific weight = 26.6 kN/m³) in free fall in (a) castor oil at 25°C, (b) water at 25°C, and (c) air at 20°C and standard atmospheric pressure. Consider the effect of buoyancy.*

17.4M *Calculate the moment at the base of a flagpole caused by a wind of 150 km/h. The pole is made of three sections, each 3 m long, of different size Schedule 80 steel pipe. The bottom section is 6-in, the middle is 5-in, and the top is 4-in. The air is at 0°C and standard atmospheric pressure.*

17.5M *A pitcher throws a baseball without spin with a velocity of 20 m/s. If the ball has a circumference of 225 mm, calculate the drag force on the ball in air at 30°C.*

17.6M *A parachute in the form of a hemispherical cup, 1.5 m in diameter, is deployed from a car trying for the land speed record. Determine the force exerted on the car if it is moving at 1100 km/h in air at atmospheric pressure and 20°C.*

17.7M *Calculate the required diameter of a parachute in the form of a hemispherical cup supporting a man weighing 800 N if the terminal velocity (see Section 2.6.4) in air at 40°C is to be 5 m/s.*

17.8M *A ship tows an instrument in the form of a 30° cone, point first, at 7.5 m/s in sea water. If the base of the cone has a diameter of 2.20 m, calculate the force in the cable to which the cone is attached.*

17.9M *A highway sign is being designed to withstand winds of 125 km/h. Calculate the total force on a sign 4 m by 3 m if the wind is flowing perpendicular to the face of the sign. The air is at −10°C. Compare the force calculated for this problem with that for Problem 16.4. Discuss the reasons for the differences.*

17.10M *Assuming that a semitrailer behaves as a square cylinder, calculate the force exerted if a wind of 20 km/h strikes it broadside. The trailer is 2.5 m by 2.5 m by 12 m. The air is at 0°C and standard atmospheric pressure.*

17.11M *A type of level indicator incorporates four hemispherical cups with open fronts mounted as shown in Fig. 17.12. Each cup is 25 mm in diameter. A motor drives the cups at a constant rotational speed. Calculate the torque that the motor must produce to maintain the motion at 20 r/min when the cups are in (a) air at 30°C and (b) gasoline at 25°C.*

17.12M *Determine the wind velocity required to overturn the mobile home sketched in Fig. 17.13 if it is 10 m long and weighs 50 kN. Consider it to be a square cylinder. The air is at 0°C.*

17.13M *A bulk liquid transport truck incorporates a cylindrical tank 2 m in diameter and 8 m long. For the tank alone, calculate the pressure drag when the truck is traveling at 100 km/h in still air at 0°C.*

(*Note:* For Problems 17.14 through 17.19, the air is at −20°F and its density is 2.80×10^{-3} slugs/ft³.

17.14E *A wing on a race car is supported by two cylindrical rods, as shown in Fig. 17.14. Compute the drag force exerted on the car due to these rods when the car is traveling through still air at −20°F at a speed of 150 mph.*

FIGURE 17.12   Problem 17.11.

FIGURE 17.13   Problem 17.12.

FIGURE 17.14 Problems 17.14 and 17.15.

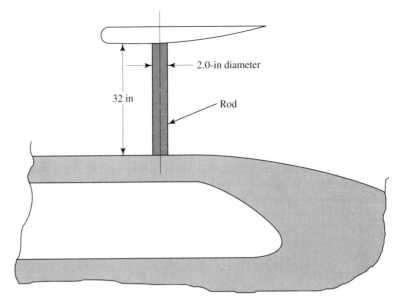

2.0-in diameter

32 in

Rod

17.15E In an attempt to decrease the drag on the car shown in Fig. 17.14 and described in Problem 17.14, the cylindrical rods are replaced by elongated elliptical cylinders having a length to breadth ratio of 8 : 1. By how much will the drag be reduced? Repeat for the *Navy strut*.

17.16E The four designs shown in Fig. 17.15 for the cross section of an emergency flasher lighting system for police vehicles are being evaluated. Each has a length of 60 in and a width of 9.00 in. Compare the drag force exerted on each proposed design when the vehicle moves at 100 mph through still air at −20°F.

17.17E A four-wheel drive utility vehicle incorporates a roll bar that extends above the cab and is in the free stream of air. The bar, made from 3-in Schedule 40 steel pipe, has a total length of 92 in exposed to the wind. Compute the drag exerted on the vehicle by the bar when the vehicle travels at 65 mph through still air at −20°F.

17.18E An advertising sign for the ABC Paper Company is shown in Fig. 17.16. It is made from three flat disks, each with a diameter of 56.0 in. The disks are joined by 4.50-in diameter tubes measuring 30 in between the disks. Compute the total force on the sign if it is faced into a 100-mph wind. The air is at −20°F.

17.19E An antenna in the shape of a cylindrical rod projects from the top of a locomotive. If the antenna is 42 in long and 0.200 in in diameter, compute the drag force on it when the locomotive is traveling at 160 mph in still air at −20°F.

17.20E A ship tows an instrument package in the form of a hemisphere with an open back at a velocity of 25.0 ft/s through seawater at 77°F. The diameter of the hemisphere is 7.25 ft. Compute the force in the cable to which the package is attached.

17.21E A flat rectangular plate, 8.50 × 11.00 in in size, is inserted into lake water at 60°F from a boat moving at 30 mph. What force is required to hold the plate steady relative to the boat with the flat face toward the water?

17.22E The windshield on an antique Stutz Bearcat automobile is a flat circular disk approximately 28 in in diameter. Compute the drag caused by the windshield when the car travels at 60 mph in still air at 50°F. The density of the car is 2.42×10^{-3} slugs/ft^3.

17.23E Assume that curve 2 in Fig. 17.4 is a true representation of the performance of a golf ball with a diameter of 1.25 in. If the Reynolds number is 1.5×10^5, compute the drag force on the golf ball and compare it to the drag force on a smooth sphere of the same diameter whose drag coefficient con-

FIGURE 17.15   Problem 17.16.

(a)

(b)

(c)

(d)

(e) Pictorial of light assembly mounted on the car

FIGURE 17.16   Problem 17.18.

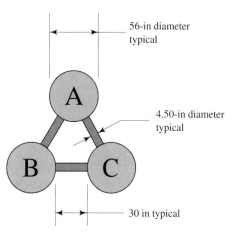

56-in diameter typical

4.50-in diameter typical

30 in typical

forms to curve 1. The air is at 50°F. The density of the air is 2.42×10^{-3} slugs/ft³ and the dynamic viscosity is 3.68×10^{-7} lb-s/ft².

17.24E In a falling ball viscometer, a steel sphere with a diameter of 1.200 in drops through a heavy syrup and travels 18.0 in in 20.40 s at a constant speed. Compute the viscosity of the syrup. The syrup has a specific gravity of 1.18. Note that the free body diagram of the sphere should include its weight acting down and the buoyant force and the drag force acting up. The steel has a specific gravity of 7.83. See also Section 2.6.4.

17.25E Compute the power required to overcome drag on a truck with a drag coefficient of 0.75 when the truck moves at 65 mph through still air at 50°F. The density of the air is 2.42×10^{-3} slugs/ft³, and the dynamic viscosity is 3.68×10^{-7} lb-s/ft². The maximum cross section of the truck is a rectangle 8.00 ft wide and 12.00 ft high.

17.26E A small, fast boat has a specific resistance ratio of 0.06 (see Table 17.2) and displaces 125 long tons. Compute the total ship resistance and the power required to overcome drag when moving at 50 ft/s in seawater at 77°F.

17.27E A passenger liner displaces 8700 long tons. Compute the total ship resistance and the power required to overcome drag when moving at 30 ft/s in seawater at 77°F.

17.28M *Assume that Fig. 17.10 shows the performance of the wing on the race car shown in Fig. 17.14. Note that it is mounted in the inverted position, so the lift pushes down to aid in skid resistance. Compute the downward force exerted on the car by the wing and the drag when the angle of attack is set at 15° and the speed is 25 m/s. The chord length is 780 mm and the span is 1460 mm.*

17.29M *Calculate the total drag on an airfoil that has a chord length of 2 m and a span of 10 m. The airfoil is at 3000 m flying at (a) 600 km/h and (b) 150 km/h. Use Fig. 17.10 for C_D and $\alpha = 15°$.*

17.30M *For the airfoil with the performance characteristics shown in Fig. 17.10, determine the lift and drag at an angle of attack of 10°. The airfoil has a chord length of 1.4 m and a span of 6.8 m. Perform the calculation at a speed of 200 km/h in the standard atmosphere at (a) 200 m and (b) 10 000 m.*

17.31M *Repeat Problem 17.30 if the angle of attack is the stall point, 19.6°.*

17.32M *For the airfoil in Problem 17.30, what load could be lifted from the ground at a takeoff speed of 125 km/h when the angle of attack is 15°? The air is at 30°C and standard atmospheric pressure.*

17.33M *Determine the required wing area for a 1350-kg airplane to cruise at 125 km/h if the airfoil is set at an angle of attack of 2.5°. The airfoil has the characteristics shown in Fig. 17.10. The cruise altitude is 5000 m in the standard atmosphere.*

■ ■ ■ ■ # 18 Fans, Blowers, Compressors, and the Flow of Gases

18.1
OBJECTIVES

Fans, blowers, and compressors are used to increase pressure and to cause the flow of air and other gases in a gas flow system. Their function is similar to that of pumps in a liquid flow system, as discussed in Chapter 15. Some of the principles already developed for the flow of liquids and the application of pumps can be applied also to the flow of gases. However, the compressibility of gases causes some important differences.

Special techniques for the design of flow systems carrying gases, such as air, have been developed by professionals after years of experience. The detailed analysis of the phenomena involved requires knowledge of thermodynamics. Because knowledge of thermodynamics is not necessary for the readers of this book, some of the methods in this chapter are presented without extensive development. New terms or concepts required for understanding the methods are, of course, described.

After completing this chapter, you should be able to:

1. Describe the general characteristics of fans, blowers, and compressors.
2. Describe propeller fans, duct fans, and centrifugal fans.
3. Describe blowers and compressors of the centrifugal, axial, vane-axial, reciprocating, lobed, vane, and screw types.
4. Specify suitable sizes for pipes carrying steam, air, and other gases at higher pressures.
5. Compute the flow rate of air and other gases through nozzles.

18.2
GAS FLOW RATES AND PRESSURES

When working in U.S. Customary System units, the flow rate of air or another gas is most frequently expressed in ft^3/min, abbreviated cfm. Velocities are typically reported in ft/min. Although these are not the standard units for the U.S. Customary System, they are convenient for the range of flows typically encountered in residential, commercial, or industrial applications.

In the SI system, m^3/s for flow rate and m/s for velocity are the units most often used. For systems carrying relatively low flow rates, the unit of L/s is sometimes used. Convenient conversions are listed below.

$$1.0 \ ft^3/s = 60 \ ft^3/min = 60 \ cfm$$
$$1.0 \ m^3/s = 2120 \ ft^3/min = 2120 \ cfm$$
$$1.0 \ ft/s = 60 \ ft/min$$

$$1.0 \text{ m/s} = 3.28 \text{ ft/s}$$

$$1.0 \text{ m/s} = 197 \text{ ft/min}$$

Pressures can be measured in lb/in² (abbreviated psi) in U.S. Customary System units when relatively large pressures are encountered. However, in most air handling systems, the pressures are small and are measured in *inches of water gage*, abbreviated as in H₂O. This unit is derived from the practice of using a pitot tube and water manometer to measure the pressure in ducts, as illustrated in Fig. 14.18. The equivalent pressure can be derived from the pressure-elevation relation, $\Delta p = \gamma h$. If we use $\gamma = 62.4$ lb/ft³ for water, a pressure of 1.00 in H₂O is equivalent to

$$\Delta p = \gamma h = \frac{62.4 \text{ lb}}{\text{ft}^3} \cdot 1.00 \text{ in} \cdot \frac{1 \text{ ft}^3}{1728 \text{ in}^3} = 0.0361 \text{ lb/in}^2$$

Stated differently, 1.0 psi = 27.7 in H₂O. In many air flow systems, the pressures involved are only a few inches of water or even a fraction of an inch.

The standard SI unit of Pascals (Pa) is itself quite small and is used directly when designing a system in SI units. Some useful conversion factors are listed below.

$$1.0 \text{ psi} = 6895 \text{ Pa}$$

$$1.0 \text{ in H}_2\text{O} = 248.8 \text{ Pa}$$

**18.3
CLASSIFICATION OF
FANS, BLOWERS,
AND COMPRESSORS**

Fans, blowers, and compressors are all used to increase the pressure of and move air or other gases. The primary differences among them are their physical construction and the pressures that they are designed to develop. A *fan* is designed to operate against small static pressures, up to about 2.0 psi (13.8 kPa). But typical operating pressures for fans are from 0 to 6 in H₂O (0.00 to 0.217 psi or 0.00 to 1500 Pa). At pressures from 2.0 psi up to approximately 10.0 psi (69.0 kPa), the gas mover is called a *blower*. To develop higher pressures, even as high as several thousand psi, *compressors* are used (see Reference 6).

Fans are used to circulate air within a space, to bring air into or exhaust it from a space, or to move air through ducts in ventilation, heating, or air conditioning systems. Types of fans include propeller fans, duct fans, and centrifugal fans.

Propeller fans operate at virtually zero static pressure and are composed of two to six blades with the appearance of aircraft propellers. Thus, they draw air in from one side and discharge it from the other side in an approximately axial direction. This type of fan is popular for circulating air in living or working spaces to improve personal comfort. When mounted in windows or other openings in the walls of a building, they deliver fresh outside air into the building or exhaust air from the building. When mounted in the ceiling or roof, they are often called *ventilators*.

Propeller fans are available from small sizes (a few inches in diameter, delivering a few hundred cfm) to 60 in or more in diameter, delivering over 50 000 cfm at zero static pressure. Operating speeds typically range from about 600 rpm to 1725 rpm. These fans are driven by electric motors, either directly or through belt drives.

Duct fans have a construction similar to that of propeller fans, except that the fan is mounted inside a cylindrical duct, as shown in Fig. 18.1. The duct could be a part of a larger duct system delivering air to or exhausting it from a remote area. Duct fans can operate against static pressures up to about 1.50 in H_2O (375 Pa). Sizes range from very small, delivering a few hundred cfm, to about 36 in, delivering over 20 000 cfm.

FIGURE 18.1   Duct fan. (Source: Hartzell Fan, Inc., Piqua, OH)

Two examples of *centrifugal fans* or *centrifugal blowers*, along with their rotors, are shown in Figs. 18.2 and 18.3. Air enters at the center of the rotor, also called the *impeller*, and is thrown outward by the rotating blades, thus adding kinetic energy. The high velocity gas is collected by the volute surrounding the rotor, where the kinetic energy is converted into an increased gas pressure for delivery through a duct system for ultimate use.

The construction of the rotor is typically one of four basic designs, as shown in Fig. 18.4. The *backward-inclined blade* is often made with simple flat plates. As the rotor rotates, the air tends to leave parallel to the blade along the vector called v_b in the figure. However, this is added vectorially to the tangential velocity of the blade itself, v_t, which gives the resultant velocity shown as v_R. The *forward-curved blades* yield a generally higher resul-

FIGURE 18.2   Centrifugal blower with straight, radial bladed rotor. (Source: Hartzell Fan, Inc., Piqua, OH)

FIGURE 18.3   Centrifugal blower with backward inclined bladed rotor. (Source: Hartzell Fan, Inc., Piqua, OH)

FIGURE 18.4 Four types of centrifugal fan rotors.

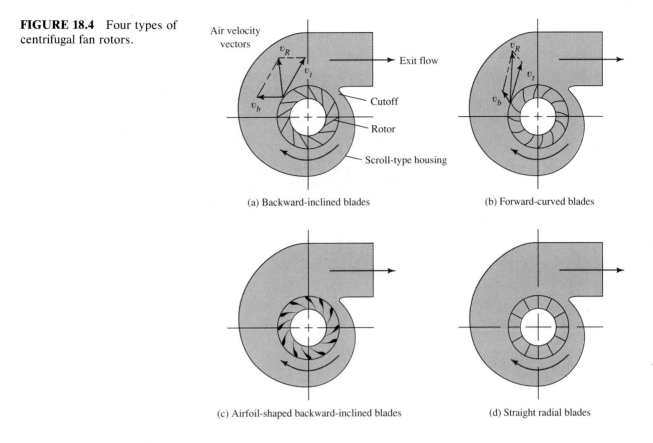

(a) Backward-inclined blades

(b) Forward-curved blades

(c) Airfoil-shaped backward-inclined blades

(d) Straight radial blades

tant air velocity, because the two component vectors are more nearly in the same direction. For this reason, a rotor with forward curved blades will run at a slower speed than a similar sized backward-inclined bladed fan for the same air flow and pressure. However, the backward-inclined fan typically requires lower power for the same service (Reference 8). *Airfoil shaped, backward-inclined fan blades* operate more quietly and efficiently than flat, backward-inclined blades. All these types of fans are used for ventilation systems and some industrial process uses. *Radial-blade* fans have many applications in industry for supplying large volumes of air at moderate pressures for boilers, cooling towers, material dryers, and bulk material conveying.

Centrifugal compressors employ impellers similar to those in centrifugal pumps (Figs. 15.7 and 15.8). However, the specific geometry is tailored to handle gas rather than liquid. Figure 18.5 shows a large, single-stage, centrifugal compressor. When a single-rotor compressor cannot develop a sufficiently high pressure, a multistage compressor, as shown in Fig. 18.6, is used. Centrifugal compressors are used for flows from about 500 to 100 000 cfm (0.24 to 47 m³/s) at pressures as high as 8000 psi (55 MPa).

FIGURE 18.5   Single-stage centrifugal compressor. (Source: Dresser-Rand, Turbo Products Divison, Olean, NY)

FIGURE 18.6   Lower case of a horizontally split multistage centrifugal compressor with rotor installed. (Source: Dresser-Rand, Turbo Products Divison, Olean, NY)

A multistage *axial compressor* is shown in Fig. 18.7. Only the lower half of the casing, the multistaged rotor, and the shaft assembly are shown. The gas is drawn into the large end, moved axially and compressed by the series of bladed rotors, and discharged from the small end. Axial compressors are employed to deliver large flow rates, approximately 8000 to 1.0 million cfm (3.8 to 470 m^3/s) at a discharge pressure up to 100 psi (690 kPa).

FIGURE 18.7 Lower case of an axial compressor with rotor installed. (Source: Dresser-Rand, Turbo Products Division, Olean, NY)

Vaneaxial blowers are similar to duct fans, described earlier, except they typically have blades that are airfoil-shaped and include vanes within the cylindrical housing to redirect the flow axially within the following duct. This results in higher static pressure capability for the blower and reduces swirl of the air.

Positive displacement blowers and compressors come in a variety of designs:

- Reciprocating—single-acting or double-acting
- Rotary—lobe, vane, or screw

The construction of a reciprocating compressor looks similar to that of an engine. A rotating crank and a connecting rod drive the piston. The piston reciprocates within its cylinder, drawing in low pressure gas as it travels away from the cylinder head and then compressing it within the cylinder as it travels toward the head. When the pressure of the gas reaches the desired level, discharge valves open to deliver the compressed gas to the piping system. Figure 15.5 shows the arrangement of both single-acting and double-acting pistons. Small versions of such compressors are seen in small shops and service stations. However, for many industrial users, they can be very large, delivering up to 10 000 cfm (4.7 m³/s) at pressures up to 60 000 psi (413 MPa).

Lobe and vane compressors appear very similar to the pumps shown in Figs. 15.1 and 15.4. Lobe-type designs can develop up to approximately 15 psi (100 kPa) and are often called blowers. Vane type compressors are capable of developing several hundred psi and are often used in pneumatic fluid power systems.

Screw compressors are used in construction and industrial applications requiring compressed air up to 500 psi (3.4 MPa) with delivery up to 20 000 cfm (9.4 m³/s). In the single screw design, air is trapped between adjacent "threads" rotating inside a close-fitting housing. The axial progression of the threads delivers the air to the outlet. In some designs, the pitch of the threads decreases along the length of the screw, providing compression within the housing as well as delivering it against system resistance. Two or more screws in mesh can also be employed. See Figure 15.2.

18.4
FLOW OF COMPRESSED AIR AND OTHER GASES IN PIPES

Many industries use compressed air in fluid power systems to power production equipment, material handling devices, and automation machinery. A common operating pressure for such systems is in the range of 60 to 125 psig (414 to 862 kPa gage). Performance and productivity of the equipment are degraded if the pressure drops below the design pressure. Therefore, careful attention must be paid to pressure losses between the compressor and the point of use. A detailed analysis of the piping system should be completed, using the methods outlined in Chapters 6 to 12 of this book, modified for the compressibility of air.

When large changes in pressure or temperature of the compressed air occur along the length of a flow system, the corresponding changes in the specific weight of the air should be taken into account. However, if the change in pressure is less than about 10 percent of the inlet pressure, variations in specific weight will have negligible effect. When the pressure drop is between 10 percent and 40 percent of the inlet pressure, we can use the average of the specific weight for the inlet and outlet conditions to produce results with reasonable accuracy (Reference 3). When the predicted pressure change is greater than 40 percent, either redesign the system or consult other references.

18.4.1
Specific Weight for Air

Figure 18.8 shows the variation of the specific weight for air as a function of changes in pressure and temperature. Note the large magnitude of the changes, particularly as pressure changes. The specific weight for any conditions of pressure and temperature can be computed from the *ideal gas law* from thermodynamics that states

IDEAL GAS LAW

$$\frac{p}{\gamma T} = \text{constant} = R \qquad \text{(18–1)}$$

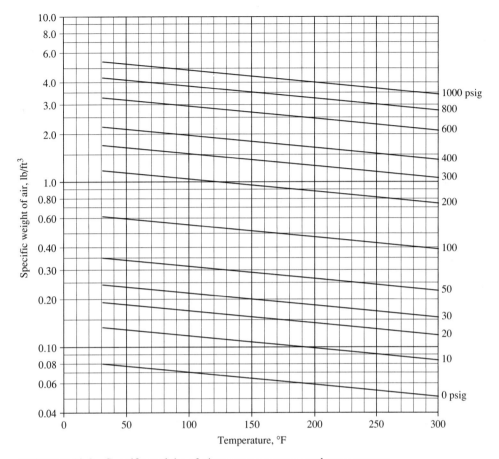

FIGURE 18.8 Specific weight of air versus pressure and temperature.

where p = absolute pressure of the gas

γ = specific weight of the gas

T = absolute temperature of the gas, that is, the temperature above absolute zero

R = gas constant for the gas being considered

Also, Eq. (18–1) can be solved for the specific weight:

$$\gamma = \frac{p}{RT} \qquad \textbf{(18–2)}$$

The absolute temperature is found by adding a constant to the measured temperature. In U.S. Customary System units,

$$T = (t\degree F + 460) \degree R$$

where °R is degrees Rankine, the standard unit for absolute temperature measured relative to absolute zero. In SI units,

$$T = (t°C + 273) \text{ K}$$

where K (Kelvin) is the standard SI unit for absolute temperature.

As presented in Section 3.2 (Eq. 3–2), absolute pressure is found by adding the prevailing atmospheric pressure to the gage pressure reading. We use $p_{atm} = 14.7$ psia in U.S. Customary System units and $p_{atm} = 101.3$ kPa absolute in SI units.

The value of the gas constant R for air is 53.3 ft·lb/lb·°R in U.S Customary System units. The numerator of the unit for R is energy in foot pounds (ft·lb), and thus the pound (lb) unit indicates *force*. The corresponding SI unit for energy is the newton meter (N·m), where

$$1.0 \text{ ft·lb} = 1.356 \text{ N·m}$$

The pound unit in the denominator for R is the weight of the air, also a force. To convert the U.S. Customary System unit °R to the SI unit K, we use 1.0 K = 1.8°R. Using these conversions, we can show that the value of R for air in the SI system is 29.2 N·m/N·K. Appendix N shows values for R for other gases.

In choosing to express the values for the gas constant R in terms of *weight*, we facilitate the computation of the *specific weight* γ, as shown in Example Problem 18.1. Later, we will use γ in calculations for the *weight flow rate* of a gas through a nozzle. It should be noted that the use of the resulting relationships based on weight should be confined to applications near the earth's surface, where the value of the acceleration due to gravity, g, is fairly constant.

In other types of analysis, particularly those in the field of thermodyamics, R is defined in terms of *mass* rather than weight. In aerospace environments, in which g and the weight can approach zero, the mass basis should also be used.

☐ EXAMPLE PROBLEM 18.1   Compute the specific weight of air at 100 psig and 80°F.

Solution   Using Eq. 18–2, we find

$$p = p_{atm} + p_{gage} = 14.7 \text{ psia} + 100 \text{ psig} = 114.7 \text{ psia}$$
$$T = t + 460 = 80°F + 460 = 540°R$$

Then,

$$\gamma = \frac{p}{RT} = \frac{114.7 \text{ lb}}{\text{in}^2} \cdot \frac{\text{lb·°R}}{53.3 \text{ ft·lb}} \cdot \frac{1}{540°R} \cdot \frac{144 \text{ in}^2}{\text{ft}^2}$$
$$\gamma = 0.574 \text{ lb/ft}^3$$

Note that the quantities 53.3 and 144 will always be used for air in this type of calculation. Then, a convenient, unit-specific equation for specific weight of air can be derived as follows:

$$\gamma = 2.70p/T \qquad \textbf{(18–3)}$$

This gives γ for air directly in pounds per cubic foot (lb/ft³) when the absolute pressure is in psia and the absolute temperature is in °R.

☐ **EXAMPLE PROBLEM 18.2**   Compute the specific weight of air at 750 kPa gage and 30°C.

Solution   Using Eq. (18–2), we get

$$p = p_{atm} + p_{gage} = 101.3 \text{ kPa} + 750 \text{ kPa} = 851.3 \text{ kPa}$$
$$T = t + 273 = 30°C + 273 = 303 \text{ K}$$

Then,

$$\gamma = \frac{p}{RT} = \frac{851.3 \times 10^3 \text{ N}}{\text{m}^2} \cdot \frac{\text{N} \cdot \text{K}}{29.2 \text{ N} \cdot \text{m}} \cdot \frac{1}{303 \text{ K}} = 96.2 \text{ N/m}^3$$

18.4.2
Flow Rates for
Compressed Air Lines

Ratings for equipment using compressed air and for compressors delivering the air are given in terms of *free air*, sometimes called *free air delivery* (f.a.d.). This gives the quantity of air delivered per unit time assuming that the air is at standard atmospheric pressure (14.7 psia or 101.3 kPa absolute) and at the standard temperature of 60°F or 15°C (absolute temperatures of 520°R or 285 K). To determine the flow rate at other conditions, the following equation can be used:

$$Q_a = Q_s \cdot \frac{p_{atm-s}}{p_{atm} + p_a} \cdot \frac{T_a}{T_s} \qquad \textbf{(18–4)}$$

where
Q_a = volume flow rate at actual conditions
Q_s = volume flow rate at standard conditions
p_{atm-s} = standard absolute atmospheric pressure
p_{atm} = actual absolute atmospheric pressure
p_a = actual gage pressure
T_a = actual absolute temperature
T_s = standard absolute temperature = 520°R or 285 K

Using these values and those of the standard atmosphere, we can write Eq. (18–4) as follows.

In U.S. Customary System units:

$$Q_a = Q_s \cdot \frac{14.7 \text{ psia}}{p_{atm} + p_a} \cdot \frac{(t + 460)°R}{520°R} \qquad \textbf{(18–4a)}$$

In SI units:

$$Q_a = Q_s \cdot \frac{101.3 \text{ kPa}}{p_{atm} + p_a} \cdot \frac{(t + 273)K}{285 \text{ K}} \qquad \textbf{(18–4b)}$$

☐ **EXAMPLE PROBLEM 18.3**　An air compressor has a rating of 500 cfm free air. Compute the flow rate in a pipe line in which the pressure is 100 psig and the temperature is 80°F.

Solution　Using Eq. (18–4a) and assuming that the local atmospheric pressure is 14.7 psia, we get

$$Q_a = 500 \text{ cfm} \cdot \frac{14.7 \text{ psia}}{14.7 + 100} \cdot \frac{(80 + 460)}{520} = 66.5 \text{ cfm}$$

18.4.3　Many factors must be considered to specify a suitable pipe size for carrying
Pipe Size Selection　compressed air in industrial plants. Some of those factors and the parameters involved follow.

- *Pressure drop.* Because friction losses are proportional to the square of the velocity of flow, it is desirable to use as large a pipe size as is feasible, to ensure adequate pressure at all points of use in a system.
- *Compressor power requirement.* The power required to drive the compressor increases as the pressure drop increases. Therefore, it is desirable to use large pipes to minimize pressure drop.
- *Cost of piping.* Large pipes cost more than small pipes, which makes the use of smaller pipes preferable.
- *Cost of the compressor.* In general, a compressor designed to operate at a higher pressure will cost more, which makes the use of larger pipes that minimize pressure drop preferable.
- *Installation cost.* Small pipes are easier to handle, but this is usually not a major factor.
- *Space required.* Small pipes take less space and provide less interference with other equipment or operations.
- *Future expansion.* To allow the addition of more air-using equipment in the future, large pipes are desirable.
- *Noise.* When air flows at a high velocity through pipes, valves, and fittings, it creates high noise level. It is desirable to use large pipes to permit lower velocities.

There is no clearly optimum pipe size for every installation, and the designer should evaluate the overall performance of several proposed sizes before making a final specification. To aid in beginning the process, Table 18.1 lists some suggested sizes.

As with other pipe line systems studied earlier, compressed air pipe systems typically contain valves and fittings to control the amount and direction of flow. We account for their effects by using the equivalent length technique, described in Section 10.10. Values for the L_e/D ratio are listed in Table 10.4.

☐ **EXAMPLE PROBLEM 18.4**　Specify a suitable size pipe for the delivery of 500 cfm (free air) at 100 psig at 80°F to an automated machine. The total length of straight pipe required between the compressor and the machine is 140 ft. The line also contains two fully open gate valves,

TABLE 18.1 Suggested compressed air system pipe sizes

Maximum Flow Rate (cfm)		Pipe Size (Schedule 40)
Free Air	Compressed Air (100 psig, 60°F)	
4	0.513	⅛
8	1.025	¼
20	2.563	⅜
35	4.486	½
80	10.25	¾
150	19.22	1
300	38.45	1¼
450	57.67	1½
900	115.3	2
1400	179.4	2½
2500	320.4	3
3500	448.6	3½
5000	640.8	4

Note: The sizes listed are the smallest standard Schedule 40 steel pipes that will carry the given flow rate at a pressure of 100 psig (690 kPa) with no more than 5.0 psi (34.5 kPa) pressure drop in 100 ft (30.5 m) of pipe. See Appendix F for pipe dimensions.

six standard elbows, and two standard tees, in which the flow is through the run of the tee. Then, analyze the pressure required at the compressor to ensure that the pressure at the machine is no less than 100 psig.

Solution As a tentative choice, let's consult Table 18.1 and specify a 1½-inch Schedule 40 steel pipe to carry the air. Then, from Appendix F we find $D = 0.1342$ ft and $A = 0.01414$ ft^2. We should now check to determine the actual pressure drop through the system and judge its acceptability. The solution procedure is similar to that used in Chapter 11. Special circumstances relative to air will be discussed.

Step 1. Write the energy equation between the outlet from the compressor and the inlet to the machine:

$$\frac{p_1}{\gamma_1} + z_1 + \frac{v_1^2}{2g} - h_L = \frac{p_2}{\gamma_2} + z_2 + \frac{v_2^2}{2g}$$

Note that the specific weight terms have been identified with subscripts for the reference points. Because air is compressible, there could be a significant change in specific weight. However, it is our intention in this design to have a small change in pressure between points 1 and 2. If this is achieved, the change in specific weight can be ignored. Therefore, let $\gamma_1 \approx \gamma_2$. The conditions at point 2 are the same as those used in Example Problem 18.1. Then, we will use $\gamma = 0.574$ lb/ft^3.

No information was given about the elevations of the compressor and the machine. Because the specific weight of air and other gases is so small, it is permissible to ignore elevation differences when dealing with the flow of gases, unless these differences are very large. As indicated in Sections 3.3 and 3.4, the pressure change is directly proportional to the specific weight of the fluid and to the change in elevation. For $\gamma = 0.574$ lb/ft^3 for air in this problem, an elevation change of 100 ft (about the height of a 10-story building), would change the pressure only 0.40 psi.

The velocity at the two reference points will be the same, because we will use the same size pipe throughout. Then, the velocity head terms can be cancelled from the energy equation.

Step 2. Solve for the pressure at the compressor:

$$p_1 = p_2 + \gamma h_L$$

Step 3. Evaluate the energy loss h_L by using Darcy's equation, and include the effects of minor losses:

$$h_L = f\left(\frac{L}{D}\right)\left(\frac{v^2}{2g}\right) + f_T\left(\frac{L_e}{D}\right)\left(\frac{v^2}{2g}\right)$$

The L/D term is the actual pipe length to flow diameter ratio.

Pipe: $L/D = (140 \text{ ft}/0.1342 \text{ ft}) = 1043$

The equivalent L_e/D for the valves and fittings are found in Table 10.4.

2 valves:	$L_e/D = 2(8)$	$= 16$
6 elbows:	$L_e/D = 6(30)$	$= 180$
2 tees:	$L_e/D = 2(20)$	$= 40$
Total:	L_e/D	$= 236$

The flow velocity can be computed from the continuity equation. It was determined in Example Problem 18.3 that the flow rate of 500 cfm of free air at actual conditions of 100 psig and 80°F is 66.5 cfm. Then,

$$v = \frac{Q}{A} = \frac{66.5 \text{ ft}^3}{\text{min}} \cdot \frac{1}{0.01414 \text{ ft}^2} \cdot \frac{1 \text{ min}}{60 \text{ s}} = 78.4 \text{ ft/s}$$

The velocity head is

$$\frac{v^2}{2g} = \frac{(78.4)^2 \text{ ft}^2/\text{s}^2}{2(32.2 \text{ ft/s}^2)} = 95.44 \text{ ft}$$

To evaluate the friction factor f, the density and viscosity of the air are needed. Knowing the specific weight of the air, we can compute the density from

$$\rho = \frac{\gamma}{g} = \left(\frac{0.574 \text{ lb}}{\text{ft}^3}\right)\left(\frac{s^2}{32.2 \text{ ft}}\right) = \frac{0.0178 \text{ lb} \cdot s^2}{\text{ft}^4} = 0.0178 \text{ slug/ft}^3$$

The dynamic viscosity of a gas does not change much as pressure changes. So, we can use the data from Appendix E, even though they are for standard atmospheric pressure. However, we must convert from U.S. Customary System to SI units.

$$t = (80°F - 32)(\tfrac{5}{9}) = 26.7°C$$

With interpolation, the dynamic viscosity is found to be $\mu = 1.84 \times 10^{-5}$ Pa·s. This must be converted back to the U.S. Customary System unit of lb·s/ft². Using Appendix K, we find

$$\mu = (1.84 \times 10^{-5} \text{ Pa·s})(2.089 \times 10^{-2}) = 3.84 \times 10^{-7} \text{ lb·s/ft}^2$$

It would be incorrect to use the kinematic viscosity listed for air in Appendix E, because that value includes the density, which is dramatically different at 100 psig than it is at atmospheric pressure.

We can now compute the Reynolds number:

$$N_R = \frac{vD\rho}{\mu} = \frac{(78.4)(0.1342)(0.0178)}{3.84 \times 10^{-7}} = 4.88 \times 10^5$$

The relative roughness D/ϵ is

$$D/\epsilon = 0.1342/1.5 \times 10^{-4} = 895$$

Then, from the Moody diagram (Fig. 9.2), we read $f = 0.021$. The value for f_T used for the valves and fittings can be found from Table 10.5 to be 0.021 for the 1½ in Schedule 40 pipe. Because this is equal to the friction factor for the pipe itself, the L/D for the pipe can be added to the total L_e/D for the valves and fittings.

$$(L_e/D)_{\text{total}} = 1043 + 236 = 1279$$

And now the energy loss can be computed:

$$h_L = f_T \left(\frac{L_e}{D}\right)_{\text{total}} \left(\frac{v^2}{2g}\right) = (0.021)(1279)(95.44) = 2563 \text{ ft}$$

Step 4. Compute the pressure drop in the pipe line.

$$p_1 - p_2 = \gamma h_L = \frac{0.574 \text{ lb}}{\text{ft}^3} \cdot 2563 \text{ ft} \cdot \frac{1 \text{ ft}^2}{144 \text{ in}^2} = 10.22 \text{ psi}$$

Step 5. Compute the pressure at the compressor.

$$p_1 = p_2 + 10.22 \text{ psi} = 100 \text{ psig} + 10.22 \text{ psi} = 110.2 \text{ psig}$$

Step 6. Because the change in pressure is less than 10 percent, the assumption that the specific weight of the air is constant is satisfactory. If a larger pressure drop had occurred, we could either redesign the system with a larger pipe size or adjust the specific weight to the average of those at the beginning and end of the system. This system design appears to be satisfactory with regard to pressure drop.

18.5 FLOW OF AIR AND OTHER GASES THROUGH NOZZLES

The typical design of a nozzle is a converging section through which a fluid flows from a region of higher pressure to a region of lower pressure. Figure 18.9 shows a nozzle installed in the side of a relatively large tank with flow from the tank to the atmosphere. The nozzle shown converges smoothly and gradually, terminating at its smallest section, called the *throat*. Other designs for nozzles, including abrupt orifices and those attached to smaller pipes at the inlet, require special treatment, as we will discuss later.

FIGURE 18.9 Discharge of a gas from a tank through a smooth convergent nozzle.

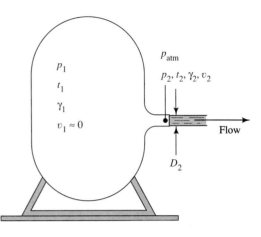

Some concepts from the field of thermodynamics must now be discussed, along with some additional properties of gases.

When the flow of a gas occurs very slowly, heat from the surroundings can transfer to or from the gas to maintain its temperature constant. Such flow is called *isothermal*. However, when the flow occurs very rapidly or when the system is very well insulated, very little heat can be transferred to or from the gas. Under ideal conditions with *no* heat transferred, the flow is called *adiabatic*. Real systems behave in some manner between isothermal and adiabatic. However, for rapid flow through a nozzle, we will assume that the flow is adiabatic.

For an adiabatic process, the equation that describes the relationship between the absolute pressure and the specific weight of the gas is

$$\frac{p}{\gamma^k} = \text{constant} \qquad (18\text{--}5)$$

The exponent k is called the *adiabatic exponent*, a dimensionless number, and its value for air is 1.40. Appendix N gives the values for k for other gases.

Equation (18–5) can be used to compute the condition of a gas at a point of interest if the condition at some other point is known and if an adiabatic process occurs between the two points. That is,

$$\frac{p}{\gamma^k} = \text{constant} = \frac{p_1}{\gamma_1^k} = \frac{p_2}{\gamma_2^k} \qquad (18\text{--}6)$$

Stated differently,

$$\frac{p_2}{p_1} = \left(\frac{\gamma_2}{\gamma_1}\right)^k \qquad (18\text{--}7)$$

or,

$$\frac{\gamma_2}{\gamma_1} = \left(\frac{p_2}{p_1}\right)^{1/k} \qquad (18\text{--}8)$$

The weight flow rate of gas exiting the tank through the nozzle in Fig. 18.9 is

$$W = \gamma_2 v_2 A_2 \tag{18-9}$$

The principles of thermodynamics can be used to show that the velocity of flow in the nozzle is

$$v_2 = \left\{ \left(\frac{2gp_1}{\gamma_1} \right) \left(\frac{k}{k-1} \right) \left[1 - \left(\frac{p_2}{p_1} \right)^{(k-1)/k} \right] \right\}^{1/2} \tag{18-10}$$

Note that the pressures here are *absolute pressures*. Equations (18–6) through (18–10) can be combined to produce a convenient equation for the weight rate of flow from the tank in terms of the conditions of the gas in the tank and the pressure ratio p_2/p_1.

$$W = A_2 \sqrt{\frac{2gk}{k-1} (p_1 \gamma_1) \left[\left(\frac{p_2}{p_1} \right)^{2/k} - \left(\frac{p_2}{p_1} \right)^{(k+1)/k} \right]} \tag{18-11}$$

Note that a *decreasing* pressure ratio p_2/p_1 actually indicates an *increasing* pressure difference $(p_1 - p_2)$ and that, therefore, it is expected that the weight flow rate W will increase as the pressure ratio is decreased. This is true for the higher values of the pressure ratio. However, it can be shown that the flow rate reaches a maximum at a *critical pressure ratio*, defined as

CRITICAL PRESSURE RATIO

$$\left(\frac{p_2}{p_1} \right)_c = \left(\frac{2}{k+1} \right)^{k/(k-1)} \tag{18-12}$$

Because the value of the critical pressure ratio is a function only of the adiabatic exponent k, it is a constant for any particular gas.

When the critical pressure ratio is reached, the velocity of flow at the throat of the nozzle is equal to the speed of sound in the gas at the conditions that prevail there. *And this velocity of flow remains constant regardless of how much the downstream pressure is lowered.* The speed of sound in the gas is

SONIC VELOCITY

$$c = \sqrt{\frac{kgp_2}{\gamma_2}} \tag{18-13}$$

Another name for c is the *sonic velocity*, the velocity that a sound wave would travel in the gas. This is the maximum velocity of flow of a gas through a converging nozzle.

Supersonic velocity, velocity greater than the speed of sound, can be obtained only with a nozzle that first converges and then diverges. The analysis of such nozzles is not presented here.

The name *Mach number* is given to the ratio of the actual velocity of flow to the sonic velocity. That is,

$$N_M = v/c \tag{18-14}$$

Equation (18–11) must be used to compute the weight flow rate of gas from a tank through a converging nozzle for values of $N_M < 1.0$ for which the pressure ratio, p_2/p_1, is greater than the critical pressure ratio. For $N_M = 1$, substituting the critical pressure ratio from Eq. (18–12) into Eq. (18–11) yields

$$W_{max} = A_2 \sqrt{\left(\frac{2gk}{k+1}\right)(p_1\gamma_1)\left(\frac{2}{k+1}\right)^{2/(k-1)}} \qquad (18\text{–}15)$$

This equation must be used when the pressure ratio p_2/p_1 is less than the critical ratio.

Figure 18.10 shows the behavior of gas flow through a nozzle from a relatively large tank, according to Eqs. (18–11) and (18–15). The graph in (a) is for the case in which the pressure in the tank, p_1, is held constant and the pressure outside the nozzle, p_2, is decreased. When $p_2 = p_1$, $p_2/p_1 = 1.00$, and there is obviously no flow through the nozzle. As p_2 is decreased, the relatively larger difference in pressure ($p_1 - p_2$) causes an increase in the weight flow rate as computed from Eq. (18–11). However, when the critical pressure ratio, $(p_2/p_1)_c$, is reached, the velocity in the throat reaches sonic velocity and the pressure remains at the critical pressure, p_2', computed from Eq. (18–12). That is,

$$p_2 = p_2' = p_1\left(\frac{2}{k+1}\right)^{k/(k-1)} \qquad (18\text{–}16)$$

Decreasing the pressure outside the nozzle any further would *not* increase the rate of flow from the tank.

Figure 18.10(b) shows a different interpretation of the variation of weight flow rate versus pressure ratio. In this case, the pressure outside the nozzle, p_2, is held constant, while the pressure in the tank, p_1, is increased. An example is discharge from a tank to the atmosphere. Obviously, when the gage pressure in the tank is zero, no flow occurs, because there is no pressure differential. As p_1 is increased, the pressure ratio p_2/p_1 is at first greater than the critical pressure ratio, and Eq. (18–11) applies. When the critical pressure ratio is reached or exceeded, the velocity in the throat will be at sonic velocity *for the condition of the gas in the throat*. Note, however, that for any given value of p_1, the critical pressure in the throat, p_2', is given by Eq. (18–16). But, because p_1 is increasing, p_2' is also increasing. Still, because the pressure ratio between the tank and the throat is at the critical value, Eq. (18–15) must be used to compute the weight flow rate through the nozzle. The weight flow rate is now dependent on p_1 and γ_1. Note also that γ_1 is directly proportional to p_1, as can be seen from Eq. (18–2). Then, after the critical pressure ratio is reached, the weight flow rate increases linearly as the pressure in the tank increases.

If the nozzle has an abrupt reduction rather than the smooth shape shown in Fig. 18.9, the flow will be less than that predicted from Eq. (18–11) or (18–15). A discharge coefficient should be applied, similar to the discharge coefficients described in Chapter 14 for venturi meters, flow nozzles, and orifice meters. Also, if the upstream section is relatively small, as in a

FIGURE 18.10   Weight flow rate of a gas through a nozzle.

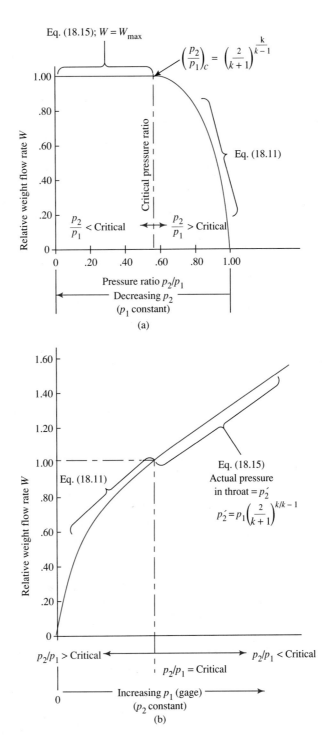

(a)

(b)

pipe, some correction for the velocity of approach should be applied (see Reference 5).

The following procedure can be used to compute the weight flow rate of a gas through a nozzle of the type shown in Fig. 18.9, assuming that the flow is adiabatic.

COMPUTATION OF ADIABATIC FLOW OF A GAS THROUGH A NOZZLE

1. Compute the actual pressure ratio between the pressure outside the nozzle and that in the tank, p_{atm}/p_1.
2. Compute the critical pressure ratio by using Eq. (18–12).
3a. If the actual pressure ratio is greater than the critical pressure ratio, use Eq. (18–11) to compute the weight flow rate through the nozzle with $p_2 = p_{atm}$. If desired, the velocity of flow can be computed by using Eq. (18–10).
3b. If the actual pressure ratio is less than the critical pressure ratio, use Eq. (18–15) to compute the weight flow rate through the nozzle. Also, recognize that the velocity of flow in the throat of the nozzle is equal to the sonic velocity, computed from Eq. (18–13), and that the pressure at the throat is that called p_2 in Eq. (18–12). The gas then expands to p_{atm} as it leaves the nozzle.

□ **EXAMPLE PROBLEM 18.5**   For the tank with a nozzle in its side, shown in Fig. 18.9, compute the weight flow rate of air leaving the tank for the following conditions:

$$p_1 = 10.0 \text{ psig} = \text{pressure in the tank}$$

$$p_{atm} = 14.2 \text{ psia} = \text{atmospheric pressure outside the tank}$$

$$t_1 = 80°F = \text{temperature of the air in the tank}$$

$$D_2 = 0.100 \text{ in} = \text{diameter of the nozzle at its outlet}$$

Solution   Using the procedure outlined above:

1. Actual pressure ratio:

$$\frac{p_{atm}}{p_1} = \frac{14.2 \text{ psia}}{(10.0 + 14.2) \text{ psia}} = 0.587$$

2. Determine the critical pressure ratio from Appendix N. For air, it is 0.528.
3a. Because the actual ratio is greater than the critical ratio, Eq. (18–11) is used for the weight flow rate. We must compute the nozzle throat area, A_2.

$$A_2 = \pi(D_2)^2/4 = \pi(0.100 \text{ in})^2/4 = 0.00785 \text{ in}^2$$

Converting to ft², we get

$$A_2 = 0.00785 \text{ in}^2 (1.0 \text{ ft}^2/144 \text{ in}^2) = 5.45 \times 10^{-5} \text{ ft}^2$$

Eq. (18–3) can be used to compute γ_1:

$$\gamma_1 = \frac{2.70 \, p_1}{T_1} = \frac{2.70(24.2 \text{ psia})}{(80 + 460)°R} = 0.121 \text{ lb/ft}^3$$

It is helpful to convert p_1 to the units of lb/ft^2.

$$p_1 = \frac{24.2 \text{ lb}}{\text{in}^2} \frac{144 \text{ in}^2}{\text{ft}^2} = 3485 \text{ lb/ft}^2$$

Then, using Eq. (18–11), with consistent U.S. Customary System units, we find the result for W in lb/s.

$$W = A_2 \sqrt{\frac{2gk}{k-1} (p_1\gamma_1) \left[\left(\frac{p_2}{p_1}\right)^{2/k} - \left(\frac{p_2}{p_1}\right)^{(k+1)/k} \right]} \tag{18–11}$$

$$W = (5.45 \times 10^{-5}) \sqrt{\frac{2(32.2)(1.4)(3485)(0.121)}{(1.4-1)} [(0.587)^{2/1.4} - (0.587)^{2.4/1.4}]}$$

$$W = 4.32 \times 10^{-3} \text{ lb/s}$$

☐ **EXAMPLE PROBLEM 18.6**   For the conditions used in Example Problem 18.5, compute the velocity of flow in the throat of the nozzle and the Mach number for the flow.

Solution   Eq. (18–10) must be used to compute the velocity in the throat. For consistent U.S. Customary System units, the velocity will be in ft/s.

$$v_2 = \left\{ \frac{2gp_1}{\gamma_1} \left(\frac{k}{k-1}\right) \left[1 - \left(\frac{p_2}{p_1}\right)^{(k-1)/k} \right] \right\}^{1/2} \tag{18–10}$$

$$v_2 = \left\{ \left(\frac{2(32.2)(3485)}{0.121}\right) \left(\frac{1.40}{0.40}\right)(1 - (0.587)^{0.4/1.4}) \right\}^{1/2}$$

$$v_2 = 957 \text{ ft/s}$$

To compute the Mach number, we need to compute the speed of sound in the air at the conditions existing in the throat by using Eq. (18–13):

$$c = \sqrt{\frac{kgp_2}{\gamma_2}} \tag{18–13}$$

The pressure $p_2 = p_{\text{atm}} = 14.2$ psia. Converting to lb/ft^2, we get

$$p_2 = \left(\frac{14.2 \text{ lb}}{\text{in}^2}\right)\left(\frac{144 \text{ in}^2}{1.0 \text{ ft}^2}\right) = 2045 \text{ lb/ft}^2$$

The specific weight, γ_2, can be computed from Eq. (18–8).

$$\frac{\gamma_2}{\gamma_1} = \left(\frac{p_2}{p_1}\right)^{1/k} \tag{18–8}$$

Knowing that $\gamma_1 = 0.121$ lb/ft^3, we get

$$\gamma_2 = \gamma_1 \left(\frac{p_2}{p_1}\right)^{1/k}$$

$$\gamma_2 = (0.121)(0.587)^{1/1.4} = 0.0827 \text{ lb/ft}^3$$

Then the speed of sound is

$$c = \sqrt{\frac{kgp_2}{\gamma_2}} \tag{18-13}$$

$$c = \sqrt{\frac{(1.4)(32.2)(2045)}{0.0827}} = 1056 \text{ ft/s}$$

Now the Mach number can be computed.

$$N_M = \frac{v}{c} = \frac{957 \text{ ft/s}}{1056 \text{ ft/s}} = 0.906$$

☐ **EXAMPLE PROBLEM 18.7**   Compute the weight flow rate of air from the tank through the nozzle shown in Fig. 18.9 if the pressure in the tank is raised to 20.0 psig. All other conditions are the same as in Example Problem 18.5.

Solution   Use the same procedure outlined before:

1. Actual pressure ratio:

$$\frac{p_2}{p_1} = \frac{14.2 \text{ psia}}{(20.0 + 14.2) \text{ psia}} = 0.415$$

2. The critical pressure ratio is, again, 0.528 for air.

3b. Because the actual pressure ratio is less than the critical pressure ratio, Eq. (18–15) should be used.

$$W_{max} = A_2 \sqrt{\frac{2gk}{k+1} (p_1 \gamma_1) \left(\frac{2}{k+1}\right)^{2/(k-1)}} \tag{18-15}$$

Computing γ_1 for $p_1 = 34.2$ psia, we get

$$\gamma_1 = \frac{2.70 p_1}{T_1} = \frac{(2.70)(34.2 \text{ psia})}{540°R} = 0.171 \text{ lb/ft}^3$$

We need the pressure p_1 in lb/ft^2.

$$p_1 = \left(\frac{34.2 \text{ lb}}{\text{in}^2}\right)\left(\frac{144 \text{ in}^2}{\text{ft}^2}\right) = 4925 \text{ lb/ft}^2$$

Then the weight flow rate is

$$W_{max} = (5.45 \times 10^{-5}) \sqrt{\frac{2(32.2)(1.4)(4925)(0.171)}{2.4} \left[\frac{2}{2.4}\right]^{2/0.4}}$$

$$W_{max} = 6.15 \times 10^{-3} \text{ lb/s}$$

The velocity of the air flow at the throat will be the speed of sound at the throat conditions. However, the pressure at the throat must be determined from the critical pressure ratio, Eq. (18–12).

$$\left(\frac{p_2}{p_1}\right)_c = \left(\frac{2}{k+1}\right)^{k/(k-1)} = 0.528$$

$$p_2' = p_1(0.528) = (4925 \text{ lb/ft}^2)(0.528) = 2600 \text{ lb/ft}^2$$

Knowing that $\gamma_1 = 0.171$ lb/ft^3, we find

$$\gamma_2 = \gamma_1 \left(\frac{p_2}{p_1}\right)^{1/k}$$

$$\gamma_2 = (0.171)(0.528)^{1/1.4} = 0.1084 \text{ lb/ft}^3$$

Then, the speed of sound and also the velocity in the throat is

$$c = \sqrt{\frac{kgp_2}{\gamma_2}} = \sqrt{\frac{(1.4)(32.2)(2600)}{0.1084}} = 1040 \text{ ft/s}$$

Of course, the Mach number in the throat is then 1.0

REFERENCES

1. Air Movement and Control Association. 1987. *Air Systems, ACMA Publication 200*. Arlington Heights, Illinois: ACMA.
2. American Society of Mechanical Engineers. 1979. *Glossary of Terms Used in the Measurement of Fluid Flow in Pipes*. Standard ANSI/ASME MFC-1M-1979. New York: Author.
3. Crane Co. 1988. *Flow of Fluids Through Valves, Fittings, and Pipe*. Technical Paper 410. Joliet, Il.: Author.
4. Hartzell Fan, Inc. 1987. *Hartzell Stock Fans and Blowers*. Piqua, Ohio: Author.
5. Idelchik, I. E. 1986. *Handbook of Hydraulic Resistance*. 2nd ed. Washington, D.C.: Hemisphere Publishing Corporation.
6. Matley, Jay, ed., and the staff of Chemical Engineering. 1979. *Fluid Movers: Pumps, Compressors, Fans and Blowers*. New York: McGraw-Hill.
7. Osborne, W. C. 1977. *Fans*. 2nd ed. New York: Pergamon Press.
8. The Trane Company. 1965. *Trane Air Conditioning Manual*. La Crosse, Wisconsin: Author.

PRACTICE PROBLEMS

Units and Conversion Factors

18.1E A pipe in a compressed air system is carrying 2650 cfm. Compute the flow rate in ft^3/s.

18.2E A duct in a heating system carries 8320 cfm. Compute the flow rate in ft^3/s.

18.3C A pipe in a compressed air system carries 2650 cfm. Compute the flow rate in m^3/s.

18.4C A duct in a heating system carries 8320 cfm. Compute the flow rate in m^3/s.

18.5C The velocity of flow in a ventilation duct is 1140 ft/min. Compute the velocity in m/s.

18.6C The velocity of flow in an air conditioning system duct is 5.62 m/s. Compute the velocity in ft/s.

18.7E A measurement for the static pressure in a heating duct is 4.38 in H$_2$O. Express this pressure in psi.

18.8C A certain fan is rated to deliver 4760 cfm of air at a static pressure of 0.75 in H$_2$O. Express the flow rate in m^3/s and the pressure in Pa.

18.9C The static pressure in a gas pipe is measured to be 925 Pa. Express this pressure in H$_2$O.

18.10C Express the pressure of 925 Pa in psi.

Fans, Blowers, and Compressors

18.11 Describe a centrifugal fan with backward-inclined blades.

18.12 Describe a centrifugal fan with forward-curved blades.

18.13 Describe a duct fan.

18.14 Describe a vaneaxial blower, and compare it with a duct fan.

18.15 Name four types of positive displacement compressors.

18.16 Name a type of compressor often used for pneumatic fluid power systems.

Specific Weight of Air

18.17E Compute the specific weight of air at 80 psig and 75°F.

18.18E Compute the specific weight of air at 25 psig and 105°F.

18.19E Compute the specific weight of natural gas at 4.50 in H_2O and 55°F.

18.20E Compute the specific weight of nitrogen at 32 psig and 120°F.

18.21M *Compute the specific weight of air at 1260 Pa(gage) and 25°C.*

18.22E Compute the specific weight of propane at 12.6 psig and 85°F.

Flow of Compressed Air in Pipes

18.23E An air compressor delivers 820 cfm of free air. Compute the flow rate of air in a pipe in which the pressure is 80 psig and the temperature is 75°F.

18.24E An air compressor delivers 2880 cfm of free air. Compute the flow rate of air in a pipe in which the pressure is 65 psig and the temperature is 95°F.

18.25E Specify a size of Schedule 40 steel pipe suitable for carrying 750 cfm (free air delivery) at 100 psig with no more than 5.0 psi pressure drop in 100 ft of pipe.

18.26E Specify a size of Schedule 40 steel pipe suitable for carrying 165 cfm (free air delivery) at 100 psig with no more than 5.0 psi pressure drop in 100 ft of pipe.

18.27E Specify a size of Schedule 40 steel pipe suitable for carrying 800 cfm (free air) to a reactor vessel in a chemical processing plant in which the pressure must be at least 100 psig at 70°F. The total length of pipe from the compressor to the reactor vessel is 350 ft. The line contains eight standard elbows, two fully open gate valves, and one swing-type check valve. After completing the design, determine the required pressure at the compressor.

18.28E For an aeration process, a sewage treatment plant requires 3000 cfm of compressed air. The pressure must be 80 psig, and the temperature must be 120°F. The compressor is located in a utility building, and 180 ft of pipe is required. The line also contains one fully open butterfly valve, 12 elbows, four tees with the flow through the run, and one ball-type check valve. Specify a suitable size of Schedule 40 steel pipe, and determine the required pressure at the compressor.

Gas Flow Through Nozzles

18.29E Air flows from a reservoir in which the pressure is 40.0 psig and the temperature is 80°F to a pipe in which the pressure is 20.0 psig. The flow is to be considered adiabatic. Compute the specific weight of the air both in the reservoir and in the pipe.

18.30M *Air flows from a reservoir in which the pressure is 275 kPa and the temperature is 25°C to a pipe in which the pressure is 140 kPa. The flow is to be considered adiabatic. Compute the specific weight of the air both in the reservoir and in the pipe.*

18.31E Refrigerant 12 expands adiabatically from 35.0 psig at a temperature of 60°F to 3.6 psig. Compute the specific weight of the refrigerant at both conditions.

18.32E Oxygen is discharged from a tank in which the pressure is 125 psig and the temperature is 75°F through a nozzle with a diameter of 0.120 in. The oxygen flows into the atmosphere, where the pressure is 14.40 psia. Compute the weight flow rate from the tank and the velocity of flow through the nozzle.

18.33E Repeat Problem 18.32, but change the pressure in the tank to 7.50 psig.

18.34E A high performance racing tire is charged with nitrogen at 50 psig and 70°F. At what rate would the nitrogen escape through a valve with a diameter of 0.062 in into the atmosphere at a pressure of 14.60 psia?

18.35E Repeat Problem 18.34 at internal pressures of 45 psig through 0 psig in 5.0 psig decrements. Plot a graph of weight flow rate versus the internal pressure in the tire.

18.36E Figure 18.11 shows a two-compartment vessel. The compartments are connected by a smooth convergent nozzle. The left compartment contains propane gas and is maintained at a constant 25.0 psig and 65°F. The right compartment starts at an equal 25.0 psig pressure, and then the pressure is

FIGURE 18.11 Vessel for Problem 18.36.

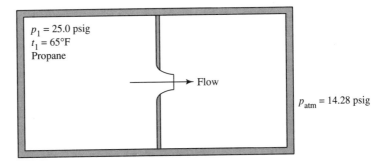

$p_1 = 25.0$ psig
$t_1 = 65°F$
Propane

Flow

$p_{atm} = 14.28$ psig

decreased slowly to 0.0 psig. The local atmospheric pressure is 14.28 psia. Compute the weight rate of flow of propane through the 0.5 in nozzle as the pressure decreases in 5.0 psi decrements. Plot the weight flow rate versus the pressure in the right compartment.

18.37M *Air flows from a large tank through a smooth convergent nozzle into the atmosphere, where the pressure is 98.5 kPa absolute. The temperature in the tank is 95°C. Compute the minimum pressure in the tank required to produce sonic velocity in the nozzle.*

18.38M *For the conditions of Problem 18.37, compute the magnitude of the sonic velocity in the nozzle.*

18.39M *For the conditions of Problem 18.37, compute the weight flow rate of air from the tank if the nozzle diameter is 10.0 mm.*

18.40M *A tank of Refrigerant 12 is at 150 kPa gage and 20°C. At what rate would the refrigerant flow from the tank into the atmosphere, where the pressure is 100.0 kPa absolute, through a smooth nozzle having a throat diameter of 8.0 mm?*

18.41M *For the tank described in Problem 18.40, compute the weight flow rate through the nozzle for tank gage pressures of 125 kPa, 100 kPa, 75 kPa, 50 kPa, and 25 kPa. Assume that the temperature in the tank is 20°C for all cases. Plot the weight flow rate versus tank pressure.*

COMPUTER PROGRAMMING ASSIGNMENTS

1. Write a program to perform the calculations called for in Eqs. (18–2) and (18–3) for the specific weight of a gas and the correction of the volume flow rate for pressure and temperatures, different from standard free air conditions.

2. Write a program for the analysis of the flow of compressed air in a pipe line system. The program should use a procedure similar to that outlined in Example Problem 18.4. Note that some of the features of the program are similar to those used in earlier chapters for flow of liquids in pipe line systems.

3. Write a program to compute the velocity and weight rate of flow of a gas from a tank through a smooth convergent nozzle. The program should use Eqs. (18–6) through (18–16), which involve the critical pressure ratio and sonic velocity.

■ ■ ■ ■ ■ 19 Flow of Air in Ducts

**19.1
OBJECTIVES**

Ventilation, heating, and air-conditioning systems distribute air at relatively low pressure through ducts. The fans or blowers responsible for moving the air can be described as high volume, low pressure devices. Knowledge of the pressures in the duct system is required to properly match a fan to a given system, to ensure the delivery of an adequate amount of air, and to balance the flow to various parts of the system.

After completing this chapter, you should be able to:

1. Describe the basic elements of an air distribution system that may be used for heating, ventilation, or air conditioning.
2. Determine energy losses in ducts considering straight sections and fittings.
3. Determine the circular equivalent diameters of rectangular ducts.
4. Analyze and design ductwork to carry air to spaces needing conditioning and to achieve balance in the system.
5. Identify the fan selection requirements for the system.

**19.2
AN EXAMPLE AIR
DISTRIBUTION SYSTEM**

Figure 19.1 is a sketch of the layout of an air distribution system. Outside air enters the building at point 1 through louvers that protect the ductwork from wind and rain. The velocity of the air flow through the louvers should be relatively low, approximately 500 ft/min (2.5 m/s), to minimize entrainment of undesirable contaminants. The duct then reduces to a smaller size to deliver the air to the suction side of a fan. A sudden contraction of the duct is shown, although a more gradual reduction would have a lower pressure loss.

The damper in the intake duct can partially close off the duct to decrease the flow, if desired. The intake duct delivers the air to the fan inlet, where its pressure is increased by the action of the fan wheel.

The outlet from the fan is carried by a main duct from which four branches deliver the air to its points of use. Dampers, shown in each of the branches, permit balancing of the system while in operation. Grilles are used at each outlet to distribute the air to the conditioned spaces (in this example, three offices and a conference room).

The ductwork shown is mostly above the ceiling. Note that an elbow is placed above each outlet grille to direct the air flow downward through the ceiling system.

It is important to understand the basic operating parameters of such an air-handling system. Obviously, the air outside the building is at the prevail-

517

FIGURE 19.1   Air distribution system.

ing atmospheric pressure. To cause flow into the duct through the intake louvers, the fan must create a pressure lower than atmospheric in the duct. This is a negative gage pressure. As the air flows through the duct, friction losses cause the pressure to decrease further. Also, any obstruction to the flow, such as a damper, and tees or wyes that direct the flow, cause pressure drops. The fan increases the pressure of the air and forces it through the supply ductwork to the outlet grilles.

The air inside the rooms of the building may be slightly above or below atmospheric pressure. Some air handling-system designers prefer to have a slight positive pressure in the building for better control and to eliminate drafts. However, when designing the ductwork, one usually considers the pressure inside the building to be the same as that outside the building.

**19.3
ENERGY LOSSES
IN DUCTS**

Two kinds of energy losses in duct systems cause the pressure to drop along the flow path. Friction losses occur as the air flows through straight sections, while dynamic losses occur as the air flows through such fittings as tees and wyes and through flow control devices.

Friction losses can be estimated by using the Darcy equation introduced in Chapter 9 for flow of liquids. However, special charts have been prepared by the American Society of Heating, Refrigerating, and Air-Conditioning Engineers (ASHRAE) for the typical conditions found in duct design. Figures 19.2 and 19.3 show friction loss h_L as a function of volume flow rate, with two sets of diagonal lines showing the diameter of circular ducts and the velocity of flow. The units used for the various quantities and the assumed conditions are summarized in Table 19.1. Reference 2 shows correction factors for other conditions. See Section 18.2 for information on air flow rates and pressures.

TABLE 19.1 Units and assumed conditions for friction charts

	U.S. Customary System Units	**SI Units**
Flow rate	ft³/min (cfm)	m³/s
Friction loss, h_L	in of water per 100 ft (in H₂O/100 ft)	Pa/m
Velocity	ft/min	m/s
Duct diameter	in	mm
Specific weight of air	0.075 lb/ft³	11.81 N/m³
Duct surface roughness	5×10^{-4} ft	1.5×10^{-4} m
Condition of air	14.7 psia; 68°F	101.3 kPa; 20°C

Although circular ducts are often used for distributing air through heating, ventilating, or air-conditioning systems, it is usually most convenient to use rectangular ducts because of space limitations, particularly above ceilings. The hydraulic radius of the rectangular duct can be used to characterize its size (as discussed in Section 8.6). When the necessary substitutions of the hydraulic radius for the diameter are made in relationships for velocity, Reynolds number, relative roughness, and the corresponding friction factor, we see that the *equivalent diameter* for a rectangular duct is

**EQUIVALENT DIAMETER FOR A
RECTANGULAR DUCT**

$$D_e = \frac{1.3(ab)^{5/8}}{(a + b)^{1/4}} \qquad \textbf{(19–1)}$$

where a and b are sides of the rectangle.

This enables you to use the friction loss charts in Figs. 19.2 and 19.3 for rectangular as well as circular ducts. Table 19.2 on p. 522 shows some results computed using Eq. 19–1.

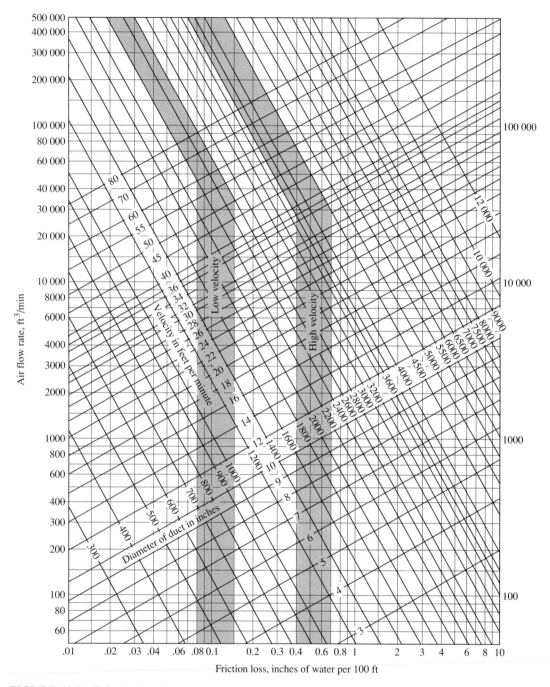

FIGURE 19.2   Friction loss in ducts—U.S. Customary System units. (Reprinted by permission from *ASHRAE Handbook: 1981 Fundamentals*).

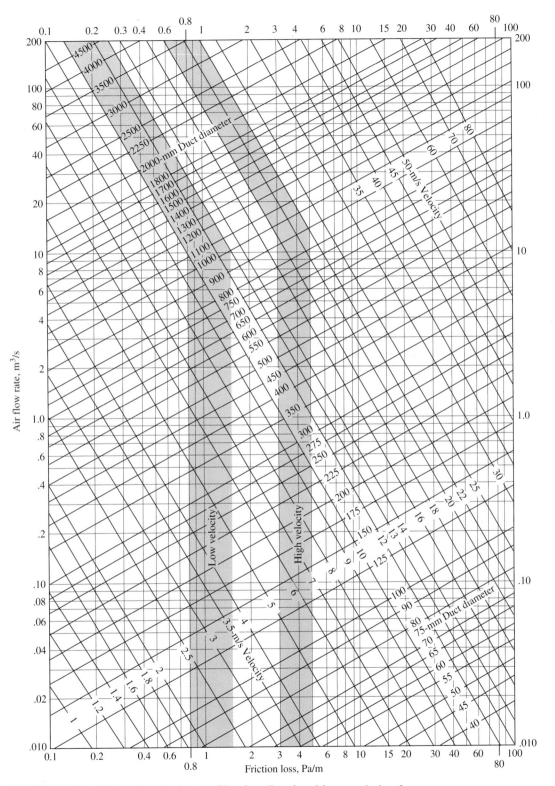

FIGURE 19.3 Friction loss in ducts—SI units. (Reprinted by permission from *ASHRAE Handbook: 1981 Fundamentals*).

TABLE 19.2 Circular equivalent diameters of rectangular ducts

Side a (in)	Side b (in)												
	6	8	10	12	14	16	18	20	22	24	26	28	30
6	6.6												
8	7.6	8.7											
10	8.4	9.8	10.9										
12	9.1	10.7	12.0	13.1									
14	9.8	11.5	12.9	14.2	15.3								
16	10.4	12.2	13.7	15.1	16.4	17.5							
18	11.0	12.9	14.5	16.0	17.3	18.5	19.7						
20	11.5	13.5	15.2	16.8	18.2	19.5	20.7	21.9					
22	12.0	14.1	15.9	17.6	19.1	20.4	21.7	22.9	24.0				
24	12.4	14.6	16.5	18.3	19.9	21.3	22.7	23.9	25.1	26.2			
26	12.8	15.1	17.1	19.0	20.6	22.1	23.5	24.9	26.1	27.3	28.4		
28	13.2	15.6	17.7	19.6	21.3	22.9	24.4	25.8	27.1	28.3	29.5	30.6	
30	13.6	16.1	18.3	20.2	22.0	23.7	25.2	26.6	28.0	29.3	30.5	31.7	32.8

☐ **EXAMPLE PROBLEM 19.1**   Determine the velocity of flow and the amount of friction loss that would occur as air flows at 3000 cfm through 80 ft of circular duct having a diameter of 22 in.

Solution   We can use Fig. 19.2 to determine that the velocity is approximately 1150 ft/min and that the friction loss per 100 ft of duct (h_L) is 0.082 in H_2O. Then, by proportion, the loss for 80 ft is

$$H_L = h_L(L/100) = (0.082 \text{ in } H_2O)(\tfrac{80}{100}) = 0.066 \text{ in } H_2O$$

☐ **EXAMPLE PROBLEM 19.2**   Specify the dimensions of a rectangular duct that would have the same friction loss as the circular duct described in Example Problem 19.1.

Solution   From Table 19.2 we can specify a 14-by-30-in rectangular duct that would have the same loss as the 22.0-in diameter circular duct. Others that will have approximately the same loss are 16-by-26-in, 18-by-22-in, and 20-by-20-in rectangular ducts. Such a list gives the designer many options when fitting a duct system into given spaces.

Dynamic losses can be estimated by using published data for loss coefficients for air flowing through certain fittings (see References 2 and 3). Also, manufacturers of special air-handling devices publish extensive data about expected pressure drops. Table 19.3 presents a few examples for use in problems in this book. Note that these data are highly simplified. For example, actual loss coefficients for tees depend on the size of the branches and on the amount of air flow in each branch. As with the minor losses discussed in Chapter 10, changes in flow area or direction of flow should be made as

TABLE 19.3 Examples of loss factors for duct fittings

Dynamic Loss Coefficient, C						
90°elbows						
Smooth, round	0.22					
5-piece, round	0.33					
4-piece, round	0.37					
3-piece, round	0.42					
Mitered, round	1.20					
Smooth, rectangular	0.18					
Tee, branch	1.00					
Tee, flow through main	0.10					
Symmetrical wye	0.30					

Damper:

Position	0° (wide open)	10°	20°	30°	40°	50°
C	0.20	0.52	1.50	4.5	11.0	29

Outlet grille: Assume total pressure drop through grille is 0.06 in H_2O (15 Pa).

Intake louvers: Assume total pressure drop across louver is 0.07 in H_2O (17 Pa).

Dynamic loss for fittings = $C(H_v)$, where H_v = velocity pressure upstream of the fitting.

Values shown are examples, for use only in solving problems in this book. Many factors affect the actual values for a given style of fitting. Refer to Reference 2 or manufacturers' catalogs for more complete data.

smooth as possible to minimize dynamic losses. The data for round, 90° elbows show the large variations possible.

The dynamic loss for a fitting is calculated from

$$H_L = C(H_v) \tag{19-2}$$

where C is the loss coefficient from Table 19.3 and H_v is the *velocity pressure* or *velocity head*.

In U.S. Customary System units, pressure levels and losses are typically expressed in inches of water, which is actually a measure of pressure head. Then,

$$H_v = \frac{\gamma_a v^2}{2g\gamma_w} \tag{19-3}$$

where γ_a is specific weight of air, v is the flow velocity, and γ_w is the specific weight of water. When the velocity is expressed in feet per minute and the conditions of standard air are used, Eq. (19-3) reduces to

▷ **VELOCITY PRESSURE FOR AIR FLOW (U.S.)**

$$H_v = \left(\frac{v}{4005}\right)^2 \tag{19-4}$$

When we use SI units, pressure levels and losses are measured in the pressure unit of Pa. Then,

$$H_v = \frac{\gamma_a v^2}{2g} \tag{19–5}$$

When the velocity is expressed in m/s and the conditions of standard air are used, Eq. (19–5) reduces to

⇨ **VELOCITY PRESSURE FOR AIR FLOW (SI)**

$$H_v = \left(\frac{v}{1.289}\right)^2 \text{ Pa} \tag{19–6}$$

□ **EXAMPLE PROBLEM 19.3**   Estimate the pressure drop that occurs when 3000 cfm of air flows around a smooth, rectangular, 90° elbow with side dimensions of 14 × 24 in.

Solution   Use Table 19.2 to find that the equivalent diameter for the duct is 19.9 in. From Fig. 19.2 we find the velocity of flow to be 1400 ft/min. Then, using Eq. (19–4), we compute

$$H_v = \left(\frac{v}{4005}\right)^2 = \left(\frac{1400}{4005}\right)^2 = 0.122 \text{ in H}_2\text{O}$$

From Table 19.3, we find $C = 0.18$. Then the pressure drop is

$$H_L = C(H_v) = (0.18)(0.122) = 0.022 \text{ in H}_2\text{O}$$

**19.4
DUCT DESIGN
EXAMPLE**

In Section 19.2, the general features of ducts for carrying the flow of air were described. Figure 19.1 shows a simple duct system, the operation of which has been described. In this section, we describe one method of designing such a duct system.

The goals of the design process are to specify reasonable dimensions for the various sections of the ductwork, to estimate the air pressure at key points, to determine the requirements to be met by the fan in the system, and to balance the system. Balance requires that the pressure drop from the fan outlet to each outlet grille is the same when the duct sections are carrying their design capacities.

The design method illustrated here uses Fig. 19.2 or 19.3 to specify a suitable velocity of flow in each duct section. The duct size, either circular or rectangular, is then decided. Then pressure drops are estimated for friction and for fittings. After all branch designs are tentatively completed, the total pressure drop to each grille is computed. An ideal design is one in which all such pressure drops are equal, and the system is *balanced*. If pressure drops are not equal, as is often the case, design changes are required. For small imbalances, such flow controls as dampers can be used to balance the system. When major deviations in pressure drops occur among the duct sections, the duct sizes or the design of fittings should be changed.

This procedure is sometimes referred to as the *equal friction method* or the *velocity reduction method*. The logic of these names is illustrated in Fig.

19.2. When we specify designs within the shaded part of the figure, the friction loss in each section will be nearly the same. Also, as the volume flow rate in each duct along a branch decreases, the recommended velocity of flow decreases. Note that the shaded part of the friction loss chart applies only to supply ducts and branches on the outlet side of fans. Proper flow of air into a fan requires a somewhat lower velocity, typically 700 to 1000 ft/min (3.5 to 5.0 m/s). Lower velocities are also used when low noise is desirable.

Other duct system design methods include the *static regain method* and the *balanced capacity method* (see References 2 and 4).

Most smaller systems for homes and light commercial applications are of the "low velocity" type, in which ductwork and fittings are relatively simple. Noise is usually not a major problem if the limits shown in Figs. 19.2 and 19.3 are not exceeded. However, the resulting sizes for ducts in a low velocity system are relatively large.

Space limitations in the design of large office buildings and certain industrial applications make "high velocity systems" attractive. The name comes from the practice of using smaller ducts to carry a given flow rate. However, several consequences arise:

1. Noise is usually a factor, and special noise-attenuation devices must be employed.
2. Duct construction must be more substantial, and sealing is more critical.
3. Operating costs are generally higher because of greater pressure drops and higher fan total pressures.

High velocity systems can be justified when lower building costs result or when more efficient use of space can be achieved.

The following design example is for a low velocity system.

☐ **EXAMPLE PROBLEM 19.4**   The system shown in Fig. 19.1 is being designed for a small office building. The air is drawn from outside the building by a fan and delivered through four branches to three offices and a conference room. The air flows shown at each outlet grille have been determined by others to provide adequate ventilation to each area. Dampers in each branch permit final adjustment of the system.

Complete the design of the duct system, specifying the size of each section of the ductwork for a low velocity system. Compute the expected pressure drop for each section and at each fitting. Then, compute the total pressure drop along each branch from the fan to the four outlet grilles and check for system balance. If a major imbalance is predicted, redesign appropriate parts of the system to achieve a more nearly balanced system. Then, determine the total pressure required for the fan. Use Fig. 19.2 for estimating friction losses and Table 19.3 for dynamic loss coefficients.

Solution   First, treat each section of the duct and each fitting separately. Then, analyze the branches.

 1. Intake duct A: $Q = 2700$ cfm; $L = 16$ ft
    Let $v \approx 800$ ft/min.

From Fig. 19.2, required $D = 25.0$ in.
$h_L = 0.035$ in $H_2O/100$ ft
$H_L = 0.035(16/100) = 0.0056$ in H_2O.

2. Damper in duct A: $C = 0.20$ (assume wide open) (Table 19.3)
   For 800 ft/min, $H_v = (800/4005)^2 = 0.040$ in H_2O
   $H_L = 0.20(0.040) = 0.0080$ in H_2O

3. Intake louvers: The 40-by-40-in size has been specified to give approximately 600 ft/min velocity through the open space of the louvers. Use $H_L = 0.070$ in H_2O from Table 19.3.

4. Sudden contraction between louver housing and intake duct: From Fig. 10.7, we know that the resistance coefficient depends on the velocity of flow and the ratio D_1/D_2 for circular conduits. Because the louver housing is a 40-by-40-in square, we can compute its equivalent diameter from Eq. (19–1):

$$D_e = \frac{1.3(ab)^{5/8}}{(a+b)^{1/4}} = \frac{1.3(40 \times 40)^{5/8}}{(40+40)^{1/4}} = 43.7 \text{ in}$$

Then, in Fig. 10.7,

$$D_1/D_2 = 43.7/25 = 1.75$$

and $K = C = 0.31$. Then,

$$H_L = C(H_v) = 0.31(0.04) = 0.0124 \text{ in } H_2O$$

5. Total loss in intake system:

$$H_L = 0.0056 + 0.0080 + 0.07 + 0.0124 = 0.096 \text{ in } H_2O$$

Because the pressure outside the louvers is atmospheric, the pressure at the inlet to the fan is -0.096 in H_2O, a negative gage pressure. An additional loss may occur at the fan inlet if a geometry change is required to mate the intake duct with the fan. Knowledge of the fan design is required, and this potential loss is ignored in this example.

Note: All ducts on the fan outlet side are rectangular.

6. Fan outlet − duct B: $Q = 2700$ cfm; $L = 20$ ft
   Let $v \approx 1200$ ft/min; $h_L = 0.110$ in $H_2O/100$ ft.
   $D_e = 20.0$ in; use 12-by-30-in size to minimize overhead space required.
   $H_L = 0.110(20/100) = 0.0220$ in H_2O
   $H_v = (1200/4005)^2 = 0.090$ in H_2O

7. Duct E: $Q = 600$ cfm; $L = 12$ ft
   Let $v \approx 800$ ft/min; $h_L = 0.085$ in $H_2O/100$ ft.
   $D_e = 12.0$ in; use 12-by-10-in size.
   $H_L = 0.085(12/100) = 0.0102$ in H_2O
   $H_v = (800/4005)^2 = 0.040$ in H_2O

8. Damper in duct E: $C = 0.20$ (assume wide open)
   $H_L = 0.20(0.040) = 0.0080$ in H_2O

9. Elbow in duct E: Smooth rectangular elbow; $C = 0.18$
   $H_L = 0.18(0.040) = 0.0072$ in H_2O

10. Grille 6 for duct E: $H_L = 0.060$ in H_2O

11. Tee 3 from duct B to branch E, flow in branch: $C = 1.00$
    H_L based on velocity ahead of tee in duct B
    $H_L = 1.00(0.090) = 0.090$ in H_2O
12. Duct C: $Q = 2100$ cfm; $L = 8$ ft
    Let $v \approx 1200$ ft/min; $h_L = 0.110$ in H_2O/100 ft.
    $D_e = 18.5$ in; use 12-by-24-in size.
    $H_L = 0.110(8/100) = 0.0088$ in H_2O
    $H_v = (1200/4005)^2 = 0.090$ in H_2O
13. Tee 3 from duct B to duct C, flow through main: $C = 0.10$
    $H_L = 0.10(0.090) = 0.009$ in H_2O
14. Duct F: $Q = 900$ cfm; $L = 18$ ft
    Let $v \approx 800$ ft/min; $h_L = 0.068$ in H_2O/100 ft
    $D_e = 14.3$ in; use 12-by-14-in size.
    $H_L = 0.068(18/100) = 0.0122$ in H_2O
    $H_v = (800/4005)^2 = 0.040$ in H_2O
15. Damper in duct F: $C = 0.20$ (assume wide open)
    $H_L = 0.20(0.040) = 0.0080$ in H_2O
16. Two elbows in duct F: Smooth rectangular elbow; $C = 0.18$
    $H_L = 2(0.18)(0.040) = 0.0144$ in H_2O
17. Grille 7 for duct F: $H_L = 0.060$ in H_2O
18. Tee 4 from duct C to branch F, flow in branch: $C = 1.00$
    H_L based on velocity ahead of tee in duct C
    $H_L = 1.00(0.090) = 0.090$ in H_2O
19. Duct D: $Q = 1200$ cfm; $L = 28$ ft
    Let $v \approx 1000$ ft/min; $h_L = 0.100$ in H_2O/100 ft
    $D_e = 14.7$ in; use 12-by-16-in size.
    Actual $D_e = 15.1$ in; new $h_L = 0.087$ in H_2O/100 ft
    $H_L = 0.087(28/100) = 0.0244$ in H_2O
    New $v = 960$ ft/min
    $H_v = (960/4005)^2 = 0.057$ in H_2O
20. Tee 4 from duct C to duct D, flow through main: $C = 0.10$
    $H_L = 0.10(0.090) = 0.009$ in H_2O
21. Wye 5 between duct D and ducts G and H: $C = 0.30$
    $H_L = 0.30(0.057) = 0.017$ in H_2O
    This loss applies to either duct G or duct H.
22. Ducts G and H are identical to duct E, and losses from steps 7, 8, 9, and 10 can
    be applied to these paths.

This completes the evaluation of the pressure drops through components in the system. Now we can sum the losses through any path from the fan outlet to the outlet grilles.

a. Path to grille 6 in duct E: Sum of losses from steps 6, 7, 8, 9, 10, and 11.

$$H_6 = 0.0220 + 0.0102 + 0.0080 + 0.0072 + 0.060 + 0.090$$
$$= 0.1974 \text{ in } H_2O$$

b. Path to grille 7 in duct F: Sum of losses from steps 6, 12, 13, 14, 15, 16, 17, and 18.

$$H_7 = 0.0220 + 0.0088 + 0.0090 + 0.0122 + 0.0080 + 0.0144$$
$$+ 0.060 + 0.090$$
$$= 0.2244 \text{ in } H_2O$$

c. Path to either grille 8 in duct G or grille 9 in duct H: Sum of losses from steps 6, 12, 13, 19, 20, 21, 7, 8, 9, and 10.

$$H_8 = 0.0220 + 0.0088 + 0.0090 + 0.0244 + 0.0090 + 0.0170$$
$$+ 0.0102 + 0.008 + 0.0072 + 0.06$$
$$= 0.1756 \text{ in } H_2O$$

Redesign to Achieve a
Balanced System

The ideal system design would be one in which the loss along any path, a, b, or c, is the same. Because that is not the case here, some redesign is called for. The loss in path b to grille 7 in duct F is much higher than the others. The component losses from steps 12, 14, 15, 16, and 18 affect this branch, and some reduction can be achieved by reducing the velocity of flow in ducts C and F.

12a. Duct C: $Q = 2100$ cfm; $L = 8$ ft
Let $v \approx 1000$ ft/min; $h_L = 0.073$ in $H_2O/100$ ft
$D_e = 19.6$ in; use 12-by-28-in size.
$H_L = 0.073(8/100) = 0.0058$ in H_2O
$H_v = (1000/4005)^2 = 0.0623$ in H_2O

14a. Duct F: $Q = 900$ cfm; $L = 18$ ft
Let $v \approx 600$ ft/min; $h_L = 0.033$ in $H_2O/100$ ft
$D_e = 16.5$ in; use 12-by-18-in size; $D_e = 16.0$ in
Actual $v = 630$ ft/min; $h_L = 0.038$ in $H_2O/100$ ft
$H_L = 0.038(18/100) = 0.0068$ in H_2O
$H_v = (630/4005)^2 = 0.0247$ in H_2O

15a. Damper in duct F: $C = 0.20$ (assume wide open)
$H_L = 0.20(0.0247) = 0.0049$ in H_2O

16a. Two elbows in duct F: Smooth rectangular elbow; $C = 0.18$
$H_L = 2(0.18)(0.0247) = 0.0089$ in H_2O

18a. Tee 4 from duct C to branch F, flow in branch: $C = 1.00$
H_L based on velocity ahead of tee in duct C.
$H_L = 1.00(0.0623) = 0.0623$ in H_2O

Now we can recompute the total loss in path B to grille 7 in duct F. As before, this is the sum of the losses from steps 6, 12a, 13, 14a, 15a, 16a, 17, and 18a.

$$H_7 = 0.0220 + 0.0058 + 0.009 + 0.0068 + 0.0049 + 0.0089 + 0.06 + 0.0623$$
$$= 0.1797 \text{ in } H_2O$$

This is a significant reduction, which results in a total pressure drop less than that of path a. Therefore, let's see if we can reduce the loss in path a by also reducing the velocity of flow in duct E. Steps 7, 8, and 9 are affected.

7a. Duct E: $Q = 600$ cfm; $L = 12$ ft
Let $v \approx 600$ ft/min
$D_e = 13.8$ in; use 12-by-14-in size.

Actual $D_e = 14.2$ in; $h_L = 0.032$ in H_2O; $v = 550$ ft/min
$H_L = 0.032(12/100) = 0.0038$ in H_2O
$H_v = (550/4005)^2 = 0.0189$ in H_2O

8a. Damper in duct E: $C = 0.20$ (assume wide open)
$H_L = 0.20(0.0189) = 0.0038$ in H_2O

9a. Elbow in duct E: Smooth rectangular elbow; $C = 0.18$
$H_L = 0.18(0.0189) = 0.0034$ in H_2O

Now, we can recompute the total loss in path a to grille 6 in duct E. As before, this is the sum of the losses from steps 6, 7a, 8a, 9a, 10, and 11.

$$H_6 = 0.0220 + 0.0038 + 0.0038 + 0.0034 + 0.060 + 0.090$$
$$= 0.1830 \text{ in } H_2O$$

This value is very close to that found for the redesigned path b, and the small difference can be adjusted with the dampers.

Now, note that path c to either grille 8 or grille 9 still has a lower total loss than either path a or path b. We could either use a slightly smaller duct size in branches G and H or depend on the adjustment of the dampers here also. To evaluate the suitability of using the dampers, let's estimate how much the dampers would have to be closed to increase the total loss to 0.1830 in H_2O (to equal that in path a). The increased loss is

$$H_6 - H_8 = 0.1830 - 0.1756 \text{ in } H_2O = 0.0074 \text{ in } H_2O$$

With the damper wide open and with 600 cfm passing at a velocity of approximately 800 ft/min, the loss was 0.0080 in H_2O, as found in the original step 8. The loss now should be

$$H_L = 0.0080 + 0.0074 = 0.0154 \text{ in } H_2O$$

But, for the damper,

$$H_L = C(H_v)$$

Solving for C gives

$$C = \frac{H_L}{H_v} = \frac{0.0154 \text{ in } H_2O}{0.040 \text{ in } H_2O} = 0.385$$

Referring to Table 19.3, you can see that a damper setting of less than 10° would produce this value of C, a very feasible setting. Thus, it appears that the duct system could be balanced as redesigned and that the total pressure drop from the fan outlet to any outlet grille will be approximately 0.1830 in H_2O. This is the pressure that the fan would have to develop.

SUMMARY OF THE DUCT SYSTEM DESIGN

- Intake duct A: Round; $D = 25.0$ in
- Duct B: Rectangular; 12 in × 30 in
- Duct C: Rectangular; 12 in × 28 in
- Duct D: Rectangular; 12 in × 16 in
- Duct E: Rectangular; 12 in × 14 in

- Duct F: Rectangular; 12 in × 18 in
- Duct G: Rectangular; 12 in × 10 in
- Duct H: Rectangular; 12 in × 10 in
- Pressure at fan inlet: −0.096 in H_2O
- Pressure at fan outlet: 0.1830 in H_2O
- Total pressure rise by the fan: 0.1830 + 0.096 = 0.279 in H_2O
- Total delivery by the fan: 2700 cfm

It is helpful to visualize the pressure changes that occur in the system. Figure 19.4 shows a plot of air pressure versus position for the path from the intake louvers, through the fan, through ducts B and E, to outlet grille 6. Similar plots could be made for the other paths.

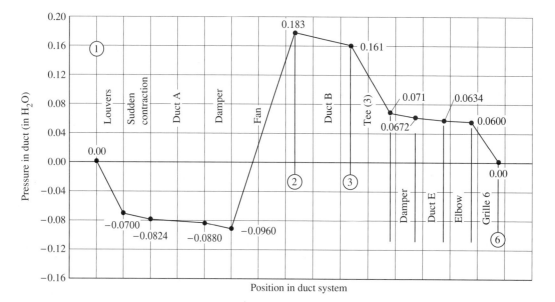

FIGURE 19.4   Pressure in duct (in H_2O) versus position for system shown in Fig. 19.1.

REFERENCES

1. American Society of Heating, Refrigerating, and Air-Conditioning Engineers. 1992. *ASHRAE Handbook: Equipment*. Atlanta: Author.
2. ———. 1989. *ASHRAE Handbook: Fundamentals*. Atlanta: Author.
3. Idelchik, I. E. 1986. *Handbook of Hydraulic Resistance*. 2nd ed. Washington, D.C.: Hemisphere Publishing Corporation.
4. McQuiston, F. C., and J. D. Parker. 1988. *Heating,*
   Ventilating, and Air Conditioning: Analysis and Design. 3rd ed. New York: John Wiley & Sons.
5. Pita, E. G. 1981. *Air Conditioning Principles and Systems: An Energy Approach*. New York: John Wiley & Sons.
6. Sheet Metal and Air Conditioning Contractors National Association. 1969. *Low Velocity Duct Construction Standards*. 4th ed. Washington, D.C.: Author.
7. The Trane Company. 1965. *Trane Air Conditioning Manual*. La Crosse, WI: Author.

PRACTICE PROBLEMS

Energy Losses in Straight Duct Sections

19.1E Determine the velocity of flow and the friction loss as 1000 cfm of air flows through 75 ft of 18-in diameter round duct.

19.2E Repeat Problem 19.1 for duct diameters of 16, 14, 12, and 10 in. Then, plot velocity and friction loss versus duct diameter.

19.3E Specify a diameter for a round duct suitable for carrying 1500 cfm of air with a maximum pressure drop of 0.10 in H_2O per 100 ft of duct, rounding up to the next inch. For the actual size specified, give the friction loss per 100 ft of duct.

19.4M *Determine the velocity of flow and the friction loss as 3.0 m^3/s of air flows through 25 m of 500 mm diameter round duct.*

19.5M *Repeat Problem 19.4 for duct diameters of 600, 700, 800, 900, and 1000 mm. Then, plot velocity and friction loss versus duct diameter.*

19.6M *Specify a diameter for a round duct suitable for carrying 0.40 m^3/s of air with a maximum pressure drop of 1.00 Pa/m of duct, rounding up to the next 50-mm increment. For the actual size specified, give the friction loss in Pa/m.*

19.7E A heating duct for a forced air furnace measures 10×30 in. Compute the circular equivalent diameter. Then, determine the maximum flow rate of air that the duct could carry while limiting the friction loss to 0.10 in H_2O per 100 ft.

19.8E A branch duct for a heating system measures 3×10 in. Compute the circular equivalent diameter. Then, determine the maximum flow rate of air that the duct could carry while limiting the friction loss to 0.10 in H_2O per 100 ft.

19.9E A ventilation duct in a large industrial warehouse measures 42×60 in. Compute the circular equivalent diameter. Then, determine the maximum flow rate of air that the duct could carry while limiting the friction loss to 0.10 in H_2O per 100 ft.

19.10M *A heating duct for a forced air furnace measures 250×500 mm. Compute the circular equivalent diameter. Then, determine the maximum flow rate of air that the duct could carry while limiting the friction loss to 0.80 Pa/m.*

19.11M *A branch duct for a heating system measures 75×250 mm. Compute the circular equivalent diame-*ter. Then, determine the maximum flow rate of air that it could carry while limiting the friction loss to 0.80 Pa/m.

19.12E Specify a size for a rectangular duct suitable for carrying 1500 cfm of air with a maximum pressure drop of 0.10 in H_2O per 100 ft of duct. The maximum vertical height of the duct is 12.0 in.

19.13E Specify a size for a rectangular duct suitable for carrying 300 cfm of air with a maximum pressure drop of 0.10 in H_2O per 100 ft of duct. The maximum vertical height of the duct is 6.0 in.

Energy Losses in Ducts with Fittings

19.14E Compute the pressure drop as 650 cfm of air flows through a three-piece 90° elbow in a 12-in diameter round duct.

19.15E Repeat Problem 19.14, but use a 5-piece elbow.

19.16E Compute the pressure drop as 1500 cfm of air flows through a wide open damper installed in a 16-in diameter duct.

19.17E Repeat Problem 19.16 with the damper partially closed at 10°, 20°, and 30°.

19.18E A part of a duct system is a 10-by-22-in rectangular main duct carrying 1600 cfm of air. A tee to a branch duct, 10×10 in, draws 500 cfm from the main. The main duct remains the same size downstream from the branch. Determine the velocity of flow and the velocity pressure in all parts of the duct.

19.19E For the conditions of Problem 19.18, estimate the loss in pressure as the flow enters the branch duct through the tee.

19.20E For the conditions of Problem 19.18, estimate the loss in pressure for the flow in the main duct due to the tee.

19.21M *Compute the pressure drop as 0.20 m^3/s of air flows through a three-piece 90° elbow in a 200-mm diameter round duct.*

19.22M *Repeat Problem 19.21, but use a mitered elbow.*

19.23M *Compute the pressure drop as 0.85 m^3/s of air flows through a damper set at 30° installed in a 400-mm diameter duct.*

19.24E A section of a duct system consists of 42 ft of straight 12-in diameter round duct, a wide open damper, two three-piece 90° elbows, and an outlet

FIGURE 19.5   Duct system for Problem 19.27.

grille. Compute the pressure drop along this duct section for Q = 700 cfm.

19.25E A section of duct system consists of 38 ft of straight 12-by-20-in rectangular duct, a wide open damper, three smooth 90° elbows, and an outlet grille. Compute the pressure drop along this duct section for Q = 1500 cfm.

19.26M *The intake duct to a fan consists of intake louvers, 5.8 m of square duct (800 × 800 mm), a sudden contraction to a 400-mm diameter round duct, and 9.25 m of the round duct. Estimate the pressure at the fan inlet when the duct carries 0.80 m³/s of air.*

FIGURE 19.6   Duct system for Problem 19.28.

FIGURE 19.7 Duct system for Problem 19.29.

FIGURE 19.8   Duct system for
Problem 19.30.

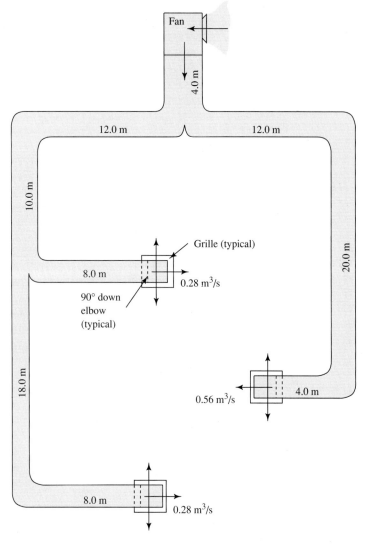

Design of Ducts

For the conditions shown in Figs. 19.5 through 19.8 on pages 532 through 534, complete the design of the duct system by specifying the sizes for all duct sections necessary to achieve a system that is balanced when it carries the flow rates shown. Compute the pressure at the fan outlet, assuming that the final outlets from the duct system are to atmospheric pressure. When an inlet duct section is shown, also complete its design and compute the pressure at the fan inlet. Note that there is no single best solution to

these problems, and several design decisions must be made. It may be desirable to change certain features of the suggested system design to improve its operation or to make it simpler to achieve a balanced system.

19.27E Use Fig. 19.5.

19.28E Use Fig. 19.6.

19.29E Use Fig. 19.7.

19.30M *Use Fig. 19.8.*

A Properties of Water

TABLE A.1 SI Units [101 kPa (abs)]

Temperature (°C)	Specific Weight γ (kN/m³)	Density ρ (kg/m³)	Dynamic Viscosity μ (Pa·s) or (N·s/m²)	Kinematic Viscosity ν (m²/s)
0	9.81	1000	1.75×10^{-3}	1.75×10^{-6}
5	9.81	1000	1.52×10^{-3}	1.52×10^{-6}
10	9.81	1000	1.30×10^{-3}	1.30×10^{-6}
15	9.81	1000	1.15×10^{-3}	1.15×10^{-6}
20	9.79	998	1.02×10^{-3}	1.02×10^{-6}
25	9.78	997	8.91×10^{-4}	8.94×10^{-7}
30	9.77	996	8.00×10^{-4}	8.03×10^{-7}
35	9.75	994	7.18×10^{-4}	7.22×10^{-7}
40	9.73	992	6.51×10^{-4}	6.56×10^{-7}
45	9.71	990	5.94×10^{-4}	6.00×10^{-7}
50	9.69	988	5.41×10^{-4}	5.48×10^{-7}
55	9.67	986	4.98×10^{-4}	5.05×10^{-7}
60	9.65	984	4.60×10^{-4}	4.67×10^{-7}
65	9.62	981	4.31×10^{-4}	4.39×10^{-7}
70	9.59	978	4.02×10^{-4}	4.11×10^{-7}
75	9.56	975	3.73×10^{-4}	3.83×10^{-7}
80	9.53	971	3.50×10^{-4}	3.60×10^{-7}
85	9.50	968	3.30×10^{-4}	3.41×10^{-7}
90	9.47	965	3.11×10^{-4}	3.22×10^{-7}
95	9.44	962	2.92×10^{-4}	3.04×10^{-7}
100	9.40	958	2.82×10^{-4}	2.94×10^{-7}

TABLE A.2 U.S. Customary System Units (14.7 psia)

Temperature (°F)	Specific Weight γ (lb/ft³)	Density ρ (slugs/ft³)	Dynamic Viscosity μ (lb-s/ft²)	Kinematic Viscosity ν (ft²/s)
32	62.4	1.94	3.66×10^{-5}	1.89×10^{-5}
40	62.4	1.94	3.23×10^{-5}	1.67×10^{-5}
50	62.4	1.94	2.72×10^{-5}	1.40×10^{-5}
60	62.4	1.94	2.35×10^{-5}	1.21×10^{-5}
70	62.3	1.94	2.04×10^{-5}	1.05×10^{-5}
80	62.2	1.93	1.77×10^{-5}	9.15×10^{-6}
90	62.1	1.93	1.60×10^{-5}	8.29×10^{-6}
100	62.0	1.93	1.42×10^{-5}	7.37×10^{-6}
110	61.9	1.92	1.26×10^{-5}	6.55×10^{-6}
120	61.7	1.92	1.14×10^{-5}	5.94×10^{-6}
130	61.5	1.91	1.05×10^{-5}	5.49×10^{-6}
140	61.4	1.91	9.60×10^{-6}	5.03×10^{-6}
150	61.2	1.90	8.90×10^{-6}	4.68×10^{-6}
160	61.0	1.90	8.30×10^{-6}	4.38×10^{-6}
170	60.8	1.89	7.70×10^{-6}	4.07×10^{-6}
180	60.6	1.88	7.23×10^{-6}	3.84×10^{-6}
190	60.4	1.88	6.80×10^{-6}	3.62×10^{-6}
200	60.1	1.87	6.25×10^{-6}	3.35×10^{-6}
212	59.8	1.86	5.89×10^{-6}	3.17×10^{-6}

B Properties of Common Liquids

TABLE B.1 SI Units [101 kPa (abs) and 25°C]

	Specific Gravity sg	Specific Weight γ (kN/m³)	Density ρ (kg/m³)	Dynamic Viscosity μ (Pa·s) or (N·s/m²)
Acetone	0.787	7.72	787	3.16×10^{-4}
Alcohol, ethyl	0.787	7.72	787	1.00×10^{-3}
Alcohol, methyl	0.789	7.74	789	5.60×10^{-4}
Alcohol, propyl	0.802	7.87	802	1.92×10^{-3}
Ammonia	0.826	8.10	826	—
Benzene	0.876	8.59	876	6.03×10^{-4}
Carbon tetrachloride	1.590	15.60	1 590	9.10×10^{-4}
Castor oil	0.960	9.42	960	6.51×10^{-1}
Ethylene glycol	1.100	10.79	1 100	1.62×10^{-2}
Gasoline	0.68	6.67	680	2.87×10^{-4}
Glycerine	1.258	12.34	1 258	9.60×10^{-1}
Kerosene	0.823	8.07	823	1.64×10^{-3}
Linseed oil	0.930	9.12	930	3.31×10^{-2}
Mercury	13.54	132.8	13 540	1.53×10^{-3}
Propane	0.495	4.86	495	1.10×10^{-4}
Seawater	1.030	10.10	1 030	1.03×10^{-3}
Turpentine	0.870	8.53	870	1.37×10^{-3}
Fuel oil, medium	0.852	8.36	852	2.99×10^{-3}
Fuel oil, heavy	0.906	8.89	906	1.07×10^{-1}

TABLE B.2 U.S. Customary
System Units (14.7 psia and 77°F)

	Specific Gravity sg	Specific Weight γ (lb/ft³)	Density ρ (slugs/ft³)	Dynamic Viscosity μ (lb-s/ft²)
Acetone	0.787	48.98	1.53	6.60×10^{-6}
Alcohol, ethyl	0.787	49.01	1.53	2.10×10^{-5}
Alcohol, methyl	0.789	49.10	1.53	1.17×10^{-5}
Alcohol, propyl	0.802	49.94	1.56	4.01×10^{-5}
Ammonia	0.826	51.41	1.60	—
Benzene	0.876	54.55	1.70	1.26×10^{-5}
Carbon tetrachloride	1.590	98.91	3.08	1.90×10^{-5}
Castor oil	0.960	59.69	1.86	1.36×10^{-2}
Ethylene glycol	1.100	68.47	2.13	3.38×10^{-4}
Gasoline	0.68	42.40	1.32	6.00×10^{-6}
Glycerine	1.258	78.50	2.44	2.00×10^{-2}
Kerosene	0.823	51.20	1.60	3.43×10^{-5}
Linseed oil	0.930	58.00	1.80	6.91×10^{-4}
Mercury	13.54	844.9	26.26	3.20×10^{-5}
Propane	0.495	30.81	0.96	2.30×10^{-6}
Seawater	1.030	64.00	2.00	2.15×10^{-5}
Turpentine	0.870	54.20	1.69	2.87×10^{-5}
Fuel oil, medium	0.852	53.16	1.65	6.25×10^{-5}
Fuel oil, heavy	0.906	56.53	1.76	2.24×10^{-3}

C Typical Properties of Petroleum Lubricating Oils

Type	Specific Gravity	Kinematic Viscosity, ν				Viscosity Index
		at 40°C (104°F)		at 100°C (212°F)		
		(m²/s)	(ft²/s)	(m²/s)	(ft²/s)	
Automotive hydraulic systems	0.887	3.99×10^{-5}	4.30×10^{-4}	7.29×10^{-6}	7.85×10^{-5}	141
Machine tool hydraulic systems						
Light	0.887	3.20×10^{-5}	3.44×10^{-4}	4.79×10^{-6}	5.16×10^{-5}	64
Medium	0.895	6.70×10^{-5}	7.21×10^{-4}	7.29×10^{-6}	7.85×10^{-5}	66
Heavy	0.901	1.96×10^{-4}	2.11×10^{-3}	1.40×10^{-5}	1.51×10^{-4}	70
Low temperature	0.844	1.40×10^{-5}	1.51×10^{-4}	5.20×10^{-6}	5.60×10^{-5}	226
Machine tool lubricating oils						
Light	0.881	2.20×10^{-5}	2.37×10^{-4}	3.90×10^{-6}	4.20×10^{-5}	80
Medium	0.915	6.60×10^{-5}	7.10×10^{-4}	7.00×10^{-6}	7.53×10^{-5}	83
Heavy	0.890	2.00×10^{-4}	2.15×10^{-3}	1.55×10^{-5}	1.67×10^{-4}	80

See also Tables 2.4 and 2.5 in Chapter 2 for properties of SAE grades of engine and transmission oils.

■ ■ ■ ■ D Variation of Viscosity with Temperature

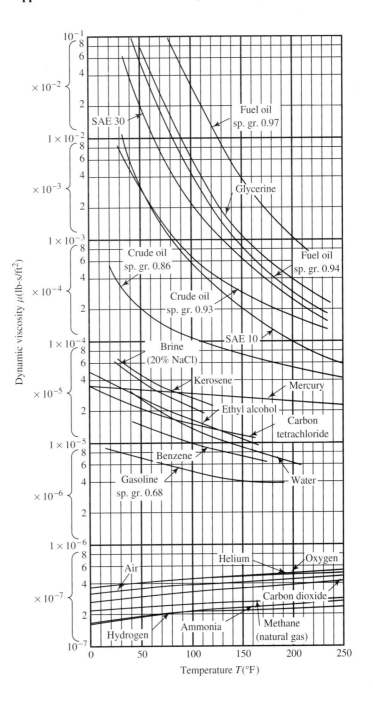

■ ■ ■ ■ E Properties of Air

TABLE E.1 Properties of air at standard atmospheric pressure

Temperature T (°C)	Density ρ (kg/m³)	Specific Weight γ (N/m³)	Dynamic Viscosity μ (Pa·s)	Kinematic Viscosity ν (m²/s)
−40	1.514	14.85	1.51×10^{-5}	9.98×10^{-6}
−30	1.452	14.24	1.56×10^{-5}	1.08×10^{-5}
−20	1.394	13.67	1.62×10^{-5}	1.16×10^{-5}
−10	1.341	13.15	1.67×10^{-5}	1.24×10^{-5}
0	1.292	12.67	1.72×10^{-5}	1.33×10^{-5}
10	1.247	12.23	1.77×10^{-5}	1.42×10^{-5}
20	1.204	11.81	1.81×10^{-5}	1.51×10^{-5}
30	1.164	11.42	1.86×10^{-5}	1.60×10^{-5}
40	1.127	11.05	1.91×10^{-5}	1.69×10^{-5}
50	1.092	10.71	1.95×10^{-5}	1.79×10^{-5}
60	1.060	10.39	1.99×10^{-5}	1.89×10^{-5}
70	1.029	10.09	2.04×10^{-5}	1.99×10^{-5}
80	0.9995	9.802	2.09×10^{-5}	2.09×10^{-5}
90	0.9720	9.532	2.13×10^{-5}	2.19×10^{-5}
100	0.9459	9.277	2.17×10^{-5}	2.30×10^{-5}
110	0.9213	9.034	2.22×10^{-5}	2.40×10^{-5}
120	0.8978	8.805	2.26×10^{-5}	2.51×10^{-5}

Note: Properties of air for standard conditions at sea level are

Temperature	15°C
Pressure	101.325 kPa
Density	1.225 kg/m³
Specific weight	12.01 N/m³
Dynamic viscosity	1.789×10^{-5} Pa·s
Kinematic viscosity	1.46×10^{-5} m²/s

TABLE E.2 Properties of the atmosphere

SI Units				U.S. Customary System Units			
Altitude (m)	Temperature (°C)	Pressure (kPa)	Density (kg/m³)	Altitude (ft)	Temperature (°F)	Pressure (psi)	Density (slugs/ft³)
0	15.00	101.3	1.225	0	59.00	14.696	2.38×10^{-3}
200	13.70	98.9	1.202	500	57.22	14.433	2.34×10^{-3}
400	12.40	96.6	1.179	1000	55.43	14.173	2.25×10^{-3}
600	11.10	94.3	1.156	5000	41.17	12.227	2.05×10^{-3}
800	9.80	92.1	1.134	10000	23.34	10.106	1.76×10^{-3}
1000	8.50	89.9	1.112	15000	5.51	8.293	1.50×10^{-3}
2000	2.00	79.5	1.007	20000	−12.62	6.753	1.27×10^{-3}
3000	−4.49	70.1	0.9093	30000	−47.99	4.365	8.89×10^{-4}
4000	−10.98	61.7	0.8194	40000	−69.70	2.720	5.85×10^{-4}
5000	−17.47	54.0	0.7364	50000	−69.70	1.683	3.62×10^{-4}
10000	−49.90	26.5	0.4135	60000	−69.70	1.040	2.24×10^{-4}
15000	−56.50	12.11	0.1948	70000	−67.30	0.644	1.38×10^{-4}
20000	−56.50	5.53	0.0889	80000	−61.81	0.400	8.45×10^{-5}
25000	−51.60	2.55	0.0401	90000	−56.32	0.251	5.22×10^{-5}
30000	−46.64	1.20	0.0184	100000	−50.84	0.158	3.25×10^{-5}

Data from *U.S. Standard Atmosphere, 1976*, NOAA-S/T76-1562. Washington, D.C.: National Oceanic and Atmospheric Administration.

FIGURE E.1   Properties of the atmosphere.

■ ■ ■ ■ F Dimensions of Steel Pipe

TABLE F.1 Schedule 40

Nominal Pipe Size (in)	Outside Diameter		Wall Thickness		Inside Diameter			Flow Area	
	(in)	(mm)	(in)	(mm)	(in)	(ft)	(mm)	(ft²)	(m²)
1/8	0.405	10.3	0.068	1.73	0.269	0.0224	6.8	0.000 394	3.660×10^{-5}
1/4	0.540	13.7	0.088	2.24	0.364	0.0303	9.2	0.000 723	6.717×10^{-5}
3/8	0.675	17.1	0.091	2.31	0.493	0.0411	12.5	0.001 33	1.236×10^{-4}
1/2	0.840	21.3	0.109	2.77	0.622	0.0518	15.8	0.002 11	1.960×10^{-4}
3/4	1.050	26.7	0.113	2.87	0.824	0.0687	20.9	0.003 70	3.437×10^{-4}
1	1.315	33.4	0.133	3.38	1.049	0.0874	26.6	0.006 00	5.574×10^{-4}
1 1/4	1.660	42.2	0.140	3.56	1.380	0.1150	35.1	0.010 39	9.653×10^{-4}
1 1/2	1.900	48.3	0.145	3.68	1.610	0.1342	40.9	0.014 14	1.314×10^{-3}
2	2.375	60.3	0.154	3.91	2.067	0.1723	52.5	0.023 33	2.168×10^{-3}
2 1/2	2.875	73.0	0.203	5.16	2.469	0.2058	62.7	0.033 26	3.090×10^{-3}
3	3.500	88.9	0.216	5.49	3.068	0.2557	77.9	0.051 32	4.768×10^{-3}
3 1/2	4.000	101.6	0.226	5.74	3.548	0.2957	90.1	0.068 68	6.381×10^{-3}
4	4.500	114.3	0.237	6.02	4.026	0.3355	102.3	0.088 40	8.213×10^{-3}
5	5.563	141.3	0.258	6.55	5.047	0.4206	128.2	0.139 0	1.291×10^{-2}
6	6.625	168.3	0.280	7.11	6.065	0.5054	154.1	0.200 6	1.864×10^{-2}
8	8.625	219.1	0.322	8.18	7.981	0.6651	202.7	0.347 2	3.226×10^{-2}
10	10.750	273.1	0.365	9.27	10.020	0.8350	254.5	0.547 9	5.090×10^{-2}
12	12.750	323.9	0.406	10.31	11.938	0.9948	303.2	0.777 1	7.219×10^{-2}
14	14.000	355.6	0.437	11.10	13.126	1.094	333.4	0.939 6	8.729×10^{-2}
16	16.000	406.4	0.500	12.70	15.000	1.250	381.0	1.227	0.1140
18	18.000	457.2	0.562	14.27	16.876	1.406	428.7	1.553	0.1443
20	20.000	508.0	0.593	15.06	18.814	1.568	477.9	1.931	0.1794
24	24.000	609.6	0.687	17.45	22.626	1.886	574.7	2.792	0.2594

TABLE F.2   Schedule 80

Nominal Pipe Size (in)	Outside Diameter		Wall Thickness		Inside Diameter			Flow Area	
	(in)	(mm)	(in)	(mm)	(in)	(ft)	(mm)	(ft²)	(m²)
⅛	0.405	10.3	0.095	2.41	0.215	0.017 92	5.5	0.000 253	2.350×10^{-5}
¼	0.540	13.7	0.119	3.02	0.302	0.025 17	7.7	0.000 497	4.617×10^{-5}
⅜	0.675	17.1	0.126	3.20	0.423	0.035 25	10.7	0.000 976	9.067×10^{-5}
½	0.840	21.3	0.147	3.73	0.546	0.045 50	13.9	0.001 625	1.510×10^{-4}
¾	1.050	26.7	0.154	3.91	0.742	0.061 83	18.8	0.003 00	2.787×10^{-4}
1	1.315	33.4	0.179	4.55	0.957	0.079 75	24.3	0.004 99	4.636×10^{-4}
1¼	1.660	42.2	0.191	4.85	1.278	0.106 5	32.5	0.008 91	8.278×10^{-4}
1½	1.900	48.3	0.200	5.08	1.500	0.125 0	38.1	0.012 27	1.140×10^{-3}
2	2.375	60.3	0.218	5.54	1.939	0.161 6	49.3	0.020 51	1.905×10^{-3}
2½	2.875	73.0	0.276	7.01	2.323	0.193 6	59.0	0.029 44	2.735×10^{-3}
3	3.500	88.9	0.300	7.62	2.900	0.241 7	73.7	0.045 90	4.264×10^{-3}
3½	4.000	101.6	0.318	8.08	3.364	0.280 3	85.4	0.061 74	5.736×10^{-3}
4	4.500	114.3	0.337	8.56	3.826	0.318 8	97.2	0.079 86	7.419×10^{-3}
5	5.563	141.3	0.375	9.53	4.813	0.401 1	122.3	0.126 3	1.173×10^{-2}
6	6.625	168.3	0.432	10.97	5.761	0.480 1	146.3	0.181 0	1.682×10^{-2}
8	8.625	219.1	0.500	12.70	7.625	0.635 4	193.7	0.317 4	2.949×10^{-2}
10	10.750	273.1	0.593	15.06	9.564	0.797 0	242.9	0.498 6	4.632×10^{-2}
12	12.750	323.9	0.687	17.45	11.376	0.948 0	289.0	0.705 6	6.555×10^{-2}
14	14.000	355.6	0.750	19.05	12.500	1.042	317.5	0.852 1	7.916×10^{-2}
16	16.000	406.4	0.842	21.39	14.314	1.193	363.6	1.117	0.1038
18	18.000	457.2	0.937	23.80	16.126	1.344	409.6	1.418	0.1317
20	20.000	508.0	1.031	26.19	17.938	1.495	455.6	1.755	0.1630
24	24.000	609.6	1.218	30.94	21.564	1.797	547.7	2.535	0.2344

■ ■ ■ ■ G Dimensions of Steel Tubing

Outside Diameter		Wall Thickness		Inside Diameter			Flow Area	
(in)	(mm)	(in)	(mm)	(in)	(ft)	(mm)	(ft^2)	(m^2)
1/8	3.18	0.032	0.813	0.061	0.00508	1.549	2.029×10^{-5}	1.885×10^{-6}
		0.035	0.889	0.055	0.00458	1.397	1.650×10^{-5}	1.533×10^{-6}
3/16	4.76	0.032	0.813	0.124	0.01029	3.137	8.319×10^{-5}	7.728×10^{-6}
		0.035	0.889	0.117	0.00979	2.985	7.530×10^{-5}	6.996×10^{-6}
1/4	6.35	0.035	0.889	0.180	0.01500	4.572	1.767×10^{-4}	1.642×10^{-5}
		0.049	1.24	0.152	0.01267	3.861	1.260×10^{-4}	1.171×10^{-5}
5/16	7.94	0.035	0.889	0.243	0.02021	6.160	3.207×10^{-4}	2.980×10^{-5}
		0.049	1.24	0.215	0.01788	5.448	2.509×10^{-4}	2.331×10^{-5}
3/8	9.53	0.035	0.889	0.305	0.02542	7.747	5.074×10^{-4}	4.714×10^{-5}
		0.049	1.24	0.277	0.02308	7.036	4.185×10^{-4}	3.888×10^{-5}
1/2	12.70	0.049	1.24	0.402	0.03350	10.21	8.814×10^{-4}	8.189×10^{-5}
		0.065	1.65	0.370	0.03083	9.40	7.467×10^{-4}	6.937×10^{-5}
5/8	15.88	0.049	1.24	0.527	0.04392	13.39	1.515×10^{-3}	1.407×10^{-4}
		0.065	1.65	0.495	0.04125	12.57	1.336×10^{-3}	1.242×10^{-4}
3/4	19.05	0.049	1.24	0.652	0.05433	16.56	2.319×10^{-3}	2.154×10^{-4}
		0.065	1.65	0.620	0.05167	15.75	2.097×10^{-3}	1.948×10^{-4}
7/8	22.23	0.049	1.24	0.777	0.06475	19.74	3.293×10^{-3}	3.059×10^{-4}
		0.065	1.65	0.745	0.06208	18.92	3.027×10^{-3}	2.812×10^{-4}
1	25.40	0.065	1.65	0.870	0.07250	22.10	4.128×10^{-3}	3.835×10^{-4}
		0.083	2.11	0.834	0.06950	21.18	3.794×10^{-3}	3.524×10^{-4}
1 1/4	31.75	0.065	1.65	1.120	0.09333	28.45	6.842×10^{-3}	6.356×10^{-4}
		0.083	2.11	1.084	0.09033	27.53	6.409×10^{-3}	5.954×10^{-4}
1 1/2	38.10	0.065	1.65	1.370	0.1142	34.80	1.024×10^{-2}	9.510×10^{-4}
		0.083	2.11	1.334	0.1112	33.88	9.706×10^{-3}	9.017×10^{-4}
1 3/4	44.45	0.065	1.65	1.620	0.1350	41.15	1.431×10^{-2}	1.330×10^{-3}
		0.083	2.11	1.584	0.1320	40.23	1.368×10^{-2}	1.271×10^{-3}
2	50.80	0.065	1.65	1.870	0.1558	47.50	1.907×10^{-2}	1.772×10^{-3}
		0.083	2.11	1.834	0.1528	46.58	1.835×10^{-2}	1.704×10^{-3}

■ ■ ■ ■ H Dimensions of Type K Copper Tubing

Nominal Size	Outside Diameter		Wall Thickness		Inside Diameter			Flow Area	
(in)	(in)	(mm)	(in)	(mm)	(in)	(ft)	(mm)	(ft²)	(m²)
⅛	0.250	6.35	0.035	0.889	0.180	0.0150	4.572	1.767×10^{-4}	1.642×10^{-5}
¼	0.375	9.53	0.049	1.245	0.277	0.0231	7.036	4.185×10^{-4}	3.888×10^{-5}
⅜	0.500	12.70	0.049	1.245	0.402	0.0335	10.21	8.814×10^{-4}	8.189×10^{-5}
½	0.625	15.88	0.049	1.245	0.527	0.0439	13.39	1.515×10^{-3}	1.407×10^{-4}
⅝	0.750	19.05	0.049	1.245	0.652	0.0543	16.56	2.319×10^{-3}	2.154×10^{-4}
¾	0.875	22.23	0.065	1.651	0.745	0.0621	18.92	3.027×10^{-3}	2.812×10^{-4}
1	1.125	28.58	0.065	1.651	0.995	0.0829	25.27	5.400×10^{-3}	5.017×10^{-4}
1¼	1.375	34.93	0.065	1.651	1.245	0.1037	31.62	8.454×10^{-3}	7.854×10^{-4}
1½	1.625	41.28	0.072	1.829	1.481	0.1234	37.62	1.196×10^{-2}	1.111×10^{-3}
2	2.125	53.98	0.083	2.108	1.959	0.1632	49.76	2.093×10^{-2}	1.945×10^{-3}
2½	2.625	66.68	0.095	2.413	2.435	0.2029	61.85	3.234×10^{-2}	3.004×10^{-3}
3	3.125	79.38	0.109	2.769	2.907	0.2423	73.84	4.609×10^{-2}	4.282×10^{-3}
3½	3.625	92.08	0.120	3.048	3.385	0.2821	85.98	6.249×10^{-2}	5.806×10^{-3}
4	4.125	104.8	0.134	3.404	3.857	0.3214	97.97	8.114×10^{-2}	7.538×10^{-3}
5	5.125	130.2	0.160	4.064	4.805	0.4004	122.0	1.259×10^{-1}	1.170×10^{-2}
6	6.125	155.6	0.192	4.877	5.741	0.4784	145.8	1.798×10^{-1}	1.670×10^{-2}
8	8.125	206.4	0.271	6.883	7.583	0.6319	192.6	3.136×10^{-1}	2.914×10^{-2}
10	10.125	257.2	0.338	8.585	9.449	0.7874	240.0	4.870×10^{-1}	4.524×10^{-2}
12	12.125	308.0	0.405	10.287	11.315	0.9429	287.4	6.983×10^{-1}	6.487×10^{-2}

I Dimensions of Ductile Iron Pipe

TABLE I.1   Class 150 for 150 psi (1.03 MPa) pressure service

Nominal Pipe Size	Outside Diameter		Wall Thickness		Inside Diameter			Flow Area	
(in)	(in)	(mm)	(in)	(mm)	(in)	(ft)	(mm)	(ft^2)	(m^2)
3	3.96	100.6	0.320	8.13	3.32	0.277	84.3	0.0601	5.585×10^{-3}
4	4.80	121.9	0.350	8.89	4.10	0.342	104.1	0.0917	8.518×10^{-3}
6	6.90	175.3	0.380	9.65	6.14	0.512	156.0	0.2056	1.910×10^{-2}
8	9.05	229.9	0.410	10.41	8.23	0.686	209.0	0.3694	3.432×10^{-2}
10	11.10	281.9	0.440	11.18	10.22	0.852	259.6	0.5697	5.292×10^{-2}
12	13.20	335.3	0.480	12.19	12.24	1.020	310.9	0.8171	7.591×10^{-2}
14	15.65	397.5	0.510	12.95	14.63	1.219	371.6	1.167	0.1085
16	17.80	452.1	0.540	13.72	16.72	1.393	424.7	1.525	0.1417
18	19.92	506.0	0.580	14.73	18.76	1.563	476.5	1.920	0.1783
20	22.06	560.3	0.620	15.75	20.82	1.735	528.8	2.364	0.2196
24	26.32	668.5	0.730	18.54	24.86	2.072	631.4	3.371	0.3132

■■■■■ J Areas of Circles

TABLE J.1 U.S. Customary
System Units

Diameter		Area	
(in)	(ft)	(in²)	(ft²)
0.25	0.0208	0.0491	3.409×10^{-4}
0.50	0.0417	0.1963	1.364×10^{-3}
0.75	0.0625	0.4418	3.068×10^{-3}
1.00	0.0833	0.7854	5.454×10^{-3}
1.25	0.1042	1.227	8.522×10^{-3}
1.50	0.1250	1.767	1.227×10^{-2}
1.75	0.1458	2.405	1.670×10^{-2}
2.00	0.1667	3.142	2.182×10^{-2}
2.50	0.2083	4.909	3.409×10^{-2}
3.00	0.2500	7.069	4.909×10^{-2}
3.50	0.2917	9.621	6.681×10^{-2}
4.00	0.3333	12.57	8.727×10^{-2}
4.50	0.3750	15.90	0.1104
5.00	0.4167	19.63	0.1364
6.00	0.5000	28.27	0.1963
7.00	0.5833	38.48	0.2673
8.00	0.6667	50.27	0.3491
9.00	0.7500	63.62	0.4418
10.00	0.8333	78.54	0.5454
12.00	1.00	113.1	0.7854
18.00	1.50	254.5	1.767
24.00	2.00	452.4	3.142

TABLE J.2   SI Units

Diameter		Area	
(mm)	**(m)**	**(mm²)**	**(m²)**
6	0.006	28.27	2.827×10^{-5}
12	0.012	113.1	1.131×10^{-4}
18	0.018	254.5	2.545×10^{-4}
25	0.025	490.9	4.909×10^{-4}
32	0.032	804.2	8.042×10^{-4}
40	0.040	1257	1.257×10^{-3}
45	0.045	1590	1.590×10^{-3}
50	0.050	1963	1.963×10^{-3}
60	0.060	2827	2.827×10^{-3}
75	0.075	4418	4.418×10^{-3}
90	0.090	6362	6.362×10^{-3}
100	0.100	7854	7.854×10^{-3}
115	0.115	1.039×10^{4}	1.039×10^{-2}
125	0.125	1.227×10^{4}	1.227×10^{-2}
150	0.150	1.767×10^{4}	1.767×10^{-2}
175	0.175	2.405×10^{4}	2.405×10^{-2}
200	0.200	3.142×10^{4}	3.142×10^{-2}
225	0.225	3.976×10^{4}	3.976×10^{-2}
250	0.250	4.909×10^{4}	4.909×10^{-2}
300	0.300	7.069×10^{4}	7.069×10^{-2}
450	0.450	1.590×10^{5}	1.590×10^{-1}
600	0.600	2.827×10^{5}	2.827×10^{-1}

■■■■■ K Conversion Factors

TABLE K.1 U.S. Customary System to International System of Units (SI)

Quantity	English Unit	SI Unit	Symbol	Equivalent Units
Length	1 foot (ft)	= 0.3048 meter	m	—
Mass	1 slug	= 14.59 kilogram	kg	—
Time	1 second	= 1.0 second	s	—
Force	1 pound (lb)	= 4.448 newton	N	$kg \cdot m/s^2$
Pressure	1 lb/in^2	= 6895 pascal	Pa	N/m^2 or $kg/m \cdot s^2$
Energy	1 lb-ft	= 1.356 joule	J	$N \cdot m$ or $kg \cdot m^2/s^2$
Power	1 lb-ft/s	= 1.356 watt	W	J/s

TABLE K.2   Other convenient conversion factors

Length
    1 ft = 0.3048 m   1 km = 1000 m
    1 in = 25.4 mm   1 cm = 10 mm
    1 mi = 5280 ft   1 m = 1000 mm
    1 mi = 1.609 km

Area
    1 ft^2 = 0.0929 m^2 1 m^2 = 10.76 ft^2
    1 in^2 = 645.2 mm^2 1 m^2 = 10^6 mm^2

Volume
    1 ft^3 = 7.48 gal   1 gal = 0.003 79 m^3
    1 ft^3 = 1728 in^3 1 gal = 3.785 L
    1 ft^3 = 0.0283 m^3 1 m^3 = 1000 L
                                    1 imperial gal = 1.201 U.S. gal.

Volume flow rate
    1 ft^3/s = 449 gal/min   1 gal/min = 3.785 L/min
    1 ft^3/s = 0.0283 m^3/s   1 L/min = 16.67 × 10^{-6} m^3/s
    1 gal/min = 6.309 × 10^{-5} m^3/s   1 m^3/s = 60 000 L/min

Temperature   *Pressure*
    $T(°C) = [T(°F) - 32]\frac{5}{9}$ 1 bar = 100 kPa
    $T(°F) = \frac{9}{5}[T(°C)] + 32$ 1 bar = 14.50 lb/in^2

Density   *Energy*
    1 slug/ft^3 = 515.4 kg/m^3 1 lb-ft = 1.356 J
                                    1 Btu = 1.055 kJ
                                    1 W-h = 3.600 kJ

Specific weight
    1 lb/ft^3 = 157.1 N/m^3

Power
    1 hp = 550 lb-ft/s   1 lb-ft/s = 1.356 W
    1 hp = 745.7 W   1 Btu/h = 0.293 W

TABLE K.3   Units and conversion factors for dynamic viscosity, μ (Multiply number in table by viscosity in given unit to obtain viscosity in desired unit.)

Given Unit → Desired Unit	lb-s/ft²	Pa·s	Poise
lb-s/ft²	1	47.88	478.8
Pa·s*	2.089×10^{-2}	1	10
poise**	2.089×10^{-3}	0.1	1
centipoise	2.089×10^{-5}	0.001	0.01

* Standard SI units
  Equivalent unit: $N \cdot s/m^2$

** $dyne \cdot s/cm^2$
  Equivalent unit: $g/cm \cdot s$

Example: Given a viscosity measurement of 200 centipoises, the viscosity in Pa·s is

$$200 \text{ centipoises} \times \frac{0.001 \text{ Pa} \cdot \text{s}}{\text{centipoise}} = 0.20 \text{ Pa} \cdot \text{s}$$

TABLE K.4   Units and conversion factors for kinematic viscosity, ν (Multiply number in table by viscosity in given unit to obtain viscosity in desired unit.)

Given Unit → Desired Unit	ft²/s	SSU	m²/s	Stoke
ft²/s	1	4.29×10^5	9.290×10^{-2}	929.0
SSU*	2.33×10^{-6}	1	2.17×10^{-7}	2.17×10^{-3}
m²/s†	10.764	4.61×10^6	1	10^4
stoke‡	1.076×10^{-3}	4.61×10^2	10^{-4}	1
centistoke	1.076×10^{-5}	4.61	10^{-6}	0.01

* Saybolt Seconds, Universal (conversions approximate for SSU > 100)
  For SSU < 100: $\nu = (0.226 \text{ SSU} - 195/\text{SSU})(10^{-6}) \text{ m}^2/\text{s}$

† Standard SI unit

‡ cm^2/s

Example: Given a viscosity measurement of 200 centistokes, the viscosity in m²/s is

$$200 \text{ centistokes} \times \frac{10^{-6} \text{ m}^2/\text{s}}{\text{centistoke}} = 200 \times 10^{-6} \text{ m}^2/\text{s}$$

■ ■ ■ ■ L Properties of Areas

Section	Area of Section, A	Distance to Centroidal Axis, $\bar{y}$	Moment of Inertia About Centroidal Axis, I_c
Square	H^2	$H/2$	$H^4/12$
Rectangle	BH	$H/2$	$BH^3/12$
Triangle	$BH/2$	$H/3$	$BH^3/36$

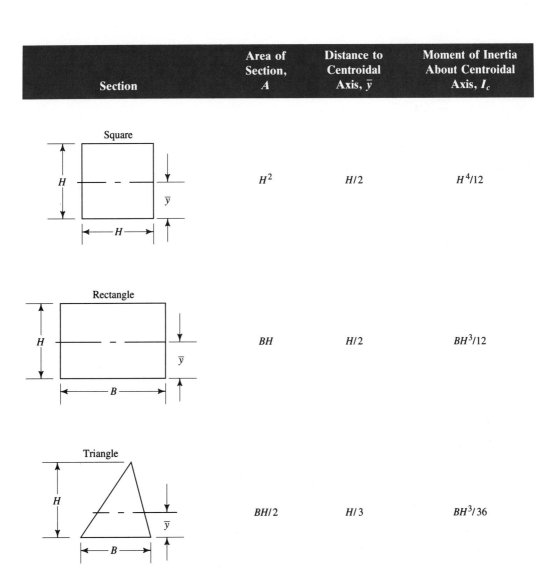

Section	Area of Section, A	Distance to Centroidal Axis, $\bar{y}$	Moment of Inertia About Centroidal Axis, I_c
Circle 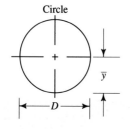	$\pi D^2/4$	$D/2$	$\pi D^4/64$
Ring	$\dfrac{\pi(D^2-d^2)}{4}$	$D/2$	$\dfrac{\pi(D^4-d^4)}{64}$
Semicircle 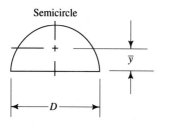	$\pi D^2/8$	$0.212D$	$(6.86\times10^{-3})D^4$
Quadrant	$\pi D^2/16$ $\pi R^2/4$	$0.212D$ $0.424R$	$(3.43\times10^{-3})D^4$ $(5.49\times10^{-2})R^4$
Trapezoid 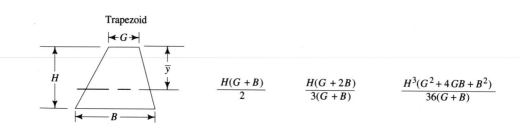	$\dfrac{H(G+B)}{2}$	$\dfrac{H(G+2B)}{3(G+B)}$	$\dfrac{H^3(G^2+4GB+B^2)}{36(G+B)}$

■ ■ ■ ■ M Properties of Solids

Shape	Volume V	Distance to Centroid, $\bar{y}$
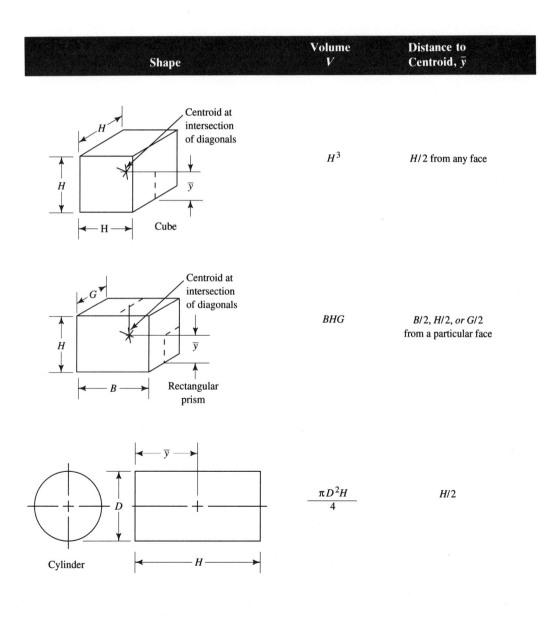 Cube	H^3	$H/2$ from any face
Rectangular prism	BHG	$B/2$, $H/2$, or $G/2$ from a particular face
Cylinder	$\dfrac{\pi D^2 H}{4}$	$H/2$

Shape	Volume V	Distance to Centroid, $\bar{y}$
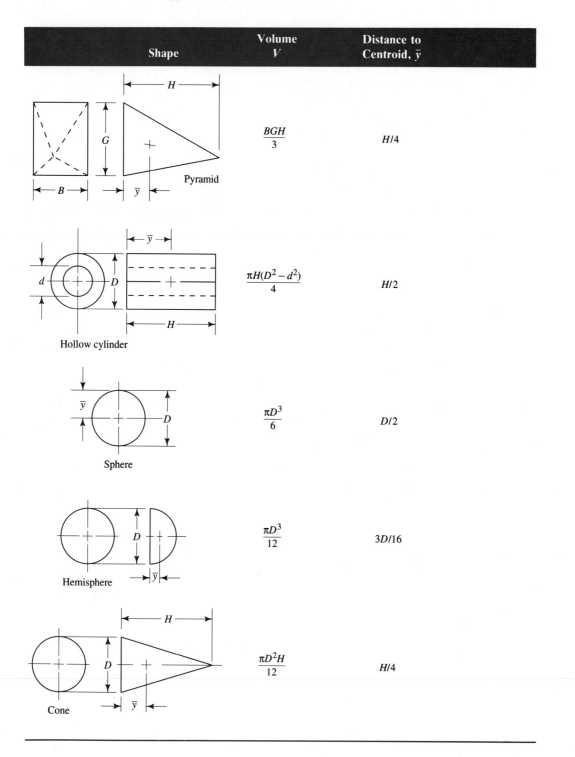		

Pyramid: $\dfrac{BGH}{3}$ $H/4$

Hollow cylinder: $\dfrac{\pi H(D^2 - d^2)}{4}$ $H/2$

Sphere: $\dfrac{\pi D^3}{6}$ $D/2$

Hemisphere: $\dfrac{\pi D^3}{12}$ $3D/16$

Cone: $\dfrac{\pi D^2 H}{12}$ $H/4$

■■■■ N Gas Constant, Adiabatic Exponent, and Critical Pressure Ratio for Selected Gases

| | Gas Constant, R | | | |
Gas	$\dfrac{ft \cdot lb}{lb \cdot °R}$	$\dfrac{N \cdot m}{N \cdot K}$	k	Critical Pressure Ratio
Air	53.3	29.2	1.40	0.528
Ammonia	91.0	49.9	1.32	0.542
Carbon dioxide	35.1	19.3	1.30	0.546
Natural gas (typical; depends on gas)	79.1	43.4	1.27	0.551
Nitrogen	55.2	30.3	1.41	0.527
Oxygen	48.3	26.5	1.40	0.528
Propane	35.0	19.2	1.15	0.574
Refrigerant 12	12.6	6.91	1.13	0.578

Answers to Selected Problems

Chapter 1

1.1 1.25 m
1.3 3.65×10^{-6} m^3
1.5 391×10^6 mm^3
1.7 22.2 m/s
1.9 2993 m
1.11 786 m
1.13 7.39×10^{-3} m^3
1.15 1.83 m/s
1.17 47.2 m/s
1.19 48.7 mi/h
1.21 8.05×10^{-2} m/s^2
1.23 0.264 ft/s^2
1.25 10.8 N-m
1.27 1.76 kN-m
1.29 37.4 g
1.31 1.56 m/s
1.33 26 700 ft-lb
1.35 6.20 slugs
1.37 4.63 ft/s
1.39 2.49 runs/game
1.41 129 innings
1.43 354 psi
1.45 2.72 MPa
1.47 119 psi
1.49 40.25 kN
1.51 2.26 in
1.57 1300 psi   8.96 MPa
1.59 1890 psi   13.03 MPa
1.61 -1.59%
1.63 884 lb/in
1.65 14 137 lb/in
1.67 62.2 kg
1.69 8093 N
1.71 0.242 slug
1.73 50.9 lb
1.75 m = 4.97 slugs
 w = 712 N
 m = 72.5 kg

1.77 9810 N
1.81 1.225 kg/m^3
1.83 0.903 at 5°C
 0.865 at 50°C
1.85 Density = 883 kg/m^3
 Specific weight = 8.66 kN/m^3
 Specific gravity = 0.883
1.87 634 N
1.89 2.72×10^{-3} m^3
1.91 Density = 789 kg/m^3
 Specific weight = 7.74 kN/m^3
1.93 w = 3.536 MN m = 360.5 Mg
1.95 3.23 N
1.97 2.38×10^{-3} slugs/ft^3
1.99 0.904 at 40°F   0.865 at 120°F
1.101 Specific weight = 56.1 lb/ft^3
 Density = 1.74 slugs/ft^3
 Specific gravity = 0.899
1.103 142 lb
1.105 2745 cm^3
1.107 1.53 slugs/ft^3; 0.79 g/cm^3
1.109 Volume = 1.16×10^5 gal
 Weight = 6.60×10^5 lb
1.111 0.724 lb

Chapter 2

2.19 1.5×10^{-3} Pa-s
2.23 1.90 Pa-s
2.25 8.9×10^{-6} lb-s/ft^2
2.29 4.1×10^{-3} lb-s/ft^2
2.31 2.8×10^{-5} lb-s/ft^2
2.33 9.5×10^{-5} lb-s/ft^2
2.35 2.2×10^{-4} lb-s/ft^2
2.55 9.3×10^{-6} m^2/s minimum
 12.5×10^{-6} m^2/s maximum
2.57 5.60×10^{-6} m^2/s 6.03×10^{-5} ft^2/s

2.59 1.36×10^{-4} lb-s/ft^2
2.61 0.402 Pa-s

Chapter 3

3.11 12.7 psia
3.13 Zero gage pressure
3.15 56 kPa (gage)
3.17 -23 kPa(gage)
3.19 384 kPa(abs)
3.21 105.4 kPa(abs)
3.23 13 kPa(abs)
3.25 8.1 psig
3.27 -3.2 psig
3.29 55.7 psia
3.31 15.3 psia
3.33 1.9 psia
3.35 1.05
3.37 5.56 psig
3.39 32.37 kPa(gage)
3.41 177.9 psig
3.43 $p_{surface} = 24.77$ kPa(abs)
       $p_{bottom} = 67.93$ kPa(abs)
3.45 61.73 psig
3.47 13.36 ft
3.49 6.84 m
3.51 70.6 kPa(gage)
3.53 110 MPa
3.55 -22.47 kPa(gage)
3.63 $p_B - p_A = -0.258$ psi
3.65 $p_A - p_B = 96.03$ kPa
3.67 $p_A = 90.05$ kPa(gage)
3.69 $p_A - p_B = 2.73$ psi
3.71 $p_A = 0.254$ kPa(gage)
3.77 30.06 in
3.81 83.44 kPa
3.83 14.99 psia
3.85 98.94 kPa(abs)

Chapter 4

4.1 1673 lb
4.3 125 lb
4.5 2.47 kN
4.7 22.0 kN
4.9 6.05 lb
4.11 137 kN
4.13 1.26 MN
4.15 $F_R = 126\,300$ lb
       $d_p = 10.33$ ft vertical depth to center of pressure
       $L_p = 11.93$ ft

4.17 $F_R = 46.8$ kN
       $d_p = 0.933$ m vertical depth to center of pressure
       $L_p = 1.32$ m
4.19 $F_R = 1.09$ kN $L_p = 966$ mm
       $L_p - L_c = 13.3$ mm
4.21 $F_R = 1787$ lb $L_p = 13.51$ ft
       $L_p - L_c = 0.0136$ ft
4.23 $F_R = 1.213$ kN $L_p = 1.122$ m
       $L_p - L_c = 5.98$ mm
4.25 $F_R = 5.79$ kN $L_p = 1.372$ m
       $L_p - L_c = 0.0637$ m
4.27 $F_R = 11.97$ kN $L_p = 1.693$ m
       $L_p - L_c = 0.0235$ m
4.29 $F_R = 329.6$ lb $L_p = 47.81$ in
       $L_p - L_c = 0.469$ in
4.31 $F_R = 247$ N $L_p = 196.5$ mm
       $L_p - L_c = 0.0465$ m
4.33 $F_R = 29\,950$ lb $L_p = 5.333$ ft
4.35 $F_R = 34\,586$ lb $L_p = 6.158$ ft
4.37 $F_R = 3.43$ kN $L_p = 3.067$ m
4.39 $F_R = 11.92$ kN $L_p = 1.00$ m
4.41 Force on hinge = 4.85 kN to left
       Force on stop = 2.95 kN to left
4.43 $F_R = 3.29$ kN
       $L_{pe} - L_{ce} = 4.42$ mm
4.45 $F_R = 1826$ lb
       $L_{pe} - L_{ce} = 1.885$ in
4.47 $F_R = 48\,58$ kN $F_V = 35.89$ kN
       $F_H = 32.74$ kN
4.49 $F_R = 120\,550$ lb $F_V = 99\,925$ lb
       $F_H = 67\,437$ lb
4.51 $F_R = 959.1$ kN $F_V = 927.2$ kN
       $F_H = 245.3$ kN
4.53 $F_R = 80.7$ kN $F_V = 54.0$ kN
       $F_H = 60.0$ kN
4.55 $F_R = 64.49$ kN $F_V = 47.15$ kN
       $F_H = 44.00$ kN
4.57 70.1 lb downward
4.59 Zero
4.61 70.1 lb downward

Chapter 5

5.1 Buoyant force = 814 N   Tension = 556 N
5.3 It will sink
5.5 234 mm
5.7 0.217 m^3
5.9 7.515×10^{-3} m^3
5.11 5.055 ft^3
5.13 0.0249 lb

5.15 1.041
5.17 1447 lb
5.19 283.6 m^3
5.21 7.95 kN/m^3
5.23 237 mm
5.25 29 mm
5.27 10.05 kN
5.29 135 mm
5.31 1681 lb
5.33 4.67 in
5.35 300 lb
5.37 14.39 lb
5.39 y_{mc} = 0.4844 m (unstable)
5.41 y_{mc} = 8.256 ft (stable)
5.43 y_{mc} = 488.8 mm (unstable)
5.45 y_{mc} = 10.55 in (unstable)
5.47 32.50 ft
5.49 y_{mc} = 436 mm (stable)
5.51 y_{mc} = 90.2 mm (stable)
5.53 y_{mc} = 410.3 mm (stable)
5.55 y_{mc} = 54.18 in (stable)
5.57 y_{mc} = 13.29 ft (stable)
5.59 y_{mc} = 467 mm (stable)
5.61 y_{mc} = 1.288 m (stable)

Chapter 6

6.1 1.89×10^{-4} m^3/s
6.3 0.550 m^3/s
6.5 2.08×10^{-3} m^3/s
6.7 0.250 m^3/s
6.9 3.30×10^4 L/min
6.11 2.96×10^{-7} m^3/s
6.13 215 L/min
6.15 1.02 ft^3/s
6.17 5.57 ft^3/s
6.19 561 gal/min
6.21 3368 gal/min
6.23 Q = 500 gal/min = 1.11 ft^3/s = 3.15×10^{-2} m^3/s
Q = 2500 gal/min = 5.57 ft^3/s = 0.158 m^3/s
6.25 2.77×10^{-2} ft^3/s
6.27 1.76×10^{-5} ft^3/s
6.29 W = 736 N/s M = 75.0 kg/s
6.31 Q = 7.47×10^{-7} m^3/s M = 8.07×10^{-4} kg/s
6.33 M = 2.48×10^{-2} slugs/s W = 2878 lb/h
6.35 5.38 ft^3/s
6.37 3.09 ft
6.39 v_1 = 0.472 m/s v_2 = 1.89 m/s
6.41 6.17 m/s

6.43 1.97×10^5 lb/h
6.45 1¼ × 0.065 steel tube for v = 8.0 ft/s min
⅞ × 0.065 steel tube for v = 25.0 ft/s max
6.47 6 in Sch 40 pipe for Q = 1800 L/min
14 in Sch 40 pipe for 9500 L/min
6.49 3.075 m/s
6.51 10.08 ft/s
6.53 ¾ × 0.065 steel tube
6.55 Q_{shell}/Q_{tube} = 2.19
6.57 Q_{tube} = 0.3535 ft^3/s Q_{shell} = 1.998 ft^3/s
6.59 v_{pipe} = 7.98 ft/s v_{nozzle} = 181.8 ft/s
6.61 34.9 kPa
6.63 25.1 psig
6.65 58.1 kPa
6.67 2.90 ft^3/s
6.69 Q = 4.66×10^{-3} m^3/s P_A = −2.82 kPa
P_B = −11.65 kPa
6.71 1.42 m
6.73 35.6 ft/s
6.75 3.98×10^{-3} m^3/s
6.77 1.48×10^{-3} m^3/s
6.79 1.035 ft^3/s
6.81 31.94 psig
6.86 6.00×10^{-3} m^3/s
6.90 1.28 ft
6.93 10.18 psig
6.95 296 s
6.97 556 s
6.99 504 s
6.101 1155 s
6.103 252 s
6.105 1.94 s

Chapter 7

7.1 34.5 lb-ft/lb
7.3 3.33×10^{-2} m^3/s
7.5 15.7 lb-ft/lb
7.7 72.7
7.9 16.2 kW
7.11 0.700 hp 70.0%
7.13 h_A = 37.46 m P_A = 0.390 kW
7.15 **(a)** p_B = 1.07 psig **(b)** p_C = 21.8 psig
(c) h_A = 48 ft **(d)** P_A = 10.9 hp
7.17 h_A = 4.68 m P_A = 43.6 W
7.19 2.84 hp
7.21 P_A = 1.25 W P_I = 2.08 W
7.23 9.15 kW
7.25 P_R = 16.79 kW P_O = 12.60 kW

7.27 2.00 hp
7.29 35.0 kPa
7.31 12.8 ft
7.33 6.88 m
7.35 21.16 hp
7.37 219.1 psig
7.39 1.01 psig
7.41 5.76 psig
7.43 4.28 hp
7.45 1.26 hp

Chapter 8

8.1 249
8.3 7.02×10^{-3} m³/s
8.5 **(a)** 3 in Type K copper tube   **(b)** 5 in tube
   (c) ¾ in tube   **(d)** ⅛ in tube
8.9 4.76×10^4
8.11 9.59×10^5
8.13 33.4
8.15 5.61×10^3
8.17 2237 (Critical zone)
8.19 1105
8.21 2.12×10^4
8.23 $Q_1 = 0.1681$ gal/min
   $Q_2 = 0.3362$ gal/min
8.25 1.06×10^4
8.27 32.4 mm
8.31 3.04×10^4
8.33 3.808×10^5
8.35 In pipe: $N_R = 6.08 \times 10^4$
   In duct: $Q = 1.39 \times 10^{-2}$ m³/s
8.37 1.58×10^5
8.39 5.90×10^3
8.41 1.61×10^7
8.43 0.750 ft³/s
8.45 6.40×10^3 2.40×10^{-3} m³/s

Chapter 9

9.1 $p_1 - p_2 = -471$ kPa
9.3 1.20 lb-ft/lb
9.5 $p_1 - p_2 = 25.2$ kPa
9.7 61.5 ft
9.9 **(a)** 12.60 ft   **(b)** 113.8 hp
9.11 46.9 psi
9.13 **(a)** 853 kPa   **(b)** 17.1 kW
9.15 89.9 kPa
9.17 $p_1 - p_2 = 39.6$ psi
9.19 $p_1 - p_2 = 411$ psi
9.21 151 hp
9.23 2.64 hp

9.25 $p_1 - p_2 = 110$ kPa
9.27 0.0273
9.29 0.0255
9.31 0.0213
9.33 0.0206
9.35 0.0289
9.37 0.713 psi
9.39 92.0 Pa
9.41 111 kPa
9.43 3.02 kPa
9.45 3.72 psi
9.47 $v = 23.05$ ft/s $Q = 187$ gal/min $h_L = 51.9$ ft
9.49 7.36×10^4
9.51 1.84 m/s
9.53 Selected values:

y(mm)	U (m/s)
10	0.530
30	0.628
50	0.674
100	0.735
300	0.833
600	$0.895 = U_{max}$

9.55 At $y = 2.44$ in, $U = v_{avg} = 6.00$ ft/s
   At $y_1 = 2.94$ in, $U_1 = 6.12$ ft/s
   At $y_2 = 1.94$ in, $U_2 = 5.85$ ft/s

9.57

N_R	f	v/U_{max}
4×10^3	0.045	0.767
1×10^4	0.040	0.778
1×10^5	0.034	0.791
1×10^6	0.033	0.794

9.59 Selected values: $v = 10.08$ ft/s

y(in)	U (ft/s)
0.05	5.83
0.15	7.98
0.50	10.35
1.00	11.71
1.50	12.51
2.013	$13.09 = U_{max}$

9.61 $h_L = 15.2$ ft
9.63 $h_L = 28.5$ ft
9.65 $h_L = 3.56$ m
9.67 **(a)** $h_L = 61.4$ ft   **(b)** $h_L = 28.3$ ft
9.69 $h_L = 14.7$ ft

Chapter 10

10.1 0.239 m
10.3 4.55 ft
10.5 $p_1 - p_2 = -0.0891$ psi
10.7 0.326 m

10.13 503.7 kPa
10.15 0.235 m
10.17 1.35 ft
10.19 False
10.21 0.224 m
10.23 4.32 ft
10.27 $K = 0.255$
10.29 (a) 0.459 m (b) 0.229 m (c) 0.115 m
(d) 0.018 m
10.31 2.04 m
10.33 1.58 psi
10.35 7.36 kPa
10.37 9.87 psi
10.39 0.340 ft
10.41 1.29 m
10.43 $h_L = 0.85$ $h_L = 1.11$
10.45 0.432 m

Chapter 11

11.1 85.1 kPa
11.3 212.8 psig
11.5 12.74 MPa
11.7 24.3 kPa
11.9 3.16 m/s
11.11 1.64×10^{-3} m³/s
11.13 (a) 49.6 ft/s (b) 129 ft/s
11.15 2.27×10^{-2} m³/s
11.17 5 in Schedule 80 pipe
11.19 2.12 ft
11.21 3.35 m
11.23 $h_A = 24.24$ ft $P_A = 0.169$ hp
11.25 6.68 ft/s
11.27 −100.0 kPa
11.29 $h_A = 276.9$ ft $P_A = 16.1$ hp
11.31 321.1 kPa
11.33 No flow
11.35 250 gal/min
11.37 270 gal/min
11.39 4.08×10^{-2} m³/s
11.41 7.32×10^{-2} m³/s

Chapter 12

12.1 0.0602 m³/s
12.3 (a) $Q_a = 518$ L/min (Upper pipe)
$Q_b = 332$ L/min (Lower pipe)
(b) 95.0 kPa
12.6 (a) 1.841 ft³/s (b) 1.385 ft³/s (c) 0.456 ft³/s

Chapter 13

13.1 75 mm
13.3 0.940 ft

13.5 40.3 mm
13.7 1.606 in
13.9 0.909 m
13.11 0.295 ft³/s
13.13 0.0125
13.15 (a) 34.7 ft³/s (b) 141.1 ft³/s
13.17 1.69 m
13.19 $Q = 15.89$ m³/s
$N_F = 0.629$ for depth = 1.50 m
$y_c = 1.16$ m
13.21 1.29 ft
13.23 $y = 0.833$ m $R = 0.270$ m
13.24 $Q = 0.0116$ m³
13.25 and **13.26**
Selected values:

Width (m)	Depth (m)	R (m)	S
0.50	1.333	0.2105	0.0162
1.00	0.667	0.2857	0.0108
1.50	0.444	0.2791	0.0111
2.00	0.333	0.2500	0.0129

13.27 $A = 7.50$ ft² $R = 0.936$ ft
13.28 $Q = 44.49$ ft³/s
13.29 $Q = 75.63$ ft³/s
13.30 and **13.31**
Selected values:

y (in)	A (ft²)	R(ft)	Q (ft³/s)
6.00	1.375	0.3616	4.33
10.00	2.708	0.5412	11.15
18.00	6.375	0.8605	35.75
24.00	10.00	1.086	65.47

13.33 $A = 0.0358$ m² $R = 0.0742$ m
13.35 $Q = 0.0168$ m³/s
13.37

	S
Rectangle	0.00519
Triangle	0.00518
Trapezoid	0.00471
Semicircle	0.00441

13.39 (a) $y_c = 0.917$ m (b) $E = 1.38$ m
(d) $E = 2.04$ m $y = 1.90$ m
(e) For y = 0.50 m:
$v = 5.5$ m/s $N_F = 2.48$
For y = 1.90 m:
$v = 1.45$ m/s $N_F = 0.34$
(f) For y = 0.50 m, S = 0.0378
For y = 1.90 m, S = 0.0011

Chapter 14

14.1 2.12×10^{-2} m³/s
14.3 0.0336 ft³/s
14.5 8.45 m/s

14.7 5.11 m/s

14.9 2.66 ft

14.12 (a) 18.8 ft^3/s   **(b)** 16.95 ft^3/s   **(c)** 6.84 ft^3/s

14.14 0.10 ft minimum
2.01 ft maximum

14.16 $H = 1.687$ ft for a 4-ft flume
$H = 1.155$ ft for a 10-ft flume

Chapter 15

15.16 Capacity is cut in half

15.17 Reduced by a factor of 4

15.18 Reduced by a factor of 8

15.19 Reduced by 25%

15.20 Reduced by 44%

15.21 Reduced by 58%

15.23 1-½ × 3–10

15.25 $Q = 280$ gal/min $P = 26$ hp
Efficiency = 53% $NPSH_R = 10.9$ ft

15.26 Head = 250 ft $Q = 220$ gal/min
$P = 24.0$ hp   Efficiency = 56%
$NPSH_R = 8.0$ ft

15.35 $Q = 390$ gal/min $N_s = 619$ $D_s = 2.94$

15.37 795 rpm

15.39 2659

15.41 2475

15.51 4.97 ft

15.53 20.48 ft

15.55 3.51 m

Chapter 16

16.1 2.76 kN

16.3 $R_x = R_y = 39.7$ lb   Resultant = 56.1 lb

16.5 $R_x = 10.13$ lb to the right $R_x = 37.79$ lb up

16.7 $R_x = 873$ N to the left $R_x = 1512$ N up

16.9 25.2 m/s

16.11 Spring force = 32.0 lb

16.13 368 lb

16.15 2676 lb

16.17 $R_x = R_y = 20.41$ kN   Resultant = 28.9 kN

16.19 $R_x = 10.17$ kN $R_y = 4.18$ kN
Resultant = 11.0 kN

16.21 $v = 45.6$ m/s

16.23 2.72 × 10^{-7} N

16.25 Moment = 0.336 lb-in

16.27 0.0307 lb

16.29 $R_x = 41.0$ N; $R_y = 19.1$N

Chapter 17

17.1 **(a)** 0.253 N   **(b)** 4.56 × 10^{-4} N

17.3 1.50 m/s   2.05 m/s   105 m/s

17.5 0.42 N

17.7 7.32 m

17.9 11.2 kN

17.11 **(a)** 2.85 × 10^{-6} N-m   **(b)** 1.67 × 10^{-3} N-m

17.13 1364 N

17.15 $F_D = 12.05$ lb

17.17 $F_D = 31.3$ lb

17.19 $F_D = 5.86$ lb

17.21 $F_D = 1414$ lb

17.23 $F_D = 0.149$ lb on golf ball
$F_D = 0.387$ lb on smooth sphere

17.25 $P_D = 137$ hp

17.27 $P_E = 4252$ hp

17.29 **(a)** $F_D = 26.5$ kN   **(b)** $F_D = 1.66$ kN

17.31 **(a)** $F_L = 26.8$ kN $F_D = 2.83$ kN
(b) $F_L = 9.24$ kN $F_D = 972$ N

17.33 $A = 90.4$ m^2

Chapter 18

18.1 44.17 ft^3/s

18.3 1.25 m^3/s

18.5 5.79 m/s

18.7 0.158 psi

18.9 3.72 in H$_2$O

18.17 0.478 lb/ft^3

18.19 0.0525 lb/ft^3

18.21 11.79 N/m^3

18.23 131 cfm

18.25 2-in Schedule 40 pipe

18.27 2½-in Schedule 40 pipe

18.29 0.2735 lb/ft^3 in the reservoir
0.198 lb/ft^3 in the pipe

18.31 1.092 lb/ft^3 at 35.0 psig
0.506 lb/ft^3 at 3.6 psig

18.33 $W = 5.76 × 10^{-3}$ lb/s
$v = 811$ ft/s

18.34 4.44 × 10^{-3} lb/s

18.37 186.6 kPa

18.39 9.58 × 10^{-3} N/s

Chapter 19

19.1 $v = 570$ ft/min $h_L = 0.0203$ in H$_2$O

19.3 $D = 17.0$ in $h_L = 0.078$ in H$_2$O

19.6 $D = 350$ mm $h_L = 0.58$ Pa/m

19.8 $D_e = 5.74$ in $Q = 95$ cfm

19.10 $D_e = 381$ mm $Q = 0.60$ m^3/s

19.12 10 × 24 or 12 × 20

19.14 0.0180 in H_2O

19.16 0.0145 in H_2O

19.18 Main duct; 1600 cfm:

$v = 1160$ ft/min; $H_v = 0.0839$ in H_2O

Main duct; 1100 cfm:

$v = 800$ ft/min; $H_v = 0.0399$ in H_2O

Branch; 500 cfm:

$v = 720$ ft/min; $H_v = 0.0323$ in H_2O

19.20 $H_L = 0.00839$ in H_2O

19.22 $H_L = 29.6$ Pa

19.24 $H_L = 0.1629$ in H_2O

19.26 $p_{fan} = -28.5$ Pa

Index

ISBN 0-02-384231-8

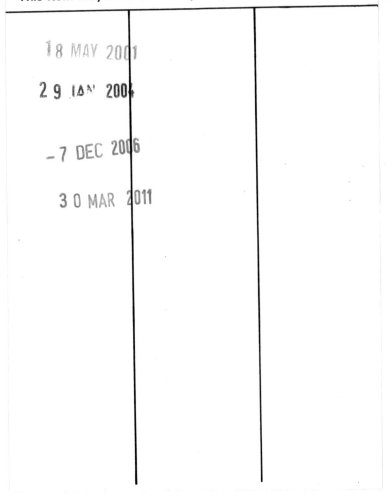

KEY EQUATIONS

FRICTION FACTOR FOR
TURBULENT FLOW

$$f = \dfrac{0.25}{\left[\log\left(\dfrac{1}{3.7\,(D/\epsilon)} + \dfrac{5.74}{N_R^{0.9}}\right)\right]^2}$$

(9–9)

REYNOLDS NUMBER FOR
NONCIRCULAR SECTIONS

$$N_R = \dfrac{v(4R)\rho}{\mu} = \dfrac{v(4R)}{\nu}$$

(9–10)

DARCY'S EQUATION FOR
NONCIRCULAR SECTIONS

$$h_L = f\dfrac{L}{4R}\dfrac{v^2}{2g}$$

(9–11)

HAZEN-WILLIAMS FORMULA
U.S. CUSTOMARY UNITS

$$v = 1.32\,C_h\,R^{0.63}\,s^{0.54}$$

(9–15)

HAZEN-WILLIAMS FORMULA
SI UNITS

$$v = 0.85\,C_h\,R^{0.63}\,s^{0.54}$$

(9–16)

HYDRAULIC RADIUS

$$R = \dfrac{A}{WP} = \dfrac{\text{area}}{\text{wetted perimeter}}$$

(13–1)

REYNOLDS NUMBER FOR OPEN
CHANNELS

$$N_R = \dfrac{vR}{\nu}$$

(13–3)

FROUDE NUMBER

$$N_F = \dfrac{v}{\sqrt{gy_h}}$$

(13–4)

HYDRAULIC DEPTH

$$y_h = A/T$$

(13–5)

MANNING'S EQUATION—SI UNITS

$$v = \dfrac{1.00}{n}\,R^{2/3}S^{1/2}$$

(13–6)

NORMAL DISCHARGE—SI UNITS

$$Q = \left(\dfrac{1.00}{n}\right)AR^{2/3}S^{1/2}$$

(13–8)

MANNING'S EQUATION—U.S.
CUSTOMARY UNITS

$$v = \dfrac{1.49}{n}\,R^{2/3}S^{1/2}$$

(13–10)

NORMAL DISCHARGE—U.S.
CUSTOMARY UNITS

$$Q = AV = \left(\dfrac{1.49}{n}\right)AR^{2/3}S^{1/2}$$

(13–11)

GENERAL FORM OF FORCE
EQUATION

$$F = (m/\Delta t)\Delta v = M\,\Delta v = \rho Q\,\Delta v$$

(16–4)